The Natures of Maps

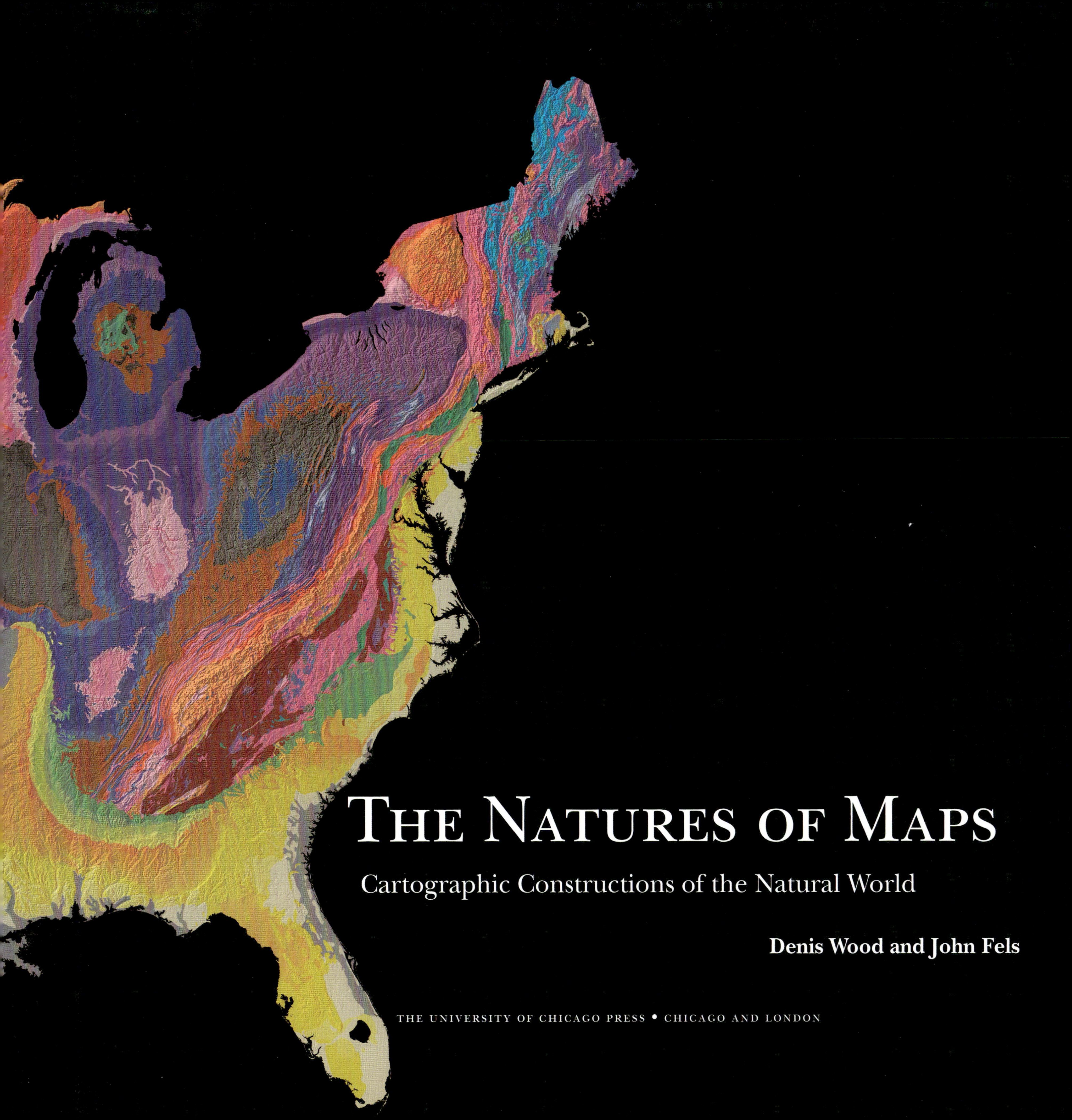

The Natures of Maps

Cartographic Constructions of the Natural World

Denis Wood and John Fels

THE UNIVERSITY OF CHICAGO PRESS • CHICAGO AND LONDON

Contents

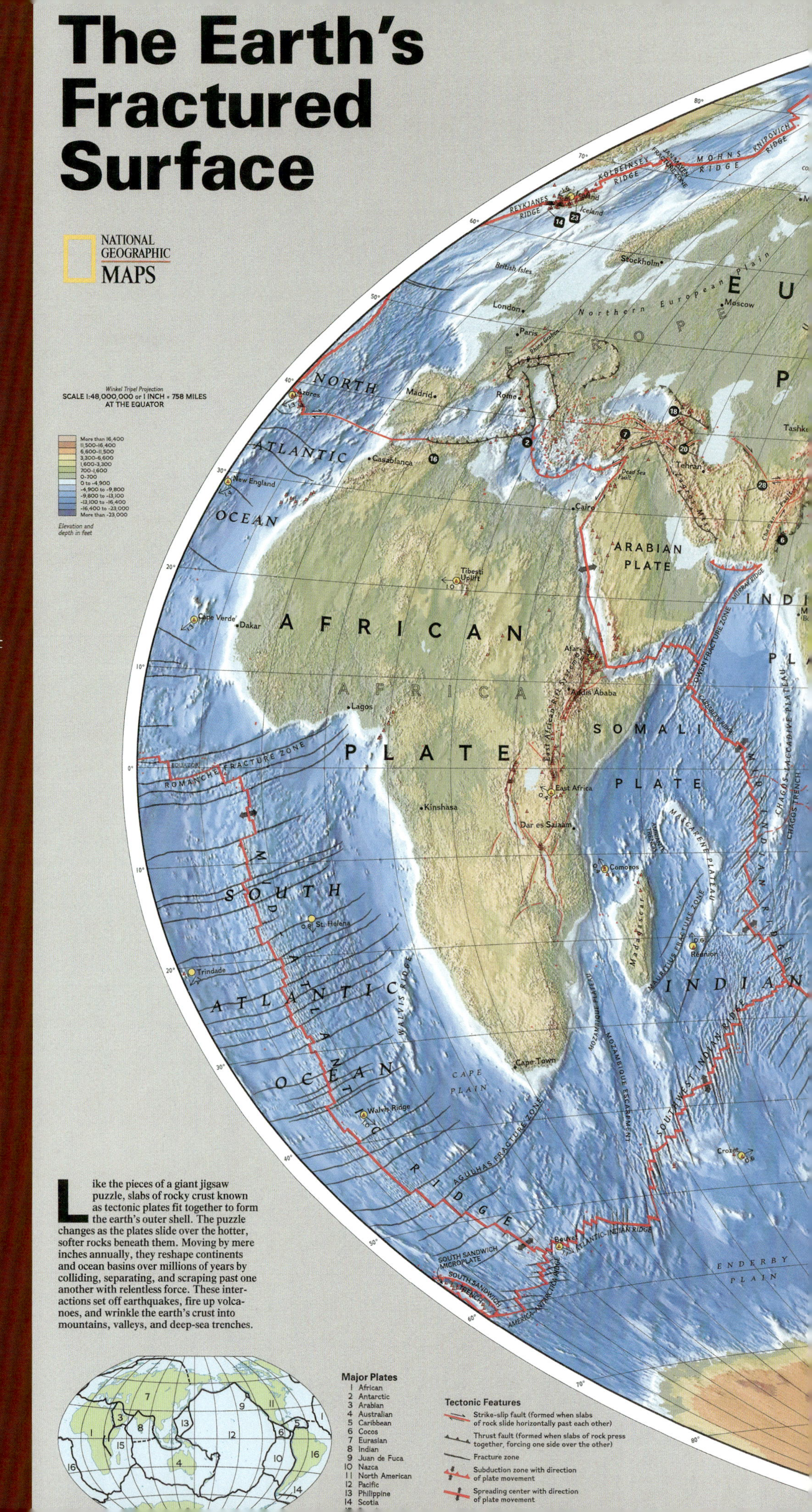

The Earth's Fractured Surface
NATIONAL GEOGRAPHIC MAPS
NORTH ATLANTIC OCEAN
AFRICAN PLATE
AFRICA
ARABIAN PLATE
SOMALI PLATE
SOUTH ATLANTIC OCEAN
INDIAN
EUROPE
Like the pieces of a giant jigsaw puzzle, slabs of rocky crust known as tectonic plates fit together to form the earth's outer shell. The puzzle changes as the plates slide over the hotter, softer rocks beneath them. Moving by mere inches annually, they reshape continents and ocean basins over millions of years by colliding, separating, and scraping past one another with relentless force. These interactions set off earthquakes, fire up volcanoes, and wrinkle the earth's crust into mountains, valleys, and deep-sea trenches.
Major Plates
Tectonic Features

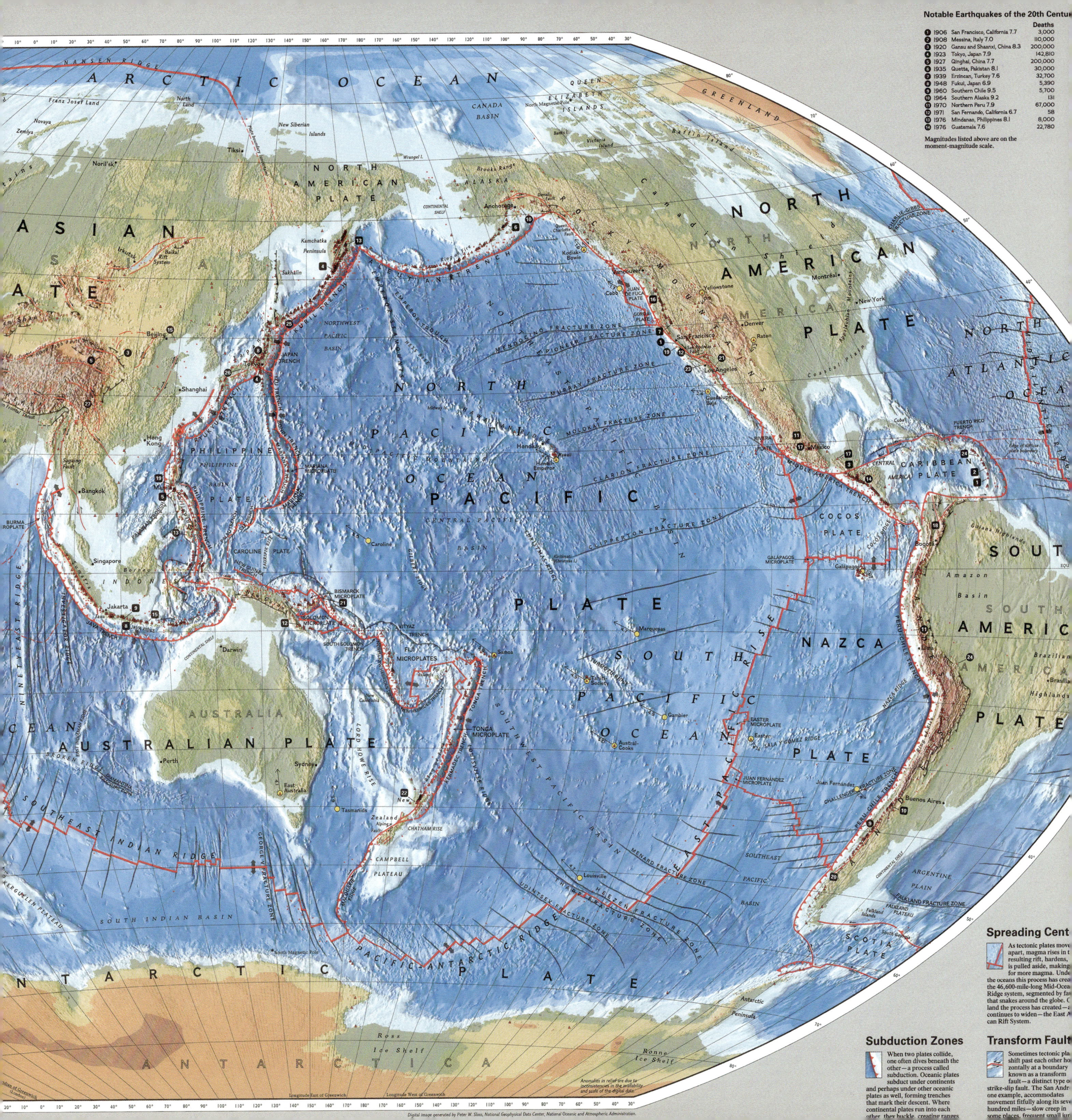

Notable Earthquakes of the 20th Centu
Deaths
1 1906 San Francisco, California 7.7 3,000
2 1908 Messina, Italy 7.0 110,000
3 1920 Gansu and Shaanxi, China 8.3 200,000
4 1923 Tokyo, Japan 7.9 142,810
5 1927 Qinghai, China 7.7 200,000
6 1935 Quetta, Pakistan 8.1 30,000
7 1939 Erzincan, Turkey 7.6 32,700
8 1948 Fukui, Japan 6.9 5,390
9 1960 Southern Chile 9.5 5,700
10 1964 Southern Alaska 9.2 131
11 1970 Northern Peru 7.9 67,000
12 1971 San Fernando, California 6.7 58
13 1976 Mindanao, Philippines 8.1 8,000
14 1976 Guatemala 7.6 22,780
Magnitudes listed above are on the moment-magnitude scale.
Spreading Cent
As tectonic plates move apart, magma rises in t resulting rift, hardens, is pulled aside, making for more magma. Und the oceans this process has crea the 46,600-mile-long Mid-Ocea Ridge system, segmented by fa that snakes around the globe. land the process has created— continues to widen—the East A can Rift System.
Subduction Zones
When two plates collide, one often dives beneath the other—a process called subduction. Oceanic plates subduct under continents and perhaps under other oceanic plates as well, forming trenches that mark their descent. Where continental plates run into each other, they buckle, creating ranges
Transform Fault
Sometimes tectonic pla shift past each other ho zontally at a boundary known as a transform fault—a distinct type o strike-slip fault. The San Andr one example, accommodates movement fitfully along its sev hundred miles—slow creep in some places, frequent small jur
ARCTIC OCEAN
NORTH AMERICAN PLATE
EURASIAN PLATE
NORTH AMERICAN PLATE
GREENLAND
NORTH PACIFIC OCEAN
PACIFIC PLATE
PHILIPPINE PLATE
CAROLINE PLATE
BISMARCK MICROPLATE
FIJI MICROPLATES
TONGA MICROPLATE
COCOS PLATE
CARIBBEAN PLATE
GALAPAGOS MICROPLATE
NAZCA PLATE
EASTER MICROPLATE
JUAN FERNANDEZ MICROPLATE
SOUTH AMERICAN PLATE
AUSTRALIA
AUSTRALIAN PLATE
SOUTH PACIFIC OCEAN
SCOTIA PLATE
ANTARCTIC PLATE
ANTARCTICA
EAST PACIFIC RISE
PACIFIC-ANTARCTIC RIDGE
SOUTHEAST INDIAN RIDGE
SOUTH INDIAN BASIN
SOUTHEAST PACIFIC BASIN
SOUTHWEST PACIFIC BASIN
ARGENTINE PLAIN
FALKLAND FRACTURE ZONE
FALKLAND PLATEAU
Falkland Islands
South Georgia
MENDOCINO FRACTURE ZONE
PIONEER FRACTURE ZONE
MURRAY FRACTURE ZONE
MOLOKAI FRACTURE ZONE
CLARION FRACTURE ZONE
CLIPPERTON FRACTURE ZONE
ELTANIN FRACTURE ZONE
CAMPBELL PLATEAU
CHATHAM RISE
Ross Ice Shelf
Ronne Ice Shelf
Digital image generated by Peter W. Sloss, National Geophysical Data Center, National Oceanic and Atmospheric Administration.

DENIS WOOD is the author of several books, including *The Power of Maps.* JOHN FELS is adjunct associate professor in the GIS graduate program in the College of Natural Resources at North Caroline State University.

The University of Chicago Press, Chicago 60637
The University of Chicago Press, Ltd., London

Printed in Canada

17 16 15 14 13 12 11 10 09 08 1 2 3 4 5

ISBN-13: 978-0-226-90604-1 (cloth)
ISBN-13: 978-0-226-90605-8 (paper)
ISBN-10: 0-226-90604-3 (cloth)
ISBN-10: 0-226-90605-1 (paper)

Library of Congress Cataloging-in-Publication Data
Wood, Denis.
The natures of maps : cartographic constructions of the natural world / Denis Wood and John Fels.
p. cm.
Includes bibliographical references.
ISBN-13: 978-0-226-90604-1 (cloth : alk. paper)
ISBN-10: 0-226-90604-3 (cloth : alk. paper)
ISBN-13: 978-0-226-90605-8 (pbk. : alk. paper)
ISBN-10: 0-226-90605-1 (pbk. : alk. paper) 1. Map reading. 2. Cartography—Philosophy. 3. Cartography—Social aspects. I. Fels, John. II. Title.
GA151 .W56 2008
912—dc22

2007043030

♾ The paper used in this publication meets the minimum requirements of the American National Standard for Information Sciences—Permanence of Paper for Printed Library Materials, ANSI Z39.48-1992.

Acknowledgments

For a number of years, the two of us met every Tuesday afternoon to talk about maps. Mostly what we did was bring maps to the table and talk them over. We had little purpose beyond that of indulging a subject we loved; but as the conversations unfolded and our thoughts matured, it became plain that we needed to get them down on paper. During the effort to realize that intention, we incurred a lot of debts.

Denis has first to thank Christine Baukus and Irv Coats. It is solely their support that made his writing of this, and a number of other books, possible. Their Raleigh bookstore, The Reader's Corner, was scarcely less critical. It was an unfailing source of books and maps, and its staff—Brian Hooley, Wayne Mann, and Todd Morman—an unfailing source of encouragement. Thanks, guys! Up the street at North Carolina State University's D. H. Hill Library, Eric Anderson and Cindy Levine provided essential resources for this project, as for so many others. Still farther afield, Tom Koch, Arthur Krim, and John Krygier each contributed in their very different ways to the final form of this book.

During our work on the book, Denis was given opportunities to present the book's central thesis—that maps are less pictures than talk—to responsive audiences at the annual meetings of the National Council on Geographic Education in 2002 and the North American Cartographic Information Society in 2003. Presentations in 2004 to visual studies students at Goldsmith College and to the geography faculty at Queen Mary's College (both of the University of London) were particularly valuable. A high point of this process was Denis's writing of the new chapter "Are Maps Talk Instead of Pictures?" for the second edition of his coauthored *Seeing Through Maps.* Both Ward Kaiser and Bob Abramms merit thanks for their role in this effort. Each iteration of the thesis made it necessary to rethink it, and even the most obtuse responses revealed deficiencies. All of this contributed to any strength the book in hand might have, and thanks are owed to all those who made these opportunities possible.

Denis was also able to develop a graphic version of the thesis with John Krygier and present it to the 13th Annual Critical Geography Mini-Conference and the North American Cartographic Information Society, both in 2006. Thinking through the issue graphically was an invaluable experience, and again John's presence in this book, entirely uncredited, is substantial.

Thanks to Maria Lane, Denis was able to read an early version of the chapter on possessable nature to a session on "American Maps, American Culture: The Cartographic Fixing of Identity, Knowledge, and Power" at the annual meetings of the American Studies Association in 2005. The audience was responsive and subsequent conversations with Maria, Stephen Hanna, and Matthew Edney were clarifying. Matthew has been helpful at many points in our work on this book.

Aileen Buckley, Edith Punt, Christian Harder, Judy Hawkins, Carmen Fye, and Michael Law, each in his or her own way, played a part in making this book possible. Kathleen Morgan helped rescue us from the permissions purgatory in which we found ourselves. Tiffany Wilkerson contributed her copyediting expertise. We especially want to thank our designer, Savitri Brant, for her strong vision, outstanding design work, and support. Anyone who likes the look of this book as much as we do has Savitri to praise. Most of all we have to acknowledge the contribution of our peerless editor, Amy Collins. Not only did Amy hold things together when it seemed they were falling apart, but her meticulous attention at once to the minutia of phrasing and to the broad sweep of argument is substantially responsible for any merit readers may find in the text. Amy, we couldn't have done it without you!

At the University of Chicago Press, Christie Henry was not only a pleasure to work with, but a beacon when everything turned dark. Dmitri Sandbeck assisted in preparing the work for final publication.

Our take on most of the maps we discuss is highly critical. Given this, the willingness with which holders of the rights to these maps gave us permission to reproduce them is remarkable and deserving of acknowledgment. We need to thank first and foremost the National Geographic Society, whose marvelous maps bear the brunt of our critique. Their maps draw our attention because they're some of the best maps ever made. We'd also like to thank the U.S. Geological Survey for their many outstanding maps. We have to thank Robert G. Bailey of the U.S. Forest Service and Tom Patterson of the U.S. National Park Service, not only for their maps and their writings about them, but for their gracious cooperation and assistance. We'd also like to thank John Harris and John Conway of WRAL-TV, Raleigh. Of course we'd *like* to thank *all* those who helped to track down and secure permissions to reproduce the maps and other images that we have, but the list would be very, very long, and inevitably we'd overlook someone whose aid really does call for recognition. Thank you all!

Most of all we want to thank Ingrid and Vicki, for their inexhaustible support and endless patience.

Foreword

Something must be afoot. If the eighteenth and nineteenth centuries were the era of exploration and mapping, and the twentieth century saw the expansion of mapping and computer-driven cartography to wider domains of life and science, the first years of the twenty-first century may be the coming into presence of an era of a "new cartography." This was the term coined by Gilles Deleuze to describe the important shift in thinking taking place in twentieth-century thought and, in this case, marked by the writings of Michel Foucault. Foucault was the "new cartographer," whose works signaled a new mode of investigation and writing that sought not to trace out representations of the real, but to show how mappings always produce worlds by combining and recombining relations and ideas; to show how knowledge and social interests always work in conjunction; how maps are always technologies that depend for their effectiveness on specific institutions, discourses, and practices; and that these specific conditions are important in constructing what we understand to be the world around us, the real.[1]

So, something is afoot. Technologies of mapping have changed rapidly, particularly as GIS, remote sensing, and graphical capacities, have transformed the resolution, richness, complexity, and speed of the data and ideas that can be presented. We now luxuriate in the creative production of new images of Earth and its worlds. As a result, the uses of maps in everyday life have grown exponentially. Maps are now ubiquitous. They adorn the dashboard, glove compartment, city directory, mobile phone, and computer screen, and from *National Geographic* to residential development to genome mapping to MRI to the Hubble to Google Earth, the practices of mapping increasingly inform and shape our daily lives.

These everyday map users know full well that maps shift scale and symbolization and are adaptable to a multiplicity of environments and uses, always more *and* less than the purposes for which they were designed. Thus, how we understand maps has also changed. The communication and transmission models of information that dominated cartographic theory from the 1970s have now given way to metaphors that better suit a globalized society of meshworks, flows, and interactions. If communication metaphors shaped thinking about maps in the 1970s and 1980s, today we are more apt to be influenced by metaphors from advertising agencies, distributed computing, or the Internet. If the earlier age was one that led cartographers to defend the accuracy of scientific representations against the growing use by the state and media of persuasive cartographies, the present age is one in which Wood and Fels's propositional logic—the map as talk, rather than picture—is much more conducive and productive.

In *The Natures of Maps,* Wood and Fels want map users to be more open to the possibilities and importance of mapping practices and the cultural politics of their use. Challenging the stuffy and suspicious epistemologies of scientific cartography, they ask their readers to take maps more seriously as tools for the creation of meaning, where meaning isn't resident in the map or the image, but instead unfolds over time, where "'the map whole' is the conclusion of a process of meaning-making," part of a conversation about how we understand the world. In contrast to scientific cartography, exercised as it is by the challenges of managing error, Wood and Fels display a much greater level of trust in their map readers. Theirs is not a policeman's concern to control the accuracy and error of representations. They are much more interested in opening the canon of cartography to the kind of cultural critique we find in film studies or literary theory. Theirs is a cultural theory of the ways in which maps gain social assent for their claims about the world, a theory of performance focused on the ways in which maps have effects and shape our understandings of place, land, and nature. Their readers do not need rules for reading, strictures and measures against which to measure real versus false representations and interpretations. They need better examples of careful map reading, conversations about maps and the work they do, the power they hold and exercise, and the many different ways in which maps produce worlds, consolidate what we take to be commonsense notions, and eventually construct what we take to be real, even true.

In the 1970s and 1980s, Wood circulated *samizdat*-like mimeographed copies of papers to friends and colleagues dealing with everyday mapping practices and practices of mapping the everyday. These were circulated with fascination among students dissatisfied with the growing consensus in cartography at the time when the map was best understood as a representational tool for communication or transmitting "information." These now-treasured collections of underground mapping prefigured what has since become one of the most important bodies of work in geography and cartography against the "*nature* of maps." In its place has emerged a pragmatics of map use and cultural politics of map reading.[2] Here the *nature* of maps has been replaced by *The Natures of Maps.*

Denis has been telling me about *The Natures of Maps* for the past two years, about the *Land of Living Fossils,* the road map, the topographic map of the Great Smoky Mountains. But I don't think I ever really grasped the scope of the project in which he and John Fels were engaged. I knew that one week he was digging around in NGS archives in Washington, another week he was devouring the *Atlas of Palestine* in the UNC Library, and the next week heading out to NACIS meetings to present on PPGIS. I knew he was visiting Israeli activists working with the Bedouin or traveling from one place to another. We knew from *The Power of Maps* the range and depth of their understanding of maps, sustained, as they say in the introduction, by the many hours spent poring over maps together. Those who have ever seen Denis at a conference or in an art gallery know how infectious is his delight in talking about maps, their producers, and their uses. But over the past few years, his has been a frenetic pace of even deeper map immersion, working in archives, at the galleries of artist friends, engaging with social movements, and presenting at a wide range of academic and professional conferences. Denis always has energy, but these two years he was driven by something special. I caught a glimpse of this drive when he came to present his work on this book to "3Cs," our counter-cartographies collective, where he presented by far the most coherent and innovative reading of maps and mapping I have ever heard. Behind it there was a special impulse. Now it is clear, that impulse was this book, and what an achievement it is.

First, there is the sheer scope and scale of the work. It is an astonishing array of full-plate color images ranging from the finest intricacies of Swiss topographic maps to the cornucopia of *National Geographic* map supplements to the tightest "reflections" of NASA's rendering of the earth.

Second, there is the renewal of their earlier engagement with the power of maps and politics of map reading. In *The Power of Maps* they introduced the reader to a pragmatic theory of map use. But here they unleash a more fully developed cognitive cartographics and art of slow map reading.

Third, they ask us to think about ways in which we can and ought to take the reading of maps just as seriously as literature and film critics and scholars read their texts. As Wood and Fels write in *The Natures of Maps*, "maps are less pictures than talk" and, like conversation, maps are part of an open exchange system through which meaning is built and changed. *The Natures of Maps* deals with "how meaning is made and expressed in maps," dealing with "maps as systems of propositions, coherent collections of things that we, or at least, the mapmaker, hold or hope to be true about the world."

Fourth, theirs is the unfettered *jouissance* of map reading. *The Natures of Maps* begins with the encouragement to readers to "tromp . . . through the text without worrying overly about whether you're getting it. You probably are, though you may not realize it until you've reached the end." Like any conversation, meaning isn't fixed in the elements of the conversation, but circulated and produced in the process of conversation, temporarily stabilized only in the ending of the conversation.

The Natures of Maps asks readers to reflect on how it is "that maps have the demonstrable power to organize social energies so as to bring into being the visions of the world they posit?" In *The Power of Maps* Wood and Fels asked how and in what ways maps have and exercise power. But now, in *The Natures of Maps* the question is no longer whether maps have power: "That's settled: they do. Now the question is where the power comes from." How do maps do work and how is the map ever more implicated in the shape our lives take?

They answer this in five ways. First, they challenge the claim that the map is a representation or a picture, instead arguing that the map is a system of propositions; an argument. The map is performative. Second, the claims made by a map receive their social assent not by how they measure up to some representational standard of accuracy, but by the way the claim they make to be valid is linked to a whole array of antecedent validations. Maps claim that "this is there" and then have to persuade us why we should believe them . . . a relatively straightforward task in one way, since the antecedents of any particular maps are reams of maps that also claimed that "this is there." We trust maps, we understand their role in proposing that this is there, and we believe we can go and check that, indeed, this is there. But this is a more complex issue than it might seem at first glance. Wood and Fels ask "what assures us that the propositions are true? That they state facts? Only the *social assent given them*. . . ." By constant processes of referencing, citing, layering, the map accumulates social assent and, historically, has established itself as an "authoritative" reference object. Third, this process of meaning creation is always carried out in a performative way, "on the fly," and over time. Cognitive cartographics is much like a conversation where meaning and understanding depend on so many other contexts that shape and condition what we understand than are contained directly in the sentences or the maps.[3] Fourth, the map depends for its meaning and effect on various aspects of its context. Maps "depend on highly structured background knowledge, associated texts and contexts, interactions with other maps and media, and are also directed towards ends in action." As a result, map reading—like literature and film—requires a critical contextual reading to show how social assent is produced and what are the many forms of commonplace on which the map depends and which it, in turn, reproduces.[4] Everything that surrounds and extends a map may be part of the context that is necessary for it to do its work; the context becomes everything that "propels it into action." What Genette called the paratext and is developed here as the *paramap* includes the *perimap* (the verbal and other productions that surround and extend the map) and the *epimap* (other verbal and material elements on which the map depends or draws but which are not materially appended to it).[5] Fifth, while cartographers have focused their attention on "the nature of maps" and critical and deconstructive cartographers like Brian Harley have focused on "the new nature of maps," Wood and Fels turn instead to "the natures of maps," where it is "through their spatialization of nature that maps have made their contribution, facilitating nature's accommodation within the spatiality of the modern state." "The interest of maps is shown to lie not so much in mimetic value but as simulacra which nevertheless may exert a profound influence upon the way space is conceptualized and organized within different societies."

Such a notion of map reading is, of course, antithetical to any interpretation that sees meaning as resident in maps, or maps as representing or "picturing" in a one-to-one manner an external world, or even what Wood and Fels call the "displacement of ideological construction from the map (as mere tool) to the map maker or user. We reject this sophistry in all its parts." It is also antithetical to any notion that there is a "nature of maps." Instead, there are many *Natures of Maps* that create propositions and claims about nature in many different ways. *The Natures of Maps* focuses on eight specific propositions about nature: nature is a victim threatened by forces that disturb its balance; the awesome power of nature itself is a threat; nature's grandeur strikes awe and powerlessness in us; nature is a cornucopia, a "cuddly" nature we embrace with fascination; it is possessable, collectible; nature is an object of science, a spatialized system of relations and flows; a mysterious unknown, a message without a code; and nature is a park where only code exists and where the map represents an inventory of coded phenomena, the archetype being the topographical map.

This, then, is a thoroughly antifoundationalist text. Maps do not contain their nature, or reflect an external nature. They produce natures through their propositional logics. There is no foundation of nature on which they rest; perhaps it is not turtles or elephants all the way down, but it is propositions all the way down, with the propositions at the bottom being the most powerful. These are the most commonplace propositions and they tend to command the most universal assent. "It is here where the map least *records* and most *builds* the world." In this sense, maps are vehicles for creating and conveying authority about the world and they shape what we understand to be factual, real, and normal. As Foucault knew so well, "this is not a pipe," and the map is not a mountain, a stream, or a natural place.[6] As Woods and Fels write, "if you've ever subscribed to the *National Geographic*, you know how map supplements work. On one side is an elaborate graphic like our painted menagerie (small type in its margin even calls it a poster). On the other side is The Map (the 'real' one), what in our introduction we called the main map . . ." Like *National Geographic* supplements, maps always depend on the *supplement*, as they claim to represent the real. Like word meanings in a dictionary that are defined by other word meanings ad infinitum, graphic images similarly take their meaning only in terms of series and series of supplements that constitute particular wholes.[7] The supplement is derived in the very practice of inscribing, annotating, citing, redrawing, formalizing, selecting, categorizing, and so on; systems of citations that are overlain and consolidated as an image of something real.[8] These supplements are not ancillary to the map, they are not merely decorative. They are the map.

The Natures of Maps is a rich reading of what elsewhere I called the everyday lives of maps and their cultural politics.[9] The map, in this reading, "is the evidence of the production of knowledge," a mangle of practice that draws on commonplace tropes, rearranges and recirculates them, constructs meaning, and, for a moment, fixes how we understand nature and how we might act in regard to it.

There is something afoot, and for readers of *The Natures of Maps* it is going to involve a rare and exciting opportunity to tromp through a spectacular gallery of full-color maps, all beautifully reproduced, parsed, and analyzed. From the National Geographic Society's (NGS) *Australia: Land of Living Fossils* and *Australia under Siege* to the WRAL *Stormtracker* to the NGS *Natural Hazards of North America* and *The Earth's Fractured Surface* to the United States Geological Survey's (USGS) *Yosemite Valley* and *The Grand Canyon* to the NGS *Grand Canyon Topographic Trail Map* and *Mount Everest* to the *Generalized Geological and Soil Maps of North Carolina* to the NGS *Portrait USA* and *Face of the Continent* and many more, *The Natures of Maps* presents us with a festival of rich deconstructive readings of maps and their consequences; glorious eye candy indeed, but of the kind that shapes real bodies. To work through *The Nature of Maps* is to rekindle the joy of maps so many people experience when they first encounter the earth through a map. It is to experience viscerally the power of maps to draw us into imaginative worlds that become so very real.

John Pickles
Earl N. Phillips Distinguished Professor of International Studies
University of North Carolina

Notes

1. G. Deleuze, *Foucault* (Minneapolis: University of Minnesota Press, 1988).

2. J. Pickles, "Texts, Hermeneutics, and Propaganda Maps," in *Writing Worlds: Discourse, Text, and Metaphor in the Representation of Landscape*, ed. T. J. Barnes and J. S. Duncan (London: Routledge, 1992).

3. Pickles, "Texts, Hermeneutics, and Propaganda Maps," 193–230.

4. D. W. Black, D. Kunze, and J. Pickles, *Commonplaces: Essays on the Nature of Place* (Lanham, MD: University Press of America, 1989).

5. Pickles, "Texts, Hermeneutics, and Propaganda Maps."

6. M. Foucault, *This Is Not a Pipe* (Berkeley: University of California Press, 1973).

7. J. Derrida, *Of Grammatology* (Baltimore: Johns Hopkins University Press, 1997).

8. B. Latour, *Science in Action: How to Follow Scientists and Engineers through Society* (Cambridge, MA: Harvard University Press, 1987).

9. J. Pickles, *A History of Spaces: Cartographic Reason, Mapping, and the Geo-Coded World* (London: Routledge, 2004).

A note to the reader

This is a book about how meaning is made and expressed in maps. It takes the position that maps are less pictures than arguments. It claims that maps are systems of propositions, coherent collections of things that we, or at least the mapmakers, hold or hope to be true about the world. While we believe this is characteristic of all maps, the maps we are most concerned with here are maps of what is widely regarded as the natural world.

We have struggled to make this book as accessible as we could to as wide an audience as we could imagine, but the fact is inescapable that parts of it will be hard going. Some people are going to boggle at the semiotics that we pretty much take for granted. The study of signs has been a part of Anglophone cartography for more than half a century, importantly so since 1983 when William Berg translated Jacques Bertin's 1967 *Semiology of Graphics*. We ourselves published a landmark text on cartographic semiology in 1986 with our "Designs on Signs: Myth and Meaning in Maps," which we incorporated in 1992 as a chapter in our widely read *The Power of Maps*. In the years since then, semiotic concepts have become so widely deployed we felt there was no need to review them or to rehash our earlier work. Still, for those unfamiliar with these concepts, they may take some getting used to.

Others, with greater reason, will boggle at the cognitive linguistics. Unlike semiotics, which traces its roots to the seventeenth century, cognitive linguistics traces its roots to the 1970s and a group of researchers frustrated by the effort to understand language from, as it were, the inside (appealing, that is, to structural properties specific and internal to language). Instead these researchers turned to things outside language. They came to believe that language was best explained by the broad principles of human cognition in general. Because they assumed that linguistic structures subserved the expression of meaning, the essential subject of their analysis became the "mappings" between meaning and form. Since we had long assumed that cartographic structures subserved the expression of meaning, these "mappings" between meaning and form made cognitive linguistics extremely attractive as a model for thinking about the expression of meaning in maps.

Because of its novelty, we have made some effort to describe the essential tenets of cognitive linguistics, especially those relevant to our pursuit; however, we have not written a primer on cognitive linguistics. Operationally, cognitive linguists assume that language opens *mental spaces* that propagate as discourse unfolds, and we assume that graphics open similar mental spaces that propagate during map reading. Here the propositional nuggets that we call *postings* get tied together into arguments about the world. Because our work proceeds largely by analogy with cognitive linguistics, we've called what we're doing *cognitive cartographics*. In doing so, we've adapted some of the conventions of cognitive linguistics. We hope we've adequately explained how the diagrams work in both the accompanying text and captions, but it may be worth noting here that the contents of mental spaces are included within brackets (< >) in our text.

Because meaning construction takes place in time, we have attempted to read maps in time, and so we *unfold* most of the maps we read, pausing, even if briefly, over each reveal. This results in a slow reading of parts initially disconnected. Only after the map has been fully unfolded, and in many cases turned over, can the integration begin that most map theorists have taken as their starting point. For us, on the contrary, this "map whole" is the conclusion of a process of meaning-making, that is, of the construction of an elaborate argument about the world out of propositional atoms concerned with the existence and locations of things. It is our hope that you will be able to construct our intended meaning by gradually integrating our structuralism, semiotics, deconstuction, propositional logic, cognitive cartographics, slow map readings, and other innovations. Like the map whole, our whole will be apparent only after its parts have been thoroughly digested; or as Humberto Maturana once put it, "The connotations of the terms used will become apparent through their usage in the text, and the complete presentation is expected to provide the foundation for its parts."

The text is copiously annotated. Some of the notes provide the source for a quotation or an observation. More notes, however, are devoted to exploring the background or context for a remark, or to the expansion of an observation, or to pointing out places where alternative points of view can be found. The notes provide, as it were, a shadow text. They may be consulted where they occur or, with greater convenience, altogether at the end of a section or chapter. We believe they enrich the text, as a walk is enriched by the consideration of things lying off the path, sometimes at a considerable remove. Those who prefer to charge straight ahead, however, may safely ignore them.

One final point: the maps we study are of many different types. Maps that are part of other publications have their titles set in quotation marks, while stand-alone maps have their titles set in italics. These maps may themselves contain many ancillary maps. We refer to the whole object as the map, to what we believe to be the backbone of this map as the primary map, and to ancillary maps by title or location. In such cases, we hope that context and the illustrations will make clear what nomenclature alone would inevitably fail to.

We heartily recommend tromping through the text without worrying overly much about whether you're getting it. You probably are, though you may not realize it until you've reached the end.

We just hope it's a good trip.

Introduction: Don't skip this

This is not a book about what maps *are*. Over the past twenty or so years, we and others have pretty convincingly demonstrated that maps are neither what they seem nor proclaim themselves to be. But having established this, we face another question: if maps *are* but partial truths masquerading as the whole story, lies layered on top of lies, nests of interests advancing one cause at the expense of others, then how is it that maps have the demonstrable power to organize social energies so as to bring into being the visions of the world they posit?

That is to say, while in a sense we and others have "decoded" the map, having done so leaves us as far as ever from understanding *how* the map does what it does. The question is no longer whether maps have power. That's settled: they do. Now the question is where the power comes from.

This book is our answer to that question.

When we began our quest in the mid-1980s we were a practicing cartographer (Fels) and a sometime historian of cartography (Wood) who had come together in mutual frustration with the state of theory in cartography. At the time this was dominated by two ideas: that maps were representations—pictures—of the world; and that they were best thought about as a problem in communications engineering—that is, as a conduit connecting a sender (ultimately the world) to a receiver (the map user). To us these ideas seemed to capture little of the nature or power of the map. They were inept when it came to capturing the roles maps had historically played in the extension of empire and the construction of national identities. Because of this, others who were equally dissatisfied with the state of cartographic theory came at the issue from a historical perspective.

Our interest, however, lay in the work done by contemporary maps. So in 1986, we took a semiological approach to what has become accepted as a classic "deconstruction" of the North Carolina state highway map, an analysis that one of us (Wood) generalized in 1992 in *The Power of Maps*, both a Smithsonian exhibition and a book. We were not alone as we traveled in this direction. Brian Harley, to whom, following his untimely death, *The Power of Maps* was dedicated, had in particular blazed a path into the thicket of map theory using tools sharpened by Michel Foucault and Jacques Derrida. Jeremy Black brought a historian's—in fact, a historiographer's—imagination to the issue. Dennis Cosgrove, Anne Godlewska, and Mathew Edney brought that of contemporary geographers with commitments that ranged from iconology to subaltern studies. Alan MacEachren also took a semiological approach, but came at the problem from within cartography. David Turnbull subjected maps to the scrutiny of the sociology of scientific knowledge.

All this work—our own especially—appealed to us as a kind of intellectual sleight. That is, all of us were saying things like, "the interests the map represents are embodied in it as presences and absences ..." (Wood) or "cartography was primarily a form of political discourse concerned with the acquisition and maintenance of power ..." (Harley) or "the space-annihilating science of cartography played an important role in the imposition of nationalist ideology ..." (Godlewska).[1]

But how? *How* did the map do these things?

Following Steven Shapin and Simon Schaffer, Turnbull suggested thinking about maps as what Ludwig Wittgenstein called "forms of life." Turnbull uses the term to mean

> ... a set of conventional linguistic practices and social structures that are "given," without which there can be no talk, knowledge, or social relations. These "givens" structure what it is possible to ask and what it is possible to answer. They lay down the criteria for what is to count as knowledge. From this constructivist perspective, knowledge can be seen as a practical, social and linguistic accomplishment, a consequence of the bringing of the material world into the social world by linguistic and practical action.[2]

We couldn't have agreed more, but for us the problem has become to say precisely what the linguistic and practical actions *are* that constitute the "form of life" we call a map.

That is, so what if the map is primarily a form of political discourse? What is it *about* this discourse that allows, even encourages people to act on it? *How* is it that maps annihilate space? So a map *does* mislead? What is it about a map that all but *obligates* people to follow it? How *can* a map embody an absence?

That the map has been increasingly conceptualized as a social construction has had no effect at all on the way maps are used. With every passing day maps are more implicated in the shape our lives take than ever before. What we're asking is, how does the map get away with it?

What follows, therefore, does *not* plow old ground. Though they may at times seem familiar, both ground and plow are decidedly new. Not that we reject the work we have been describing. Far from it. But we hope we integrate it in a more productive way by giving it an epistemological foundation capable of bearing weight. Our plow has five blades:

1. We start by replacing the whole idea of the map as a representation with that of the map as a system of propositions. Too long

has the eye reigned over cartographic theory. The map is not a picture. It is an argument. Basic here is what we call a *posting*, the fundamental cartographic proposition that *this is there*. Each posting encapsulates a powerful existence claim—*this is*—that gains enormous power by being *posted*, that is, from the indexicality vouchsafed by the sign plane of the map.[3] Multiple postings participate in the construction of a territory, which facilitates the transmission of authority. *Everything* about a map, from top to bottom, is an argument. In taking this position we are not only following a course dictated by our earlier work, but aligning ourselves with the gradual shift from representational to performative idioms broadly characteristic of the work of Ian Hacking, Bruno Latour, Andrew Pickering, and others.[4]

2. What gives the posting its uniquely powerful ability to make existence claims is the social assent that is given to the proposition—*this is there*—that it embodies. This happens because every instance of map use constitutes an act of validation. This validation—all but automatic—is shaped by antecedent validations that have been performed in situations ranging from map learning exercises in school, through successfully using a map to find your way around, to watching Colin Powell point to sites of weapons of mass destruction on a map of Iraq. (The whole process is similar to what Steven Shapin and Simon Schaffer write about as the production of facts by witnessing and reporting.[5]) To claim that *this is there* is to make a powerful claim precisely because it implies the ability to perform an existence test: *you can go there and check it out*. Having done so in the past, you know the outcome. (Besides, who would fake such a challenge?) The assent given to the postings spreads to the territory that the postings mutually construct, and this endows the map with an intrinsic factuality whose social manifestation is the authority the map carries into public action.
3. Complementing this reconstruction of the map as a system of propositions is our adoption of a cognitive linguistics model for map reading. Cognitive linguistics is an exciting new approach to language with powerful connections to cognitive neuroscience. Instead of imagining that people somehow *perceive* map *information*, we show how they actually *construct* the meaning of the map *on the fly* in a process analogous to the way they construct meaning in conversation, in both cases directed toward an end *in action*, that is, toward doing something. This "cognitive cartographics" makes it clear that the principles underlying the graphic design of maps, far from being essentially aesthetic, are wholly at the service of the map's construction of knowledge, a construction built in *real time* by the map readers and typically validated on the spot (as evidenced by its use).
4. Clarifying our understanding of the graphic design of the map is our adoption of what, by analogy with Gerard Genette's clarifying distinction between *text* and *paratext*,[6] we propose to call the *map* and the *paramap*. As a para*text* is everything that surrounds and extends a *text* in order to present it (title, table of contents, foreword, etc.), so the para*map* consists of everything that surrounds and extends a map in order to present it (title, legend, text, photos, ancillary maps, associated texts, etc.). As the distinction illuminates the rhetorical power of the paramap, it underscores the logical primacy of the map (or maps) proper, for it is only there where postings, with their unique power to establish, can be made. Ultimately, it is the *interaction* between map and paramap that propels the map into action.
5. Finally, we focus our analysis on maps of nature to make our case as strongly as we can. Instead of working with the maps that have been the mainstay of "social construction of the map" analysis—that is, with maps of exploration and discovery, maps of empire and nation building, political maps, and others in which the interests of the map (their involvement in the acquisition and maintenance of power) is comparatively straightforward—we attempt to answer our question about how it is that maps do what they do—lead to social action—by looking at maps of species habitat, the Grand Canyon, the migration of birds, plate tectonics, and the Milky Way. Here we see that it is through their spatialization of nature that maps have made their contribution, facilitating nature's accommodation within the spatiality of the modern state.

This nature that maps spatialize turns out not to be unitary, but a many-faced subject of unceasing social negotiations. Indeed we quickly realized that maps proposed the existence of no fewer than eight *different* natures—these related to, but different from, the "ways of knowing" described by John Pickstone.[7] While we have listened to Clarence Glacken, T. J. Clark, William Cronon, Steven Shapin and Simon Schaffer, Bruno Latour, Paula Findlen, and others in our efforts to make sense of the propositions about nature thrown at us by our maps, we have listened even more carefully to the arguments the maps made themselves. What do our maps propose nature to be? What existence claims do our maps advance with their powerful indexicality? What connections do they propose through the natural territories they produce? What social actions do the maps incite?

The spatialized nature we found in our maps is a nature with many faces. It is a nature that can be threatened and that can threaten; a nature that can inspire awe and be cuddled; a nature that can be collected and systematized; a nature that is unknowably remote and underfoot. These are the natures, respectively, of conservationists and disaster relief organizations, IMAX and stuffed toy companies, The Nature Company and university science departments, space-age portraits of the earth and topographic surveys. Our maps, through the connections they propose between nature and society, indict us and caution us, humble us and inflate us, arouse our cupidity and our curiosity, fan our anxieties and our sense of competence. At once they incite us to change our ways before we wipe nature out *and* to buy insurance against its next attack; to stand in open-mouthed humility before the majesty of nature *and* to swaddle ourselves in its warmth; to buy specimens of rocks and butterflies *and* to advance environmental science; to tremble in anticipation of nature's future *and* to visit it in the parks we have set aside for it. In these shifting perspectives nature and humankind seem alternately masters of each other. Bully and victim, owner and pet, subject and object, doubt and assurance exchange position with the folding up of one map and the unfolding of another. In the end, what the maps seem to reveal most is our profound ambivalence about our place in the universe.

Our book falls into two parts. The first opens with a chapter that lays out this argument in greater detail, passes through a second chapter that spells out the propositional logic of the map, and concludes with a third in which we try to read the nature of an individual map. In the second part of the book, we devote a chapter to each of the eight natures proposed by our maps, as we do so exploring in greater depth the power of the propositional logic of the map, and the cognitive construction of nature that maps promote.

A note about the maps

Because we're interested in the natures of map*s*, we attend to as wide a variety of maps as we can. We concern ourselves with everything from the little thumbnail maps that can appear in magazine advertisements to the large sheets issued by specialist scientific organizations. Each is a part of the larger map culture and has been made with the existence of others in mind.

Our interest is overwhelmingly, if not exclusively, on popular maps, and so we devote a lot of attention to maps published by the U.S. Geological Survey and the National Geographic Society. There are a couple of reasons for this. In the first place, the National Geographic Society is the largest geographic society in the world, by a very great margin, and its magazine is read around the world. For years subscribers have received a beautiful large map folded up in every other issue. Among sheet maps these are, without competition, the most widely distributed—and probably the most broadly influential—in the world.

Furthermore, they are available everywhere. In the pages that follow we attempt to describe the maps we deal with as well as we can, and have included figures of most. Although the figures will illuminate the discussion a great deal, they cannot take the place of the real thing. One important reason for dwelling on *National Geographic* maps is because you can so easily find them, for fifty cents or a dollar, in nearly any Goodwill, Salvation Army, or other thrift store. Used book dealers sell them too. Some even give them away. In the back of this book we've provided a list of every map we discuss. Take it with you to the nearest thrift shop, used-book store, or weekend flea market, stock up on the *National Geographic* maps we treat, and follow our discussions with the real maps in hand. U.S. Geological Survey maps are also readily available. With the maps in hand, not only will you find everything we say more interesting, but you'll be in a much better position to second-guess, query, or disagree with the positions we take.

And speaking of the National Geographic Society...

Because of its dominant position in popular discourse about geography, the National Geographic Society has endured its share of criticism. Though we are both members of the National Geographic Society—Wood has been a subscriber, with brief interruptions, for over forty years—we are in full agreement with Catherine Lutz and Jane Collins' reading of *National Geographic* as a conservative arbiter of taste, wealth, and power.[8] While we do extend their analysis of *Geographic* photography with ours of cartography, our interest is focused on neither the magazine nor the society. We are interested in either only insofar as either informs our reading of their maps.

Gilles Fauconnier, the cognitive linguist, has observed that "Language data suffers when it is restricted to language."[9] This is not just because language depends on highly structured background knowledge, conversational meaning, negotiations, and so on, but because it leads to action. The same has to be said of maps. Map study suffers when it is restricted to maps. Maps, too, depend on highly structured background knowledge, associated texts and contexts, interactions with other maps and media, and are also directed toward ends in action. Most of the maps we deal with are enormously complicated, and our readings of them are deep and extended. Given the central role maps play in our society, we are dismayed that maps aren't routinely given the rich, critical reviews that novels and films receive every day, and we advance our readings as examples of what might be accomplished.

Notes

1. For Wood, the relationship between absences and presences was the key to the map's power to articulate a claim. See *The Power of Maps* (New York: Guilford Press, 1992), especially the introduction and first chapter ("Maps Work by Serving Interests"). Harley's understanding of "discourse" was comparatively narrow, limited to "those aspects of a text which are appraisive, evaluative, persuasive, or rhetorical," an emphasis he got from reading Robert Scholes. The notion that map discourse could be concerned with the acquisition and maintenance of power he got from Michel Foucault. See Harley's early formulation of these ideas in his "Maps, Knowledge and Power," in *The New Nature of Maps: Essays in the History of Cartography* (Baltimore, Md.: Johns Hopkins University Press, 2001), 51–81. Godlewska at this time was especially concerned with the role played by geography in the expansion of empire, especially with its ability to harness tools, like maps, capable of annihilating the greater and greater distances, geographic and social, that empire was obligated to span. See Godlewska, "Napoleon's Geographers (1797–1815): Imperialists and Soldiers of Modernity," in *Geography and Empire*, ed. Godlewska and Smith (Oxford: Blackwell, 1994), 34.

2. Turnbull, *Maps Are Territories: Science is an Atlas* (Geelong, Victoria: Deakin University Press, 1989), 10.

3. With Ferdinand de Saussure (*Course in General Linguistics*. New York: McGraw-Hill, 1966) we take a sign to be the union of a concept of some kind—that is, of an element from the plane of content (say "house")—with a mark of some kind—that is, with an element from the plane of signification (say a ■). This union of "house" and ■ takes place on the plane of the sign or sign plane. Only on this sign plane is a ■ understood as a house. The *cartographic* sign plane differs from other sign planes by virtue of the convention that locations on the cartographic sign plane are themselves signs. Their content is "location *x,y* in the world," their mark nothing other than their location x,y on the map. When a ■ is located on the cartographic sign plane the ■ is understood not as any house but as the house in the world at that location. That is, the cartographic sign plane acts as an index to which house in the

world the ■ is a sign of, meeting in this way the conditions for the indexical sign function set by Charles Sanders Peirce. See chapter 2, notes 1 and 2.

4. While almost anything by these three is worth reading, we have been especially influenced in our work here by Hacking's *The Social Construction of What?* (Cambridge, Mass.: Harvard University Press, 1999); Latour's *We Have Never Been Modern* (Cambridge, Mass.: Harvard University Press, 1993) and his *Pandora's Hope* (Cambridge, Mass.: Harvard University Press, 1999); and especially by Pickering's *The Mangle of Practice: Time, Agency, and Science* (Chicago: University of Chicago Press, 1995).

5. Shapin and Schaffer, *Leviathan and the Air-Pump: Hobbes, Boyle, and the Experimental Life* (Princeton: Princeton University Press, 1985). See especially the section "Witnessing Science," pp. 55–60.

6. Genette, *Paratexts: Thresholds of Interpretation* (Cambridge: Cambridge University Press, 1997).

7. Pickstone isolates natural history, analysis, and experimentalism as three distinct "ways of knowing" characteristic of the modern period. He pulls his thoughts on these themes together in his *Ways of Knowing: A New History of Science, Technology and Medicine* (Chicago: University of Chicago Press, 1995).

8. The notion among certain classes that reading *National Geographic* gave one social cachet has long been understood. The *Geographic* marketed the magazine on this assumption from shortly after its inauguration, and commentators like Paul Fussell, for example, took it for granted (as in Fussell's *Class: A Guide Through the American Status System.* New York: Simon & Schuster, 1983, 144–45). Lutz and Collins put the assertion on a firm statistical foundation in their *Reading National Geographic* (Chicago: University of Chicago Press, 1993). Also see the chapters, "Science, Culture, and Expansionism in the Making of the *National Geographic*" and "Negotiating Success at the *National Geographic*" in Susan Schulten's *The Geographical Imagination in America, 1880–1950* (Chicago: University of Chicago Press, 2001), 45–59 and 148–75.

9. Fauconnier, *Mappings in Thought and Language* (Cambridge, Mass.: MIT Press, 1999)

The Natures of Maps

Part I

Map courtesy of Ng Maps/National Geographic Image Collection.

NORTHERN
TERRITORY
QUEENSLAND
SOUTH
AUSTRALIA
NEW
SOUTH
WALES
GULF OF
CARPENTARIA
CORAL SEA
CORAL SEA
ISLANDS
TERRITORY
GREAT BARRIER REEF
ARNHEM LAND
GREAT DIVIDING RANGE
GREAT ARTESIAN BASIN
Tanami Desert
Simpson Desert
Darwin
Alice Springs
Brisbane
Gold Coast
Adelaide
SYDNEY
Canberra
Newcastle
Wollongong
Broken Hill
Mount Isa
Townsville
Cairns
Rockhampton
Mackay
Port Augusta
Eyre Peninsula
Cape York Peninsula
BASS STRAIT

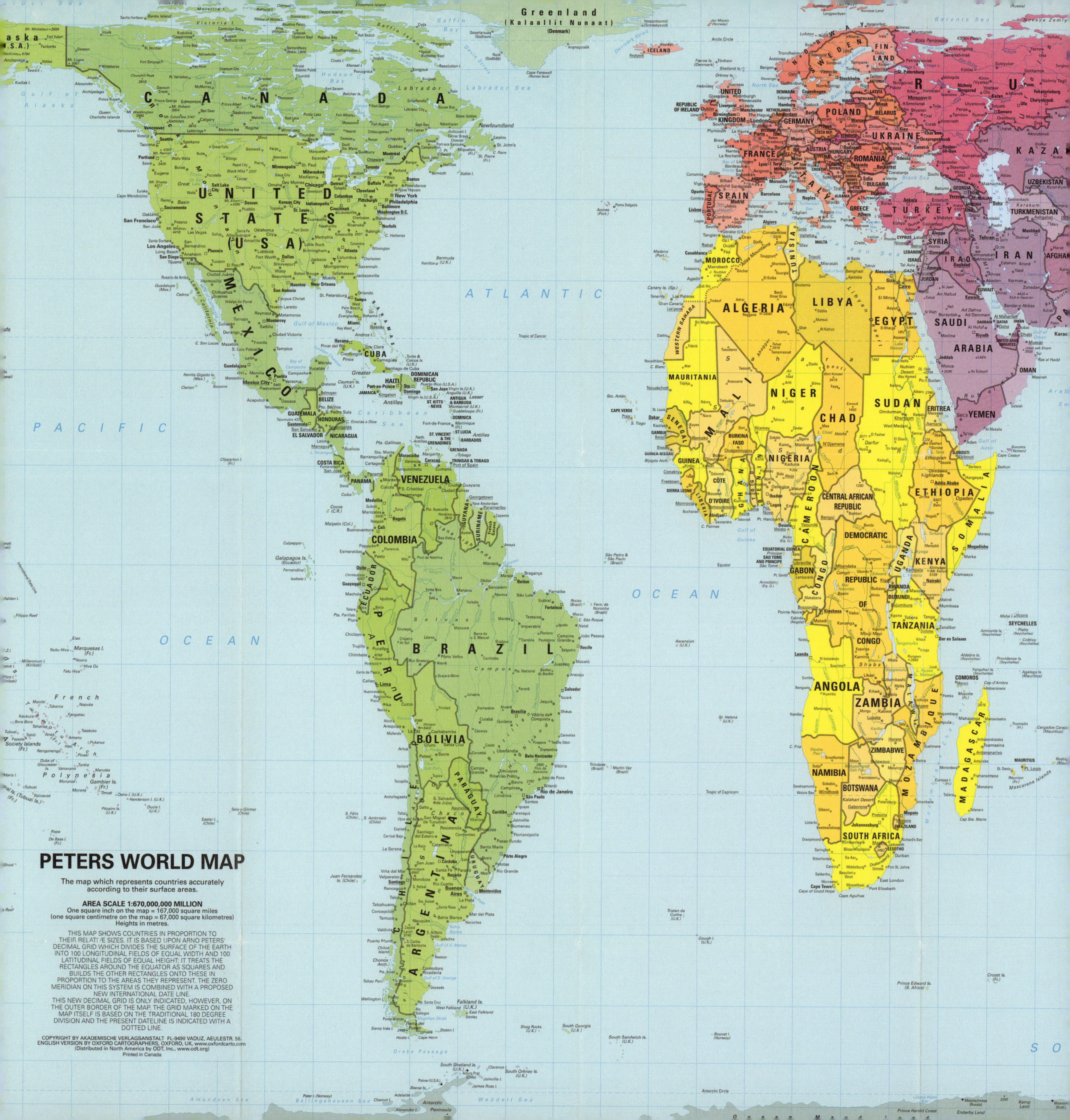
PETERS WORLD MAP
The map which represents countries accurately according to their surface areas.
AREA SCALE 1:670,000,000 MILLION
One square inch on the map = 167,000 square miles
(one square centimetre on the map = 67,000 square kilometres)
Heights in metres.
THIS MAP SHOWS COUNTRIES IN PROPORTION TO THEIR RELATI'E SIZES. IT IS BASED UPON ARNO PETERS' DECIMAL GRID WHICH DIVIDES THE SURFACE OF THE EARTH INTO 100 LONGITUDINAL FIELDS OF EQUAL WIDTH AND 100 LATITUDINAL FIELDS OF EQUAL HEIGHT; IT TREATS THE RECTANGLES AROUND THE EQUATOR AS SQUARES AND BUILDS THE OTHER RECTANGLES ONTO THESE IN PROPORTION TO THE AREAS THEY REPRESENT. THE ZERO MERIDIAN ON THIS SYSTEM IS COMBINED WITH A PROPOSED NEW INTERNATIONAL DATE LINE.
THIS NEW DECIMAL GRID IS ONLY INDICATED, HOWEVER, ON THE OUTER BORDER OF THE MAP. THE GRID MARKED ON THE MAP ITSELF IS BASED ON THE TRADITIONAL 180 DEGREE DIVISION AND THE PRESENT DATELINE IS INDICATED WITH A DOTTED LINE.
COPYRIGHT BY AKADEMISCHE VERLAGSANSTALT FL-9490 VADUZ, AEULESTR. 56.
ENGLISH VERSION BY OXFORD CARTOGRAPHERS, OXFORD, UK, www.oxfordcarto.com
(Distributed in North America by ODT, Inc., www.odt.org)
Printed in Canada
CANADA
UNITED STATES (U.S.A.)
MEXICO
Greenland (Kalaallit Nunaat) (Denmark)
ICELAND
PACIFIC
OCEAN
ATLANTIC
OCEAN
CUBA
VENEZUELA
COLOMBIA
BRAZIL
PERU
BOLIVIA
PARAGUAY
ARGENTINA
CHILE
UNITED KINGDOM
FRANCE
SPAIN
POLAND
UKRAINE
ROMANIA
TURKEY
IRAQ
IRAN
SAUDI ARABIA
YEMEN
MOROCCO
ALGERIA
LIBYA
EGYPT
MAURITANIA
MALI
NIGER
CHAD
SUDAN
NIGERIA
ETHIOPIA
CENTRAL AFRICAN REPUBLIC
CAMEROON
DEMOCRATIC REPUBLIC OF CONGO
KENYA
TANZANIA
SOMALIA
ANGOLA
ZAMBIA
MOZAMBIQUE
NAMIBIA
BOTSWANA
SOUTH AFRICA
MADAGASCAR

One

The nature of maps

The nature of maps

The nature of maps: an ambiguous phrase.

Furthermore, a comparatively famous one. In 1976, Arthur Robinson and Barbara Bartz Petchenik used it for the title of a book they subtitled *Essays Toward Understanding Maps and Mapping.* In 1991, J. B. Harley added "New" to the phrase to give the book he was proposing the title, *The New Nature of Maps: Essays in the History of Cartography.*[1]

Harley's was an explicitly subversive gesture. Although Harley died before he was able to write the introduction that would have justified his title, he gave his publisher the following description of his intentions:

> The dominant view of modern Western cartography since the Renaissance has been that of a technological discipline set on a progressive trajectory. Claiming to produce a correct relational model of terrain, maps are seen as the epitome of representational modernism, rooted in the project of the Enlightenment, and offering to banish subjectivity from the image. Cartographers have thus promoted a standard scientific model for their discipline, one in which it is claimed that a mirror of nature can be projected through geometry and measurement. Furthermore, this model for maps has colored the critical values of historians of cartography; they often assess early maps by this modern yardstick, thereby excising from the accepted canon of mapping not only maps from the pre-modern era but also those from other cultures that do not match Western notions of accuracy.
>
> The essays in this book—through historical examples and by a critical examination of the practices of modern cartography—seek to offer an alternative view of maps. Drawing on ideas in art history, literature, philosophy, and the study of visual culture, they subvert the positivist model of cartography, replacing it with one that is grounded in iconological and semiotic theory of the nature of maps. The interest of maps is shown to lie not so much in mimetic value but as simulacra which nevertheless may exert a profound influence upon the way space is conceptualized and organized within different societies. The theme of power is central to many of the essays. The way in which power—whether military, administrative, religious, or economic—is inscribed on the land through cartography is dissected and the nature of the political unconscious in maps is explored and illustrated. In new introductory and concluding essays aspects of this debate will be updated. The conclusion addresses the ultimate cartographic paradox: the map is not the territory, yet it often precedes, and even becomes that territory.[2]

Despite their differences, Harley's and Robinson and Petchenik's ideas about the nature of maps—and certainly Harley intended his first paragraph to be a description of Robinson and Petchenik's nature of maps—refer to the nature of *maps,* that is, to the nature, or inherent character, of *maps* as distinguished from the nature of *painting, sports,* or *small dogs.* But with equal grace the phrase can refer to the *nature* of maps, that is, to concepts of the natural—as distinguished from the cultural—figured by and brought into being on and by maps.

It is our intention both to insert ourselves into this history of ideas about the nature of the map *and* to embrace the ambiguity of the phrase, to explore the nature of *maps* by exploring the *nature* of maps and the *nature* of maps by exploring the nature of *maps.* We contend there can be little understanding of the one project except in the light of the other. We will show that the *nature* maps bring into being is one—actually it is a multitude—dependent on the nature of *maps,* while the nature of *maps* is best understood through its mapping of *nature.* This follows from the very idea of nature, which is about the intrinsic, the essence, the physical, the out-of-doors, the forces of the physical world, the primitive, the untouched-by-civilization, the uninfluenced-by-artificiality: the real. Nature wants to be the just-born, the innate, the native, the naïve, the untutored, the untaught, the unsophisticated, the unpolluted, the apolitical, the above-all-else *non*ideological. This is the one-word way Harley described what he'd been writing in those essays of his—an "inquiry into ways in which maps are ideological constructions and have been used as a classic form of power/knowledge in past societies."[3]

In the years since Harley wrote these words—and since we published "Designs on Signs" and *The Power of Maps*[4]—it has grown apparent that many people (if by no means all) are willing to accept maps as ideological constructions when it comes to zoning, school attendance districts, legislative districts (people love to say "gerrymander"), and national boundaries. But, then, the subjects of such maps are understood to be human constructions *in the first place.* There is nothing (it is said) natural about political boundaries; all are ideological creations. In this way, the ideological construction gets displaced from the map to its subject. The map itself remains uncontaminated; it is recovered as (what it claimed to be all along) no more than a conduit through which the ideological content—as *all* map content—passes undistorted, or if at all, then by no more than the "white lies" necessitated by the difficulties of printing the world on paper.

We reject this sophistry in all its parts.

The structure of the map's construction of knowledge

By focusing our attention on the *nature* of maps, that is, on what above all is supposed to be free of ideological construction—mapped wildlife, earthquakes, hurricanes, mountains, canyons, birds, butterflies, pinnipeds, ecosystems, landforms, vegetation, topography—we show that it is the *map,* hardly alone, in collaboration with other sign systems, which *creates* ideology, transforms the world *into* ideology, and by printing the world on paper *constructs the ideological.* It doesn't matter what has the map's attention. Whatever its subject is will be turned into something it isn't and in the process inescapably, unavoidably made ideological. At a minimum, at the most atomistic, it will be a construction, an invention, a conception, something drawn not from the world but from the mind of men and women; for maps are made not of wildlife, earthquakes, hurricanes, mountains, canyons, birds, but of *signs*—these themselves composed of marks and concepts.

The map: a field of concepts. There can be no escaping this.

But it's worse, much worse, for as slippery as these conceptual atoms may be, to make a map they must be aggregated into molecules and macromolecules of meaning in which constructions, interests, and ideologies enter at every point. But no sooner have we realized this than we find ourselves dealing with the nature of the *map.* We will show that the map is nothing more than a vehicle for the creation and conveying of authority about, and ultimately over, territory. We will demonstrate that the authority the map claims is the social manifestation of what the map presents as its "intrinsic" and "incontrovertible" factuality. We will spell out the way this factuality is constructed through the social assent given to the propositions maps embody. We will show how these propositions take the form of connections made among conditions, states, processes, and behaviors. Finally, we will make clear the way these connections are realized through the fundamental spatial/meaning propositions we propose to call *postings.* The posting is a proposition of the form, "*this is there.*"

By uniting an existence claim and a location, the posting locks together the *nature* of the map and the nature of the *map.* It is here, at the level of the posting where it is claimed that *this* of nature *is*—a waterfall or cliff, sequoia or syncline, high pressure cell or coral reef, mountain range or river—*and* that it is *there*—at this bend in the river or on that face of the mesa, in this grove or beside that anticline, in this system of winds or surrounding that island, rising above that plain or draining that basin—that the *this* takes on its *there* form, and the *there* takes on its *this* form. It is with the posting that nature is made spatial. The claims, that it *is,* and that it is *there,* reinforce each other. The *there* claim implies a reality test, that you can go there and look, a test that rises to the level of a challenge: "Why would we put it there if it weren't so? Check it out if you want!" Insisting that something is *there* is a uniquely powerful way of insisting that something *is.* Mapped things—no matter how conceptually daunting—possess such extraordinary credibility that they're capable of propelling into popular discourse abstruse abstractions cantilevered from abstruse abstractions: high pressure cells, El Niño, seafloor spreading, thermohaline circulation.

"You don't believe it? Check it out."

This is *there*—that tree—and *this* is *there* and *this* is *there*: through spatial magic the existence of the tree is transmuted into the existence of a forest, the existence of the forest is transfigured into the existence of an ecosystem, the existence of the ecosystem is transmogrified into the existence of nature. Nature. In space. As a spatial thing.

But the map can't leave well enough alone. It wouldn't be a map if it did. If it stopped at this atomic level—at the level of spatialized thing—the map would amount to a kind of spatial ontology. What makes the map a *map* is its exploitation of spatialized things—themselves propositions (this is there)—as the subjects of yet higher order propositions (this is there and *therefore it is also. . .*). The map *is* these propositions. Technically, a proposition is a statement in which the subject is affirmed or denied by its predicate (this is there). Take this ginseng plant. The map affirms of this ginseng plant (the proposition's subject) that it *is,* and therefore that it is also *in,* which is to say *of,* the Great Smoky Mountains National Park (the proposition's predicate). It could be the other way around (there is this). The map equally affirms of the park (the new proposition's subject) that it *is,* and therefore that it also *contains* ginseng (the new proposition's predicate). Either way the map *links* the plant and the park.

In so doing it connects the plant to the system of rules and regulations that is just another way of saying "national park." The park is not a collection of trees, shrubs, and other wildlife. That would just be a forest. The park is a way of *relating* to trees, shrubs, and other wildlife. These ways of relating are codified in rules and regulations. Some of these forbid the culling of ginseng. To cull ginseng in the Great Smoky Mountains National Park is therefore to poach. To cull ginseng outside the park, say across the road in a national *forest* (Pisgah or Nantahala), or on private land, is either to harvest or to steal, depending on how the map in question links the *theres* of the plants in question to the relevant systems of rules and regulations, codes and laws (to the relevant property rights—see figure 1.1). In the national forest, where trees can be cut, animals hunted, and plants gathered and sold, anyone can get a permit to cull ginseng. Poaching from private land, on the other hand, is a larceny.[5]

Note how that at this point a territory has been invoked. It has a national park, national forests, and parcels of private property. These are all equivalently subjects of different propositions made by the maps that invoke the territory. It is through the simultaneous affirmation of these propositions that the territory *as such* is brought into being. What assures us that the propositions are true? That they state facts? Only the *social assent given them,* the confirmation by the courts and by the court of public opinion, the voice of newspapers, and friends: "You shouldn't have been in the park. You should have stayed in the forest on the other side of the road."

Social assent and reference authority

The continual assent given to the propositions made by maps endows them with the authority that is uniquely that of *reference objects.* These include catalogs, calendars, concordances, encyclopedias, directories, phone books, dictionaries (*Merriam-Webster's,* the *OED* [look it up!]), thesauruses (*Roget's*!), glossaries (at the end of every textbook), textbooks (*Organic Chemistry*—no subtitle), the *National Geographic,* the *Times* (*New York, London, Los Angeles*), *TV Guide,* style guides (*The Chicago Manual of Style* [*fifteenth edition*!], Turabian, Strunk and White), cookbooks, field guides, travel books ("What does the *Mobil Guide* say?"), footnotes, citations, legal citations, priests, eye witnesses, constitutions, parliamentary procedures.[6] All of these constitute objectifying resources that permit a claimant to insist that, "It is not I, not I who says this, but—" before dropping, like a tombstone, the name of some revered reference object (*Langenscheidt's, Groves,* the *Britannica, Larousse,*

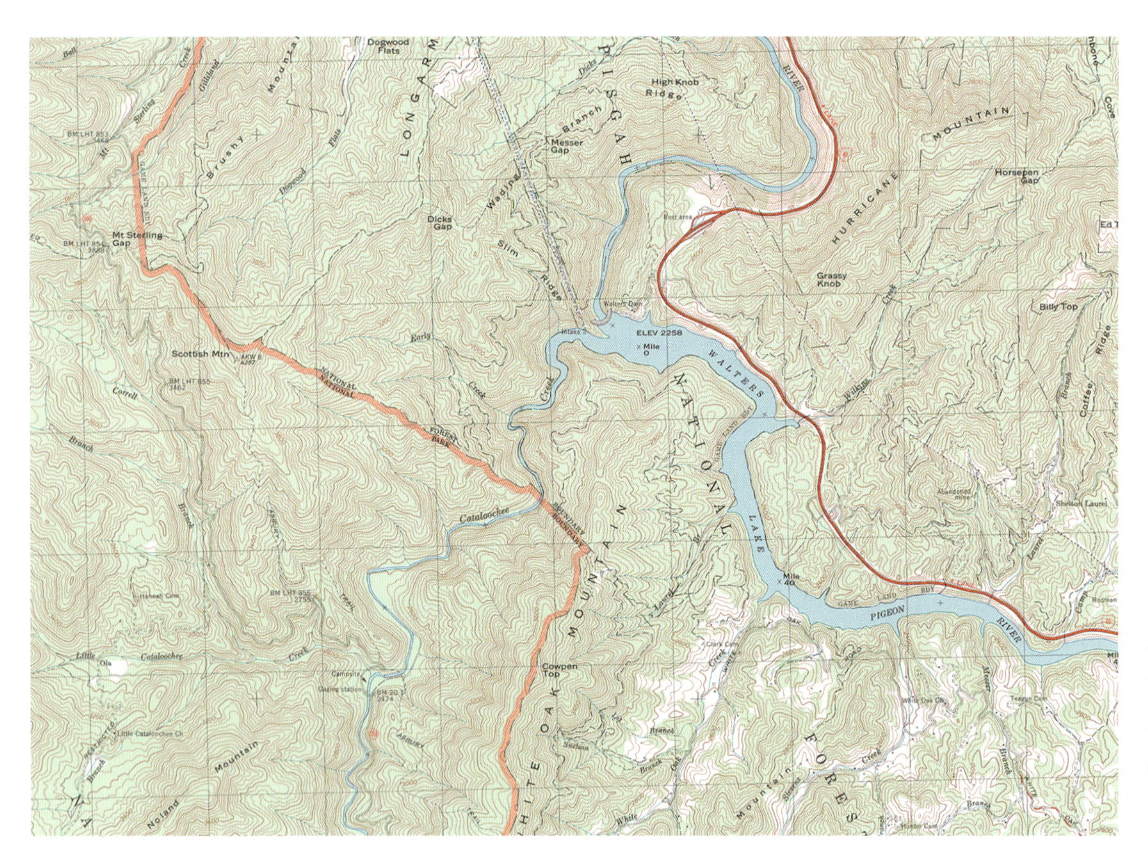

Figure 1.1
This detail of the *Cove Creek Gap*, NC, topographic quadrangle map describes, among much else, the border between the Great Smoky Mountains National Park and the adjacent national forest land (the thick orange line). With a permit, you may harvest ginseng in the national forest, but harvesting in the national park (or on private land) is against the law. The map not only invokes territory, but also civil code. (See the whole map in figure 11.1 on page 208.)

Courtesy of USGS, 1997.

Merck). Maps, too, are objectifying resources: the maps of Hammond, Bartholomew, Rand-McNally, Esselte, the National Geographic Society, AAA, Mobil, Michelin, the United States Geological Survey, other national mapping services, state highway maps, the Thomas Guides, Falk's, bus maps, maps of metro lines. Maps objectify by winnowing out our personal agency, replacing it with that of a reference object so constructed by so many people over so long a time that it might as well have been constructed by no one at all ("It is not I who says this, but . . . *the entire human race*"). Citation enhances a source's authority but also the authority of the one who cites it. The reflected light is blinding. Opposition is extinguished.

"You don't believe the map? Check it out!"

This authority, apparently descriptive, is inherently prescriptive. The phone book is not a guide to numbers from which one may feel free to pick and choose (though plenty evidently do): it *tells* you what to dial, it *prescribes* the number. A street directory *gives* you the address. There is no "Hmmm" here as there is over the choices a thesaurus offers or among the shades of meaning provided by decent dictionaries, where even so there is little hemming or hawing over spelling. The dictionary is *absolutely* prescriptive about spelling, a social fact we acknowledge—that we *dramatize*—in the annual rite of the National Spelling Bee. Among the mutual validations—spellers validating the authority of the dictionary, dictionary validating the speller's spelling—the prescriptive, the authoritative, is hard to miss.

Here: in this morning's paper there is an article about the new legislatively mandated North Carolina social studies curriculum.[7] The large, colorful photo illustrating the story is an overhead shot of an eighth-grade girl crouched over the state's transportation map (figure 1.3). Her left hand, forefinger extended, is on the transportation map, while her right hand transfers features—interstate highways and state and national forests—to a small outline map of the state. She is a human pantograph, literally reproducing—and by reproducing affirming—the existence (the *this-ness*) of state and national forests. As she traces their location (their *there-ness*), she simultaneously reproduces—and by reproducing affirms—the existence of North Carolina as a *state* of state and national forests. North Carolina's thereness is established later, in an exercise caught in another color photo on an inside page, where another student uses a globe to establish the state's coordinates. In all of this the map's authority is absolutely taken for granted.

The newspaper validates, with its literally glowing presentation, this power of the map to establish, almost in the religious sense, the following: the world as a sphere; North Carolina as a state of roads and forests; and the state and national forests as enclaves of green (the students color them green). It is these validations—the newspaper's, the curriculum's, the school's, the girl's—repeated uncountable times (hundreds and hundreds of times in this classroom alone)—that makes the map the potent vehicle it is for the creation and conveyance of authority about, and ultimately over, territory.

The paramap tells us how to read the map

The map itself—the piece of paper covered with ink—*insists* on this authority. Rare is the map that fails to advertise *in itself* its claims to be taken authoritatively. This advertisement takes the form of what, by analogy with Gerard Genette's coinage of "paratext," we propose to call the *paramap*. Genette distinguishes paratext into *peritext* and *epitext* (thus, the *perimap* and *epimap*). "In other words," Genette says, "for those who are keen on formulae, *paratext = peritext + epitext*."[8]

Figure 1.2
Reference objects are presumed to tell the truth. We learn to trust them from an early age.
Courtesy of Comstock Images—Education 1 Disc/Jupiterimages

Figure 1.3
An eighth-grade girl uses the state highway map as a reference object in her social studies class, affirming both the authority of the map and the subjects of the map's attention.
Courtesy of *The News & Observer*, Raleigh, N.C.

The *peritext* consists of all the verbal and other productions that surround and extend a text in order to present it: the quality of the paper, the quality of the binding, the character of the type, that of the printing, the dust jacket copy, the series indication (if any), the author name (anonymous, pseudonymous, with titles, without, etc.), and the work's title, together with whatever dedications, inscriptions, epigraphs, prefaces, forewords, intertitles, notes, and illustrations there may be.[9] The *epitext* consists of all the paratextual elements "not materially appended to the text within the same volume, but circulating, as it were, freely, in a virtually limitless physical and social space"—for example, advertisements, the letters publishers send out with review copies, promotional appearances by the author, interviews, lectures and so on, again, *surrounding* the text in order to present it, in order to shape its reception.[10] We have in hand, for example, a book club flyer, "Bonus Book Selections: Choose from a Wide Range of Reader Favorites" advertising *The Smithsonian Atlas of the Amazon.*[11] Copy promising an "exhaustively researched volume" is decorated with a cover shot, an inset map, and a blue bubble enthusing, "More than 150 Color Maps!" The exclamation point, the large number, the assurances that the book is a "Reader Favorite," that it's been "exhaustively researched," and its institutional affiliation with the Smithsonian conspire to position the atlas as authoritative and desirable. In Genette's terms, the flyer is a piece of the epitext; it also happens to be a piece of the *epimap* of every map in the atlas.[12]

The *perimap* carries out its labor closer in. "Australia Under Siege," a map supplement from the National Geographic Society, smothers its primary map (equivalent to Genette's "text") with seventeen ancillary maps, a timeline, a chart, four graphs, five photographs, twenty-seven blocks of type, several dozen call-outs, legends, titles, scales, and credits.[13] The map's construction of Australia as a biological horn of plenty besieged by its human inhabitants is largely a function of this rich perimap, though the epimap—the accompanying article in the *National Geographic* and *its* paratext (the title, "Australia—A Harsh Awakening," the blurb on the contents page with its "now barren fields of salt and dwindling marsupial populations," the note "From the Editor," the photographs with *their* titles and captions ["A graveyard of skeletons with gray arms raised in good-bye"], the "Behind the Scenes" item, and the later letters to the editor)—contributes to the construction substantially.

Ignoring the paramap, as contemporary cartography textbooks do (except for titles, legends, and scale bars, it's like the paramap doesn't exist), makes it much easier for such texts to ignore the claims of ideological construction to which the paramap is the essential guide. As a way of suggesting what's at stake we ask, "Would the projection promoted by German historian Arno Peters have stirred an iota of interest had it not been for its paramap?"

This is easy to answer, since *except for its paramap* the Peters projection (figure 1.5) is identical to James Gall's 1885 Orthographic Projection (figure 1.6), which never attracted any attention at all. But then Gall's perimap said, "Gall's Orthographic Projection/ Equal-area Perfect/ for Physical Maps, chiefly statistical," and its epimap more of the same at greater length.[14] Whereas Peters' perimap said, among much, *much* else (in large type along the margins of the map):

> Five thousand years of human history have brought us to the threshold of a new age. It is an age typified by science and technology, by the end of colonial domination, by a growing awareness of the interdependence of all nations and all peoples.
>
> Such a moment in history demands that we look critically at our understanding of the world. This understanding is based, to a significant degree, on the work of map-makers of the age when Europe dominated and exploited the world. Surprisingly, maps still reflect that bygone era.
>
> The new map, the work of German historian Arno Peters, provides a helpful corrective to the distortions of traditional maps. While the Peters Map is superior in its portrayal of proportions and sizes, its importance goes far beyond questions of cartographic accuracy. Nothing less than our world view is at stake.

Paramap	
Perimap	Epimap
Titles	Accompanying article(s)
Photographs	Advertisements that refer to the map
Illustrations	Marketing copy
Charts, graphs, timelines	Letter from the editor
Legends, scale bars, north arrows, other standard cartographic elements	Letters to the editor about the map
Callout text, blurbs	Behind the scenes info (how the map was created)
Credits	
Borders, decorative elements	

Table 1.1
The paramap can be broken down into perimap elements and epimap elements.

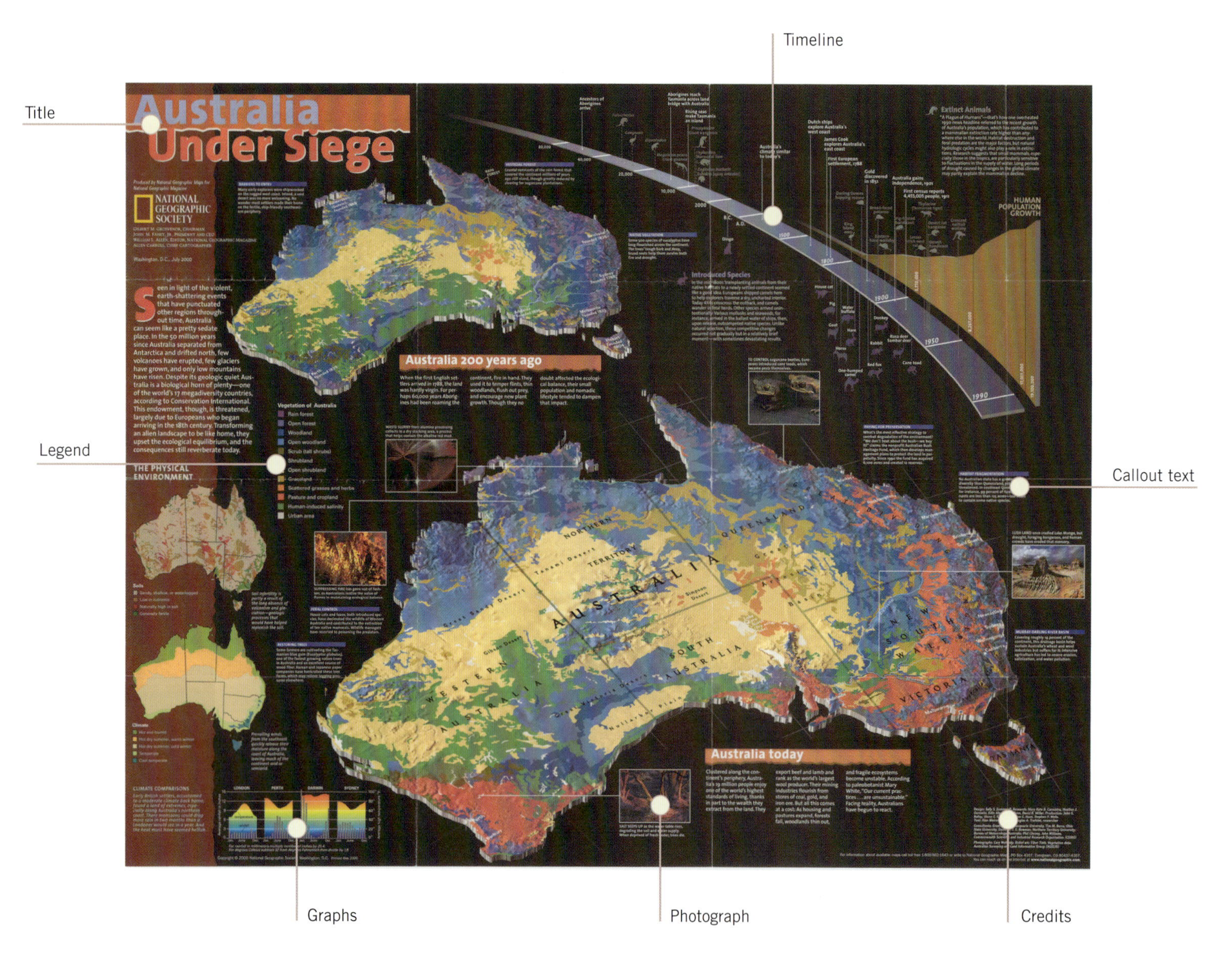

Figure 1.4
Perimap elements of "Australia Under Siege," a *National Geographic* map scrutinized in chapter 4.

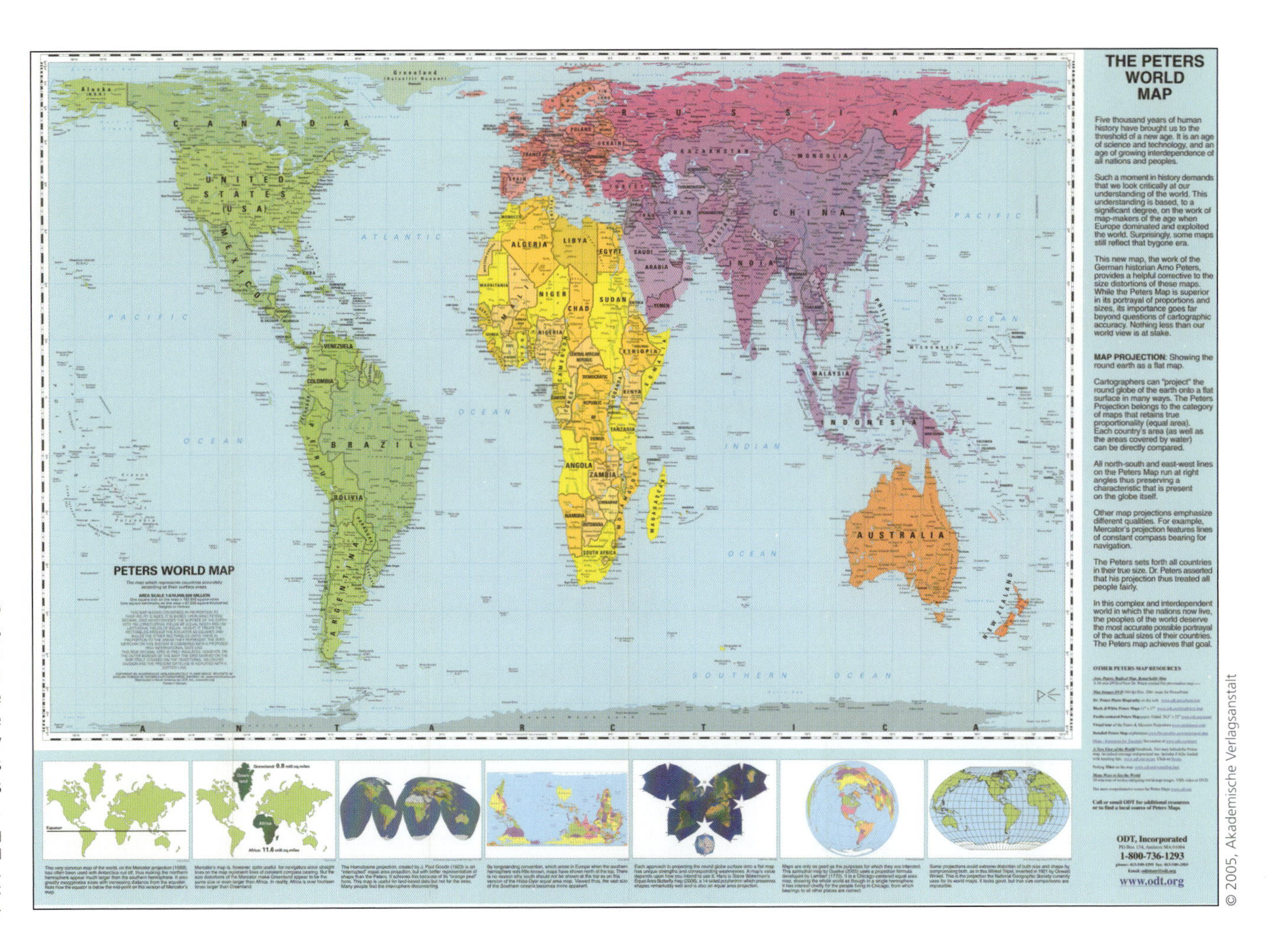

Figure 1.5
Arno Peters' map of the world. The notorious press conference in Bonn at which Peters announced his map was held in May 1973. The German Lutherans first published versions of the full-size map in 1974 in German. The first English language version was in 1983. The map shown here, heavily revised following Peters' death in 2002, was released in December 2005. Map dimensions (WxH): 50in. x 35in.

The Peters World Map was produced with the support of the United Nations Development Programme. For maps and other related teaching materials contact ODT, Inc., P.O. Box 134 Amherst, Mass. 01004, USA (800-736-1293; Fax 413-549-3503; e-mail odtstore@ODT.org.

> . . . In the complex and interdependent world in which the nations now live, the peoples of the world deserve the most accurate possible portrayal of their world. The Peters Map is that map for our day.[15]

Peters' perimap essentially accused cartographers of producing distorted maps in the service of a discredited European colonialism —of being ideologists in a bad cause—and positioned his map as a unique antidote. Next to the UN seal in the map's lower right-hand corner it said, "This map is produced with the support of the United Nations Development Program."

Cartographers flipped! Driving them even more insane was an epitext, Peter's inflammatory book, *The New Cartography*.[16] The most reputable review of *The New Cartography*—Arthur Robinson's—opened with, "The review of a book such as *The New Cartography* would ordinarily be short since much of it is misrepresentation, is illogical and erroneous, and one's initial reaction is simply to dismiss it as being worthless." The review nevertheless proceeded to eviscerate Peters for another eight pages.[17] As a scholar and a gentleman, Robinson did not stoop to mudslinging, but characterizations like, "Arno Peters, the German architect of this novel map, was in fact not a cartographer at all but a journalist and propagandist for leftist causes who had mastered 'the art of writing press releases,'" etc., etc.,[18] by other critics made them sound like right-wing ideologues on an AM talk show. An entrenched profession attacked *everything*—especially the claim that the map was new (Peters hadn't known about Gall)—but remarkably, the critics didn't confine themselves to Peters' paramap or even the rechristened Gall-Peters projection: they launched an attack against *rectangular world maps in general.*

Like Mercator's and many others, the Gall-Peters projection produces a rectangular world, unlike those of, say, Robinson and Mollweide, which are curved (figure 1.7). At the very height of the controversy, the American Congress on Surveying and Mapping adopted a "sternly worded resolution condemning [rectangular maps] for 'showing the round earth as having straight edges and sharp corners.'"[19] This preposterous (and wholly ineffectual) resolution was endorsed by the American Cartographic Association, the American Geographical Society, the Association of American Geographers, the Canadian Cartographic Association, the National Geographic Society, etc.,[20] and all because of the *paratext*—which few of those endorsing the resolution would even consider part of the map—of a map they universally dismissed.[21]

That the paramap should have this power is no surprise. Rare is the image that can dispense with words. Roland Barthes wondered whether *any* system of signs could do without them: "Is there," he asked, "any system of object-signs which can dispense with articulated language? Is not speech the inevitable relay of any signifying order?"[22] By *relay* Barthes always understood a second-order message, a connotation parasitic on a first-order message, as a caption to a photograph (say in a fashion magazine), or the text on a map (say in the title or legend). Among what Barthes called the relay effects of speech were its ability to fix—to immobilize—perception at a given level, first of

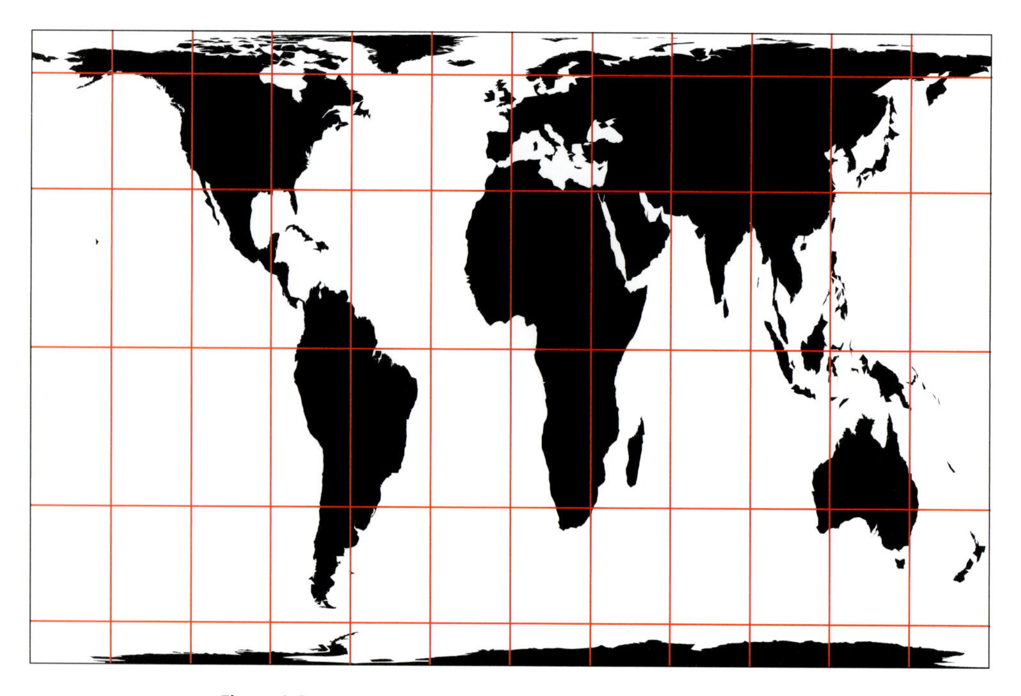

Figure 1.6
The Gall orthographic projection, originated by James Gall in 1885. As you can see, the projection is identical to Peters'. It's the perimap of the Peters map that sets it dramatically apart.

all at the level of the photo or map, say, rather than at that of the paper, the printing screen, or the typeface; but then to draw attention, as to the collar or hemline (in the fashion photo) or to the system of highways (in a road map). Other relay effects of speech include its ability to go beyond the image, to interpret, to say what the narrow collar *means* (it's sexy) or the tint of red in the legend (the road is limited access), and its ability to direct attention, to emphasize ("Pay attention to *this*!").[23] In Peters' case, the paramap attempts to keep us focused on the equal-area property of his map, to force us to compare it *along this dimension* to the unequal-area Mercator, and to pretty much ignore everything else. His paramap immobilizes our perception on his chosen ground.

Every map does this. Consider *Seasonal Land Cover Ecoregions*, published by the United States Geological Survey (USGS) in 1993 and distributed as a supplement to the *Annals of the Association of American Geographers* in 1995[24] (figure 1.8). The title doesn't add, "of the coterminous United States" but then the familiar shape obviates the need. This is an extremely fine-grained mosaic. Colors change within minute parts of millimeters, shifting almost continuously, but are dominated in the west by acid pinks and purples (blotchy and pimpled with bright and sallow greens), in the plains states and Midwest by intense yellows and oranges, and in the east by mottled opal and jade greens. It's pretty amazing looking (*it's a nightmare*), and absent its perimap, *completely* unintelligible. Oh, sure, if you know your geography you can pick out the Mississippi Delta and the Appalachians, the Corn Belt and the Great Plains, the Black Hills and the Central Valley of California; but if you know your geography, you can find these on an outline map, and none of them are actually shown on this map anyway. What *are* shown are "159 polythetic seasonal land-cover classes in 159 different colors, each keyed to a unique combination of vegetable/land-cover types, seasonal properties, and relative primary production." You learn all this on the back of the map where a text is supplemented by twenty-six smaller maps (including a simpler "monothetic land-cover map" of only twenty-six classes), as well as in the epimap, the article that accompanied the map in the *Annals*. Indeed it would not be too much to insist that *Seasonal Land Cover Ecoregions'* epimap also included all of the literature cited in the article, so concentrated within the map is the density of previously concentrated material (for example, "These maps were derived from an AVHRR normalized difference vegetation index [NDVI] time series . . ."). Absent this supplemental literature, there is little true understanding of the map.

Besides stabilizing our attention on the map as a map, the *Seasonal Land Cover Ecoregions'* paramap *forces* us to see the map as one of polythetic seasonal land-cover classes, to compare it along this dimension to a less subtle monothetic classification, and to pretty much ignore everything else. The paramap immobilizes our perception on the map's chosen ground.

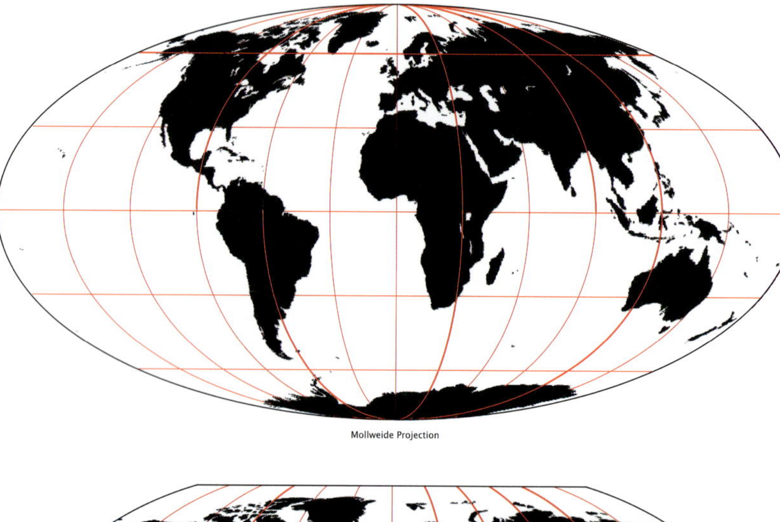

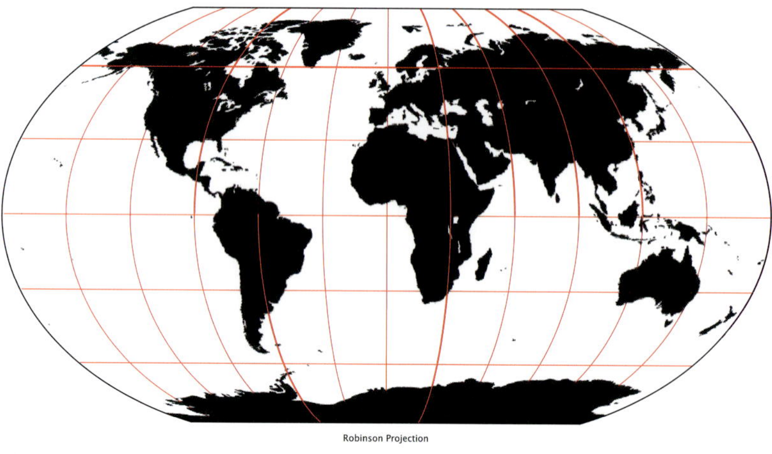

Figure 1.7
Examples of curved map projections. Shown here are those of Karl Mollweide (top) and Arthur Robinson (bottom).

Cognitive cartographics

Given: a nightmarishly encrypted main map, a 159-item key, a text (with citations to many other texts), twenty-six supplementary maps,

Figure 1.8
Seasonal Land Cover Ecoregions. What does it all mean? Map dimensions (WxH): 27in. x 32in.
Courtesy of USGS, 1993.

titles, scales, and credits as in *Seasonal Land Cover Ecoregions*; or a main map, legend, texts, seventeen ancillary maps, a timeline, a chart, four graphs, five photographs, several dozen call-outs, legends, titles, scales, and credits as in "Australia Under Siege"; or even a main map, text, seven ancillary maps, titles, scales, and credits as in *The Peters World Map*; given this heterogeneity, what is one to do? That is, what sense is one to make of it? How to assemble it, pull it all together?

Contemporary cartography textbooks treat this as a problem in graphic design: "Titles, legends, scales, and insets may be arranged in various ways in the graphic organization of a map," say the authors of *Elements of Cartography*, sixth edition, where none of their examples comes near to approaching the complexity of ours:

> Nothing should seem out of place. Layout is the process of arriving at proper balance. In a well-balanced design, nothing

3 Map layout: balance

Balance refers to the stability of a map layout. When balance is poor, map readers may be distracted. When balance is achieved, map readers will focus on the content of the map. Balance can be symmetrical or asymmetrical.

balance

Balancing map elements is complicated and intuitive. The map elements to balance vary in weight. ***Heavier*** *elements include those that are larger, darker, brightly colored, simpler and more compact in shape, and closer to the map edge (particularly the top).* ***Lighter*** *elements include those that are smaller, lighter, dully colored, complex or irregularly shaped, and closer to the map center.*

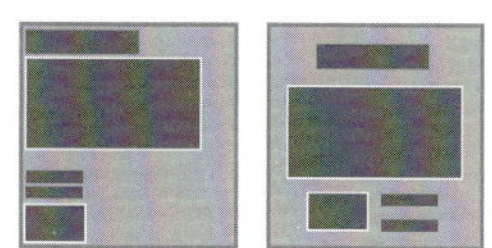

Poor balance:

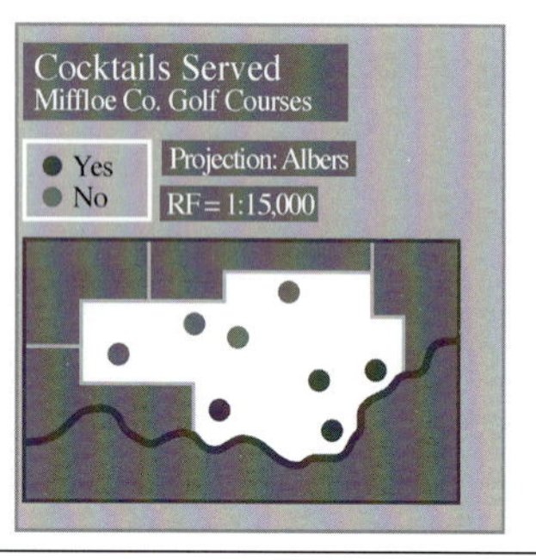

Better balance:

Figure 1.9
Achieving good balance is easy when there's a handful of elements to be laid out. But how relevant are diagrams like this to the real-world problems faced by designers of complicated maps like the ones we're looking at?
Derived from Krygier and Wood, *Making Maps*, (New York: Guilford, 2005).

> is too light or too dark, too long or too short, too small or too large, in the wrong place or too close to the edge. . . . The cartographer's job is to balance visual items so that they "look right."[25]

In the illustration accompanying these remarks, circles and squares balance or unbalance a beam depending on their size and distance from the fulcrum. Another illustration displays differently proportioned rectangles with the admonition that those in the ratio of three to five make "the most stable and pleasing map format." Yet another illustration shows different arrangements of title, legend, and locator inset, all *three* of which mapmakers are encouraged to retain no matter how difficult the design. The text does *not* contemplate a 159-item key, a text (with citations to other texts), twenty-six supplementary maps, titles, scales, and credits. Another text, Borden Dent's *Cartography: Thematic Map Design* is more sophisticated, but again *nothing* like the complexity of our examples is contemplated.[26] This whole tradition of thinking about maps as graphics comes out of an illustration, out of an advertising tradition. Indeed, the text in *Elements of Cartography* (whose first edition came out in 1953) could have been lifted from something like William Longyear's *Advertising Layout* (whose first edition came out in 1946). For example, Longyear says the following:

> Balance is most important in a layout. The various sizes and shapes of the elements in the layout must have good artistic composition. There are few, if any, distinct formal rules to guide the layout man in deriving good balance. Balance has some of the qualities of a seesaw. By setting a vertical line through the center of the layout to serve as a fulcrum, elements may be balanced for both size and weight. . .[27]

Given the prevalent idea that maps amount to a kind of "seeing," none of this is surprising. Committed as most cartographers are to the idea that maps "present information," cartographers rather appropriately approach map design as they would the design of an advertisement, or a smorgasbord, where the aim is to make everything as attractive as possible to draw the grazing eye.

Doubtless this is all sound advice (though what heart a designer is to take from knowing that in a well-balanced design nothing is "too light or too dark, too long or too short" is open to question) but, given that *we* see maps as systems of propositions (as *arguments*), nothing could be further from what we have in mind. The question is *not* for us how things are arranged for the eye, but how the design promotes and constrains, how it directs, the construction of meaning. It is not about the "presentation of information." It is about the construction of meaning as a basis for action. It is for us a question of cognition.

The discipline that has contributed most substantially to our thinking is the new and rapidly evolving one of cognitive linguistics. We're proposing that cognitive linguistics is a good model for thinking about cartography, for thinking about *cognitive cartographics.*

Why cognitive linguistics? Because it is a nonrepresentational approach to language that is concerned with how we think, act, and communicate. Unlike historical forms of linguistics, which were essentially concerned with the nature of the *signal,* cognitive linguistics is concerned with the *meaning construction* upon which language operates. For cognitive linguists, "meaning construction refers to the high-level, complex mental operations that apply within and across domains when we think, act, and communicate."[28]

This makes it a form of linguistics analogous in intent to the theorizing we're doing about cartography, which is directed toward the thinking, acting, and communicating that maps facilitate (i.e., cognitive cartographics). No surprise then that cognitive linguistics critiques historical forms of language theorizing in much the same way that we have critiqued traditional theories of cartography. For example, cognitive linguistics critiques traditional forms of language theory for their predisposition to sharply separate components (syntactic, semantic, pragmatic), and to study these in isolation, especially independent of their use in the world for reasoning and communication.[29] This parallels traditional cartographic thinking, which not only compartmentalized mapmaking from map use, but within mapmaking, compartmentalized projection, generalization, symbolization, design, and the rest. In its interest in understanding the role of, say, grammar in discourse configuration, cognitive linguistics is a model of appropriate procedure for, to give one example, understanding the role that the choice of map projection plays in shaping world view. As we've already quoted Gilles Fauconnier in the introduction, "Language data suffers when it is restricted to language,"[30] not just because language depends on highly structured background knowledge, conversational meaning, negotiations, and the like, but because it is directed toward an end in action. The same has to be said of maps: map study suffers when it is restricted to maps.

Furthermore, unlike historical forms of language analysis, including semiotics (which we nonetheless hang on to), cognitive linguistics is dynamic, committed to understanding the way meaning is constructed *on the fly*, which is certainly the way we propose to understand—and model—map reading, as a process in time, which encourages the construction of certain kinds of meaning and ultimately behavior.[31] We're not interested in maps as pictures. We're interested in maps as the significant players they are in the world of action. Maps—let us acknowledge this—are not just *of* the world, but in it, very much a part of it.

At the heart of cognitive linguistics is what its developers think and write about as *mental spaces*. Mental spaces, says Fauconnier, "are partial structures that proliferate when we think and talk." Since these constructions take place on a *cognitive* level, they are partial *cognitive* structures. This is to mark their distinction from the structure of language. Such a cognitive structure "is *not* an 'underlying form,' it is *not* a 'representation' of language or of language meaning, it is *not* bijectively associated with any particular set of linguistic expressions." Such a cognitive structure is not a representation of the world either, but it relates language *to* the world by providing "real-world inferences and action patterns." Fauconnier and Mark Turner characterize these mental spaces as "small conceptual packets constructed as we think and talk, for purposes of local understanding and action." These small conceptual packets (or partial cognitive structures), "correspond," Fauconnier and Turner elaborate, "to activated neuronal assemblies," which are linked or link themselves to other activated neuronal assemblies. [32] Cognitive linguists think about these neuronal linkages as *mappings*. For example, the configurations of words you're reading right now are opening up thinking spaces in your brain, that is, activating assemblies of neurons, which are connected to, project to, are mapped onto, other thinking spaces in the process of constructing meaning.

These *mental space mappings* are the essential subject of cognitive linguistics (giving rise to an alternative name—*space grammar*):

> In terms of processing, elements in mental spaces correspond to activated neuronal assemblies, and linking between elements corresponds to some kind of neurobiological binding, such as co-activation. On this view mental spaces operate in working memory but are built up partly by activating structures available from long-term memory. Mental spaces are interconnected in working memory, can be modified dynamically as thought and discourse unfold, and can be used generally to model dynamic mappings in thought and language. Spaces have elements and, often, relations between them. When these elements and relations are organized in a package that we already know about, we say that the mental space is framed and we call that organization a "frame."[33]

George Lakoff says that these frames can be structured by idealized cognitive models (ICMs). ICMs are descended from the earlier *plans* and *scripts* of Roger Shank and Robert Abelson's "script theory," where a *script* was a hypothetical knowledge structure capable of generalizing about a socially appropriate sequence of events. A script was a sort of ideal, an ideal you attempted to follow, or that you expected others to follow. Schank and Abelson's best known example was the Restaurant Script. Script theory, in turn, made powerful connections to Steven Toulmin's theory of logic as "generalized jurisprudence" and to Mikhail Bakhtin's ideas about speech genres.[34]

One of the appealing things about cognitive linguistics is the way it absorbs, integrates, and updates so many worthwhile concepts from the past, while at the same time promising to connect them to neurophysiologic evidence being developed tomorrow from PET scans and functional MRIs.[35] Via cognitive linguistics, yesterday's hypothetical knowledge structures promise either to disappear into the junkyard of failed models (still always worth braving the junkyard dogs to visit) or to transform themselves into actual knowledge structures. It's all very heady.

"The dynamics of mental space construction and space linking are technically abstract, but conceptually straightforward," Fauconnier and Eve Sweetser write. "The basic idea is that, as we think and talk, mental spaces are set up, structured, and linked under pressure from grammar, context, and culture. The effect is to create a network of spaces through which we move as discourse unfolds."[36] Similarly, as we read the main map and the various elements of the paramap—text, ancillary maps, title, photos, scale bar, graphs—one or more mental spaces open up that are structured (frequently by ICMs or frames) and linked under pressure from the graphic structure, context, and culture to create a network of spaces—one space opening up after another—through which we move as we read and make sense of the map.

In figure 1.10, motion through this network starts from a *base space*, which establishes the initial *viewpoint* (the space from which, at a given point in the reading, other spaces can be accessed or created) and *focus* (the space to which structure is actively being added); and then *shifts* viewpoint and focus as the reading unfolds. In natural languages, it is grammar that helps answer questions such as the following: Where is the starting point (the base space)? What space is currently the viewpoint? What space is currently in focus? What is the relationship of the viewpoint to the base? What is the nature of the connections between spaces? In maps it is graphic structure—the design—that helps answer these questions.

Contemporary cartography texts are not entirely unaware of this parallelism. For example, *Elements of Cartography* says the following:

> The task of map design has much in common with writing. An author—a literary designer—must employ words with due regard for many important structural elements of the written language, such as grammar, syntax, and spelling, in order to produce a first-class written communication. Likewise the cartographer—a map designer—must pay attention to the principles of graphic communication.[37]

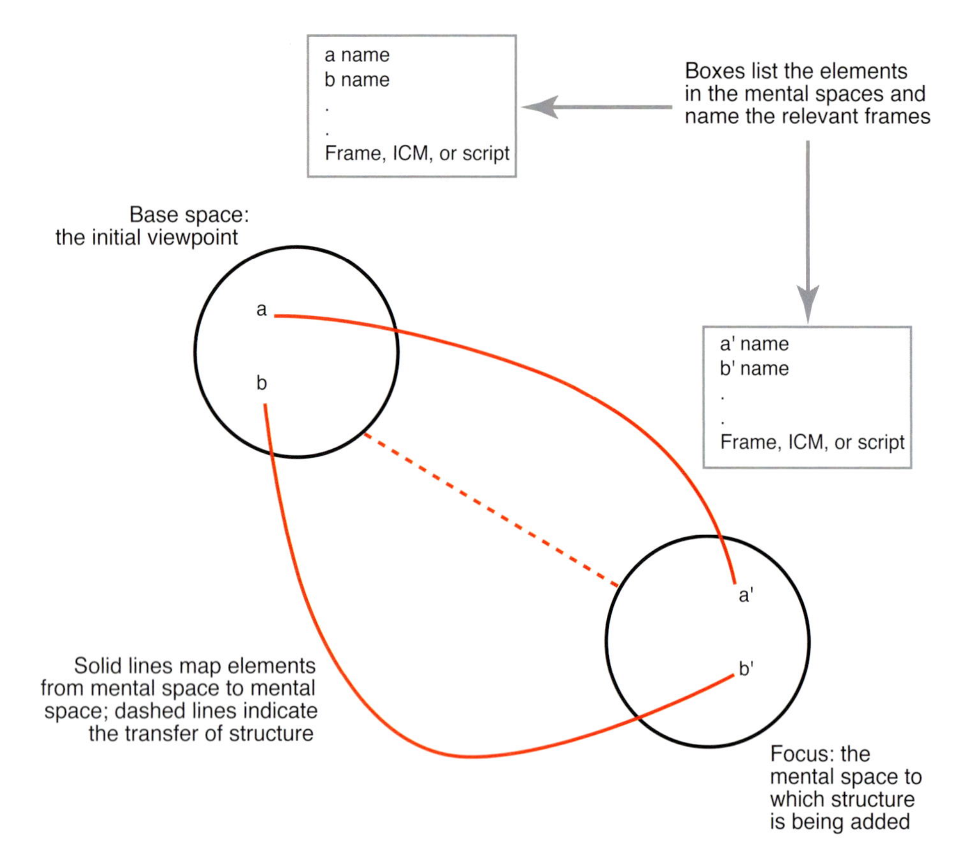

Figure 1.10
Cognitive linguistics diagrams represent mental spaces with circles, a network of which is propagated as discourse unfolds. The first space that a discourse opens is considered its *base space* (here upper left). The elements in the space are lettered. If a frame, ICM, or script structures these elements, it appears as a box containing the elements and naming the framing structure. For example, the frame "buying and selling" with its buyer, seller, consumables, money, price, and rich set of inferences about owning, exchange, and so on; or the frame "vegetation map" with its locative field and vegetation classes, and inferences relating to hierarchic relationships, adjacency expectations, and the like. Continuing discourse spawns further spaces. The space to which structure is being added is the *focus* (here lower right). Dashed lines indicate the transfer of structure from space to space, while solid lines map the movement of elements. The diagrams are a graphic way of keeping track of what's going on.

Of these so-called principles we have seen a sample ("nothing too dark or too light, too long or too short"), and despite infusions of psychophysics over the years this remains state of the art.[38] As such, these "principles" bear little relationship to the structure provided by grammar and indeed, absent explicit scaling arguments, offer no guidance to mapmakers—and so no guidance to map readers—whatsoever. Yet however unarticulated, implicit principles masked by the chatter about aesthetically pleasing appearance and "looking right" must in fact be structuring the elements of the map, that is, guiding the creation of spaces through which we move as we read and make sense of the map.

Space mapping has convinced us—and we are convinced it will convince you—that the principles underwriting the graphic design of maps are wholly at the service of *the structure of the map's construction of knowledge*. That is, the principles of map design are concerned with the straightforward display of postings amenable to consumption by propositions appearing on the plane of the map as incontrovertible characteristics of the territory the map thereby evokes and over which it exhibits its authority. The essential goal of these principles is not "looking right" but the preservation and enhancement of authority, and nothing supports this goal more strongly than the pretense, and so the impression, that all maps do is "present information."

Did we mention how preliminary the work in cognitive linguistics is, how tentative its conclusions? Even more preliminary are our proposals, which nonetheless we advance as a model for understanding how maps hoist themselves off the page into our brains, spawning world views, images of the city, and a spatialized, a regionalized nature; a nature plucked equally from the vagaries of veneration and from the toils of taxonomy; a nature capable of being isolated as a region, capable of coming into conflict with other regions, and capable of being legislated and commercialized. This spatialized nature can threaten and be threatened; it can awe and it can be cuddled; it can be collected and it can be systematized; it is unknowably remote and it is underfoot. It is a nature, ultimately, *quietly put in its place.*

Eight natures of maps

Which is our question: the place of nature, *what is it?* Our contention is that, today, maps play a significant role in the way we frame this question and in the answers we give to it. Since what nature is taken to be affects the possibilities of its being mapped, and since what mapping is taken to be affects the nature we can imagine being mapped, there has been a continuous evolution in the mapping of nature over the half millennium during which maps have played a significant role in human affairs.[39] This evolving history, being eagerly explored, has not yet been written—nor do we propose, despite its importance, to write such a history ourselves. Our interest lies elsewhere, in the present, in the ways in which everyday map readers, encountering maps throughout the course of their lives, find maps participating in the construction and reconstruction of their ideas of nature.

Nature, as we suggested earlier, is a powerful concept, circling as it does around ideas of the real and the nonideological. It can be used as a heavy hammer to attack the "unnatural" and as a powerful flag around which to rally the "natural." So it has been interesting, as we have worked our way through the maps that came to hand, to discover so many *different* natures. There is the nature that is threatened, but there is also the nature that threatens. There is a sublime, awe-inspiring nature, but there is also a pretty, endearing, and bounteous nature. There is a nature that we collect, which may be different from the nature that we study. There is an unfathomable, mysterious nature, but there is also a nature in which we can picnic.

1. *Threatened nature.* Nature as victim, susceptible to countless threats, is inescapable these days. This is nature harassed by man. It is nature on the ropes. "Wildlife as Canon Sees It" is the headline in a series of full-page advertisements that Canon has run for years in a broad range of magazines with an enormous readership: *Scientific American, National Geographic, Natural History, The Smithsonian* (figure 1.11). A photograph of an animal (doubtless taken with a Canon camera) fills the top half of the page. A text—one

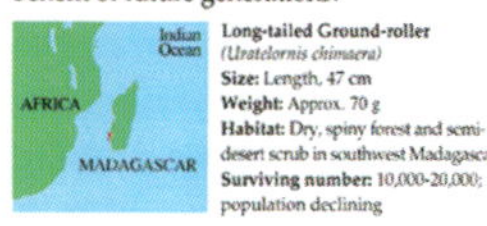

Figure 1.11
A Canon wildlife ad (2001) that includes a map illustrating the minuscule—almost nonexistent—habitat of the terrestrial long-tailed ground-roller. One reading of this might be "It's a good thing someone had a Canon in the right place at the right time, because unless we do something about it, this photograph may be the last chance we have to see the bird in its threatened, shrinking home."
Courtesy of Canon, Inc.

of Barthes' relays—says (in the case at hand), "In the relative cool of early morning, a terrestrial long-tailed ground-roller probes among leaf litter and around thorny thickets, hunting for insects and their larvae. The shy bird stands quietly for extended periods surveying an area, slowly lifting and lowering its long tail. Then, with a few quick hops, it disappears into the scrub."[40] Another sentence sketches the bird's domestic economy ("stays with its mate while nesting"), and another its imminent peril: "Confined to a small strip of unprotected coastal forest, the long-tailed ground-roller is threatened by loss and degradation of habitat." A map is invariably appended: ours shows, in green and blue, the Indian Ocean, southeast Africa, and Madagascar. An aging eye can hardly discern the miniscule dot (in red) on the southwest coast of Madagascar that signifies the bird's remaining—shrinking, threatened—habitat. By translating "habitat" into space, the map gives the habitat real credibility at the same time it dramatizes how small this habitat is. Beleaguered nature. Canon wants to help. Canon wants us all to help.

2. *Threatening nature.* Yet every bit as common are maps of a beleaguering nature: nature threatening man, nature on the rampage. Every summer newspapers in our part of the country publish inserts with titles like, "Stormtracker 2005, Your Official Hurricane Survival Guide." A joint effort of Raleigh's *News and Observer* and a local television station, this one was widely distributed and "proudly sponsored" by Jiffy Lube and North Carolina's Electric Cooperatives with paid advertisements from local companies sprinkled throughout. Stuffed with sound advice ("Prepare a Family Disaster Plan"), these inserts are really all about the maps. There are usually two of them. One describes areas prone to flooding and sketches the evacuation routes. The other—typically a couple of feet across—is a hurricane tracking map showing the East Coast and Atlantic Ocean, and extending to 30° west (figure 1.12). The water area is gridded in one-degree increments. Inset is a graph for you to record facts about the storm, the time, its latitude and longitude, and other statistics, sort of like a line score in baseball. Transferring the storm's latitude and longitude to the map lets you keep track of the storm.[41] As you keep updating its location, you transform the hurricane into something spatial. *You* spatialize it. It's a short step from this to synoptic hurricane maps (like the widely reproduced satellite map NASA's Goddard Laboratory made of 1989's Hurricane Hugo), maps compiling tracks of hurricanes, and maps of hurricane regions. On the National Geographic's "Hurricanes: Where Ill Winds Blow" map, gradations of blue demarcate the frequency of hurricanes per hundred years in steps of forty.[42] Hurricaniana: it's now a region—a place—like any other.[43]
3. *Nature as grandeur.* What can threaten also can awe, and the sense of powerlessness and personal insignificance that hurricanes inspire is not unrelated to what people experience standing on the rim of the Grand Canyon, looking up at Everest, down on Victoria Falls, or across the Amazon. With their majesty, their sublimity, each inspires a sense of the *power* of nature, less its strength (hurricanes are strong), than its boundlessness, its magnanimity, its *glory*. As we write these words, a new *National Geographic* map of Everest arrives, an extraordinary image, photographic in detail. Here Everest, vast beyond understanding, is caught at a resolution of nineteen inches (May 2003 supplement). But . . . didn't *National Geographic* just publish a map of Everest? Wholly different but just as awesome? A joint production of the *Geographic,* the Boston Museum of Science, and the governments of Nepal and China? Actually, that was fifteen years ago (November 1988, figure 1.13), and it came in a long line of powerful Himalayan images. *The Kingdom of Sikkim,* glorious mountains from north to south, appeared as a supplement to the *Annals of the Association of American Geographers* in 1969. Four years earlier the *Annals* had published *The Kingdom of Bhutan,* twelve square feet of Himalayas folded up and shipped along with the journal.[44] Before that . . .

   But the list is long. Each of the great sublimities has been mapped, the maps as extraordinary in their way as their subjects, the efforts invariably daunting (so high, so deep, so far away). This is not a nature we can threaten (not one we can *dream* of threatening), nor yet is it one that threatens. This is a nature *beyond us.*
4. *Nature as cornucopia.* There is yet another nature, the nature that we embrace, that we cuddle. This is the nature of the small and the soft, the fuzzy and the warm. This is the nature of fur and

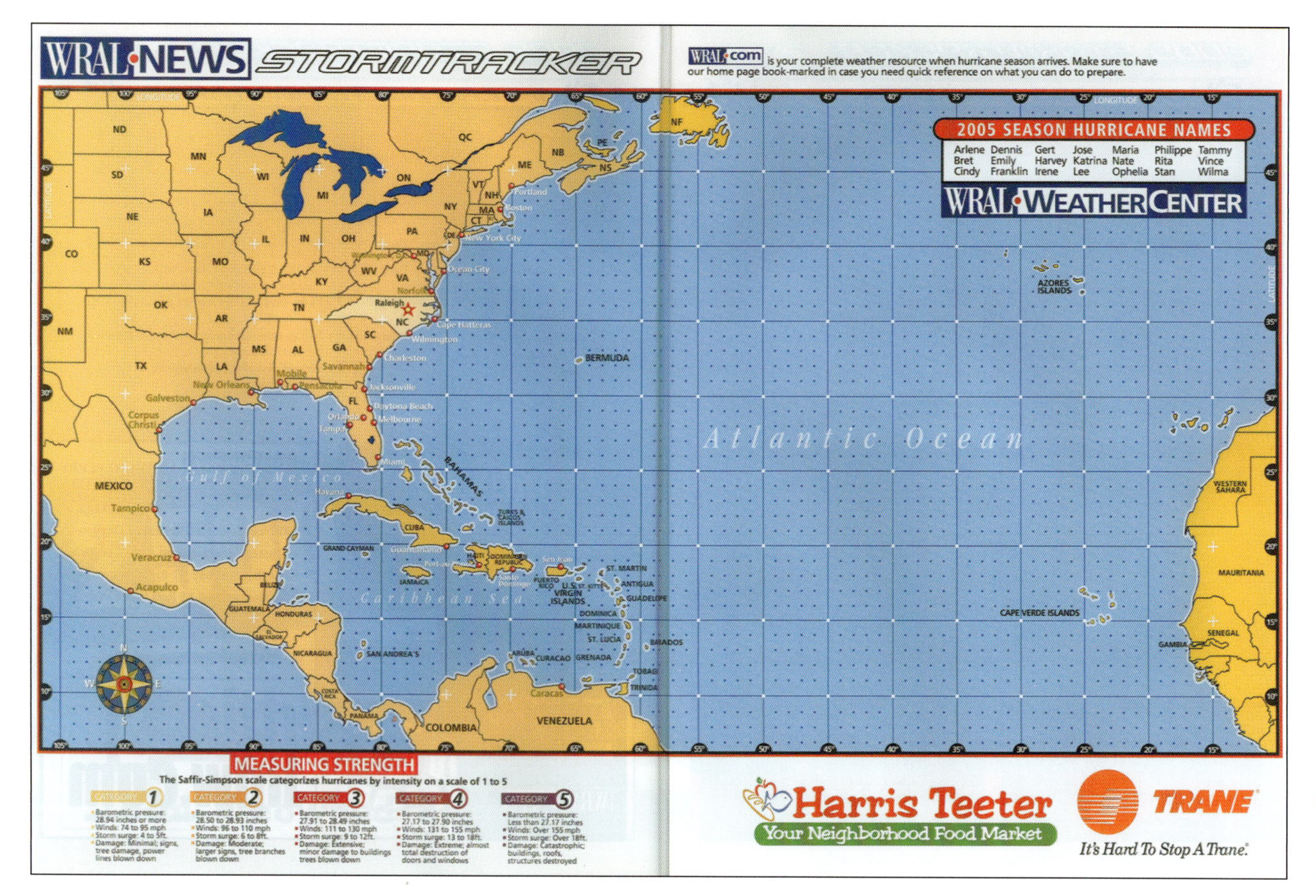

Figure 1.12
2005 Stormtracker map.

Courtesy of WRAL-TV5, Capitol Broadcasting Company, Raleigh, North Carolina.

Figure 1.13
A detail of "Mt. Everest" (supplement to *National Geographic*, November 1988). The full map showing Everest in all her glory can be seen in figure 6.16 on page 117.

Courtesy of Ng Maps/National Geographic Image Collection.

feathers, birds and bees, flowers and seed. If the mountain is awesome, its flower-strewn meadows are beautiful. If the oak is sublime, the dogwood is sweet. Anything but austere, this nature is giving, prodigal. It is a gigantic cornucopia, an unceasing gush of bounty: flowers, fruit, berries, nuts. "The sublime *moves*," Immanuel Kant wrote, "the beautiful *charms*."[45] But it also feeds, also nurtures, and the soul no less than the stomach: "Emblazoned with beauty, this floral map shows the origins of 117 of man's favorite flowers" begins the perimap of "The World of Flowers" (figure 1.14). Beguiling bouquets burst from the hearts of continents. A clump of tulips sprouts in Turkey. A branch of a *flamboyán* flowers in Madagascar. Oriental poppies bloom in Pakistan. The theme of profligacy mingles with that of beauty. The abundance of this nature is inexhaustible: in yet another *National Geographic* map, individual portraits of sixty-seven birds—from "hundreds of kinds"—festoon a map of migratory routes in the Americas.[46] The routes lace the continents from pole to pole. The numbers are insane: the arctic tern may travel twenty-five thousand miles a year! *Ain't nature something else*! The maps demonstrate that this nature—flowers, trees, birds, seals, furry friends—is everywhere.

5. *Possessable nature*. The beautiful, the profligate (and so the exotic) is also the collectible. We yearn to tally it, catalog it, photograph it, and perhaps even own a small piece of it. Maps of this collectible, possessable nature—bird sightings, birds' nests, rocks and minerals, gemstones, big game animals, highest points, stars—are less interested in display than they are in inventory. At stake here are lists, head counts, censuses, catalogs, statistics.

Figure 1.14
A detail of "The World of Flowers" (supplement to *National Geographic*, May 1968, by artist Ned Seidler). The full map is shown in figure 7.1 on page 127.
Courtesy of Ned Seidler/National Geographic Image Collection.

PIN OAK takes its name from the many short, pinlike twigs that clutter the horizontal or downward-sloping branches. These make identification easy in winter. The leaf has five to seven deep lobes with long teeth; it is dark green above, lighter and smooth below. The gray-brown bark remains smooth for some time, gradually breaking into scaly ridges. Pin Oak is partial to moist soil. It is a hardy tree, widely used in street and ornamental planting. Pin Oak is widely cultivated in Europe. Acorns are small (about ½ in. long), rounded, with a shallow cup.

Height: 70 to 80 ft. Beech family

Figure 1.15
This range map of the pin oak tree is the spatial equivalent of its species identity. Range maps are the spatial face of our inventorying of nature. The map shows us where we might find a pin oak, so we might check it off our lists, snap a picture, pocket a leaf.

We're holding in our hands *A Bird Lover's Life List and Journal,* a luxurious, hardbound volume, based on the checklist of the American Ornithological Union, in which birdwatchers can keep score. It lists 715 species and is decorated with illustrations by John James Audubon. While life lists rarely include maps, field guides almost always do. There are 362 maps, for example, in Peterson's *Birds of Britain and Europe,* each map distinguishing breeding and winter ranges for an individual species. Here the maps are corralled into an "atlas" in the back of the book, but in *The Audubon Society Field Guide to North American Birds (Western Region)* the maps accompany the text, one per species, each with its textual relay: "Southeastern Arizona, southern New Mexico, and western Texas, where it breeds at the northern fringes of its otherwise all-Mexican range."[47]

Historically, the construction of spatial identities for species led to the construction of synthetic regions composed in different ways of numbers of species,[48] and these syntheses, too, appear in the field guides, as in *Trees of North America* (a Golden Field Guide), where hundreds of thumbnail maps (such as the one shown in figure 1.15) are preceded by a map of forest regions. Here, for example, we read that in the Northern Forest region "far northern tree associations" consist of conifers, birches, and willows. In Hugh Johnson's *The Principles of Gardening* we find maps not only of where domesticated plants originally grew wild, but of plant hardiness zones which pretty much amount to maps of zones of consistent annual average minimum temperature.[49] Maps like these hint at the systematization—that is, at the science—that consumes the collectible nature.

6. *Nature as system.* The nature of science, of system, is anything but collectible, for it is a nature that exists less in its parts than in the whole. It is an inherently spatialized nature, and maps are a primary way of knowing it. Here individual outcrops metamorphose into strata and strata into geologic formations; soil series aggregate into soil associations and these into soil groups; plant species fall into plant associations and associations combine into plant communities; variations in barometric readings grow into weather systems and these merge into climate. It is a paradigmatic nature. With *Seasonal Land Cover Regions* we have already

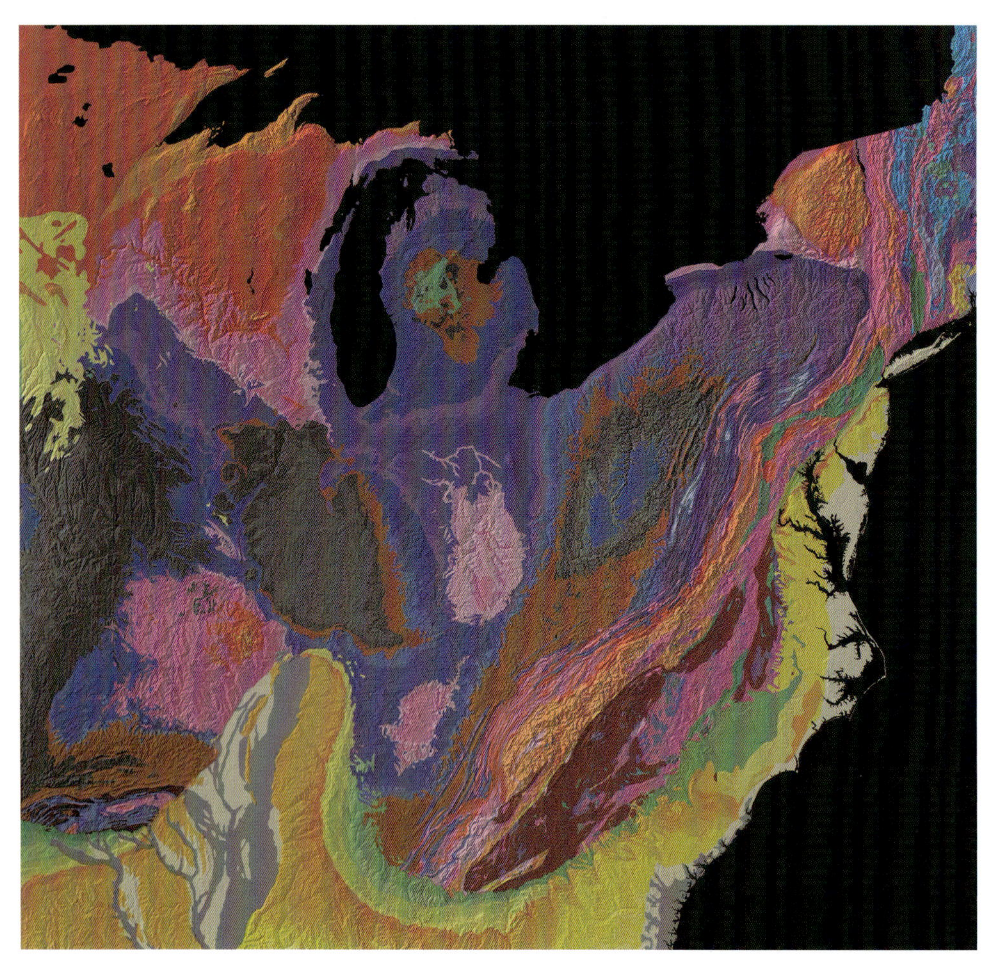

Figure 1.16
A Tapestry of Time and Terrain shows us an illuminating synthesis of two natural systems—physiography and its underlying geology. (See the whole map in figure 9.9 on page 177.)
Courtesy of USGS, 2000.

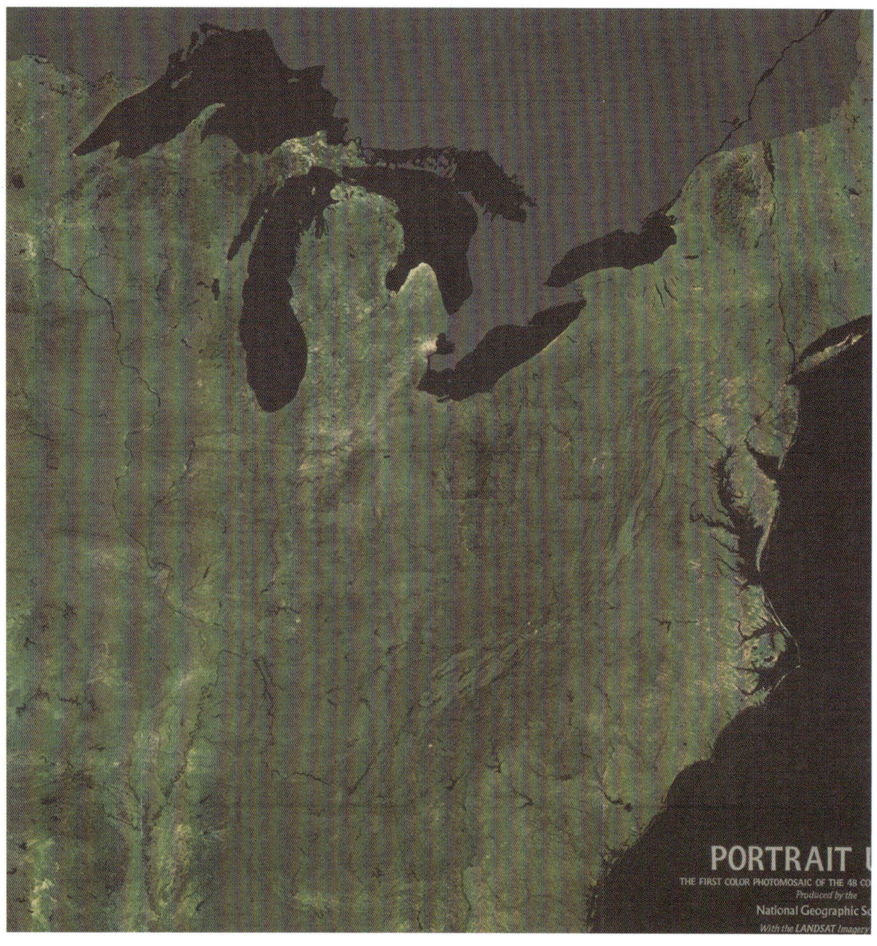

Figure 1.17
This is a detail of "Portrait USA" (*National Geographic* map supplement, July 1976): what does it say to you? (See the whole map in figure 10.2 on page 189.)
Courtesy of Cartographic/National Geographic Image Collection.

glanced at a representative of the genre, but other examples abound: Robert Bailey's *Ecoregions of North America,* USGS's *A Tapestry of Time and Terrain* (figure 1.16—a *truly* mind-boggling geologic/physiographic map of the U.S.), Simon and Fels' *Plant Associations of the Chattooga River Basin,* the endless suites of thematic maps (of landforms, climate, temperature, winds, precipitation, ocean currents, natural vegetation, soils) that stand in the front of so many atlases. This nature is neither threatened nor does it threaten. It does not awe nor is it cute. It is anything but collectible. It is nature that is *known.* It is that of science.

7. *Nature as mystery.* Out of science a new nature has lately risen: it is a nature seen but mysterious, *unknown.* It is that from space. Its construction reverses the usual process through which careful measurements are compiled over time to reveal, for example, a continent (as in the gradual emergence of the Americas on European maps in the fifteenth and sixteenth centuries), an ocean current (as on Ben Franklin's map of the Gulf Stream), a hole in the ozone layer (as with the TOMS data from the Nimbus 7 satellite). This new seen-but-unknown nature emerges whole, apparently unscarred by conceptual categories. Maps of this nature pass for photos, of which Barthes famously remarked, "the feeling of 'denotation', or, if one prefers, of analogical plentitude, is so great that the description of a photograph is literally impossible." This special status of a photograph? *It is that of a message without a code.* At least it *appears* to be without a code. Barthes showed that photos did have a code, but one developed on the basis of a message without one: "It is *read,* connected more or less consciously by the public that consumes it to a traditional stock of signs."[50] It also turns out that these new maps are not photographs, they are maps after all, the connotation—the code, the concepts—has been imposed in their production (looking *like* a photograph is part of this code). This new genre of "portrait" maps (figure 1.17) presents a nature of gradations without distinctions. "What is *that*?" The map does not answer. It is whatever you wish to make of it. This nature is fragile. It is threatened. Or it is tough, resilient. It is enduring. It is distant. It is somewhere else. It is unknowable. It is a vehicle for our anxiety; a recipient of our admiration.
8. *Nature as park.* One final nature, the *intimately known,* that nature mapped at a scale of two and a half inches to a mile, with a contour interval of ten feet (or less). This is the nature of the USGS topographic quad (figure 1.18) and other national mapping surveys. Here again is the sense that everything can be seen, but here *everything* is coded. In fact, here *only* the coded exists, anything not on the (admittedly capacious) leg-

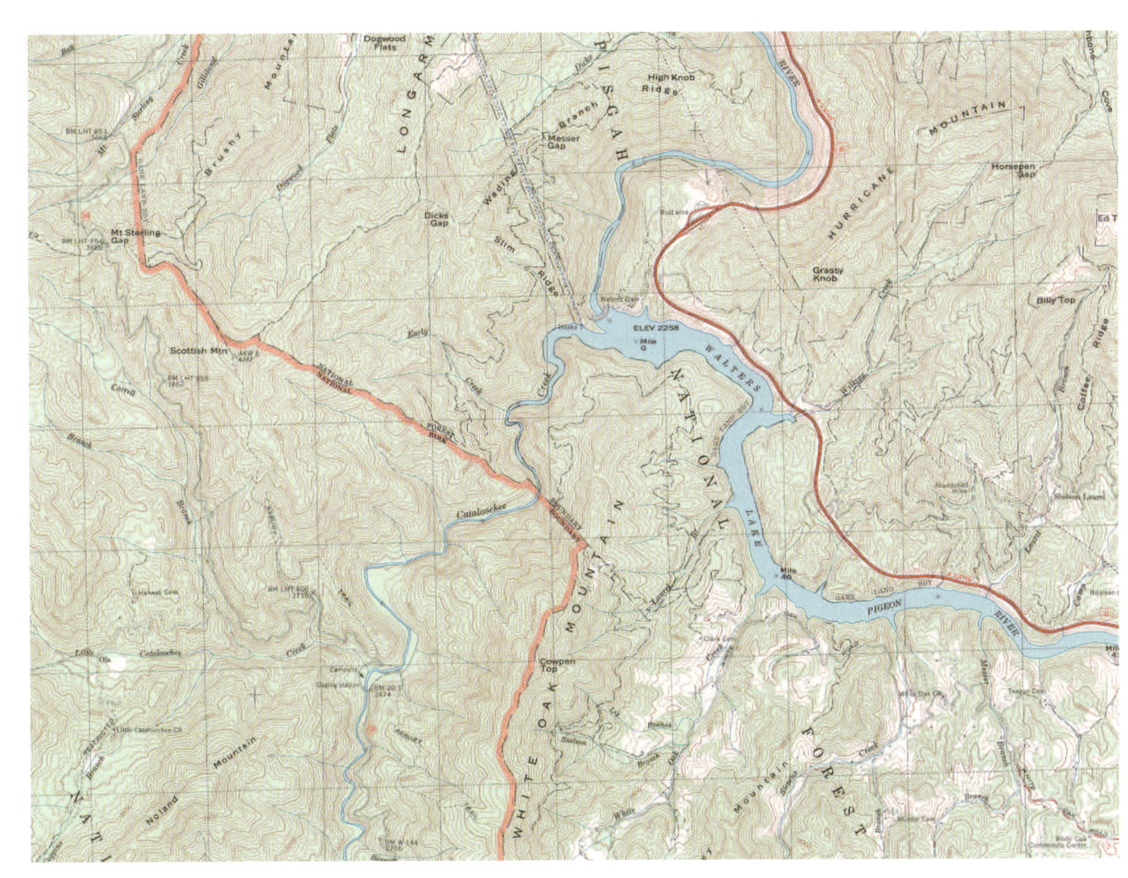

Figure 1.18
To return to where we started, *Cove Creek Gap* quadrangle: the border of Great Smokey Mountain National Park protects nature by, among other things, criminalizing the poaching of ginseng within it.
Courtesy of USGS, 1997.

end doesn't. "What is *that*?" The map returns an answer. It is an intermittent stream; it is a mangrove; it is a dry lake; it is a sunken rock; it is scrub; it is a gravel beach. But nature is not brought to the foreground here. The map is as loquacious about exposed wrecks, landing strips, railroads under construction, vineyards, gauging stations, built-up areas, and dams as it is about glaciers and permanent snowfields, shorelines, mountains, swamps, and rivers. Here nature is subject to no rhetorical flourish, no isolation, no highlighting. It is not the *theme* of these maps. It is along for the ride. This is the nature of the *phenomenological* inventory. At this level nature lies so deep in the conceptual frame that it manifests itself in things instead of attitudes. But the things it manifests itself in are not hills, rivers, or trees, which, undifferentiated from culture, here lie *below* the level of nature. Here nature shows up as parks, monuments, sanctuaries, and preserves. It is a fenced-in nature that we can visit, that we can protect, that we have to protect. . . because it is threatened.

So we have come full circle. Only it is not a circle. It is a multidimensional space of contradictions. It is a dialectical space ripe with the interpenetration, struggle, and unity of opposites. Eight natures —doubtless there are others—each spatialized, each areal, each hoisting itself off the page, taking shape in the mental spaces of cognitive linguistics as we read the map, as we unfold it, turn it over, and refold it; as we bring it closer to our eyes or move it away; as we scale its distances with our fingers: nature as victim, bully, spectacle, cornucopia, collectable, paradigm, mystery, park.

Ours is not a systematic survey. We have made no effort to search for maps of nature but taken as examples those that came to hand in our grappling with the nature of maps. We shall proceed by unfolding in each chapter a map or maps of a different nature, and to use this reading as an opportunity for probing one component or another of our model of the map—the logical structure of the map's construction of knowledge; the physical structure of the paramap; the intellectual structure of the act of map reading itself—as well as probing the nature of the nature in question. Inescapably, we attend closely to the concept of nature as it intertwines itself with economic structures, class formations (nature is above all else a construction of class), and official systems of construal. As we scan sheet after sheet, more and more the maps appear as players in a complicated social game defining the relationship of our species to the rest of existence. Pretending to be no more than scorekeepers, maps stand revealed as more like the ball, the very medium through which the game's moves are made.

Notes

1. Robinson and Petchenik, *The Nature of Maps: Essays toward Understanding Maps and Mapping* (Chicago: University of Chicago Press, 1976); Harley, *The New Nature of Maps.* Note that although Harley's book wasn't published until 2001, the manuscript, lacking the promised introductory and concluding essays, was submitted for publication ten years earlier. The title originally bandied about was *Maps and Society*, which Harley had found "a bit tame: could we devise something more arresting?" (letter to George Thompson, 15 October, 1991). In a postscript a month later Harley wrote, "Please note the new title which

is final as far as I'm concerned: *The New Nature of Maps: Essays in the History of Cartography*" (letter to George Thompson, 26 November 1991).

2. This is from the questionnaire prospective authors were asked to file with the Johns Hopkins University Press. Harley dated it 11/25/91.

3. Ibid.

4. Wood and Fels, "Designs on Signs," *Cartographica* 23 (Autumn 1986): 54–103; Wood, *Power of Maps,* where "Designs on Signs" appears as the fifth chapter.

5. In fact picking ginseng without a permit anywhere but on your own property is larceny under the North Carolina General Statute 14-79: Larceny of ginseng. This statute links the spatialized entity North Carolina with ginseng. It's punishable by up to six months in prison and a five-thousand-dollar fine. Poaching in the national park carries a maximum sentence of a year, though, according to Burkhard Bilger ("Wild Sang: Rangers, Poachers, and Roots that Cost a Thousand Dollars a Pound," *The New Yorker* [July, 15, 2002]: 38–47), no one has ever received it.

6. For one window into the fascinating subject of reference objects, try John Willinsky's *Empire of Words: The Reign of the OED* (Princeton: Princeton University Press, 1994), which struggles with the source of the authority of the Oxford English Dictionary.

7. Hui, "Social Studies Squeeze," *News and Observer* (August 13, 2003): 1B, 9B.

8. Genette, *Paratexts,* 5.

9. Jacques Derrida plows related ground in his treatment of the *parerga,* those elements about, outside, or around a work—in short, the frame; but also the columns of a building, the drapery on a statue—in short, *hors d'oeuvres.* See especially pp. 53–82 and the whole section "Cartouches," (183–253) in his *Truth in Painting* (Chicago: University of Chicago Press, 1987). Be forewarned: "Parergon" is a reading of Kant's *Critique of Aesthetic Judgment,* and "Cartouches" is a catalog essay for a show of Gérard Titus-Carmel's drawings. Where Derrida and Genette are closest is in their understanding of the paratext/parerga as liminal, as threshold. See also (always) Goffman's *Frame Analysis: An Essay on the Organization of Experience* (Cambridge, Mass.: Harvard University Press, 1974) .

10. Genette, *Paratexts,* 344.

11. Scientific American Book Club, August 2003.

12. While rarely discussed in the cartographic literature as such, the epimap has become an issue in the liminal area between the history of cartography and the history of science. Jane Camerini, for instance, is explicit about her interest in "the notion that the meaning of a map resides not only in the map, but in relation to the written text of which it is a part." See her PhD dissertation "Darwin, Wallace and Maps" (PhD diss., University of Wisconsin, Madison, 1987), or her "Evolution, Biogeography, and Maps: An Early History of Wallace's Line," *Isis* 84 (1993): 700–27. The quotation comes from this latter, p. 702.

13. We propose to treat freestanding maps such as this as though they were the independent reference objects they so frequently and so rapidly become. At the same time we need to observe their connection to their epimap: here, that the map was a supplement to *National Geographic,* July 2000.

14. Gall, "Use of Cylindrical Projections for Geographical, Astronomical, and Scientific Purposes," *Scottish Geographical Magazine* 1 (1885): 119–23.

15. There are many versions of this map in circulation, with more or less inflammatory perimaps. This is from a copy distributed by ODT in 2000.

16. Peters, *Die Neue Kartographie/The New Cartography* (Klagenfurt and New York: Universitätsverlag and Friendship Press, 1983).

17. Robinson, "Arno Peters and His New Cartography," *American Cartographer* 12, no. 2 (1985): 103–11. Although clearly a review of Peters' book, the piece appeared in the Views and Opinions section. Considering how completely batty Peters' book is, Robinson's treatment is surprisingly temperate.

18. From a review by Jonathan Yardley of Mark Monmonier's *Drawing the Line* (New York: Henry Holt, 1995). Source unknown (*Washington Post?*). Yardley developed his jibe by stringing together phrases of Monmonier's. The tone, however, cannot be attributed to Monmonier whose treatment of the Peters affair in *Drawing the Line* is thorough and thoroughly scrupulous (9–44). Monmonier's list of sources for both sides is the best available (301–2).

19. This is from *The Wall Street Journal's* front page story about the resolution, June 8, 1989. The ACSM's resolution was a way of dealing with Peters' use of the Mercator as a straw map: excommunicate both projections!

20. This list—and it goes on—is from Robinson, "Rectangular World Maps-No!," *The Professional Geographer* 42, no. 1 (1990): 101–4. Robinson concludes by predicting the resolution's futility.

21. All this commentary, the reviews, the ACSM resolution, its coverage by *The Wall Street Journal,* and the rest (and see Monmonier's treatment for an idea of how extensive this "rest" was) constitute what Genette calls the "metatext." See the presentation of the five varieties of "transtextuality"—intertextuality, paratextuality, metatextuality, hypertextuality and architextuality—in Genette's *Palimpsests: La littérature au second degré* (Paris: Seuil, 1982). Each of these forms of transtextuality plays an important role in the world of maps.

22. Barthes, *The Fashion System* (New York: Hill and Wang, 1983), 11.

23. Barthes uses the word "relay" constantly. An explicit discussion can be found in "The Photographic Message" (of 1961) in Barthes' *Image—Music—Text* (New York: Hill and Wang, 1977). If you pick up the book, "Rhetoric of the Image" and "The Third Message" are also relevant.

24. Loveland, et al., "Map Supplement: Seasonal Land-cover Regions of the United States," *Annals of the Association of American Geographers* 85, no. 2 (1995): 339–55. This is then a second example of a map supplement that can be experienced as a freestanding map; only in this case, the magazine and map were published by unrelated but cooperating entities.

25. Robinson, et al., *Elements of Cartography,* 6th ed. (New York: John Wiley & Sons, Inc., 1995), 333.

26. Dent, *Cartography: Thematic Map Design,* 3rd ed. (Dubuque: William Brown Publishers, 1993). Among major texts in cartography, only Jacques Bertin's *Semiology of Graphics: Diagrams, Networks, Maps* (Madison, Wis.: University of Wisconsin Press, 1983) deals—vigorously—with displays on our order of complexity, but then *Semiology of Graphics* is analytical, not prescriptive.

27. Longyear, *Advertising Layout* (New York: Ronald Press, 1946), 11.

28. Fauconnier, *Mappings,* 1.

29. About the limitations of this kind of "modularity" see, among many others, the pointed comments in Dancygier and Sweetser's *Mental Spaces in Grammar: Conditional Constructions* (Cambridge: Cambridge University Press, 2005), throughout, but most emphatically on p. 15.

30. Fauconnier, *Mappings,* 7.

31. We are not unaware of Roland Barthes' semiotic construction of meaning on the fly in *S/Z* (New York: Hill and Wang, 1974), his slow rereading of Balzac's *Sarrasine,* but it is the only example that comes to mind. And besides, it was actually a rereading of Balzac's text: "We must accept one last freedom: that of reading the text as if it had already been read" (15). Nonetheless, Barthes' reading here is very much in the spirit of the cognitive linguists he preceded by a generation, and very much a model for our own unfolding of the map.

32. Fauconnier, *Mappings,* 11, 36 (emphasis ours). Fauconnier and Turner, *The Way We Think: Conceptual Blending and the Mind's Hidden Complexities* (New York: Basic Books, 2002), 40, 102.

33. Fauconnier and Turner, *Way We Think,* 102.

34. Lakoff, *Women, Fire, and Dangerous Things* (Chicago: University of Chicago Press, 1987), chap. 4. Schank and Abelson, "Scripts, Plans, and Knowledge" in *Thinking: Readings in Cognitive Science,* ed. P. N. Johnson-Laird and P. C. Wason (Cambridge: Cambridge University Press, 1977), 421–32. A cool, accessible, book-length treatment of the restaurant script is Schank's *The Connoisseur's Guide to the Mind* (New York: Summit Books, 1991). Toulmin, *The Uses of Argument* (Cambridge: Cambridge University Press, 1958). Bakhtin, *Speech Genres and Other Late Essays* (Austin, Tex.: University of Texas Press, 1986).

35. See Lakoff and Johnson's discussion of these points in their *Philosophy in the Flesh: The Embodied Mind and Its Challenges to Western Thought* (New York: Basic Books, 1999), chap. 3.

36. Sweetser and Fauconnier, "Cognitive Links and Domains: Basic Aspects of Mental Space Theory," in Fauconnier and Sweetser, *Spaces, Worlds and Grammar* (Chicago: University of Chicago Press, 1996). 11.

37. Robinson et al., *Elements,* 316.

38. In fairness to *Elements,* it must be acknowledged that its discussion of the principles of graphic design—following the caveat that "There is little agreement among professional designers about what they mean by graphic design"—is relatively more sophisticated and revolves around the concepts of legibility, visual contrast, figure-ground, and hierarchical structure (324–38). Discussions of these issues occupy hundreds of subsequent text pages. Much of this, however, is the development of technical production vocabularies. The principles themselves remain at the level of "If these visual relationships coincide with the cartographer's intentions, effective communication can take place. If not, the map design is likely to fail" (324) which is—effectively—meaningless.

39. More or less. Not that maps played no role in human affairs prior to, say, 1400, but that after that time they begin to play the role they continue to play today. Our decision to draw the line here is akin to Ian Hacking's drawing the line for the birth of statistics at 1660. It's not that there weren't all kinds of precursors but that "We do not ask how some concept of probability became possible. Rather we need to understand a quite specific event that occurred around 1660: the emergence of our concept of probability. If there were Indian concepts of probability 2,000 years ago, they doubtless arose from a transformation quite different from the one we witness in European history," and so on (Hacking, *Emergence of Probability,* Cambridge: Cambridge University Press, 1975, 9). Similarly, we are not concerned with the host of potential precursor map-like things, but with the map as we know it, and have known it for five or six hundred years. See Wood's "P. D. A. Harvey and Medieval Mapping: An Essay Review" *Cartographica* 31 (Autumn 1994): 52–59; and his "Maps and Mapmaking" in *Encyclopedia of the History of Science, Technology and Medicine in Non-Western Cultures,* ed. Helaine Selin. Dordrecht, Boston: Kluwer Academic, 1997), 549–54.

40. Our example comes from the May, 2001, issue of *National Geographic,* among the unpaginated front matter.

41. Americans have been known to similarly track the war front. For more about the practice of publishers to package maps in kits with little flags and pins during World War II, see the section, "War Is God's Way of Teaching Us Geography," in Schulten's *Geographical Imagination,* 206–14.

42. An element of the "Great Disasters: Nature in Full Force" poster, supplement to *National Geographic,* July 1998.

43. Marita Sturken observes this and more in her "Desiring the Weather: El Niño, the Media, and California Identity" *Public Culture* 13, no. 2 (2001): 161–89." She focuses on TV not maps, but her paper is all but a disquisition on the spatialization of the weather in which all such media collaborate.

44. Karan, Pradyumna P. *The Kingdom of Sikkim,* supplement to *Annals of the Association of American Geographers* 59 (March 1969), and *The Kingdom of Bhutan,* supplement to *Annals of the Association of American Geographers* 55 (December 1965).

45. Kant, *Observations of the Feeling of the Beautiful and Sublime* (Berkeley: University of California Press, 1960), 47.

46. "Bird Migration" was a supplement to the August, 1979, issue.

47. Boucher, *Bird Lover's Life List and Journal* (Boston: Museum of Fine Arts, 1992). Peterson et al., *A Field Guide to the Birds of Britain and Europe,* 4th ed. (Boston: Houghton Mifflin, 1983); Udvardy, *The Audubon Society Field Guide to North American Birds (Western Region)* (New York: Knopf, 1977), 647. John Law and Michael Lynch compare and contrast a number of these field guides in their "Lists, Field Guides, and the Descriptive Organization of Seeing: Birdwatching as an Exemplary Observational Activity" in *Representation in Scientific Practice,* ed. Michael Lynch and Steve Woolgar, 267–99. (Cambridge, Mass.: MIT Press, 1990).

48. See Camerini, "The Physical Atlas of Heinrich Berghaus: Distribution Maps as Scientific Knowledge" in *Non-Verbal Communication in Science Prior to 1900,* ed. R. G. Mazzolini, 479–512. (Florence: Olschki, 1993).

49. Brockman, *Trees of North America: A Field Guide to the Major Native and Introduced Species North of Mexico* (New York: Golden Books, 1968); Johnson, *Principles of Gardening* (New York: Simon & Schuster, 1979). The maps in *The Principles* came from the Mitchell Beazley studio.

50. Barthes, *Image—Music—Text,* 18, 19.

ASHE
ALLEGHANY
SURRY
STOKES
ROCKINGHAM
WATAUGA
WILKES
YADKIN
FORSYTH
GUILFORD
CALDWELL
ALEXANDER
IREDELL
DAVIE
DAVIDSON
RANDOLPH
BURKE
CATAWBA
ROWAN
LINCOLN
CABARRUS
GASTON
CLEVELAND
STANLY
MONTGOMERY
MECKLENBURG
UNION
ANSON
RICHMOND
BLUE RIDGE MTS.
Winston-Salem
Greensboro
High Point
Thomasville
Lexington
Asheboro
Statesville
Hickory
Lenoir
Morganton
Boone
Wilkesboro
Salisbury
Kannapolis
Concord
Mooresville
Davidson
Cornelius
Huntersville
Charlotte
Gastonia
Kings Mountain
Shelby
Lincolnton
Newton
Albemarle
Monroe
Matthews
Mint Hill
Pineville
Indian Trail
Stallings
Wadesboro
Rockingham
Hamlet
Eden
Reidsville
Clemmons
Kernersville
Mocksville
Conover
Cherryville
Belmont
Mount Holly
Rock Hill
Gaffney
Cheraw
Lancaster
N. CAROLINA
S. CAROLINA
UWHARRIE NATIONAL FOREST
South Mountains State Park
Morrow Mountain St. Pk.
Crowders Mtn. St. Pk.
Kings Mtn Natl Mil. Pk.
Andrew Jackson Historical St. Pk.
N. C. Zoological Park
Town Creek Indian Mound
PEE DEE NAT'L WILDLIFE REFUGE
Sandhills Rec. Area
Hanging Rock St. Pk.
Pilot Mtn. St. Pk.
Stone Mountain St. Pk.
Doughton Park
Cumberland Knob
Lake Norman
Badin Lake
Lake Tillery
High Rock Lake
Tuckertown Lake
Lake Hickory
Lake Rhodhiss
Lake Wylie
Mountain Island Lake
Blewett Falls Lake
ASHEBORO
ASHEVILLE
BLOWING ROCK
BOONE
BURLINGTON
CAPE HATTERAS
CHAPEL HILL
CHARLOTTE
CHEROKEE

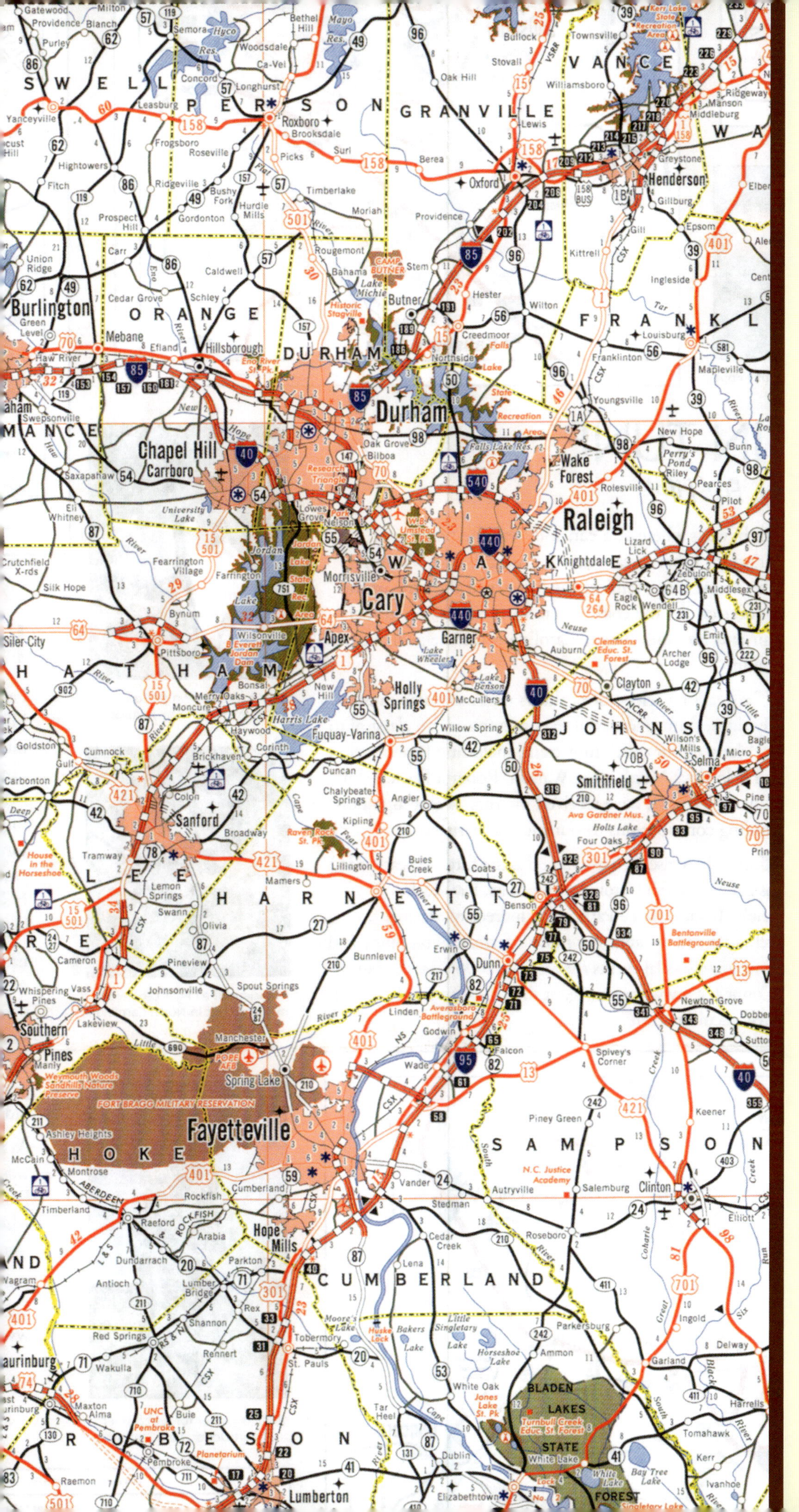

Source: 2007 ESRI Data & Maps. Basemap: Tom Patterson.

Two

The propositional logic of the map

The propositional logic of the map

We're looking at the newspaper, at a map of legislative districts the North Carolina General Assembly is proposing for future elections. Its propositional character is obvious. It is on its face. It is in its name. (It is labeled, "Proposed for 2004.")

The interests the map embodies—those of the Democratic Party—are scarcely less obvious. No surprise, then, that the Republican Party has taken the map to court and elicited from a Superior Court judge an alternative map, one that embodies—and has already advanced—the interests of the Republican Party (see figure 2.1).

Both political parties have argued that the maps they propose embody broader, nonpartisan interests, specifically those embedded in the North Carolina and U.S. constitutions. This is to say they have argued that their maps embody common interests which amount to, or so they imply, no interests at all.

Few to date have taken these arguments seriously. The interest is too naked. The propositional character of the maps is too pronounced. But most would agree that this propositional character is essentially superficial, that is, that it pertains only to the proposed legislative districts but does not reach the counties which these districts variously divide. That is, it is the *legislative districts* that are proposed for 2004, not the counties that also appear on the map. Most would agree that counties, far from being propositions, are *facts*.

The cruisers of their sheriffs' deputies and the big yellow school buses of their school systems do endow counties with an existential quality lacking in the proposed legislative districts, though once the legislative districts have been fixed they too will share this quality (it will be materialized as ballots). But, just as the districts are debatable, so, too, are county boundaries. Wilson and Greene counties, for instance, have been fighting for the last twenty or so years over 240 acres that Greene County claims it lost to Wilson in 1902. More generally, as with the legislative districts, it has been political expediency that has driven the creation of counties in North Carolina, for much of whose history representation in the State Assembly was county-based. Because of this, each time a new county was created in the state's growing Piedmont and western regions, a "new" county had to be carved out of an *existing* eastern county to maintain the regional balance. Similar remarks can be made about North Carolina itself, and about the United States. What gives any political unit its factuality is the assent people give to its existence. When this is withdrawn, the propositional nature of the unit once more becomes apparent. (Recent national examples include the former Czechoslovakia, the former Yugoslavia, the former Soviet Union.)

Except for the rivers implicit in the county boundaries, then, these maps are composed of nothing *but* propositions to which varying degrees of assent have been granted. Actually this is true of the rivers as well, especially in the coastal plain, and even of the state's coast, where no more than arbitrary lines can be drawn between water and land in its morass of sounds, estuaries, swamps, and marshes, to say nothing of along the relentlessly shifting Outer Banks. As for "absolute location," it's *never* been more than arbitrary grids and highly conventional datums, proposed, debated, and adopted until new datums and grids are proposed.

Stephen Hawking tells a story about a little old lady rejecting an astronomer's description of the universe with, "What you have told us is rubbish. The world is really a flat plate supported on the back of a giant tortoise." To the astronomer's superior, "And what is the tortoise standing on?" The little old lady replies, "You're very clever, young man, very clever. But it's turtles all the way down." If the picture of a tower of turtles makes you laugh, it nevertheless gets at the nature of the map in a way few other images do, for a map is a system of propositions "all the way down." In fact, it is the propositions at the bottom that are the most powerful, because these command almost universal assent. These fundamental spatial/meaning propositions are the building blocks out of which the common-sense view of the world is constructed. It is here where the map least *records* and most *builds* the world.

A conceptual scaffold

This argument of ours is built around a conceptual scaffold consisting of five propositions (table 2.1). We propose, first, that maps are vehicles for creating and conveying authority about, and ultimately over, territory. We then argue that this authority the map wields is the social manifestation of its intrinsic and incontrovertible factuality. Next, we argue that the factuality of the map is a function of the social assent given to the propositions the map embodies. We argue, fourth, that these propositions take the form of linkages among conditions, states, processes, and behaviors conjoined through the territory. Finally, we argue that these linkages are realized through fundamental spatial/meaning propositions expressed in the sign plane of the map. We call these fundamental spatial/meaning propositions *postings*.

To stay with our example—which is simply one that is in the newspaper after every new census—what is the proposed legislative district map but a vehicle for the creation and conveying of authority over the territory of North Carolina? The map is explicitly the vehicle for the creation of the districts, but it is also the way the agreed-upon new districts will be *communicated, conveyed* to the electorate, the boards of elections, the political parties, and the campaigns of candidates for election. With the new map, county boards of elections will print ballots containing the names of only some candidates and not others (the ballots will carry the names of only those candidates running for office in a given district). The authority the map has to bring about these behaviors is the manifestation in social action of the map's factuality. This factuality, like all factuality, arises from the broad (if not universal) assent given to the propositions the map embodies (the map will have been voted on and finally accepted by the state legislature, and it will have survived judicial scrutiny). The map *is* these propositions.

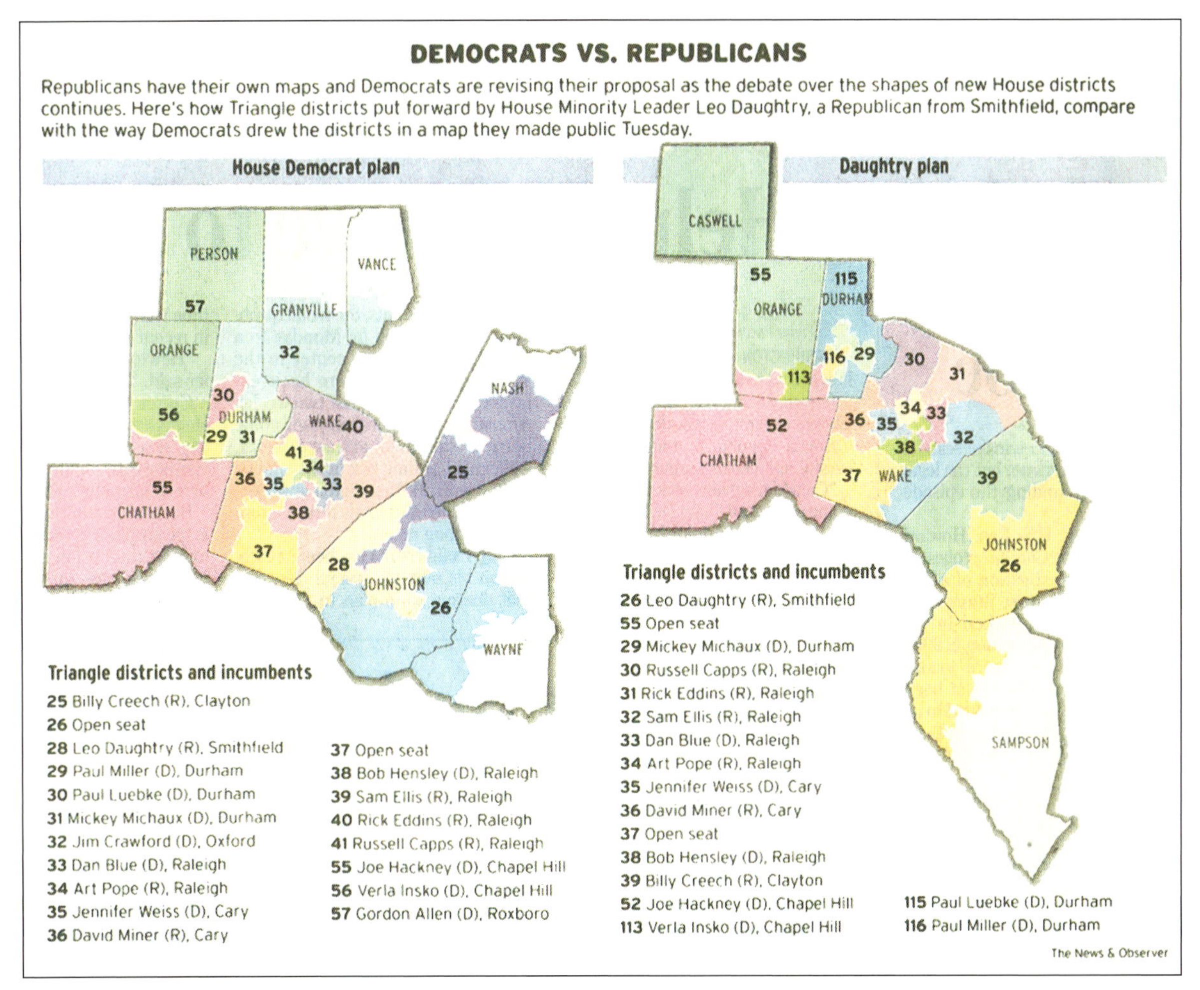

Figure 2.1
Maps of North Carolina legislative districts proposed by Democrats and Republicans. While the propositional character of the legislative districts may be obvious, that of the counties is less marked, and that of the coastlines is not at all obvious to most map readers. It is "natural" things like coastlines with which we are mostly concerned in this book.

Courtesy of *The News & Observer*, Raleigh, N.C.

A conceptual scaffold
1. The map is a vehicle for creating and conveying authority about, and ultimately over, territory.
2. The authority of the map is the social manifestation of its intrinsic and incontrovertible factuality.
3. The factuality of the map is given through social assent to the propositions it embodies.
4. These propositions assume the form of linkages among conditions, states, processes, and behaviors conjoined through the territory.
5. These linkages are realized through fundamental spatial/meaning propositions expressed in the sign plane of the map.

Table 2.1
At the top of our conceptual scaffold is the map's ability to create and convey authority over territory. This ability is the social manifestation of the map's factuality, which is secured through social assent to the propositions the map advances. These propositions assume the form of linkages among things conjoined in the territory, and are realized through the fundamental spatial/meaning propositions we call postings.

Technically, a proposition is a statement in which the subject is affirmed or denied by its predicate (*this is there*). The propositions advanced by maps always take the form of linkages among conditions, states, processes, and behaviors conjoined through the territory; in the case at hand, a linkage between places of residence and electoral districts (that is, between residents and candidates). The map says, *if you live here, you will be able to vote there* (that is, for the candidates appearing on your precinct's ballot). In turn, these linkages are realized through fundamental spatial/meaning propositions, or postings. An example here might be that of your residence on the map maintained by your county board of elections. Another might be that of your precinct (which determines where you actually go to cast your vote). Another might be that of the legislative district your residence is *in*, that is, that it is linked to.

The authority, factuality, assent to propositions, and linkages are all relatively transparent in this example. We believe every map has this form. Only the nature of the authority and its construction varies—here political, but in other cases scientific. The nature of the posting, however, is perhaps less plain, and it is that to which we turn.

A spatial/meaning calculus

All cartographic propositions, however elaborate, are ultimately realized through fundamental spatial/meaning propositions expressed in the sign plane of the map. The most fundamental of these cartographic propositions—*this is there*—establishes an equivalence between a *this*, a specific instance of a preexisting conceptual type, and a *there*, a specific location in the cartographic sign plane. Similar *thises* may be located at more than one *there*, and a *there* may be populated by many *thises*. Multiple *this is there* propositions (cartographic *postings*) comprise and create the territory of the map. This territory becomes the subject of the map's social and political action.

Relationships established among *thises* and *theres* facilitate the exchange of meaning within the cartographic sign plane. The coincidence of *theres* affords and affirms authority (political, cultural, religious, scientific) over territory and its constituent *thises*, potentially at many levels. The coincidence of *thises* raises the opportunity for constructing new *theres* (say, in the present, past, or imagined future), and consequently new territories. We attempt to unfold these operations here through a provisional calculus of space and meaning propositions as they are expressed in the sign plane of the map. This calculus is presented through diagrams and algebra to illustrate how cartographic propositions are formed and how they function in creating both space and meaning. In the special logic of map signs, space and meaning are indivisible.

Every map is about something—some things and some place—and what are the *things* that populate maps if not content types, ready-made to be realized, to be instantiated in the cartographic sign plane? There will be a house on the map, and a church, and a fire station, and a park, and some roads, and . . . the list goes on. The house, the church, the fire station, the park, the roads, are all categorical types residing (in some way we don't pretend or need to define) in a conceptual space (or "conceptual universe" or "content space" or "content plane" or "semantic field" or "semantic cloud"). Any of these *types* can be given existential claim in the map insofar as it is differentiated from other types and is able to be expressed in the map (figure 2.2). The universe of preexisting (but ever-changing) conceptual types is essentially a *system of differences* within which the differentiation of each specific type from all others is necessary to its very existence

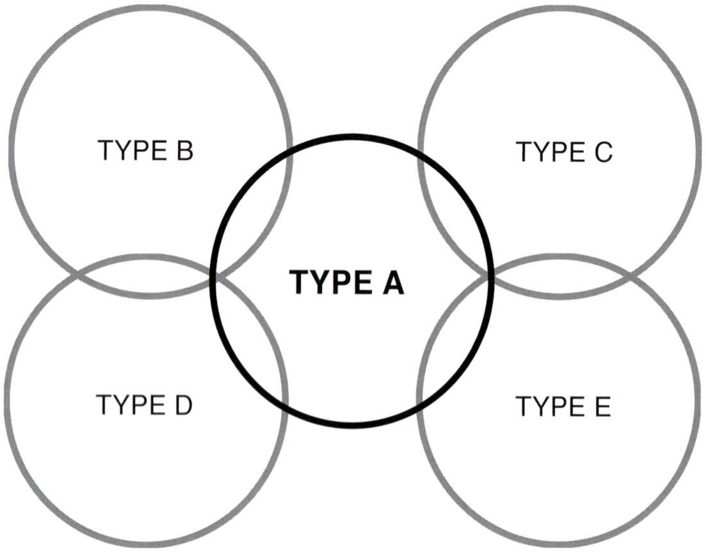

Figure 2.2
A cluster of types in conceptual space. At any level in hierarchical conceptual space, the attributes, and therefore conceptual boundaries, of similar types inevitably overlap to a greater or lesser degree. There are any number of preexisting conceptual types: trees, rivers, voting districts, shores, buildings, roads, houses, and so on. They exist on a content plane, in a concept space, or in a semantic field, or are created by humans in communication.

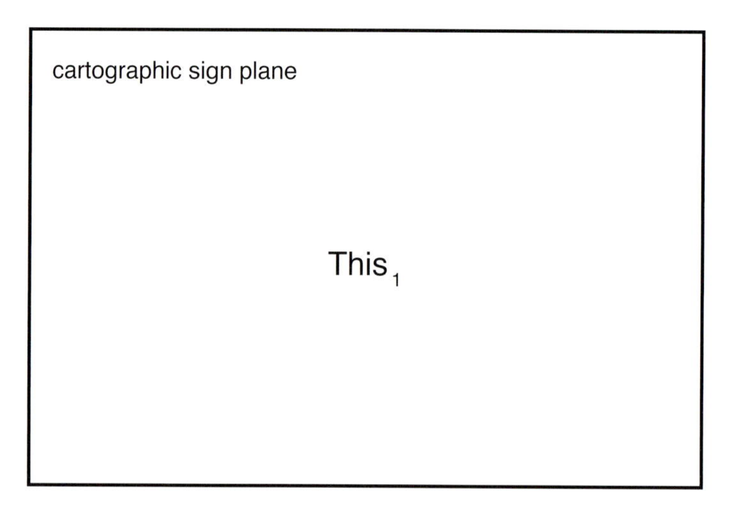

Figure 2.3
The precedent existential proposition. In order for an instance of a type to enter into the cartographic sign plane, its existence must be proposed both as a valid conceptual type and as an actual and undeniable instance of that type. Hence, not simply *This* but *This$_1$*. Types are manifested as signs. As types they only exist in concept space. To be communicated they must be wed to a mark. This happens on the plane of the sign, or sign plane. This plane could be a map (the sign would be a map sign), but it could also be a page of text (the sign would be a word), a geometric diagram (the sign would be a figure or a part of a figure), and so on.

in conceptual space. And this differentiation can be —indeed, nearly always is—hierarchical. For example, white pine can be differentiated from pitch pine, and both from longleaf or shortleaf pine, but these are all *types* of pine. Pine in turn can be differentiated from spruce or hemlock, but these are all types of needleleaf trees, as needleleaf trees can be differentiated from broadleaf trees and trees can, in turn, be differentiated from cacti and grasses.

Type A might be differentiated from Type B, Type C, Type D, and Type E in denoting a dwelling type "house" as distinct from a "condominium," an "apartment," a "mobile home," or a "hut." But the deployment of cartographic signs entails much more than semantic differentiation and denotation. Each of these signs also carries an enormous connotative load, waiting to be dropped into the map at the stroke of a pen, or a keystroke. The precedent existential proposition not only entails a distinctive, if abstract, existential claim but also requires certain territorial allowances for its realization. In the city of Raleigh, for example, there can be houses, condominiums, and apartments but no mobile homes (at least not legally). As for huts, those only exist in Third World countries, don't they?

When a content element is expressed in the cartographic sign plane it is simultaneously given a location in that sign plane.[1] We call this a *posting*. *This* is no longer *a* house or *a* church or *a* fire station but *that* house, or *that* church, or *that* fire station. *This* is no longer a type but a concrete and specific instance of a type. In figure 2.3, This_1 is an instance of Type A, a house, let's say, as distinct from a condominium or apartment; that is, a specific structure expressible in the sign plane with a distinct mark such as a small black square. In the map, the small black square already expresses several assertions: that This_1 exists, that it is located There_1, that This_1 is a legitimate instance of Type A, and that Type A has a viable claim to existence in the first place. Moreover, and the point is usually overlooked, it equally asserts that There_1 exists, that it is the unique locale of This_1, that There_1 is a legitimate instance of another concept—*location*—and that that concept has a viable, if even more abstract, existential claim as well.

In proposing that *this is there* (see figure 2.4), the term "is" may be taken as a statement of equivalence. The posting establishes an equivalence between a *this*, an instance of a type, and a *there*, an instance of a location, expressed jointly in the sign plane. (In this and the following diagrams, *thises* will be indicated by the label This_n and corresponding *theres* by their bounding circles.) Through the posting, *this* acquires *thereness*, a quality or condition of being somewhere, as *there* acquires *thisness*, a quality or condition of being something. Thisness and thereness are now inseparable. The same mark (expression) that denotes a conceptual type in the legend signifies, when posted in the cartographic sign plane, a material (or in some other way "real") and spatial instantiation of that type. It's not that This_1 is located at some abstract position There_1. This_1 *is* There_1 and vice versa.[2]

The cartographic posting—the instantiation of a This/There or There/This is the elemental building block of the map's territory (see figure 2.5). Since each posting embodies a *there* as well as a *this*, each invests its own space in the map. Collectively these define the territory of the map (This_4). A park occupies part of the territory, and a road traverses some of it, and the small black square is there as well. While the spatiality of the park and the road is evident, the small black square may appear to be a simple location marker, without any intrinsic space or territory. But that's hardly the case. If the small black square instantiates a house, then its parental concept dictates that it occupies and contains space and has other spaces attached to it—a front yard, a back yard, a sidewalk, a driveway, probably a garage or carport, maybe a garden or a gazebo or a swimming pool (not to mention all of the social and political attributes of a dwelling or a home).

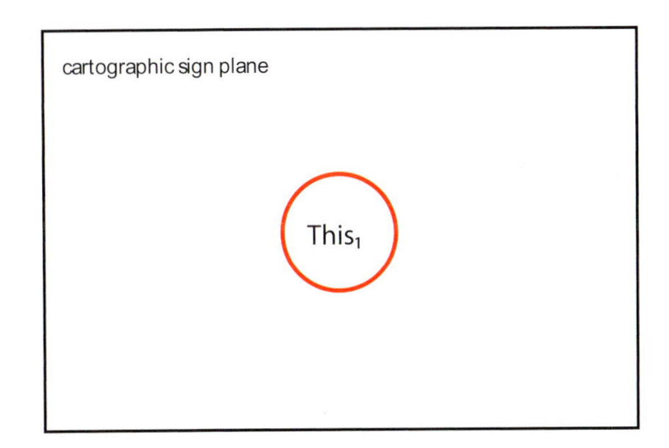

Figure 2.4
The fundamental cartographic proposition—the posting. The posting establishes an equivalence between a *this* and a *there*. It is the attribution of *thisness* to *thereness*, and vice versa, cementing their inseparability. In these diagrams, each cartographic posting will be symbolized by both a label (e.g., This_1) that indicates its *thisness*—its claim to existence as an instance of a type—and a bounding circle that indicates its *thereness*—its claim to existence at a particular place in real space. The cartographic sign plane signifies not only through the mark, which realizes the concept type, but through location, which specifies an instance of the type: that house *there*.

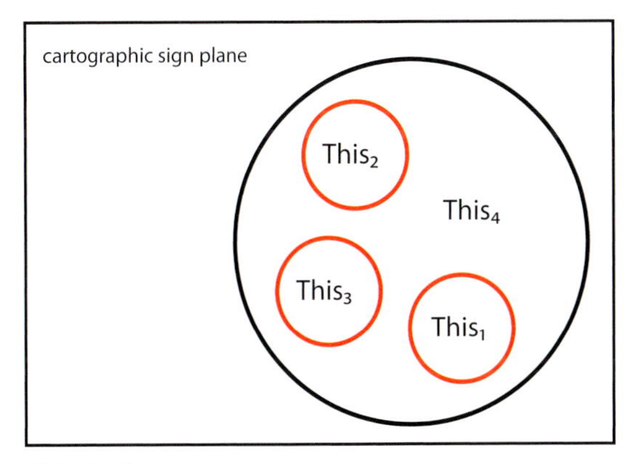

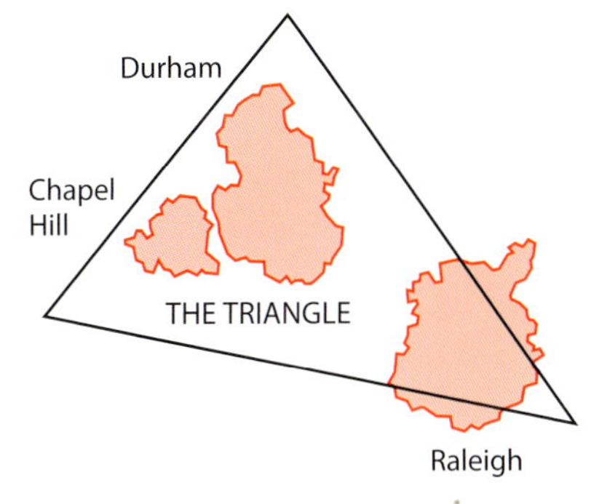

Figure 2.5
Multiple postings participate in the construction of the territory. In order for the territory to materialize in the map as a legitimate posting in its own right, it has to be built from many constituent postings. This is simply to say that the territory cannot materialize out of—or consist of—thin air.

When the "house" is posted in the map, all of those other spaces are posted there as well. At some level, the instantiation of the house is very clearly a territory in its own right.[3] Post enough of these, along with some roads, and we have a neighborhood, another form of territory; post enough of those, along with some commercial and institutional postings, and we have a town, another form of territory; post enough towns (Raleigh, Durham, Chapel Hill) and we have a region (The Triangle). Even if the territory of the map appears *a priori* and fully formed, it still had to be *built* in some map or, more realistically, maps—probably lots of them—as, for example, the territory called the City of Raleigh is built not just from maps proclaiming its official limits but also maps of school districts, fire districts, police districts, city council districts, zoning overlays, among other things.

In figure 2.6, This$_1$/There$_1$ is again an instantiation of a type "house," with its unique material and spatial realization, and domestic function (stripped of its *thisness*, it is simply an address) and This$_2$/There$_2$ is the city of Raleigh or, more specifically, a portion of the city of Raleigh carrying a residential zoning or, more specifically still, the residential zoning overlay itself (There$_2$) along with its associated restrictions and covenants (This$_2$). This$_1$ is There$_1$ and This$_2$ is There$_2$, but This$_1$ is subordinate to There$_2$ as This$_2$ is supraordinate to There$_1$. Because There$_1$ is subordinate to There$_2$, certain behavioral rules embodied in This$_2$ are applied to This$_1$ as well. We cite, for example, the story of the Preddys who grew tomatoes in their back yard and sold them out of the garage of their home outside the northern limits of the city of Raleigh. Eventually, the Preddys' property and those of their neighbors were annexed by the city and overlain by a residential zoning that prohibited, among other things, any commercial enterprise. No longer could the Preddys sell their tomatoes from their garage. They could still grow their tomatoes of course, and eat them, and give them away, and even sell them elsewhere—they just couldn't sell them *there.* And that's hardly all. Prohibited also are open fires, the discharging of firearms, the playing of loud music after a specified hour, the running of pets at large, the keeping of potbellied pigs (they're typified as livestock rather than pets), and many other behaviors deemed inappropriate to civilized urban life in North Carolina. One's first instinct might be to regard the Preddys' fate as the visible consumption of This$_1$/There$_1$ (the Preddys' property) by This$_2$/There$_2$ (the city of Raleigh) but, in the map, there is seldom any sign of such calculated and conspicuous aggression. This$_1$/There$_1$ is simply coincident with—coinstantiated with—This$_2$/There$_2$. Not apparently consumed or subsumed, not subjugated—just a part of the territory. If the City *is* aggressive in its corporate intentionality, that is occluded by the dispassion of the map.

The posting This$_1$/There$_1$ may be—indeed, always is—coincident with many other postings, many other *This is also There* propositions: a school district, a fire district, a voting district, a municipality, a county, a state or province or territory, a nation (figure 2.7). The various authorities invested in This$_2$, This$_3$, This$_4$, etc., are simultaneously imposed on This$_1$ through the coincidence of their respective Theres. If This$_1$/There$_1$ is coincident with a political This$_2$/There$_2$, such as Bolivia, it is also coincident with a This$_3$/There$_3$ of language (Spanish), a This$_4$/There$_4$ of currency (the bolivar), and so on. Constituency, as opposed to mere coincidence, emerges from a hierarchy of authority. Raleigh, North Carolina, and the United States are all coincident but North Carolina is constituent to the United States and Raleigh is constituent to both. In this way, coincidences of location entail constituencies to multiple hierarchical territories and their resident authorities.[4] Can such hierarchies of territorial authority be constructed without maps? We doubt that they can. Maps are the indispensable instruments of their social construction.

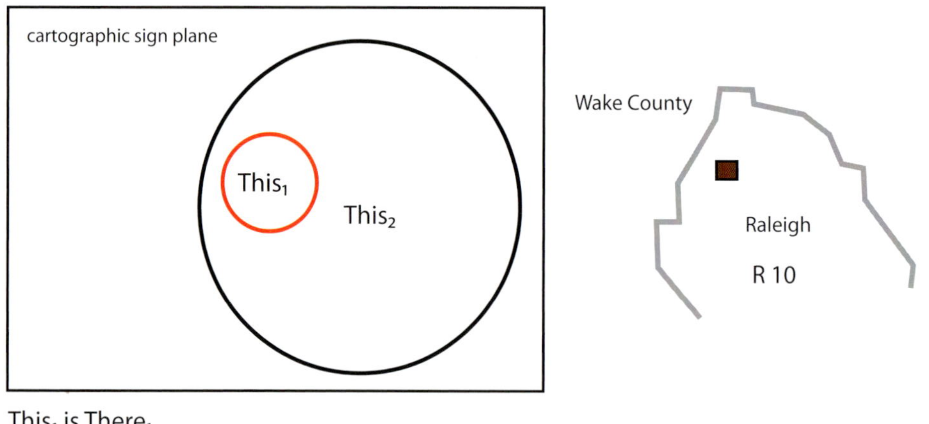

This$_1$ is There$_1$
This$_2$ is There$_2$
There$_1$ < There$_2$
This$_1$ < This$_2$

Figure 2.6
Two *thises* are linked through a common *thereness* invoked by the map, facilitating the transmission of authority through the coincidence and subordination of territory. In this example, a house (This$_1$) is encompassed by Wake County but also by the City of Raleigh (This$_2$).

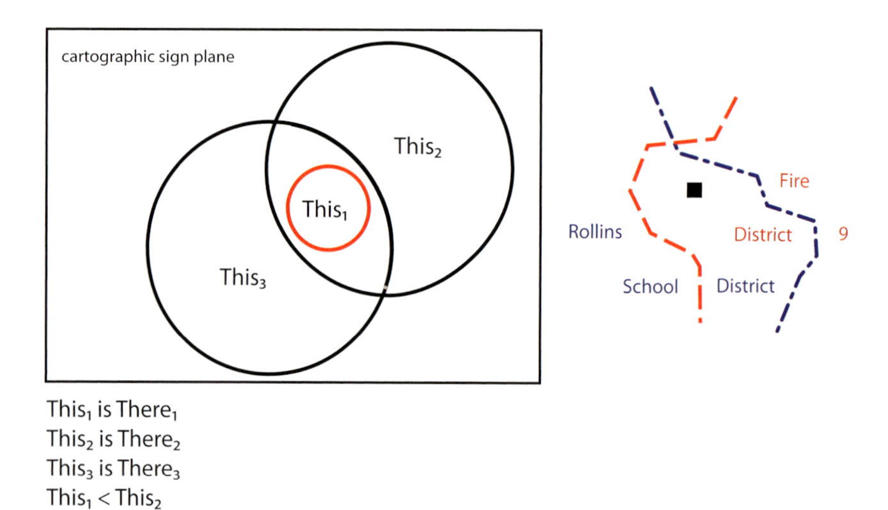

Figure 2.7
A posting can be simultaneously coincident with, and constituent to, many other postings, such as school districts or fire districts. These multiple, and multileveled, coincidences facilitate, through the map, the transmission of multiple tiers of authority.

In figure 2.8, This_1 and This_3 are proposed as disjunct instances of the same type (Type A). Their kinship suggests a greater posting —This_5—consuming both (figure 2.9). Since This_1 is constituent to There_2 and This_3 is constituent to There_4, the corresponding theres are entrained in their union to form a new There_6. That syntactical procedure has been employed countless times in nation building and empire building enterprises. Consider the tug-of-war for Labrador between Newfoundland and Quebec, or the contestation of Gibraltar or Cyprus or Nepal or the Holy Land, or Adolf Hitler's vision of an Aryan supernation. The process can operate in reverse as well, as in the dissolution of the Soviet Union or Yugoslavia. (When journalists invoke "the former Soviet Union" or "the former Yugoslavia," they are still referring to recognizable postings, but posted in a *then* rather than a *now*.)

One of our favorite examples is Alfred Wegener's argument for continental drift and the existence of an ancient continent consisting of Africa and South America. Wegener's argument, based on empirical observation and evidence, proposed the instantiation This_5 of a recognizable geologic unit (Type A) coincident with, and constituent to both $\text{This}_2/\text{There}_2$—the South American continent (Type B)—and $\text{This}_4/\text{There}_4$—the African continent (Type B). Wegener offered, in addition, several corroborating cartographic propositions invoking geomorphic features, contemporary fauna, and paleofauna. If $\text{This}_1/\text{There}_1$ and $\text{This}_3/\text{There}_3$ are brought together, then so are $\text{This}_2/\text{There}_2$ and $\text{This}_4/\text{There}_4$. The union of $\text{This}_2/\text{There}_2$ and $\text{This}_4/\text{There}_4$ is a new territory $\text{This}_6/\text{There}_6$ (figure 2.9), and Wegener called this new territory "Gondwana."[5]

Map logic and its semiotic expression

Our objective in attempting a spatial/meaning calculus has been to formalize the description of potential relationships into which postings can enter. It is these relationships that permit the exchange of meaning within the sign plane of the map. And it is out of this exchange of meaning that the higher-order propositions that facilitate the creation and conveyance of authority about, and ultimately over, territory arise.

Our calculus, like every calculus, is no more than a method or system of calculation. That is, it is a logical structure. For the calculus to have anything to work with, the postings have to be realized in the sign plane of the map, and this can only be done as signs, that is, as unions of physical marks (signifiers) with some sort of ideational content (the thing signified). Traffic lights provide the classic example. The red light provides the physical mark, the signifier, while the necessity of stopping provides the ideational content, the thing signified. The collection of all the traffic signs (green light/go, yellow/caution, and so on) comprises the highway code. A code is an assignment scheme—a rule—that couples signifiers and signifieds; and codes vary from situation to situation. A given signifier can be assigned by different codes to a whole range of signifieds, while a given signified can be assigned to any number of signifiers. Depending on the code, for example, a line (the signifier) can be assigned to a river (one signified), a road (a second signified), a contour line (a third signified), an isobar (a fourth signified), and so on. Some of the code in maps is expressed in the map's legend, but most of it is not. Furthermore, every sign in a map is under the control of more than a single code.

In fact, as we have shown elsewhere in a formulation that has been widely adopted, map signs are under the control of at least ten codes.[6] Five of these codes, the intrasignificant ones, facilitate the exchange of meaning *among* the postings. Here signs are subject to an

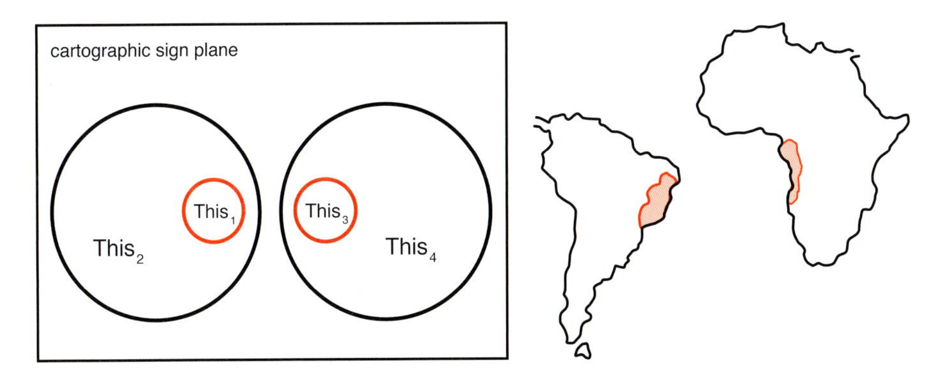

Figure 2.8
Two *thises*—This_1 and This_3—are proposed to be, or to have been, disjunct instances of the same type, in this example two similar geologic formations found on the African and South American continents. To bring them together would conjoin not only those two *this/theres*, but also any other *this/theres* entrained along with them.

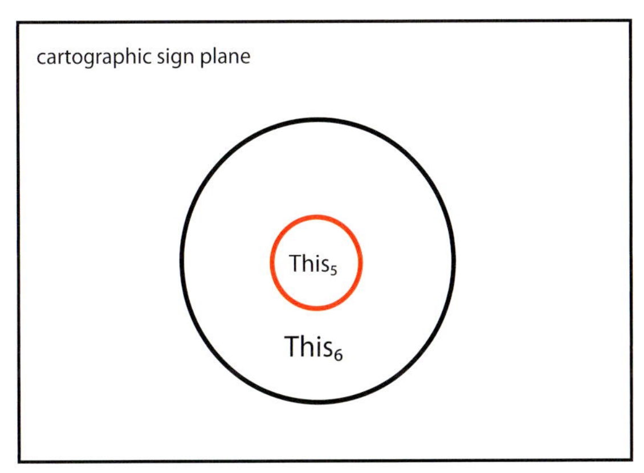

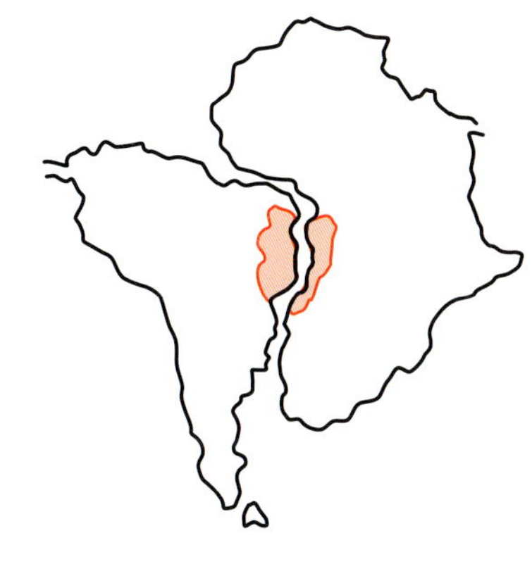

This_1 : Type A
This_3 : Type A
$\text{This}_1 + \text{This}_3 = \text{This}_5$
This_5 : Type A
$\text{This}_1 < \text{There}_2$
$\text{This}_3 < \text{There}_4$
This_2 : Type B
This_4 : Type B
$\text{This}_2 + \text{This}_4 = \text{This}_6$
This_6 : Type B

Figure 2.9
Two *theres* are linked through the commonality of their constituent *thises*, facilitating the construction of a new territory. In this way, new territories—past, present, or imagined—are invoked through the union of their constituent postings.

iconic code concerned with their *whatness* (say, streets and schools); a linguistic code concerned with their names; a *tectonic* code concerned with their spatial relationships (within which *scalar* and *topological* codes can be differentiated); a temporal code concerned with their temporal relationships (within which codes of duration and tense can be distinguished); and a *presentational* code concerned with the style and structure of their ensemble. The other five, the extrasignificant codes, facilitate the creation and conveyance of authority about, and ultimately over, territory. Here signs are subject to the *thematic* code that organizes the signs of the iconic code into a theme (it's a map of school districts); the *topic* code that organizes their spatial relationships into a place (they turn into a county, say, Wake); the *historical* code that organizes their temporal relationships into an epoch, into an era (for example, the coming school year); the *rhetorical* code that organizes their presentation into a style (that most supportive of the point that these are the school districts); and the *utilitarian* code that organizes the whole for the uses to which the map is intended (to achieve the complicated goals of the school board, i.e., whether it will appear on screen or be printed, in what fashion and in what numbers, and to whom the map will be addressed, given, sent, or sold).

While maps may in fact be subject to codes we have failed to recognize, no map can do without the ten we've identified; and all will follow the propositional logic of the map that we've attempted to formalize in our spatial/meaning calculus. In the end, of course, the map's meaning will be constructed by the map reader. This construction, however, is by no means free. It is grounded in the signs and the codes that govern them, and is constrained by the map's propositional logic. It unfolds, furthermore, in a dynamic process easier to exemplify than characterize, and it is to an example that we therefore now turn.

Notes

1. What do we mean by the *cartographic sign plane?* We refer simply to that part of the cartographic presentation commonly recognized as "the map" or "the map part," which is to say, the integral planar display of map signs often bounded by a neatline but, in any case, set apart from ancillary displays such as supportive maps, graphics, text, documentation, etc. The distinction being drawn here is between the map image as opposed to paramap elements of the larger cartographic discourse. (In the GIS environment, ESRI makes this distinction functionally explicit: our cartographic sign plane is called a "view" or "data frame" and the larger cartographic presentation a "layout.") We emphasize "cartographic sign plane" rather than simply "sign plane," since as opposed to other graphical sign planes—the photograph, painting, poster, comic strip—the sign plane of the map is unique in the degree and nature of its indexicality of planar location. Compare the Cartesian plot where location is also explicitly indexical but points to a plane of numbers instead of a geographic plane.

2. The second of Charles Sanders Peirce's three trichotomies of signs (see his "Logic as Semiotic" in *Philosophical Writings of Peirce,* ed. Justus Buchler [New York: Dover, 1955], 98–119), which has propagated within semiotic literature for a century, recognizes three general sign functions: iconic, symbolic, and indexical. Contemporary interpretations of these three functions distinguish among icons, signs that manifest some perceptible quality or qualities of their referents (Peirce's "objects"); symbols, signs that represent through some explicit rule or law; and indices, signs that refer, in Peirce's words, through a "dynamical (including spatial) connection." Peirce goes on to say that indexical signs or indices "have no significant resemblance to their objects" but that they "direct attention to their objects." Cartographic signs—the postings asserting that this is there—clearly actualize these functions simultaneously.

Most, perhaps all, cartographic signs are iconic to one degree or another, offering up graphical resemblances of their referents. At one extreme, a sign for "church" may take on the shape of an actual church facade. Toward another, the sign for "road" may pick up on a road's linearity. Unlike the sounds of spoken language, few cartographic signifiers are entirely free of some tincture of the signified. Anything but arbitrary instantiators of a there, cartographic signs fully participate in the map's construction of its thises (and so a sign for "road" that implies a channel for vehicular movement also defines that road as something different from, say, an interstate or walking or bicycle trail).

At the same time, all map signs are also symbolic in that they refer through some convention, either one established within the cartographic presentation by a legend or a key (where a circle of a specific diameter may be stated to represent a quantity of a certain size) or through some widely shared cultural convention (as green refers to forest on U.S. topographic maps but to Crown Lands on Canadian topographic maps).

Iconic and symbolic signing functions bring the this into the map, but what about the there? That is precisely the indexical function of every cartographic sign—to connect a location in the sign plane of the map with a corresponding location in the "real world." Can any sign be purely indexical? Peirce didn't think so and neither do we, but the graticule crosshair demarking the intersection of a line of longitude and a line of latitude may seem to offer no *this is there* proposition at all, only the proposition that *there is there*. It becomes evident, though, that the crosshair does indeed post the instantiation of not one but two conceptual types (it is a symbol after all): the line of longitude and the line of latitude. Exactly what *thises* might be manifest there is left to be discovered. The indexical function of the crosshair is foregrounded; it is less a mark than a marking. A fine example is the vegetation survey, which entails the fixing of random locations within a given territory and then their subsequent visitation to witness and enumerate just which floral *thises* are actually at the given *theres*. The cartographic plan of the vegetation survey, and other surveys of that kind, is the posting of multiple empty *theres* awaiting population by as yet unknown *thises* of some prescribed type, or multiple types (fauna, soils, environmental conditions).

3. If there remains any doubt about the intrinsic spatiality of the posting, one need only consider the index to *Goode's World Atlas* (2005) where, for example, the location of California is given as 38.10N, 121.20W (about twenty miles north of Stockton), the location of the Mississippi River as 31.50N, 91.30W (somewhere between Natchez and Vicksburg), and the location of the United States as 38.00N, 110.00W (the northeastern arm of Lake Powell, Utah).

4. In our "Designs on Signs" we addressed these hierarchies from a semiotic perspective, distinguishing among cartographic signs (individual postings), sign systems (assemblages of signs of similar type), and sign syntheses (juxtapositions of signs and sign systems of dissimilar types). Also at issue here, though, are concepts of scale intrinsic to cartographic signs.

5. The propositional construction of the Third Reich would have been an equally illuminating example, but that would have serviced the notion that, if maps have agendas at all, these must inevitably be sinister. Of course maps have agendas. If they didn't, no one would take the time and trouble to make them. Does acknowledging the simple but critical fact that maps are motivated imply their ethical or moral corruption? Not necessarily. Clearly, Wegener deployed cartographic arguments in an attempt to construct new intellectual territory, new knowledge of the planet's processes and prehistory; and it would be impossible to imagine contemporary planetary geology without his contributions. The determined opposition of Wegener's American (not European or South African) peers had almost nothing to do with the viability of his arguments and almost everything to do with the perceived challenge to their own authority (and consequently their prestigious and lucrative careers). Whose ethics color the map in a case like that?

6. Wood and Fels, "Designs," and Wood, *Power of Maps*. Alan MacEachren discusses these codes at length in his *How Maps Work: Representation, Visualization, and Design* (New York: Guilford Press, 1996), 298–353; Peter Wollen reads Guy Debord's maps through our codes in "Mappings: Situationists and/or Conceptualists" in *Rewriting Conceptual Art*, ed. Michael Newman and Jon Bird (London: Reaktion, 1999), 27–46; Irit Rogoff uses the codes in her work on what she calls "geography's visual culture" in her *Terra Infirma: Geography's Visual Culture* (London: Routledge, 2000), especially pp. 74–76; and Vincent Del Casino and Stephen Hanna make exemplary use of the whole approach in "Mapping Identities Reading Maps: The Politics of Representation in Bangkok's Sex Tourism Industry" in *Mapping Tourism*, ed. Hanna and Del Casino, 161–85 (Minneapolis, Minn.: University of Minnesota Press, 2005).

AUSTRALIA
LAND OF LIVING FOSSILS
Produced by the Cartographic Division
National Geographic Society
ROBERT E. DOYLE, PRESIDENT
NATIONAL GEOGRAPHIC MAGAZINE
GILBERT M. GROSVENOR, EDITOR
RICHARD J. DARLEY, CHIEF CARTOGRAPHER
JOHN F. SHUPE, ASSOCIATE CHIEF CARTOGRAPHER
Design: John F. Dorr, J. Robert Teringo
Paintings: Roy Andersen; Map Art: Tibor G. Toth
Text: John Eliot; Research: Merrill Clift,
Jeanne E. Peters; Principal Consultant: Dr. T. L. Riggert
WASHINGTON
FEBRUARY 1979
Red-tailed Black Cockatoo
Gouldian Finches
Crimson Finch
Sugar Glider
Lumholtz's Tree Kangaroo
Freshwater Crocodile
Sulphur-crested Cockatoo
White Ibis
Wedge-tailed Eagle
Rainbow Lorikeet
Bustard
Galah
Emu
Agile Wallaby
Brolga
Frilled Lizard
Red Kangaroo (female)
Red Kangaroo (male)
Euro
Dingo
Rabbit-eared Bandicoot
Melville Island
Van Diemen Gulf
Darwin
Daly
Katherine
Joseph Bonaparte Gulf
Victoria
Ord
Kununurra
L. Argyle
King Sd.
Dampier Land
Kimberley Plateau
Mt. Ord 937
Roebuck Bay
Great Sandy Desert
Tanami Desert
NORTHERN
Port Hedland
Hamersley Range
Mt. Bruce
Mt. Meharry
North West Cape
Exmouth Gulf
NORTH WEST BASIN
Lake Mackay
WESTERN
1510 Mt. Zeil
Macdonnell
L. Disappointment
Gibson Desert
Mt. Augustus 1106
WESTERN AUSTRALIA
PLATEAU
Amadeus Depression
Ayers Rock 868
L. McLeod
Shark Bay
Dirk Hartog I.
Mt. Woodroffe 1440

Three

Reading Land of Living Fossils

Courtesy of Roy Anderson/National Geographic Image Collection.

Reading Land of Living Fossils

Maps do not appear in our hands by magic, but as the result of a sequence of actions. A book has to be removed from a shelf; pages have to be turned. The map is the third in a series of maps, or it is the seventh. Or it is surrounded by text to which it is related. The way we approach the map—or it approaches us—constructs a frame, and this frame encourages us to see the map this way or that. Perhaps we had to open the newspaper. The map accompanies this article. It is on that page; it is attached to this ad. It is a locator map; it is a weather map. Or the map was handed out at a public hearing—an usher counted the number for each row, they were passed down. The map pertains to the hearing. It is a legal document. Or it was handed out at a briefing. It is the map of a target; it shows the sites to be bombed. In our case, the map had to be unfolded. We came across it riffling through a bin of maps in a used bookstore. The map had been divorced from its original context—its epimap (the pages of an issue of the *National Geographic*)—and any it had, it now carried with itself, in its perimap and in the way it allowed itself to be revealed as it was opened. This is true of most *National Geographic* maps: they tell you how they want to be read by the way they unfold themselves.

AUSTRALIA
LAND OF
LIVING FOSSILS
Produced by the Cartographic Division
National Geographic Society
ROBERT E. DOYLE, PRESIDENT
NATIONAL GEOGRAPHIC MAGAZINE
GILBERT M. GROSVENOR, EDITOR
RICHARD J. DARLEY, CHIEF CARTOGRAPHER
JOHN F. SHUPE, ASSOCIATE CHIEF CARTOGRAPHER
Design: John F. Dorr, J. Robert Teringo
Paintings: Roy Andersen; Map Art: Tibor G. Toth
Text: John Eliot; Research: Merrill Clift,
Jeanne E. Peters; Principal Consultant: Dr. T. L. Riggert
WASHINGTON
FEBRUARY 1979
Wedge-tailed Eagle
Galah
Emu

Figure 3.1
Cover of "Australia: Land of Living Fossils," supplement to *National Geographic*, February 1979.

Figures 3.1–3.4 courtesy of Roy Anderson/National Geographic Image Collection.

Unfolding "Land of Living Fossils"

"Australia: Land of Living Fossils" (figure 3.1) begs to be unfolded. One attraction is the whiff of the aboriginal wafting from "Australia." It comes from the way the letterforms whisper *hand drawn* and the way the word is printed in a "natural dye" brown (hinting of nut shells). Inside each letter its form is picked out again in white as if to say "Primitive X-ray Art."

Below the title, credits, and publication information (National Geographic Society, 1979), three birds have been posed: an emu (still exotic in 1979), a galah, and a wedge-tailed eagle. The galah blazes in phosphoric pink. It has a white comb and a faintly yellow nib. The neck of the emu glows with an opalescent blue.

Living fossils!

Everything here—title, letterforms, birds—plays with this paradox which, *simply by being paradoxical*, is removed from our everyday, our hum-drum world. It would be hard not to be thinking, "Australia—not an everyday place!" The tip of the wedge-tailed eagle's wing wraps out of sight around the right fold. The feet of the emu dip around the fold below. Like a barker at a carnival the cover pleads, "Unfold me!"

The first fold gives up a pair of cockatoos (one sulphur-crested), a gaggle of multicolored Gouldian finches, a rainbow lorikeet, and an agile wallaby; the next fold a colony of koalas, the superb lyrebird, a bustard, a brolga, a white ibis, a royal spoonbill, a pied goose, a sugar glider, a spotted cuscus, a tree kangaroo, and a freshwater crocodile (figure 3.2). Do *all* these illustrate the paradox? Certainly they're exotic, colorful. Only the next fold, unveiling the central portion of the supplement, reveals . . . the *map*, this of a tan and green Australia surrounded, *guarded*, by totemic kangaroos, dingoes, a black swan, a variety of other birds, a hairy-nosed wombat, a truly prehistoric-looking frilled lizard, and, falling off the bottom of the page (inviting the final unfolding), a kookaburra with a common black snake twisting in its bill.

With the final fold, a platypus, an echidna, a Tasmanian devil, more kangaroos, another wallaby, another ibis (figure 3.3).

Wow!

Swarmed by the fury of fur and feathers, the map comes off as little more than an excuse for the animal portraiture. And indeed there is that about the map too, a suggestion of portraiture, as if the continent had sat for, perhaps even commissioned, this lush, gorgeous, almost tactile rendering in tawny shades of khaki and sand and lightly done toast. The colors slip through old ivory and olivesheen and citron to conclude in a deep grass-green, minty, almost viridian in the shadows of the Atherton Tableland (the northeast highland region, figure 3.4). There is a ripeness about the rendering, a swelling, a fullness. A production house notorious for the lavishness of its type has here restrained itself as though type ill comported with portraiture: seven retiring province names; a few features picked out, but in the smallest type imaginable; in a larger face, but so widely spaced as to sink into the land, the names of a single plateau, lowland, and range of mountains. This *virginal land* (of living fossils) has yet to be conquered by nomenclature.

There is a text, but it's unlikely that it's read before the map is flipped over. If you've ever subscribed to the *National Geographic*, you

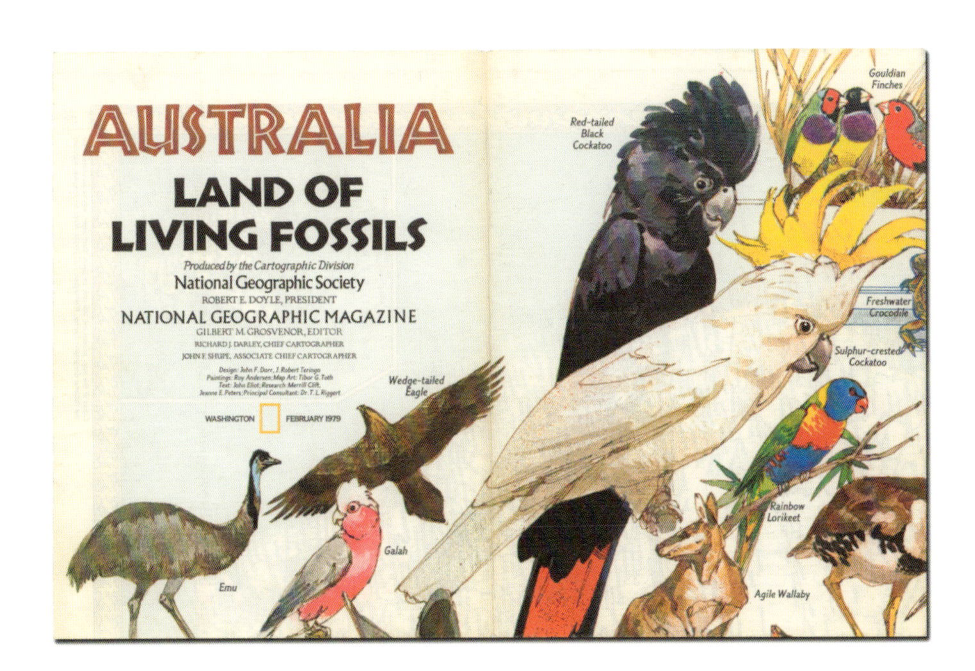

Figure 3.2
Unfolding "Australia: Land of Living Fossils."

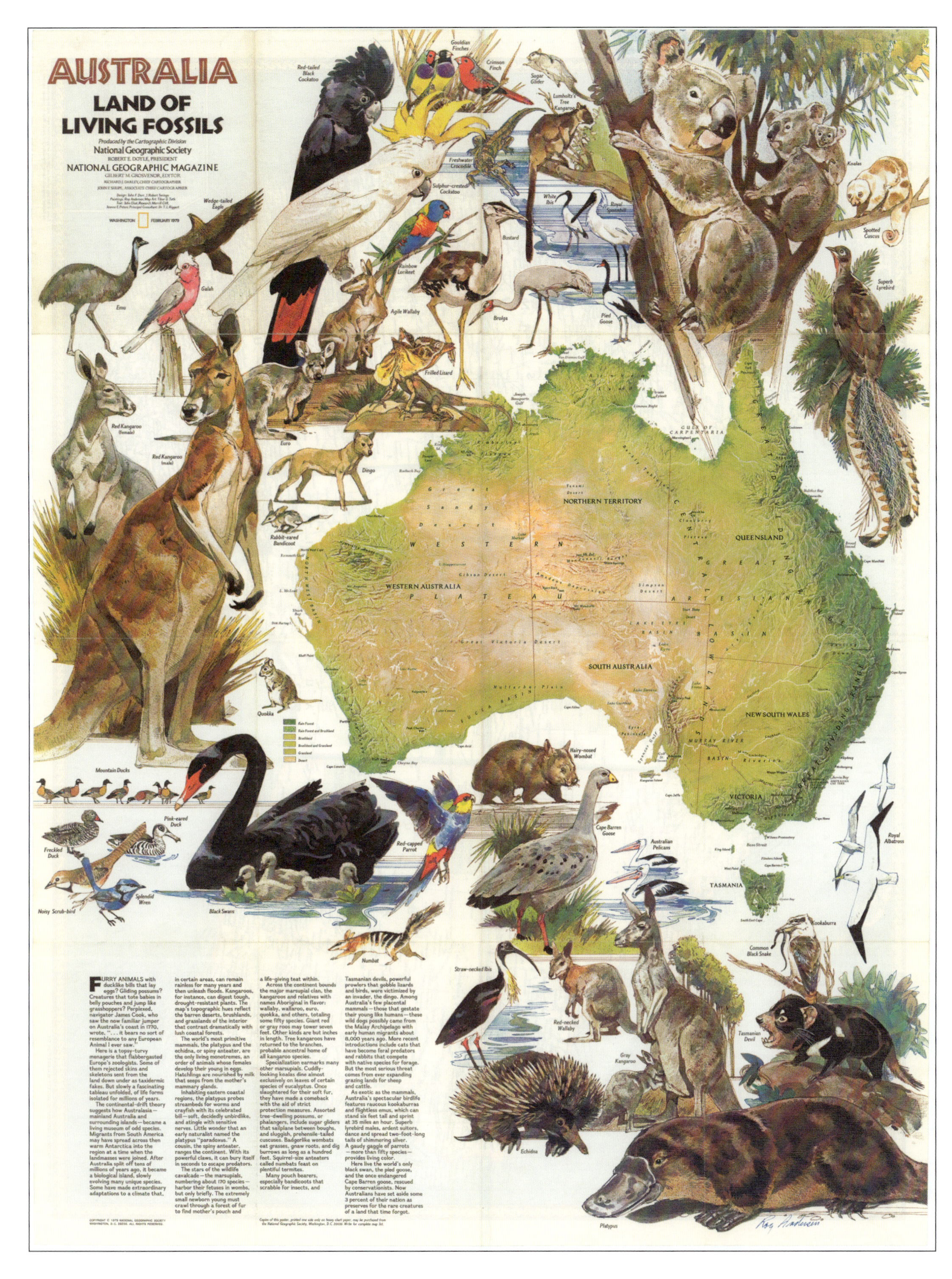

Figure 3.3
"Australia: Land of Living Fossils," full map. But "map" is used loosely here, as the animals are clearly meant to overshadow the subtle map in the center. Even the *Geographic* concedes it's better called a "poster."

Figure 3.4
Detail of "Australia: Land of Living Fossils."

know how these map supplements work. On one side is an elaborate graphic like our painted menagerie (small type in its margin even calls it a poster). On the other side is The Map (the "real" one), what in our introduction we called the main map (small type in *its* margin even calls it a "map"). This map *is* in the house style. An all but white Australia—Easter-egg tones shadow the borders—is washed by the palest of blue waters and just *covered* with type (figure 3.5). There's not a kookaburra in sight: in its place, a dense spiderwork of roads and railroads, homesteads, airports, oil fields, water holes, ports, stock routes, towns, cities, metropoli, *stippled in type*, nomenclature nailing everything. Even the boundless waters are named: the Timor Sea is distinguished from the Arafura, the Tasmanian from the Coral, and all from the Indian Ocean.

What modulates smoothly on the poster-side map from the palest of Caucasian winter-skin "whites" to watermelon-rind green, goose-steps on the main-map side from Western-Australian pink to South-Australian purple to Queensland yellow. What on the poster-side map was a "self-distinguishing" landscape of gradually varying land*forms* (mountains, valleys, plains), is severed, on the main map, into land *status types*: aboriginal lands (bounded by a black line shadowed in gray); and wildlife sanctuaries, nature reserves, and national parks (bounded by dashed lines and filled with green). What on the poster side explodes from the map in abundant profusion (animals, which is to say, nature) is on the main map corralled, bordered, set apart (in parks). Everything else is white, which is to say . . .

Which is to say . . . everything else is, precisely, *what*?

The topography that forms the *surface* of the poster-side map slips *beneath* it on the main map. The main map's surface is made of type. Beneath it the signs for swamp, desert, and dry salt lake lie like a layer of dust. In barely distinguishable grays only ghosts of relief can be made out. Yet the maps are not contentious. Instead of suggesting a state of schizophrenia, the day face of a Dr. Jekyll on the one side and the night face of a Mr. Hyde on the other, the map as a whole—that is, the two-sided sheet taken as a unit ("*Hand me that map of Australia, will you?*")—asks us to imagine that these worlds are separated for no more than (technical) reasons of legibility, that in fact they somehow *imply* each other.

Correspondingly the map's text—ten paragraphs on the poster side (figure 3.7)—situates its "topsy-turvy menagerie" in a historical context of loss and survival among introduced species (dingoes, cats, rabbits, sheep, and cattle) that very much *takes for granted* the Australia mapped on the nomenclatural side. In describing the "fascinating tableau" of this "living museum of odd species," the text does not hesitate to refer to the slaughter of the koala for its soft fur, the victimization of the Tasmanian devil by the introduced dingo, and the competition given native species for forage by the more recently introduced rabbits, sheep, and cattle. Indeed, tragic as this history might have been, today koalas "have made a comeback with the aid of strict protection measures," and "the once endangered Cape Barren goose" has been "rescued by conservationists." Without a trace of irony the text concludes that "Now Australians have set aside some 3 percent of their nation as preserves for the rare creatures of a land that time forgot."

Though it wasn't forgotten altogether, as the text's introductory paragraph reminds us, recovered as Australia was for history (for *time*) at least as long as 230 years ago by James Cook, if not in fact 8,000 years before that, as the text further reminds us, by the human migrants from the Malay Archipelago who introduced the dingo. "Forgot" only means to say forgotten in the mainstream of *our* history, which is why

Figure 3.5
"Australia" (verso of "Land of Living Fossils") is what we call the main map of the supplement. It grounds the fur-covered map on the front side.

Figures 3.5 and 3.6 courtesy of Ng Maps/National Geographic Image Collection.

Figure 3.6
Detail view of "Australia."

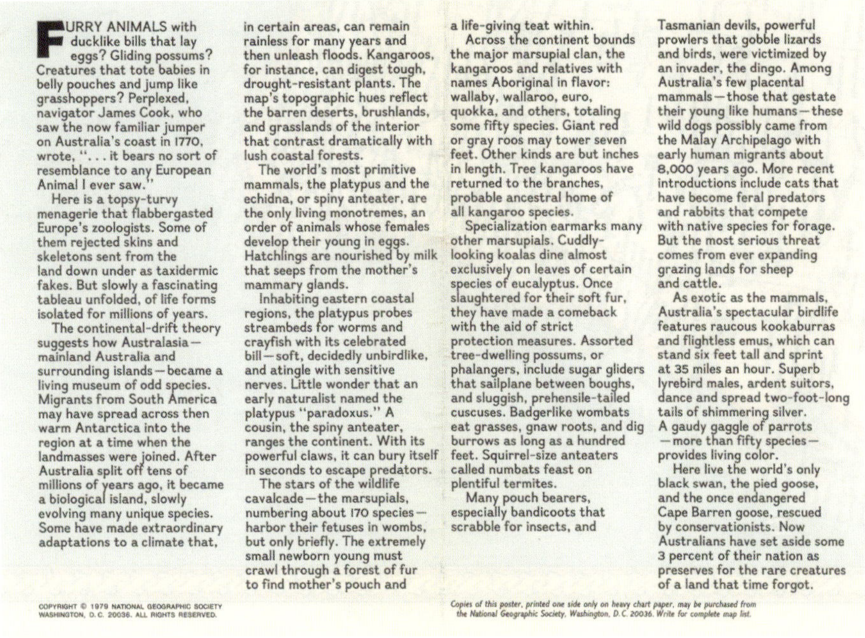

FURRY ANIMALS with ducklike bills that lay eggs? Gliding possums? Creatures that tote babies in belly pouches and jump like grasshoppers? Perplexed, navigator James Cook, who saw the now familiar jumper on Australia's coast in 1770, wrote, ". . . it bears no sort of resemblance to any European Animal I ever saw."

Here is a topsy-turvy menagerie that flabbergasted Europe's zoologists. Some of them rejected skins and skeletons sent from the land down under as taxidermic fakes. But slowly a fascinating tableau unfolded, of life forms isolated for millions of years.

The continental-drift theory suggests how Australasia—mainland Australia and surrounding islands—became a living museum of odd species. Migrants from South America may have spread across then warm Antarctica into the region at a time when the landmasses were joined. After Australia split off tens of millions of years ago, it became a biological island, slowly evolving many unique species. Some have made extraordinary adaptations to a climate that, in certain areas, can remain rainless for many years and then unleash floods. Kangaroos, for instance, can digest tough, drought-resistant plants. The map's topographic hues reflect the barren deserts, brushlands, and grasslands of the interior that contrast dramatically with lush coastal forests.

The world's most primitive mammals, the platypus and the echidna, or spiny anteater, are the only living monotremes, an order of animals whose females develop their young in eggs. Hatchlings are nourished by milk that seeps from the mother's mammary glands.

Inhabiting eastern coastal regions, the platypus probes streambeds for worms and crayfish with its celebrated bill—soft, decidedly unbirdlike, and atingle with sensitive nerves. Little wonder that an early naturalist named the platypus "paradoxus." A cousin, the spiny anteater, ranges the continent. With its powerful claws, it can bury itself in seconds to escape predators.

The stars of the wildlife cavalcade—the marsupials, numbering about 170 species—harbor their fetuses in wombs, but only briefly. The extremely small newborn young must crawl through a forest of fur to find mother's pouch and a life-giving teat within.

Across the continent bounds the major marsupial clan, the kangaroos and relatives with names Aboriginal in flavor: wallaby, wallaroo, euro, quokka, and others, totaling some fifty species. Giant red or gray roos may tower seven feet. Other kinds are but inches in length. Tree kangaroos have returned to the branches, probable ancestral home of all kangaroo species.

Specialization earmarks many other marsupials. Cuddly-looking koalas dine almost exclusively on leaves of certain species of eucalyptus. Once slaughtered for their soft fur, they have made a comeback with the aid of strict protection measures. Assorted tree-dwelling possums, or phalangers, include sugar gliders that sailplane between boughs, and sluggish, prehensile-tailed cuscuses. Badgerlike wombats eat grasses, gnaw roots, and dig burrows as long as a hundred feet. Squirrel-size anteaters called numbats feast on plentiful termites.

Many pouch bearers, especially bandicoots that scrabble for insects, and Tasmanian devils, powerful prowlers that gobble lizards and birds, were victimized by an invader, the dingo. Among Australia's few placental mammals—those that gestate their young like humans—these wild dogs possibly came from the Malay Archipelago with early human migrants about 8,000 years ago. More recent introductions include cats that have become feral predators and rabbits that compete with native species for forage. But the most serious threat comes from ever expanding grazing lands for sheep and cattle.

As exotic as the mammals, Australia's spectacular birdlife features raucous kookaburras and flightless emus, which can stand six feet tall and sprint at 35 miles an hour. Superb lyrebird males, ardent suitors, dance and spread two-foot-long tails of shimmering silver. A gaudy gaggle of parrots—more than fifty species—provides living color.

Here live the world's only black swan, the pied goose, and the once endangered Cape Barren goose, rescued by conservationists. Now Australians have set aside some 3 percent of their nation as preserves for the rare creatures of a land that time forgot.

Copies of this poster, printed one side only on heavy chart paper, may be purchased from the National Geographic Society, Washington, D.C. 20036. Write for complete map list.

Figure 3.7
The descriptive text found in the lower left corner of "Land of Living Fossils."

Courtesy of Roy Anderson/National Geographic Image Collection.

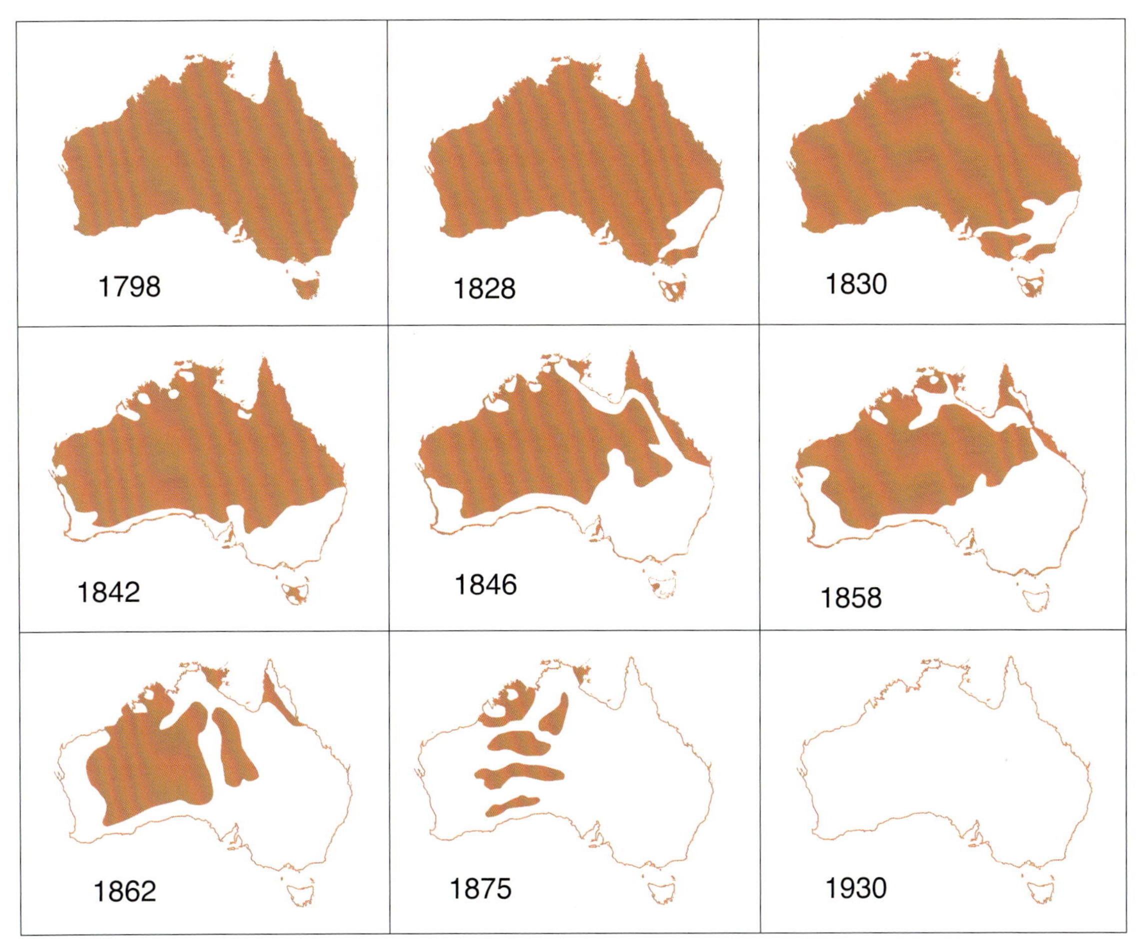

Figure 3.8
These maps track the expansion across Australia of the British, who ultimately arrogated to themselves the entire continent. It is this arrogated land that appears in white on "Australia."

our zoologists were "flabbergasted" by the "topsy-turvy menagerie" of "furry animals with ducklike bills"; even as "forgot" ignores the fact that today Australia is very much in history, and populated by recent, European immigrants who have arrogated to themselves a continent they initially claimed by . . . naming.

The white on the main map we asked about a page ago? That white is all the land arrogated by the British through the magic of naming not "returned" or "set aside" for the rare creatures (including aboriginal humans) of this land that time forgot.

Comparing the two sides of the map

The two sides of the map are connected in precisely this way: the land of living fossils—the poster side, the wild and wacky animals, and the wild and wacky terrain they imply (the lushly painted land of the poster's map)—are subsumed within the bounded and often rectilinear areas of green on the—how to say this?—more "comprehensive," more "real-world" main map of Australia, which while not a USGS topo quad is almost as authoritative. In the terminology of our introduction, nature as cornucopia (theme of profligacy) is subsumed within nature as park. Simultaneously, nature as park takes on the color of the profligate. In other words, nature as park *authorizes* nature as cornucopia. It says, "*I* am authoritative, so *that* is authoritative." Nature as cornucopia *colors*, *perfuses* nature as park. It says, "In Australia nature is bountiful, it is extravagant," and so it becomes hard to see the shape of Australia without seeing koalas, wallabies, kangaroos. But this effect occurs in our heads. It's not on the paper.

It is the purity of each of the maps—their complete independence—which potentializes this effect. The two sides, in fact, seem to have been produced within the Geographic's cartographic division by entirely separate units. The main map, again, *called* a map ("Copies of this map, printed one side only on heavy chart paper, may be purchased . . ."), carries in its upper right the designation, "Supplement to the National Geographic, February, 1979, Page 152A, Vol. 154 No. 2—AUSTRALIA." There is no hokey type, funky colors, or hanky-panky X-ray art. Among the credits are none for art or design. None but the society president, magazine editor, and chief and associate chief cartographers are named; they and the projection: the Chamberlin Trimetric. The scale is given in four forms. The graticule is marked every four degrees.

None of this is true of the poster, which, again, is *called* a poster ("Copies of this poster, printed one side only on heavy chart paper, may be purchased . . ."). The poster carries no designation as a supplement to the magazine. The type of the title is *cute*. Beyond the names of president, editor and chief, and associate chief cartographers, credit is given for *design*, *painting*, *map art*, *text*, *research*, and *principal consultant*. The projection is *not* named. There is no scale or graticule. There is no ocean either, and therefore no reefs. There is no sea life among that of the land that time forgot. The *poster's* Australia stops at the water's edge.

The Australia of the main map slips *into* the water. Here the "outer limit of the continental shelf" is marked. The offshore waters are thick with reefs and banks and shoals. A block of text describes the Great Barrier Reef as, "Containing the widest variety of marine life found anywhere, the barrier is the world's most extensive stamping ground for coral fancier, shell collector, underwater explorer, and student of marine biology." Another text on the map describes the 400,000 square miles of Australia's Coral Island Sea Territory. These waters *connect* Australia to the rest of the world—Indonesia dips into the map at the upper left, Papua New Guinea into the upper right—and so *imply* the rest of the world. This map's border less demarcates *an entity* than frames *a piece of the world*. Indeed, this *is* the world, we have merely zoomed in on one part of it for a closer look, we could pull out, the Pacific would reveal itself, it's all *real*.

It is precisely the absence of these features that permits the poster map to be a portrait, an essay, an opinion piece (to be cute). This map's border (which stops at *land's* end) less frames a piece of the world (a piece that *contains* Australia) than brings a world into being ("the land of living fossils"). It is this exemption from the burden of "objective reality" (from the burden of the authoritative) that licenses the map "art" with its "living" color and somewhat breathless topography (any "art" on the main-map side is confined to the elaborate decorative mini-kangaroo border that emphasizes the *un*decorative quality of the main map itself—see figure 3.12). It is this same exemption from the burden of "objective reality" that licenses the riot of animals that surrounds the poster map, especially the "unscientific," indeed *popular* emphasis on birds and mammals (the main map is coolly dispassionate, anything but popular).

There is nothing unusual about this structure. State highway maps, for example, the ones produced by state transportation departments, have two sides too. On one side is the highway map itself (this is its main map). On the other side (the "poster" side) is a heavily illustrated inventory of points of interest. As the main map tells us how to get there, the poster side tells us where to go. As in "Australia: Land of Living Fossils," the main map *authorizes* the rest.[1]

A cuddly cornucopia

In introducing nature as cornucopia we characterized it as the nature of the small and the soft, the fuzzy and the warm. We said it was the nature of fur and feathers, and in fact the poster has it that the land of living fossils is largely inhabited by birds. Forty have their portraits here, forty birds, twenty-two mammals, and three reptiles, one of which is hanging from the bill of a kookaburra. Australia may have "slowly evolved many unique species" but the evidence here is that these confined themselves to a couple of classes of a single phylum of animals. The birds are colorful, the mammals are furry (the text even refers to "cuddly" koalas), but . . . *what do they eat?* Where are the echinoderms, the gastropods, the crustacea and the insects, the numerous worms and sponges? How about the corals for which Australia is so renowned? And this is to enumerate only a few of the *animals*, it's not to get started on the other kingdoms.

Admittedly, microbial mats are less picturesque than such "stars of the wildlife cavalcade" as the koalas and giant red kangaroos. Microbial mats are hard to embrace, to pet. They're less obviously responsive, they don't sit up and beg. (They're bacterial scums.) But their absence from the "the land of living fossils" gives precisely the lie to the phrase, reveals it for the cute "hook" that it is. For if anything in existence merits the description "living fossil," certainly it is the microbial mats—the stromatolite reefs (figure 3.9)—in Shark Bay several hundred miles north of Perth on Australia's west coast.[2] Here the stromatolites flourish as they have for more than three and a half *billion* years, their appearance little changed since the Isuan Era of the Archean Eon. Comparatively rare in the earlier Archean, stromatolites owned the Proterozoic Eon, "the Age"—in Steven Stanley's phrase—"of Prokaryotes."[3]

This Age of Prokaryotes lasted from somewhere around two and half billion years ago until the onset of the Paleozoic nearly two billion years later. (The Paleozoic began approximately 570 million years ago.) It is a large chunk of the earth's history, and the stromatolites were a big part of it. Not only were stromatolites the most abundant form of life throughout the Precambrian, they were also responsible for building up the oxygen content of the atmosphere to its present levels; that is, for making modern life possible:

> The clearest evidence of the lives of ancient and extensive bacterial confederacies, however, are stromatolites. Stromatolites were to the Proterozoic landscape what coral reefs are to the present ocean: rich and beautiful collectives of intermingled, interdependent organisms. These domed, conical, columnar, or cauliflower-shaped rocks, found throughout the fossil record and still in existence today, are composed of rock layers that were once microbial mats. Communities of bacteria, especially photosynthetic cyanobacteria, lived and died atop one another . . . . Some of the ancient stromatolites exceeded thirty feet in height.
>
> Today—in restricted parts of the world—we can see that the top layers, only a few centimeters in width, are dominated by photosynthetic blue-green bacteria . . . . Below the top layer are thriving populations of anaerobic purple photosynthesizers, which are sulfur depositers. Beneath them are dependent microbes, living on the produce of the bodily remains of the others.[4]

The cyanobacteria (blue-green algae) precipitate calcium carbonate, which binds with sediments trapped in the microbial mat to build up the rock base. The organisms continually migrate upward through the sediments to maintain their access to the sunlight, which fuels the life of the cyanobacteria. These are the oldest lifeforms extant. That they have no place in "Australia: Land of Living Fossils" simply begs the question, why not?

A possible answer, that when our map was published in 1979 the significance of stromatolites was unrecognized, does not bear much scrutiny. The *Geographic* itself had published a photo of stromatolites in its May 1978 issue;[5] and J. William Schopf's article, "The Evolution of the Earliest Cells"—largely about stromatolites and indeed including a photograph of the living stromatolites in Shark Bay—was published in *Scientific American* in September of 1978, which is to say, their significance was already acknowledged in the popular press. Schopf

Figure 3.9
Stromatolite reefs are a true example of "living fossils."

had been writing about stromatolites since the early 1970s, but the groundwork for this recognition had been laid as long ago as the 1950s with the realization that even the most ancient sedimentary rocks contained fossils, and in the 1960s with the growing recognition that the greatest division among organisms was not between plants and animals, but between those with and without cellular nuclei. We recognize this distinction today as that between eukaryotes (with nuclei) and prokaryotes (without). Among the prokaryotes are both eubacteria and archaebacteria and what used to be called blue-green algae, but are now better known as cyanobacteria.[6]

To seriously entertain this answer is, however, to overlook the absence on our map of yet another "living fossil," Australia's great saltwater crocodile (figure 3.10), which the *Geographic* itself had called—and only the year before—"Survivor of the Dinosaur Age," noting that "crocodiles have been around for nearly 200 million years," and that "crocodiles survived while their close kin the dinosaurs died out."[7] If these too are not exemplars of "living fossils," it is hard to imagine what could be. It is even harder to understand why the *Geographic* would include among its "topsy-turvy menagerie" a very small image of the far less significant *freshwater* crocodile when a year before it had called the Australian *saltwater* croc "the biggest and some say the most dangerous of crocodiles." In the 1978 article, author Rick Gore goes on to say that "Fishermen in Queensland once hauled in one that reportedly measured 33 feet," adding that the crocodiles are "revered as a totem in parts of northern Australia" by contemporary Aborigines. Such an animal would certainly seem to qualify as a *Geographic* totem.

Except that it's *not* cuddly. "They're not cuddly," Wayne King said in his epigraph to Gore's article. "They don't have big soulful

Figure 3.10
Australia's great saltwater crocodile, another living fossil. Again, none can be found in our *National Geographic* map!
Courtesy of St. Augustine Alligator Farm, 2004.

Figure 3.11
"Australia: Land of Living Fossils" detail. Imagine these as stuffed animals.
Courtesy of Roy Anderson/National Geographic Image Collection.

eyes like seals. Most of the animals the world is concerned with are beautiful, or they tug at your heartstrings. Crocodiles have a pretty toothy leer. They eat dogs in Florida—sometimes even people. Who could love them?" Yet at least crocodiles *have* eyes. Stromatolites lack even those, don't move, just sit there, in the shallows, like the cold, black, near-rocks that they are.

This marginalization, not to say rejection, of the cold and dark and hard by our map serves to construct an idea of nature as *warm and bright and soft*, though as a glance at the "living museum of odd species"—forty birds, twenty-two mammals, three reptiles—makes clear, still more is at stake than this. The issue, as King understood, is *cuddlesomeness*, which may be reduced to the question: *what would it come to as a stuffed animal* (as a Teddy bear)? Filling the upper-right corner of our map is a family of koalas (figure 3.11). A baby clings to its mother's shoulders. Completely dominating the left side of the sheet, and better than a third its height, are a pair of red kangaroos, a male and a female. Commanding the lower right? A platypus, its fur so lushly rendered you can almost sink your fingers into the plush.

In fact, the reigning images on the sheet *are* stuffed animals, and in each corner there is as well an added hint of *domesticity* (in addition to the family of koalas in the upper right, the platypus partially obscures a gray kangaroo and her nursing baby). The cuddly, the caressable, the kissable. These are all, or are all presented as being, things to hold close for warmth or comfort or in affection, which is to say, things to care for. No surprise, then, to recall that "Australians have set aside some 3 percent of their nation as preserves for the rare creatures of a land that time forgot." As for the crocs? In 1978, Australian biologist Gordon Grigg said, "If a child is taken at a beach, say, it will become almost impossible to defend our efforts to conserve the crocs."[8]

It is *not* about conservation. (How can we conserve animals when there is not a hint of the ecosystems required to sustain them?) It is *not* about rare creatures. (It is about our furry and feathered friends.) It is about . . . what *is* it about?

The Geographic's construction of society

This is *not* about the National Geographic Society. Or, rather, it is. But only because it was the Society that produced "Australia: Land of Living Fossils." Without diminishing the (enormous) significance of the Society as a force in American geographic consciousness, it needs to be said that what we are attempting to trace here is a very general state of mind, almost a Foucauldian episteme or, more precisely, an aspect of an episteme. The deeply conflicted ideas of nature that legitimate only certain discourses are not the product of any single institution, however dominant, but are broadly developed. We could be looking at the greeting card industry—which has its own instincts for conceptualizing nature—but we happen to be looking at maps, of which the National Geographic Society is an important, but as our canvassing of the terrain in our introduction must have suggested, by no means unique producer. That said, there can be no coming to grips with this map or the natures it proposes without some understanding of the organization that produced it.

The late 1970s nearly coincided with a peak in the *National Geographic*'s circulation of close to eleven million.[9] Some eleven million copies of "Land of Living Fossils" were distributed to the magazine's readership. After slipping the map from the magazine's pages, these readers, as we did, unfolded it to the koalas, unfolded it again to the kangaroos and the simple map image—over which they lingered for a second—and finally unfolded it to the platypus. They glanced at the text in the lower left, and turned the map over. Their eyes roved over the main map. Maybe they mouthed the names of the states —"Western Australia," "Queensland"—and looked to make sure they knew where Sydney was before turning the map back over. Here they glanced again at the "stars of the wildlife cavalcade" and maybe read a paragraph or two of the text before folding the map up and slipping it back into the pages of the magazine, where it remained until the used-book dealer extracted it to add to his map bin.[10]

This was the fate of almost every one of these maps. What in the world were they produced for?

Frankly, they were produced as the primary form of the cultural capital whose possession certified the class position of the *Geographic* family; that is, the readership and the staff that produced the magazine and, with it, the effect of a certain kind of cultured life, a life with maps and globes, pipes and booklined-studies, knowledge, and therefore wisdom, and since wisdom, the right to wield social power that, despite the changes afoot at the *Geographic* in the late 1970s, amounted nonetheless to a powerful conservatism.[11] The magazine, with its unmistakable yellow border, was the lever, but the fulcrum was the map, for the pretense here was far from a simple consumption of knowledge, but the *participation*—as a supporting member—in a scientific organization that *produced* geographic knowledge that, as everyone knew, was ultimately codified in the form of a map.[12]

Even today, when the *National Geographic* can be purchased at newsstands, one cannot *subscribe* to the magazine. One *joins* the Society and receives the magazine as a benefit of membership. Members receive membership cards, which they are encouraged to "keep handy" (a signature line implies that another might want to steal one's membership rights). The card, says the Society, "distinguishes you as a member of the world's largest nonprofit scientific and educational organization." The Society spells this out again and again: "You have the satisfaction of supporting," it elaborates, "the Society's program for improving the geographic literacy of youngsters; important worldwide research and exploration, exemplified by the work of scientists and explorers such as Jane Goodall, Sylvia Earle, Will Steger, and Robert Ballard; quality family programming on public, network, and cable television; [and] the development of educational materials for schools in the United States and Canada."[13] That this is all good is taken for granted—family, quality, education, public television, Jane Goodall—but nothing certifies its reality like the map.

This is to say, *the map is the evidence of the production of the knowledge* that transforms Society membership from a magazine subscription into a really significant form of cultural capital (the map is a knowledge fetish). The magazine, whatever its éclat, does not have this power, because every magazine subscription brings with it a magazine. Only *Society* membership brings with it a map, and this map is the fulcrum that enables the magazine to be leveraged to another level—ultimately to that of reference authority.

One can imagine the Cartographic Division of the National Geographic Society bristling at the imputation, but it is exceedingly difficult to imagine to what use the main map of one of the Society maps can be put. The poster-side is easier to understand: it can go up on the bulletin board of a classroom. In the case at hand, the wildlife cavalcade if nothing else is highly decorative. Even if they fall into no other category, the animals all live in Australia where many, if not all, are emblematic. The connection between the large forms of the koalas, kangaroos, and the platypus and the lushly portrayed Australia would be clear even from the back of a classroom, and this is precisely the sort of middlebrow knowledge treasured by classroom teachers and exploited in popular games and game shows like *Trivial Pursuit*, *Jeopardy!*, and *Who Wants To Be a Millionaire?*

But the main map is devoid of animal portraiture, and as we have seen not only is its land buried in type, but the whole is burdened with an elaborate "scientific" apparatus not only invisible from the back of the classroom—or for that matter any distance at all—but rarely the subject of quiz games. It is indeed of an intimidating and arcane nature: a graticule marked every four degrees (mysterious references lines); the scale in four (occasionally incomprehensible) forms; and the name of the projection (completely opaque). Fourteen symbols are distinguished, and it is noted that elevations and soundings are shown in meters. The miles used on one of the scales are *statute* miles.

Earlier, imagining how this map might be handled by Society members, we wrote, "Their eyes roved over the main map. Maybe they mouthed the names of the states—'Western Australia,' 'Queensland' —and looked to make sure they knew where Sydney was before turning the map back over." As their eyes roved over the map what they took in were all these signs of the admirable science that their membership underwrote, *the more admirable precisely the less readily understood, the less evidently useful.* Consider the graticule. What are we to make of it? It pretends to be an aid to location, but vis-à-vis what? Is it credible that "Longitude East 140° of Greenwich" is meaningful to many when only about every other American can identify New York State on a map of U.S. states?[14] (Indeed in more recent years the locator function of the graticule has been taken over by inset maps of the globe.) Certainly the graticule cannot help find places *on the map*, for despite the division of the surface into a grid, there is no index of names to take advantage of it. In fact, there is no way to access any of the plethora of names, hundreds, perhaps thousands, which make the main map a nomenclatural gray. The lack of an index renders the maps useless. (Literally: what can you do with it?[15])

The kangaroos around the edge of the map that constitutes its border—there are 1,080 of them—emphasize this (figure 3.12). At the same time that they point out the undecorativeness of the map they circumscribe, they thoroughly undercut any "scientific" pretension the graticule might have claimed. They are an element of the playful that turns science into a plaything—into a Thing of Science[16] —in company with the lettering of "Indian Ocean" and the signature pastels of the state borders. Each of these elements must open a cognitive space, cue a context into which the map must be mentally slotted, or draw on a different domain of general knowledge. The pastels, type, and lettering of the oceans, for example, must be slotted into a "history of cartography." The signs will speak with greater force the better this history is understood, but it is not necessary that the reader know any of it for the signs to work. All that is required is that the reader recognize them as "antiquarian" vis-à-vis design norms established, among other things, in the layout of the magazine, the design

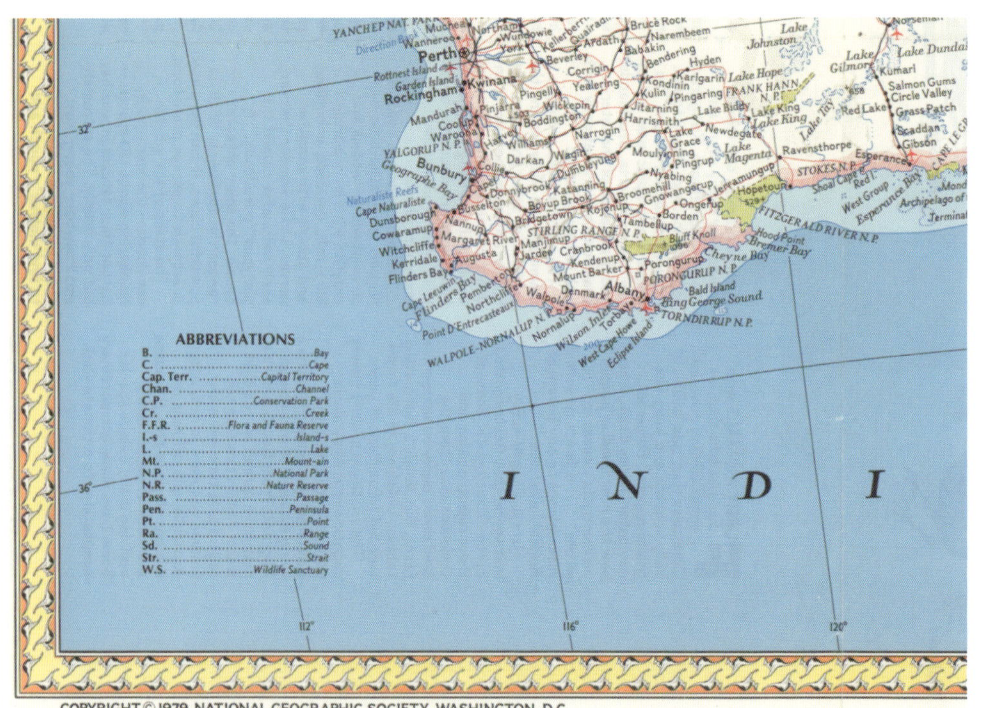

Figure 3.12
"Australia" border detail (verso of "Land of Living Fossils"). There are 1,080 kangaroos hopping around the map's edge.
Courtesy of Ng Maps/National Geographic Image Collection.

of the advertisements, and graphic "life" in general. In this context, the pastels recall the hand watercoloring of maps; the type over the land the era of wax engraving; the letterforms used for the oceans an even earlier calligraphic age.[17] This echo of a history draws attention to the present, to the map in hand—to the wealth of names, the precision of the graticule, the luxury of the four forms of the scale; and this comparison bespeaks progress, foregrounding the progressivism of the *National Geographic* of which the reader is a part, a supporting member (and therefore progressive), a foregrounding that makes of this particular map an emblem of progress, and for the reader a badge of his or her progressiveness.[18]

Though the reader can't use it, because it is a progressive map, it is a map that *could* be used for science *even though it is not intended for science*, but rather for the schoolroom or for the study of the professional who supports the Society (the study of the reader, the schoolroom of his or her children). That is, the map is "scientized," not scientific. It is schoolbookish. It is a middlebrow fetish. It says of the members, "We are the sort of people who have these kinds of things, who think these kinds of thoughts; we are not frivolous." So the graticule, the names, the colors, the typeface, even Australia, finally, come to be understood not as content, but as instructions about how this content is to be construed. All the content of this map, the carefully accumulated locational details, the endlessly rechecked spellings of exotic names, all is no more than a peg on which to hang a mass of procedural code. "Australia: Land of Living Fossils" makes concrete the reader's seriousness and the reasonableness of taking his or her opinions seriously ("I read *National Geographic.*"). Any subject would have served this purpose.

In the end, the map is nothing but a set of instructions about how to think about oneself: *with satisfaction.* (You deserve a Cadillac [and there used to be an ad for one right up front, though these days it's a Toyota, a Range Rover].)

Nature in its narrow place

Implicit in all of this is an attitude toward nature. Not that it is articulated as such. On the contrary, the word "nature" does not appear on "Australia: Land of Living Fossils," anymore than it does in the contemporaneous statement, "Reaffirmation of Editorial Policy," that the *Geographic's* then-editor, Gilbert M. Grosvenor, agreed to print by way of concluding the contretemps his publication of stories about Harlem, Cuba, and South Africa had stirred up among the more conservative members of his Board of Trustees: "The mission of *National Geographic* is to increase and diffuse geographic knowledge. Geography is defined in a broad sense: the description of land, sea, and universe; the interrelationship of man with the flora and fauna of earth; and the historical, cultural, scientific, governmental, and social background of people."[19] Unnoticed in the anxiety aroused by the political situation the statement resolved was the *Geographic's* taxonomy of nature: land, sea, universe, flora, fauna.

Though we have already run into the limitations imposed by "flora and fauna" (stromatolites are neither) it was less in the taxonomy per se than in its selective presentation that the *Geographic* revealed, indeed insinuated, its pervasive ideology. En route to concluding that the phrase "Land of Living Fossils" was a cute hook for a schoolroom poster intended to teach students to associate selected mammals with Australia, we observed: (a) a profound bias toward animals, and among these, birds and furry mammals; (b) a marginalization, among "living fossils," of the cold, dark, and hard (stromatolites, crocodiles);[20] (c) a valorization of the warm, bright, and soft, of the cuddly (birds, koalas, kangaroos); (d) a commitment to the domestic; and (e) a complete absence of ecosystemic consciousness. We characterized this as no more than an *idea* of nature, but if an ideology is the body of ideas reflecting the social needs and aspirations of an individual, group, class, or culture, as most dictionaries have it, then this idea was an important component of the Society's *ideology* of nature.

How else to make sense—other than as an ideology—of the bias toward birds and mammals? It was not, after all, as though there were anything unusual about the display on "Land of Living Fossils." It actually represents *Geographic* thinking with uncanny precision. Birds account for approximately 62 percent of the animals illustrated on "Land of Living Fossils," mammals for 34 percent, and reptiles for 5 percent. In the *National Geographic* index to its first hundred years (1888–1988), birds account for 69 percent, mammals for 27 percent, and reptiles for 4 percent of the entries devoted to those classes.[21] With respect to these, then, our "topsy-turvy menagerie" nearly mirrors the magazine's biases during its first hundred years,[22] a period, incidentally, not only renowned for the magazine's extraordinary circulation, but notorious for its abundance of articles about birds, as C. D. B. Bryan notes:

> Through the decade that began with the Depression and ended in the midst of world war, Geographic readers could

still take comfort in birds - "Birds of the High Seas," "Birds of the Northern Seas," and "Birds that Cruise the Coast and Inland Waters." There were "Canaries and Other Cage-Bird Friends," "Crows, Magpies, and Jays," "The Eagle, King of Birds, and His Kin," "Far-Flying Wild Fowl and Their Foes," "Game Birds of Prairie, Forest, and Tundra," "Humming-birds, Swifts, and Goatsuckers," "Parrots, Kingfishers, and Flycatchers," "The Shorebirds, Cranes, and Rails." There were "Thrushes, Thrashers, and Swallows," "Winged Denizens of Woodland, Stream, and Marsh," there were "Sparrows, Towhees, and Longspurs: Those Happy Little Singers Make Merry in Field, Forest, and Desert Throughout North America," and "The Tanagers and Finches: Their Flashes of Color and Lilting Songs Gladden the Hearts of American Bird Lovers East and West."[23]

The problem, from the perspective of "increasing and diffusing geographic knowledge," or even from that of "the interrelationship of man with the flora and fauna of earth," is that birds ill represent the fauna, much less the flora and fauna or anything more comprehensive. The diversity of life on earth is a vexed subject, but it is broadly acknowledged that one and a half million species are currently known to exist, and that these represent no more than a tenth of those alive.[24] E. O. Wilson, a well-known biologist, is on record—often—as believing the number could be ten times larger, that is, that 100 million species could be living on the planet.[25] Because the smaller the organism the less we know about it, a vast portion of this unknown number is undoubtedly microbial. Wilson observes that "There may be up to 5,000 species of bacteria in a single gram of forest soil, almost all of which are unknown to science."[26] You never would have guessed this from reading *National Geographic,* and certainly not from looking at "Australia: Land of Living Fossils," but "in terms of metabolic impact and numbers, prokaryotes still dominate the biosphere, outnumbering all eukaryotes combined."[27] Prokaryotes are the most numerous organisms, and the most pervasive and necessary. Were prokaryotes wiped out, it would mean the end of us and all the rest of the eukaryotes; but were the eukaryotes to disappear, the prokaryotes would continue on much as they have for the past three and half billion years. Confining ourselves to known species, half are insects (we know 750,000 species of insects). Of the million known animal species, barely 4 percent are chordates, that is, animals with notochords like us. (Most animal phyla are aquatic worms of one kind or another.[28]) Of these chordates, not even a quarter are birds (bony fishes are far more numerous[29]). That is, birds account for less than a single percent of the known animal species. This is to take nothing from birds, but it does beg the question why the emphasis in the pages of the *National Geographic.*

A family portrait

It may not be irrelevant that mammals, birds, and reptiles all develop an amnion in embryonic life. This is the membrane that surrounds the sac filled with the fluid—the amniotic fluid—in which the embryo is suspended during gestation. The amnion's development was the essential adaptation enabling terrestrial vertebrates—again, most vertebrates are bony fishes—to sever the remaining ties with their aquatic origins. Indeed reptiles, birds, and mammals make up a monophyletic group, which is to say they share an exclusive and unique ancestor. One could say of "Land of Living Fossils"—and indeed of *National Geographic* in general—that it constitutes a *family* portrait.[30]

This still doesn't explain the emphasis on birds. There may well be nearly twice as many bird as mammal species (8,600 species of birds to 4,500 of mammals)—and so the "Land of Living Fossils" has that right—but there are also nearly as many reptile species (7,000), so the map—and the *National Geographic* in general—has that very, very wrong.[31] If it is a family portrait, it is one in which the "black sheep" have been pretty much kept out of sight. Indeed, it is even clearer in this "family analysis" precisely how marginalized the cold, dark, and hard have been.[32]

Comparison with a natural history is instructive. Right now we're flipping through *Australia's South East: A Natural History of Australia,* [2]. It's a big, illustrated volume, beautifully printed, meant for the general reader, and if there is a flock of crested terns on the title page, there is a pair of male lace goannas "fighting with sharp and powerful claws" on the contents page. Facing the first text page is an orchid. The next photo features a grass tree. Then follow a blue-winged parrot, a mountain grasshopper, a flowering wattle, a smooth-barked eucalypt, a satin bowerbird, a mouse Sminthopsis, a long-horned grasshopper, the common wombat, a yellow-faced honeyeater, and another lace goanna. (The goanna is a large, marauding lizard.) Pages are devoted to sawfly larvae. There's a superb lyrebird, but *lots* of snakes. There are pages of flowers and butterflies. Shrubs. Ants. Wasps. The lungfish, a damselfly, a loggerhead turtle.[33]

Yes, the emphasis remains on the macrobiota, but there seem to be no more birds than reptiles, there is an abundance of insects, in addition to plants—pages and pages devoted to the eucalypt alone—there are fungi. Everything's connected. There's a strong sense of the whole, of an ecosystem.

What was it with the *National Geographic?* [34]

One thing was that the magazine's longtime editor Gilbert Hovey Grosvenor (from 1899 to 1949) loved birds, and since he felt that "what I liked, the average man would like," he ran articles about them. A 1943 *New Yorker* writer profiling Grosvenor wrote

> This war, like the last one, has made the editor of the *Geographic* go easy on birds in his magazine, recent issues of which have featured articles on United States food production, the Alaska highway, the Coast Guard, army dogs, convoys, aircraft carriers, women in uniform, and military and naval insignia, but he knows that wars are fleeting affairs compared to birds and he does not propose to wait for the end of the conflict before again doing justice, editorially, to his favorite topic. "The Chief has been begging us to run a series of color photographs on the wrens of Australia for the past three years," a Geographic editor recently told an acquaintance. "Everyone conspires to keep him from using these goddam birds. We keep putting him off, but he'll sneak them in any month now."[35]

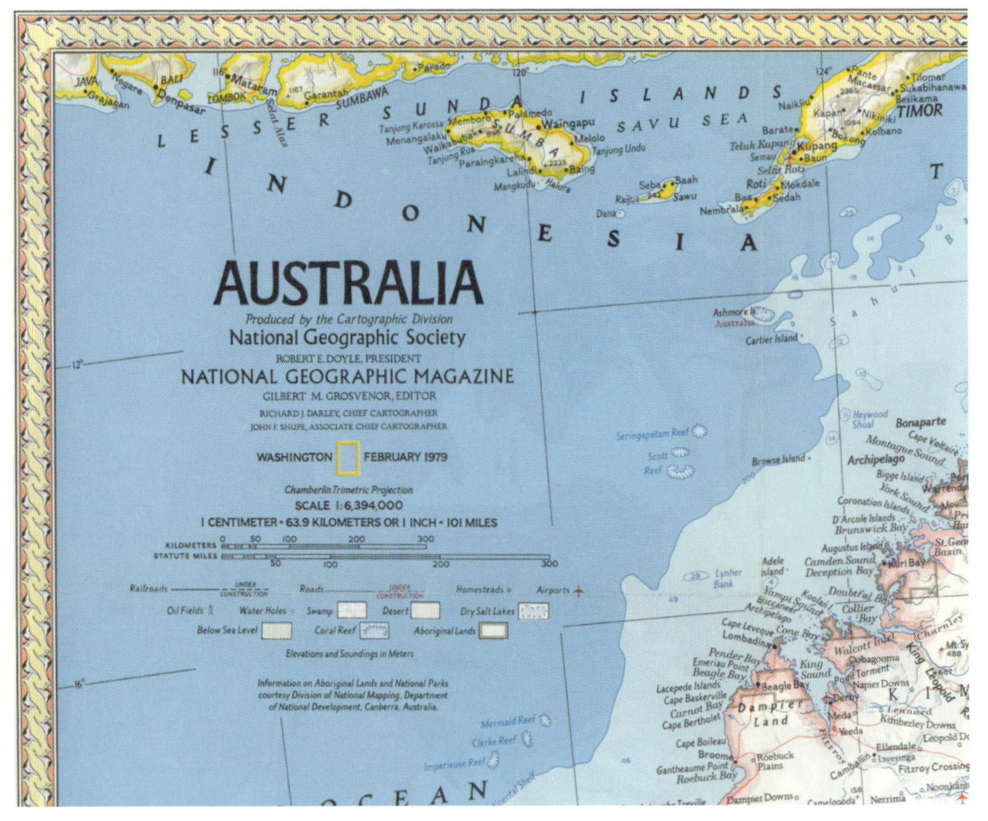

Figure 3.13
"Australia" detail. The presence of the Lesser Sunda Islands assures us we are looking at but a piece of a much bigger picture, a piece of the world.
Courtesy of Ng Maps/National Geographic Image Collection.

The wrens ran in the October 1945 issue ("The Fairy Wrens of Australia: The Little Longtailed 'Blue Birds of Happiness' Rank High Among the Island Continent's Remarkable Birds")—but the real fact is that the *Geographic* had not been about diffusing geographic knowledge since shortly after its inception. Rather, as Catherine Lutz and Jane Collins have put it, it was about "*popularizing and glamorizing* geographic and anthropological knowledge,"[36] both of which are more readily achieved with our furry and feathered friends than with bugs. Even today, with our "improved" understanding, prokaryotes remain better known as germs; crocodiles and sharks as mankillers; snakes, spiders, and bats as menaces; and marine worms—ribbon worms (900 species), gnathostomulids (probably a thousand species), nematomorphs (240 species), spirunculans (or peanut worms, 300 species), spoon worms (140 species), annelids (5,500 polychetes alone), among many others—hardly at all. The *Geographic* does not bedeck its solicitations for membership renewals with pictures of priapulids (Latin *priapulus*, little penis), with their spines and warts and their retractable mouths flush with the bottoms of estuaries, but with polar bear mommas frisking with their cubs.

We like our nature as close to us as possible (we like to keep it in the family), and if we can't have it close, then decorative (butterflies, roses, sunsets). Key words are pretty, warm, bright, abundant, soft, cuddly. It is an *inescapable* idea of nature, and by playing into and nurturing it, the *Geographic* was able to position itself as "an arbiter of national culture."[37]

The function of the map? To invest this idea (this ideology) of nature with the authority of the cartographic. In "Australia: Land of Living Fossils" what is in fact a highly arbitrary, limited, and patronizing vision of nature is passed off as *a straightforward fact of geography*. The map achieves this by securing the patent ideology within the ostensibly factual, by subsuming *nature as cornucopia* within *nature as park*: park authorizes cornucopia even as cornucopia tinctures park. Underwriting this effort, a fundamental claim: the world *is*. Of this, the main map frames a piece. We have said this before, but it is worth observing again that the pretension of the main map is that it *no more than* frames a piece (that's *all* it does). Here is the world: here is a frame around a piece of it. This says, "Yes, we have imposed a *border*, but no *filter*." The main map no more than *presents the facts*, these as postings, that is, as existence claims that *this is there*.

Because these claims are essential—*every*thing depends on them (the Society's claim to being more than a popular magazine, our acceptance of its ideology of nature)—evidence is marshaled to defend them. We have already observed the way things incidentally caught in the frame of the main map—pieces of Indonesia, a snippet of Papua New Guinea—are emphasized rather than being eliminated (as they are from the poster's map). Thanks to this emphasis, the title, "Australia," is surmounted by golden flakes of the Lesser Sunda Islands (figure 3.13). This presence of the Lesser Sundas *guarantees* the continuity of the world, it *assures* the reader that we have merely zoomed in. Another sign of good faith: the water lapping against the map's border. The graticule offers a third: the acute angle at which it slices into the border articulates the border's irrelevance (the graticule's orientation is global, the map's frame a local, impertinent interruption). The *signs of science*—the multiple versions of the scale, the name of the projection, the statute miles—further assure us of the map's factuality. So, too, the endless postings: of internal borders, railroads, mountain heights (in meters), cities, towns, reefs, ranges, sounds, straits, homesteads, airports, roads, and names, names, names.

The facticity *bleeds* through the sheet, it *infects* the map on the other side, forcing its reading as yet another *thing of science*. The logic is simple: if the main map is factual—and who could doubt it? —so is the poster's map. And if the poster's *map* is factual, so is the rest of it: "Why should they start lying now?" And indeed, they're not *lying*, per se, the picture is just (fatally) incomplete.

But . . . *bleed*? *infect*? Doesn't this language recall T. S. Eliot's "periphrastic study in a worn-out poetical fashion"? Doesn't it remind us of his, "Words strain, crack and sometimes break under the burden"? If they haven't snapped already, "bleed" and "infect" are about to.

But how else to put it? Well, cognitive linguists would view each of these "infections"—of the fauna by the poster map, of the poster map by the main map—as examples of *spreading*, which is a powerful cognitive mechanism that builds structure in mental spaces by allowing the transfer of large amounts of structure without explicit specification (a sort of wholesale transfer of knowledge). "Underlying forces in the discourse construction," writes Fauconnier, "have the aim of spreading structure across [cognitive] spaces, using minimal linguistic effort, through powerful default procedures. We find that Spreading happens in both directions, top to bottom, and bottom to top."[38] When the structure (or presuppositional knowledge) spreads from the top of

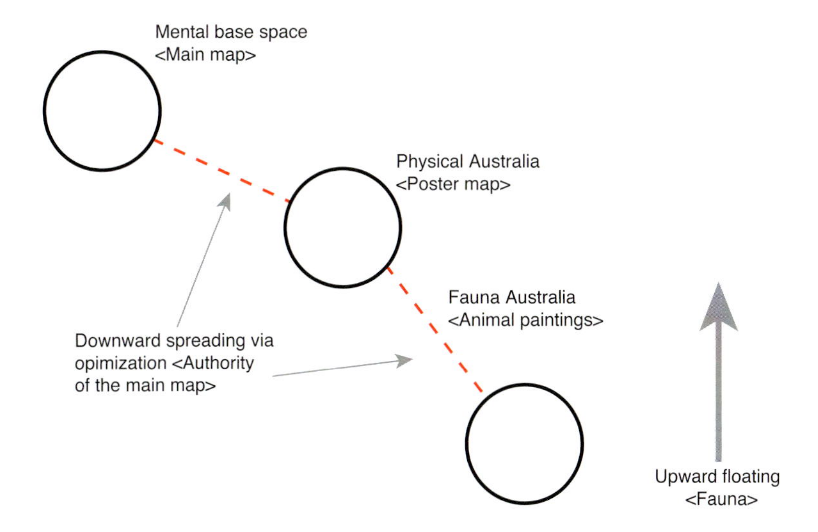

Figure 3.14
Cognitive structure of "Australia: Land of Living Fossils." The circle at the upper left symbolizes the mental space opened in the minds of readers who've unfolded and flipped over the map. In this space, they construct Australia as a nation and continent, drawing on all their general knowledge about Australia, and using it to structure their "reading" of the map. The middle circle symbolizes the mental space propagated by the first when readers flip the map back over and contemplate the poster map. Everything that's been built in the first space gets transferred down to this new space by optimization, while new structure from the poster map is being built. The third circle in the lower right symbolizes the mental space propagated by the second mental space as the readers' eyes move out to deal with the paintings of animals. These animals get "Australianized" by the structure that optimization builds here. The resulting new Australia, with all its animals, then rises back through the second to the first space via floating, embedding the animals in the authoritative main map.

the discourse construction (or *base*), it's called *optimization.* When it spreads upward from the, as it were, conclusion, it's called *floating.* The top-down bottom-up terminology reflects space grammar's formalism, which sites the base space (in our case the main map) at the top of the diagram (see figure 3.14). In our case, what spreads is the presuppositional knowledge embodied (and encoded) in the main map, which first spreads down to ("bleeds through the sheet" into) the poster map; and then spreads down to (out to) the faunal display (giving it the implicit authoritative structure of the main map), and all this through optimization. Then, through floating, the presuppositional knowledge embodied (and encoded) in the faunal display (living fossils) spreads bottom to top (infusing the main map with the cornucopial structure of the poster map).

We'll lay this out in greater detail as we go along. For the moment, we just want you to savor the grace with which an "infected" terminology can be supplanted by that of a model that promises a link straight into neuronal assemblies (the mental spaces *are* neuronal assemblies). However, "bleeds" or "spreads," the concepts constructed are identical:

Nature: soft, cuddly.

Australia: wacky but wonderful. (What else to expect of a land "down under"?)

Notes

1. See our analysis of the 1978–79 North Carolina Transportation Map & Guide to Points of Interest—the North Carolina state highway map—in "Designs" (also available in Wood's *Power of Maps*). Incidentally, this map was also produced by two different units of state government: the map side by the Department of Transportation, the poster side by the Department of Commerce.

2. Grazed into extinction almost everywhere else, stromatolites also survive in similarly hypersaline environments—salt flats, shallow embayments—in the Persian Gulf, the Bahamas, and the west coast of Mexico. See Margulis and Schwartz, *Five Kingdoms: An Illustrated Guide to the Phyla of Life on Earth.* 2nd ed. (New York: Freeman, 1988), 48.

3. Stanley, *Life and Earth through Time.* 2nd ed (New York: Freeman, 1989), 258. See also the treatment of the Proterozoic era and stromatolites in his *Exploring Earth and Life through Time* (New York: Freeman, 1993), 210–37, especially 210–20. Margulis and Schwartz, in *Five Kingdoms,* refer to the Proterozoic era as "the golden age of cyanobacteria," 48.

4. This from Margulis and Sagan's indispensable *Microcosmos: Four Billion Years of Microbial Evolution* (New York: Summit, 1986), 106–7.

5. Where it referred to them as "fossils formed from algae" in the caption to the photo, in Robin Davidson's "*Alone: Daring the harsh and beautiful Australian outback,* a young woman makes a remarkable journey across half a continent. This account is based on her diaries," 581–611.

6. As put, this is consistent both with the five-kingdom system almost universally used in contemporary biology texts, and Carl Woese's three domain system, prevalent in microbiology texts. In the realignment of our understanding of life represented by these systems (not to mention the six- and the eight-kingdom systems), the significance of microbial mats would be hard to overestimate. Read Margulis's rapturous paean in her *Symbiotic Planet* (New York: Basic Books, 1998), 69–70, or the enthusiastic description in Wilson's *The Diversity of Life* (New York: Norton, 1992), 183–6. The mats are the subject of an academic growth industry. See, among others, Zavarzin, "Cyanobacteria Mats in General Biology" in *Microbial Mats: Structure, Development and Environmental Significance.* Vol. 35 of NATO ASI Series G: Ecological Sciences, ed. Lucas Stal and Pierre Caumette, 443–52. (Berlin: Springer-Verlag, 1994), and indeed the whole volume in which it is found (*Microbial Mats*). (Among other things, Zavarzin is concerned with distinguishing cyanobacterial mats and stromatolites. Why do the former not lithify? Why did the latter become extinct—or nearly extinct—near the end of Proterozoic era? Or are these the same question? That is, do they both point toward a change in biogenic carbonate precipitation?)

7. Gore, "A Bad Time to Be a Crocodile," *National Geographic* (January 1978), 90–115.

8. The stromatolites, too, are threatened. Referring to those at Lake Clifton, also along Australia's southwestern coast, biologist Linda Moore

was quoted in a 1998 *Geographic* article as saying, "Their fate is hanging in the balance. If nutrients [from sewage and agricultural runoff] increase any more in the lake, the stromatolites could die. And once they're gone, they're gone for good. It's almost like watching their decline at the end of the Precambrian all over again" (in Monastersky, "The Rise of Life on Earth," National Geographic [March 1998], 54–81).

9. Lutz and Collins, *Reading National Geographic,* 37. This is a book about "the magazine and the Society as a key middlebrow arbiter of taste, wealth and power in America." For a related view, but one more tightly focused on the magazine's "imperialist" imagery, see Rothenberg's "Voyeurs of Imperialism: *The National Geographic Magazine* before World War II" in *Geography and Empire,* ed. Anne Godlewska and Neil Smith (Oxford: Blackwell, 1994) 155–72. Abramson's *National Geographic: Behind America's Lens on the World.* (New York: Crown Publishing, 1987) may have been the first extended critical take on the magazine. Despite their differences (and Lutz and Collins provide the most nuanced reading), the emphasis in all these is on the Geographic's photographs (especially of naked brown women) and their captions, and so on the magazine's construction of human culture. Our interest is in the magazine's construction of nature, especially as spatialized in its maps.

10. What the Society imagined members did with these maps is not known to us, but we do know that it was taken for granted that the articles were not read: "Aware that many readers simply looked at the *Geographic's* photographs and ignored the articles, M[elville]B[ell]G[rosvenor] structured photograph captions to provide distillations of the text they illustrated" (Bryan, *National Geographic Society: 100 Years of Adventure and Discovery.* Rev. ed. [New York: Abrams, 1997], 337). That class fraction of Americans capable of building cultural capital with a Geographic membership is not particularly well educated: they have had more formal education than the average American, and 30 percent of the readers can be categorized as upper and upper-middle class, but 43 percent of its readers have not attended college. See Lutz and Collins, *Reading National Geographic,* 221–3, who derived their statistics from the Simmons 1987 "Study of Media and Markets," a sample of over 19,000 Americans. To reach a similar conclusion, Paul Fussell needed no statistics. See his right-on characterization of the *Geographic* as middle-class, "nonideological," and "nice" in his *Class,* 144–5.

11. For a Society-approved history of these changes see Bryan, *National Geographic Society,* especially 378–99. But also see Lutz and Collins, *Reading National Geographic,* especially 41–46; and for an attempt at integrating a related reading of the magazine into the "geographical imagination of America," see Schulten's *Geographical Imagination.*

12. In their analysis, Lutz and Collins (*Reading National Geographic*) see this construction as a function of the Geographic's sponsored research: "From the institution's second decade, the funding and conduct of research was always marginal to the institution's main role of popularizing and glamorizing geographic and anthropological knowledge, yet it was sufficient to establish and retain its reputation as a scientific and educational organization. This made it possible for the Geographic to speak with the voice of scientific authority, while remaining outside of and unconstrained by the scientific community" (24). We don't doubt this, but believe that this sponsorship was fetishized in the map, which the member could unfold, fold back up, file, collect, and display. It is hard to do any of this with a heavily mediated sponsorship.

13. All this from the form conveying the membership card to one of us. Wood has been a member, with a brief lapse while he was getting his PhD in geography, since the late 1950s. In graduate school, far from being a form of cultural capital, membership in the National Geographic Society was . . . well, it would have been a stigma, but either no one was a member or no one admitted it. Certainly it was understood that the National Geographic had nothing to do with geography, which was (and is) published in the *Annals of the Association of American Geographers,* the *Geographical Review,* and other even more specialized journals. To a substantial extent, the Geographic's style was forged in direct opposition to the style of these journals (see the mythologizing surrounding the elder Grosvenor's publication of an "unintelligible" article by William Morris Davis [Bryan, *National Geographic Society,* 90]).

14. Only 55 percent could identify New York on an outline map of U.S. states in a survey Gallup conducted for the National Geographic Society in 1988. Regular readers of *National Geographic* did outperform nonreaders by a significant margin (and this despite the fact that the readership is not markedly well educated). From the 1988 *Geography: An International Gallup Survey* (Princeton: Gallup,1988), 4, 33.

15. The one thing readers have regularly claimed they could do with Geographic maps was follow war on them. The following from a letter from Annabel Girard is only one example in a lineage of such readers: "Congratulations on the map of Afghanistan. The map stays where I watch television, and it is a pleasure to pick it up and document where the action is" (*National Geographic* [April 2002]: letters column). We hypothesize that the news stories—with their accompanying maps—supply the missing but necessary index function.

16. Things of Science was something Wood was involved with as a child. Every month a small, blue, cardboard box arrived in the mail. Inside were "things of science," small mineral samples, simple lenses, minute quantities of chemicals, and texts describing experiments. It was a like an "experiment of the month" club. It was science, but it was play, too.

17. These issues were frequently a concern for the magazine (especially given its deep and cherished conservatism), as when, for example, its photographic layouts began to be perceived as "old-fashioned" (see Bryan, *National Geographic Society,* 330–51).

18. We're arguing (very casually at this point) that the various map elements—graticule, pastels, four forms of scale, type—open mental spaces just as words or phrases do, to, in this case, construct the meaning "Scientific Map at the leading edge of Progress." From a broader perspective, this map is but part of a larger meaning construction involving the magazine, the article to which the map was attached, the Society, and indeed the context of other magazines, and so on.

19. The statement appeared on the contents page of the same issue that carried Rick Gore's article about the crocodile (January 1978). Grosvenor wrote the statement, although a Board committee wrangled with some of the wording, according to the account in Bryan, *National Geographic Society,* 395–9.

20. Part of the problem with marginalizing the cold, dark, and hard is that "living fossils" tend to be reptiles. A *Geographic* article about New Zealand—but from 2002 (October)—opens with a double-page portrait of a large lizard: "Clinging to life on an offshore crag, the tuatara wears the moniker 'living fossil,' its appearance little changed since the Jurassic" (75), not something that can be said of kangaroos and koalas.

21. There were 196 entries for the three classes according to *National Geographic: Index 1888–1988* (Washington, D.C,: National Geographic Society, 1989). The entries included all National Geographic products. Reptiles fared slightly better in the number of cross references under See also: there were 78 for birds (68 percent), 29 for mammals (25 percent), and 7 for reptiles (6 percent). The fact is, birds completely dominated the magazine's attention for years.

22. This is an admittedly crude measure of the magazine, one that fails, for example, to acknowledge the 65 entries under "fishes," the 38 under "insects," and so on. Nor does it take into account the length of the articles and the nature of their illustration. What such further probing does underscore, however, was the extraordinary emphasis during the early era on "birds," with more entries than all other nonhuman vertebrates combined, indeed than all other chordates combined.

23. Bryan, *National Geographic Society,* 219.

24. Numbers like these are only part of the story. For an ecological perspective on the issue of biodiversity see, for example, Ricklefs, *Ecology,* 3rd ed. (New York: Freeman, 1990) 708–27.

25. Wilson, *Diversity of Life,* 132–3 ff.

26. In an interview in Campbell, *Biology,* 4th ed. (Menlo Park, Calif.: Benjamin/Cummings, 1996), 485. The number could be much higher. Soil scientists feel lucky to be able to culture even 1 percent of the microbes found in a typical soil sample, of which the total number is truly astronomical. For a very readable introduction to the biota of the soil, see Wolfe's *Tales From the Underground: A Natural History of Subterranean Life* (Cambridge, Mass.: Perseus, 2001).

27. Campbell, *Biology,* 498.

28. Wilson, *Diversity of Life,* 136; Margulis and Schwartz, *Five Kingdoms,* 170.

29. "Of all vertebrate classes, bony fishes, of the class Osteichthyes, are the most numerous, both in individual and in species (about 30,000)," reports Campbell in *Biology,* 637.

30. It is probably not irrelevant that taxonomically birds and mammals should be subsumed in the class Reptilia, since a monophyletic taxon includes a common ancestor and all of its descendents. The distinction of birds and mammals into separate classes is generally justified on the grounds of convenience. See Campbell, *Biology,* 644.

31. The numbers of species are from Campbell, ibid., 643, 648, 649. Other figures are regularly cited. In Morell's "Variety of Life," (*National Geographic* [February, 1999], 6–87) the figures given are birds, 10,000; mammals, 4,500; reptiles and amphibians, 10,500 (22). (But why group reptiles with amphibians, which are not amniotes?)

32. We originally chose these adjectives with stromatolites in mind, but they turn out to have wide currency with respect to reptiles. Here is Judith Thurman—reviewing a Versace retrospective at the Victoria and Albert Museum—writing about the use in his clothing of Oroton, a brass and aluminum alloy: "Its 'intrinsic qualities,' Chiara Buss writes in a catalog essay, are 'symbols of the invincible woman.' They are also rather pointedly the attributes of a reptile: slinkiness, hardness, and 'impenetrability'" ("String Theory"). Of course, the correlation of reptiles and invincible women is preposterous.

33. Breeden and Breeden, *Australia's South East: A Natural History of Australia.* Vol. 2. Sydney: Collins, 1972. This was the second in a multivolume natural history intended for the general reader.

34. Here is the characteristic flavor: "The uniqueness of Australia is most strongly felt in the eucalypt forests and in the heaths . . . . In Australia alone eucalypts grow in forests, and these forests are totally unlike those of any other continent. Associated with these eucalypts is a fauna which, equally, has no counterpart in any other part of the world" (Ibid., 25). (Nor is it that the magazine's editors had never heard of the Breedens: they had been responsible for twenty-five pages of the February 1973 issue, "Eden in the Outback" and "Rock Paintings of the Aborogines.") In more recent years, the Geographic has striven for a greater sense of integration. Compare "Land of Living Fossils" to "Africa's Natural Realms," a map insert to the September 2001 issue. Running down the left side of the poster-side of the sheet are forty-two lifeform portraits, but they include plants (eleven) as well as insects (two) along with the mandatory birds (seven) and mammals (twenty-one). In keeping with the ecoregional theme of the poster, the portraits—which are only partial and overlapping, and so imply a kind of interdependence—are keyed to four "natural realms." Note the continued avoidance of the cold, dark, and hard: among the paneled lifeforms, a single reptile, a geometric tortoise. An earlier example—with five reptiles!—would be "Amazonia: A World Resource at Risk" with South America on the map side from the August 1992 issue. To the objection that natural history wasn't the Geographic's brief, it needs to be recalled that when, in the first decade of the last century, the elder Grosvenor inaugurated the magazine's turn to natural history, "two distinguished geographers on the Board resigned, stating emphatically that 'wandering off into nature is not geography.' They also criticized me for 'turning the magazine into a picture book'" (Gilbert Hovey Grosvenor, as quoted in Bryan, *National Geographic Society,* 121). See also the discussion of the turn to natural history in Lutz and Collins, *Reading National Geographic,* 22–24. Our comparison, in fact, is exactly appropriate.

35. Quoted in Bryan, *National Geographic Society,* 118, 259.

36. Lutz and Collins, *Reading National Geographic,* 24. Our emphasis. See our note 121 for context.

37. Ibid., 22.

38. Fauconnier, *Mappings,* 63. See his further discussion on 112–26. See also Kay, "The Inheritance of Presupposition," *Linguistics and Philosophy* 15 (1992): 333–81.

Part II

Courtesy of USGS, 1972.

FOREST
KAIBAB NATIONAL FOREST
POWELL PLATEAU
WALHALLA PLATEAU
TIONAL FOREST
COCONINO PLATEAU

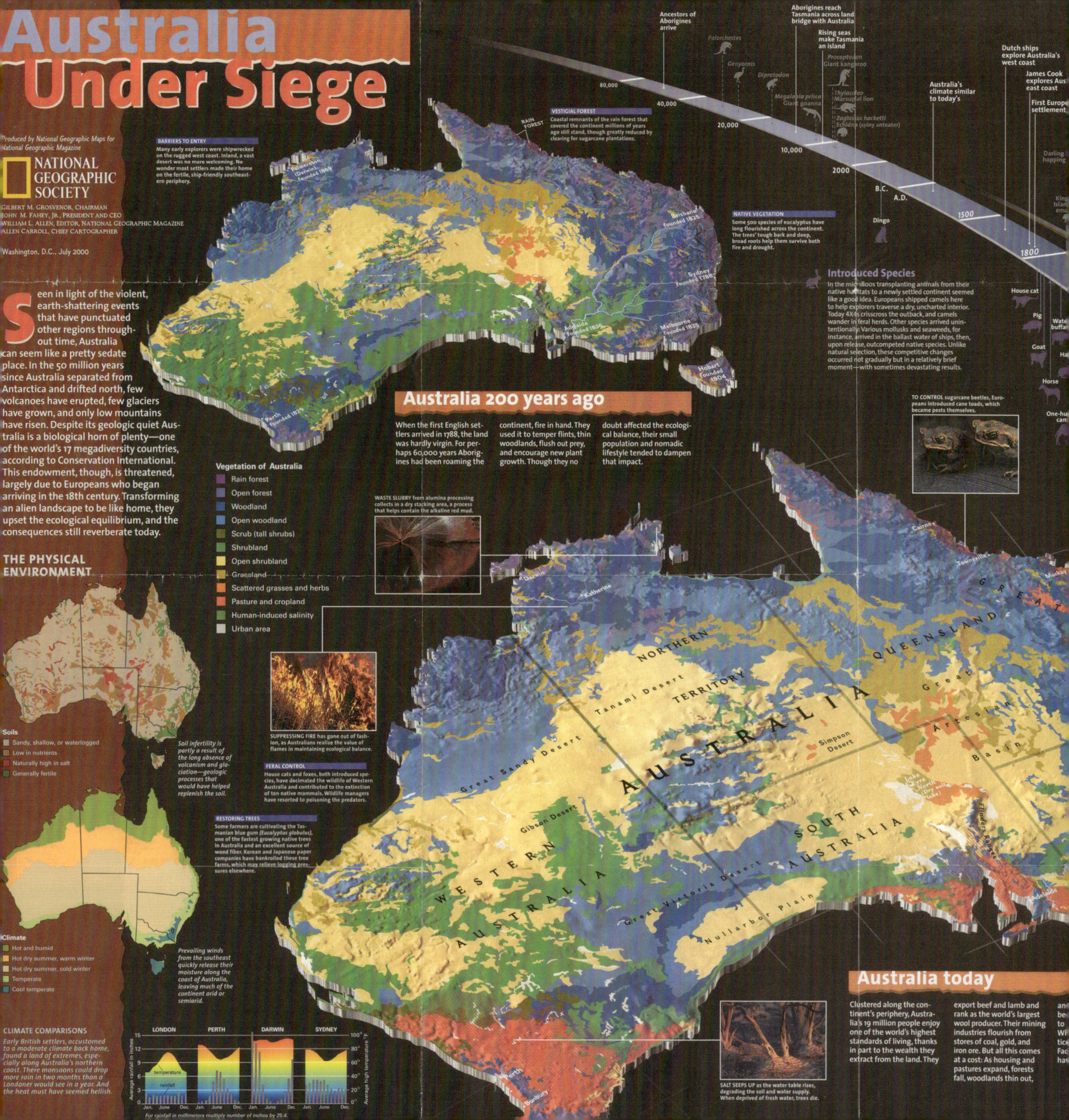

Australia
Under Siege
Produced by National Geographic Maps for National Geographic Magazine
NATIONAL GEOGRAPHIC SOCIETY
GILBERT M. GROSVENOR, CHAIRMAN
JOHN M. FAHEY, JR., PRESIDENT AND CEO
WILLIAM L. ALLEN, EDITOR, NATIONAL GEOGRAPHIC MAGAZINE
ALLEN CARROLL, CHIEF CARTOGRAPHER
Washington, D.C., July 2000
Seen in light of the violent, earth-shattering events that have punctuated other regions through-out time, Australia can seem like a pretty sedate place. In the 50 million years since Australia separated from Antarctica and drifted north, few volcanoes have erupted, few glaciers have grown, and only low mountains have risen. Despite its geologic quiet Australia is a biological horn of plenty—one of the world's 17 megadiversity countries, according to Conservation International. This endowment, though, is threatened, largely due to Europeans who began arriving in the 18th century. Transforming an alien landscape to be like home, they upset the ecological equilibrium, and the consequences still reverberate today.
BARRIERS TO ENTRY
Many early explorers were shipwrecked on the rugged west coast. Inland, a vast desert was no more welcoming. No wonder most settlers made their home on the fertile, ship-friendly southeastern periphery.
RAIN FOREST
VESTIGIAL FOREST
Coastal remnants of the rain forest that covered the continent millions of years ago still stand, though greatly reduced by clearing for sugarcane plantations.
Brisbane founded 1825
Sydney founded 1788
Melbourne founded 1835
Adelaide founded 1836
Hobart founded 1804
Perth founded 1829
NATIVE VEGETATION
Some 500 species of eucalyptus have long flourished across the continent. The trees' tough bark and deep, broad roots help them survive both fire and drought.
Ancestors of Aborigines arrive
Palorchestes
Genyornis
Diprotodon
Megalania prisca
Giant goanna
Aborigines reach Tasmania across land bridge with Australia
Rising seas make Tasmania an island
Procoptodon
Giant kangaroo
Thylacoleo
Marsupial lion
Zaglossus hacketti
Echidna (spiny anteater)
Australia's climate similar to today's
Dutch ships explore Australia's west coast
James Cook explores Australia's east coast
80,000
40,000
20,000
10,000
2000
B.C.
A.D.
1500
1800
Dingo
Introduced Species
In the mid-1800s transplanting animals from their native habitats to a newly settled continent seemed like a good idea. Europeans shipped camels here to help explorers traverse a dry, uncharted interior. Today 4X4s crisscross the outback, and camels wander in feral herds. Other species arrived unintentionally: Various mollusks and seaweeds, for instance, arrived in the ballast water of ships, then, upon release, outcompeted native species. Unlike natural selection, these competitive changes occurred not gradually but in a relatively brief moment—with sometimes devastating results.
House cat
Pig
Goat
Horse
TO CONTROL sugarcane beetles, Europeans introduced cane toads, which became pests themselves.
Australia 200 years ago
When the first English settlers arrived in 1788, the land was hardly virgin. For perhaps 60,000 years Aborigines had been roaming the continent, fire in hand. They used it to temper flints, thin woodlands, flush out prey, and encourage new plant growth. Though they no doubt affected the ecological balance, their small population and nomadic lifestyle tended to dampen that impact.
Vegetation of Australia
Rain forest
Open forest
Woodland
Open woodland
Scrub (tall shrubs)
Shrubland
Open shrubland
Grassland
Scattered grasses and herbs
Pasture and cropland
Human-induced salinity
Urban area
WASTE SLURRY from alumina processing collects in a dry stacking area, a process that helps contain the alkaline red mud.
THE PHYSICAL ENVIRONMENT
Soils
Sandy, shallow, or waterlogged
Low in nutrients
Naturally high in salt
Generally fertile
Soil infertility is partly a result of the long absence of volcanism and glaciation—geologic processes that would have helped replenish the soil.
SUPPRESSING FIRE has gone out of fashion, as Australians realize the value of flames in maintaining ecological balance.
FERAL CONTROL
House cats and foxes, both introduced species, have decimated the wildlife of Western Australia and contributed to the extinction of ten native mammals. Wildlife managers have resorted to poisoning the predators.
RESTORING TREES
Some farmers are cultivating the Tasmanian blue gum (Eucalyptus globulus), one of the fastest growing native trees in Australia and an excellent source of wood fiber. Korean and Japanese paper companies have bankrolled these tree farms, which may relieve logging pressures elsewhere.
Climate
Hot and humid
Hot dry summer, warm winter
Hot dry summer, cold winter
Temperate
Cool temperate
Prevailing winds from the southeast quickly release their moisture along the coast of Australia, leaving much of the continent arid or semiarid.
CLIMATE COMPARISONS
Early British settlers, accustomed to a moderate climate back home, found a land of extremes, especially along Australia's northern coast. There monsoons could drop more rain in two months than a Londoner would see in a year. And the heat must have seemed hellish.
LONDON
PERTH
DARWIN
SYDNEY
Average rainfall in inches
Average high temperature °F
temperature
rainfall
Jan. June Dec.
For rainfall in millimeters multiply number of inches by 25.4.
Darwin
Katherine
Cairns
Townsville
NORTHERN TERRITORY
QUEENSLAND
AUSTRALIA
Tanami Desert
Great Sandy Desert
Gibson Desert
Simpson Desert
Great Artesian Basin
Lake Eyre North (Dry salt lake)
WESTERN AUSTRALIA
SOUTH AUSTRALIA
Great Victoria Desert
Nullarbor Plain
Flinders Ranges
Adelaide
Perth
Bunbury
SALT SEEPS UP as the water table rises, degrading the soil and water supply. When deprived of fresh water, trees die.
Australia today
Clustered along the continent's periphery, Australia's 19 million people enjoy one of the world's highest standards of living, thanks in part to the wealth they extract from the land. They export beef and lamb and rank as the world's largest wool producer. Their mining industries flourish from stores of coal, gold, and iron ore. But all this comes at a cost: As housing and pastures expand, forests fall, woodlands thin out,

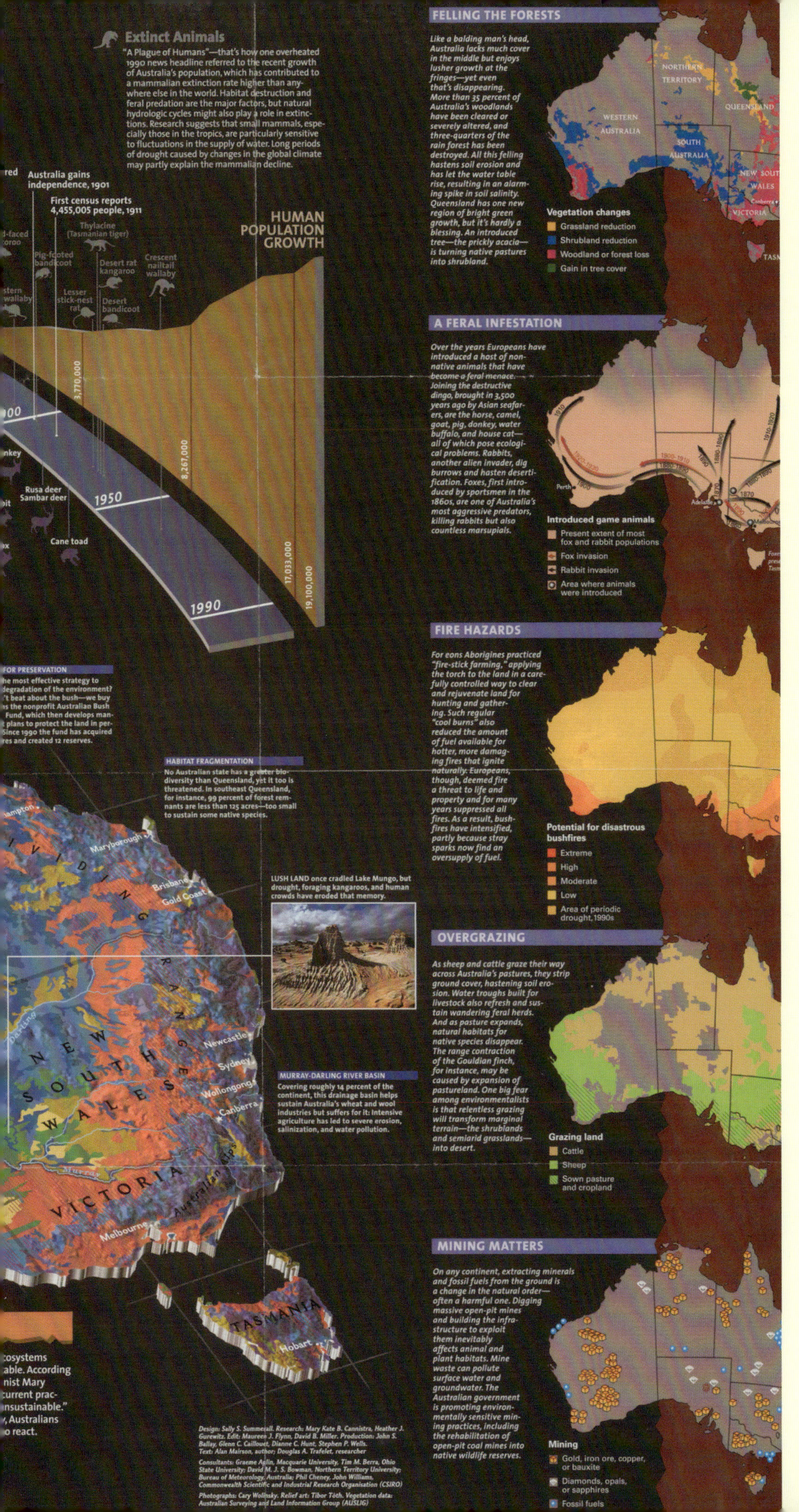

Four

Threatened nature

Courtesy of Ng Maps/National Geographic Image Collection.

Threatened nature

A couple of decades later, this all looked very different. Inserted in the pages of the July 2000 *Geographic* is a map of Australia of a wholly different character: "Australia Under Siege" (figure 4.1) It, too, just begs to be unfolded, but the whiff here is of danger, of a conflagration, and it isn't wafting toward you: you are enveloped in it. The title, in metallic gray and burning-fire orange is printed over, and sort of under, an orange-brown fragment seemingly torn from its source (the edge is ragged), the tops of the risers and the dots over the "i"s cut off (presumably in the haste with which it was pasted up). The sense is that it was just ripped from the teletype machine. The news is urgent, and it's *bad.* The background's in black, *its* torn edge ravaging a deeply chocolate brown.

Below the title, and beside the publication information, a corner of what one takes to be Australia—in acid yellows and blues—warps out of sight around the right and bottom folds. "Under Siege" is not folded like "Land of Living Fossils," which opened to the right to reveal, after the first two folds, the top quarter of the sheet. "Under Siege" opens down (figure 4.2). The first fold gives up half a map of Australia, perhaps of vegetation, whose acid tones are imperative, and a block of text jittery with phrases like, "threatened largely due to Europeans" and "transforming an alien landscape to be like home." This text is dropped out in white across a rip that severs what is gradually being revealed as a brown border from the black background of most of the poster. The next fold gives up two smaller maps of Australia, a portion of the largest version yet—again in acid tones—and four charts which, blue-gray-purple below and yellow-orange above, seem to be on fire. A small photo inset just below the second fold *is* of a fire, and indeed, there is about all of this a sense of watching a house go up in flames, and the urgency compels the unfolding of the rest of the poster (figure 4.3).

The background is black, matte. In addition to the four maps previously glimpsed, this unfolding reveals five more falling down the right side of the page. There is a feeling that the central continents are flying toward you *out* of the map—both are perspective views—the smaller one above and "behind" the larger one that commands bottom center. Reinforcing this perspective is a timeline, barely perceptible in the upper left, which swells in presence as it crashes right and center.

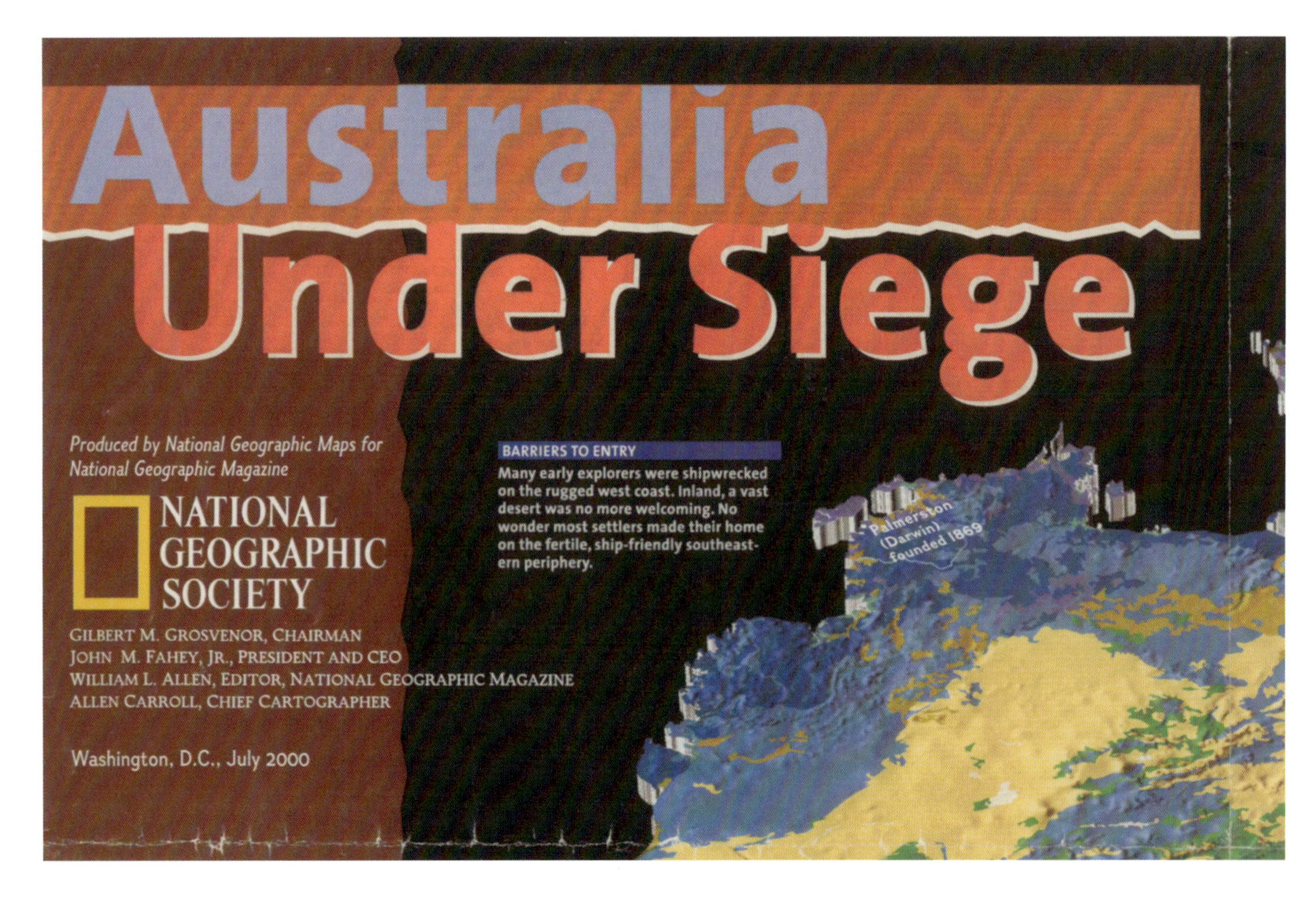

Figure 4.1
Cover of "Australia Under Siege," supplement to *National Geographic*, July 2000.

Figures 4.1–4.10 courtesy of Ng Maps/National Geographic Image Collection.

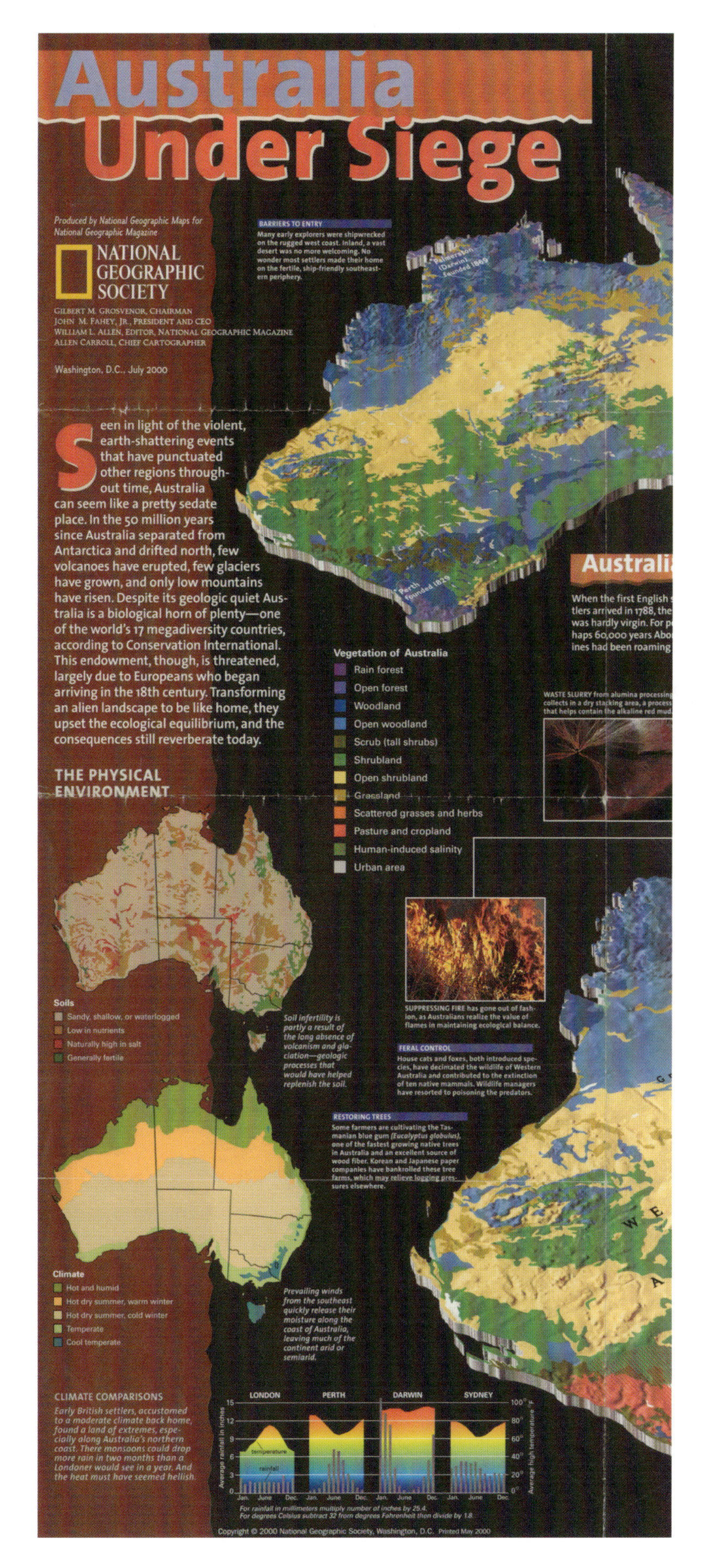

Figure 4.2
As "Australia Under Siege" is unfolded, soils and climate maps are revealed.

The smaller map toward the upper left is "Australia 200 years ago," that in the foreground "Australia today." The timeline charges from a hundred thousand years ago to the present. Yet the dynamism transcends any idea of the simple passage of time. The situation here is *in play*, everything's in motion, everything's threatened, and you just *have* to turn the map over to see what's on the other side (figure 4.4).

It's *some* sort of refuge. Essentially, this is the main map from "Australia: Land of Living Fossils"—instantly recognizable (the same all-but-white, the same Easter-egg tones along the borders, the same pale blue waters, the same immersion in type)—except that this time there's no border, the image bleeds right off the page, it really *is* just a piece of the world, and there's a little inset image of the globe with Australia distinctly colored to make sure you get the picture (figure 4.5).[1] Java, the Lesser Sundas, Papua New Guinea dip into the picture, the graticule rushes off, and the sense is strong of having just zoomed in. To make it even stronger, we can zoom in further to close-ups of Brisbane, Sydney, Canberra, Melbourne, Adelaide, and Perth (figure 4.6).

Again the reader no more than scans this sheet (maybe mouthing the names of the towns honored with inserts), takes in the "signs

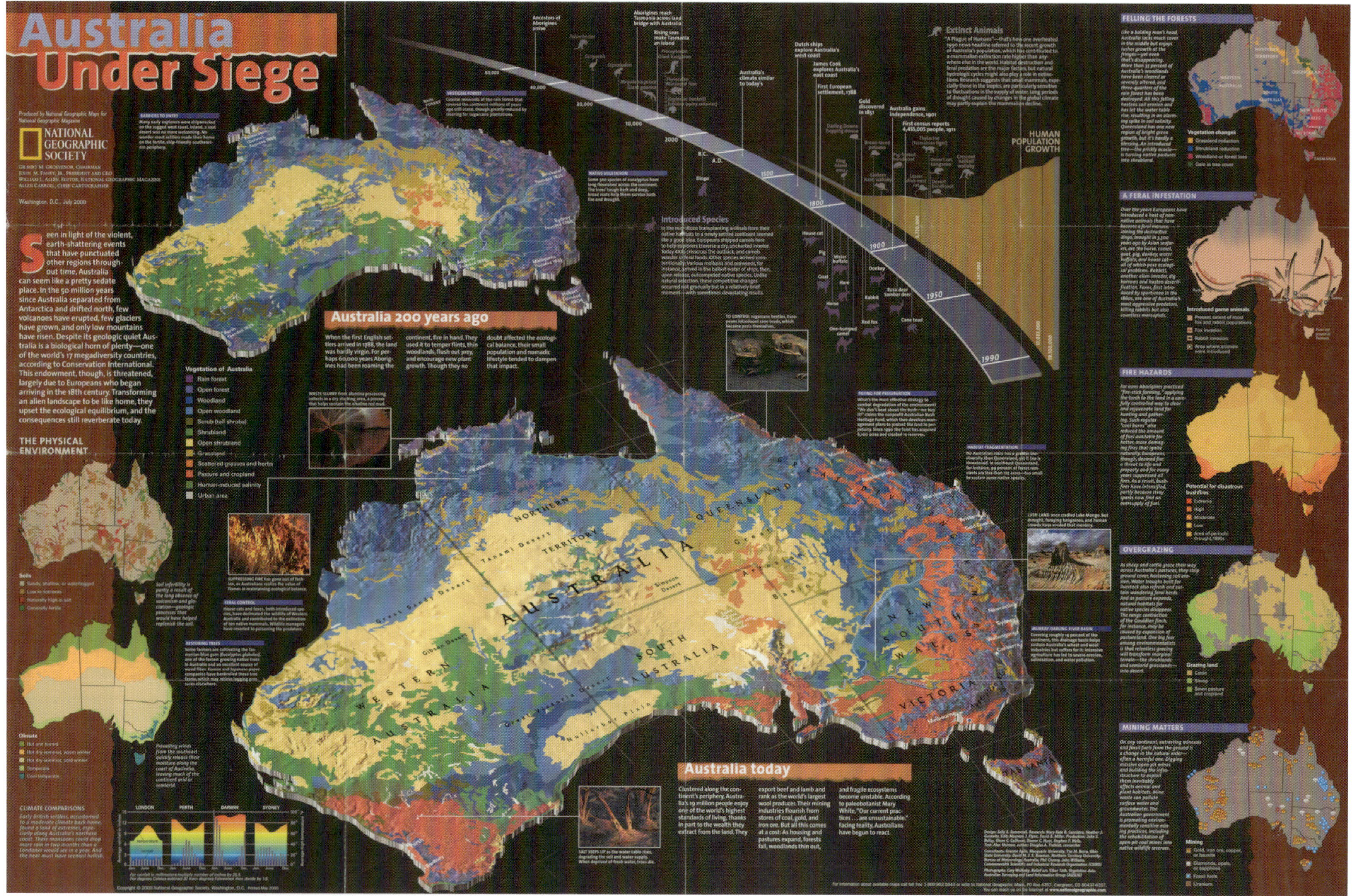

Figure 4.3
Fully unfolded, "Australia Under Siege" bombards us with facts, figures, maps, and photos, all pointing to the same conclusion: Australia is at risk. Map dimensions (WxH): 31in. x 20in.

of science" (the azimuthal equidistant projection has replaced the Chamberlin Trimetric), and turns the sheet back over. Whatever we have on this side, it is not the totemic mammals garlanding the portrait-map of "Living Fossils." That image and the simple idea it encoded, legible from the back of the classroom, reduced to the simple formula, AUSTRALIA = KOALA + KANGAROO + PLATYPUS (= NATURE = SOFT/CORNUCOPIA).

"Australia Under Siege" may reduce to a similarly simple formula, but it *cannot* be read from the back of the classroom. Where "Living Fossils" is decorative, "Under Siege" is informative. Where "Living Fossils" is smothered with bright feathers and lushly rendered fur, "Under Siege" invigorates with facts, prompts with cautionary lessons, stirs with warnings. Where "Living Fossils" is innocuous, "Under Siege" is alarming.

An explicit lesson instead of a celebration

Both maps start from the same premise: Australia was for a time isolated from other landmasses. Therefore, it developed into either a "living museum of odd species" (in the words of "Living Fossils") or "a biological horn of plenty—one of the world's seventeen megadiversity countries" (in the words of "Under Siege"). The maps develop these conclusions into substantially different ideas of nature.

Without denying that its "topsy-turvy menagerie" was "once slaughtered for [its] soft fur," "victimized by an invader, the dingo," competed with by introduced cats and rabbits, and threatened by "ever expanding grazing lands," "Living Fossils" celebrated both "the stars of the wildlife cavalcade" *and* Australia's conservation activities, the "comeback" of the koala thanks to "strict protection measures," "the

Figure 4.4
"Australia" (verso of "Australia Under Siege") is the main map of the supplement, designed in the typical house style. This map, considered by most to be an unbiased reference object, lends credibility to the other side.

Cape Barren goose, rescued by conservationists," and the "3 percent set aside" for the preserves. These (tepid and qualified) equivocations and complications were buried in the text. The lush continent embowered in its furry mammals sent the less equivocal message summarized in our formula. It needs to be added that the overwhelming impression was that everything was *extremely simple.*

The poster side of "Australia Under Siege" is anything but simple, and the fact that it's *not* simple is clearly part of the message. Besides the nine maps, the timeline, and the four graphs, there's a chart of human population growth, five photos, and twenty-seven blocks of type: nine map captions, eight paragraphs floating around the big maps, five photo captions, two timeline captions, a chart caption, an intro text, and a credit block. Each map has a legend, of three or more items. "Vegetation of Australia" has twelve. Furthermore, each map has a title. The five along the right side are printed over strips of purple that disappear beneath the chocolaty border. "Australia 200 years ago" and "Australia today" are printed over "torn" strips of the same orange-brown as the main title (figures 4.7 and 4.8). All three strips have drop shadows. The eight paragraphs floating around the big maps have titles too, these printed over strips of purple. Thirty-nine call-outs are associated with the timeline alone, some in white, some in gray, some in white, gray, and purple. Only a very sure sense of hierarchy keeps the poster from spinning completely out of control. It can still be hung in the classroom (where else?), but it has to be studied up close; it has to be taught.

Resembling a giant piece of a jigsaw puzzle, "Australia today" dominates this sheet. The name "Australia" sprawls across it, the states are delineated and named, and a number of physical features are picked out (the Nullarbor Plain, the Great Sandy Desert, the Darling River). Eighteen cities are named. Blotched with what could

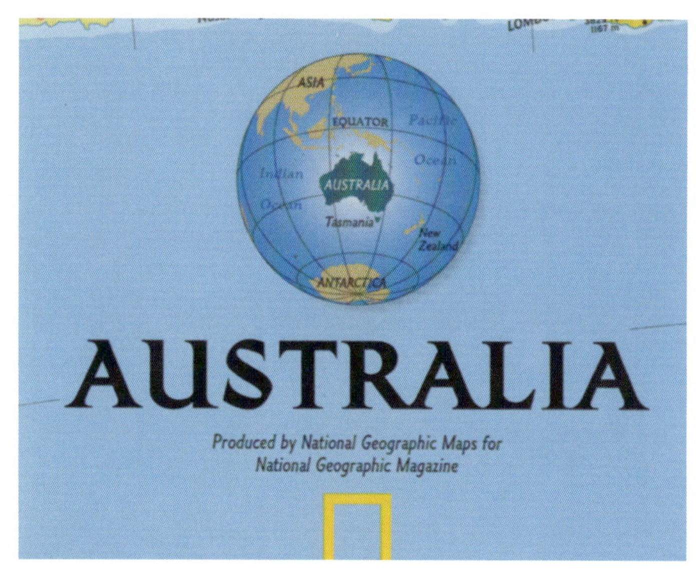

Figure 4.5
The inset globe locator map in "Australia" makes the point that the map is no more than a window onto a part of the whole.

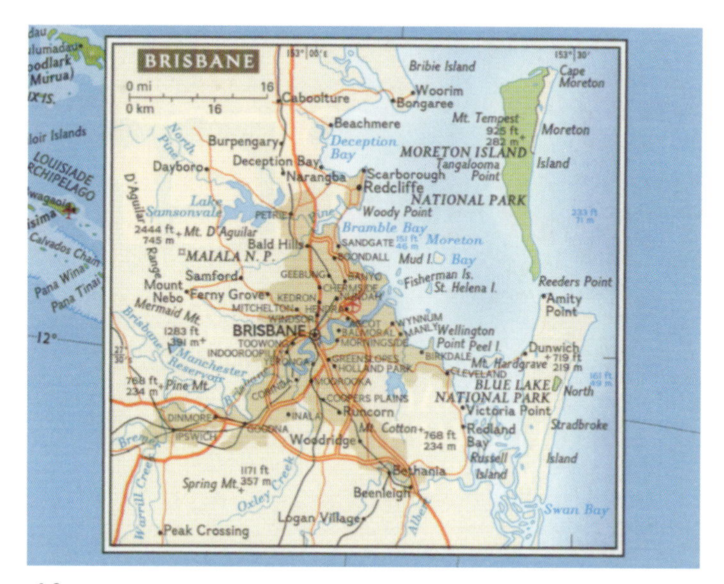

Figure 4.6
The sense of having zoomed in from the locator globe to Australia is increased by the city insets, which let us zoom in still further

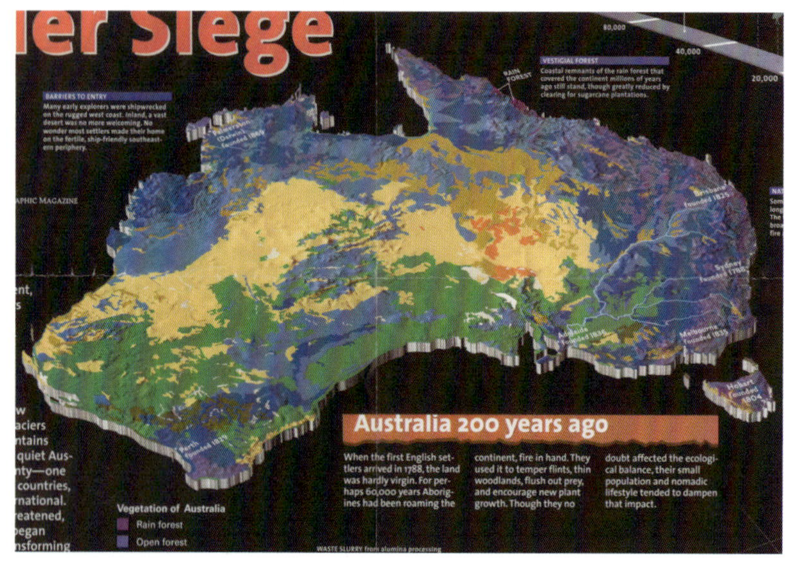

Figure 4.7
"Australia 200 years ago," one of two maps featured on the poster side of "Australia Under Siege," shows us an older Australia in its "natural" state.

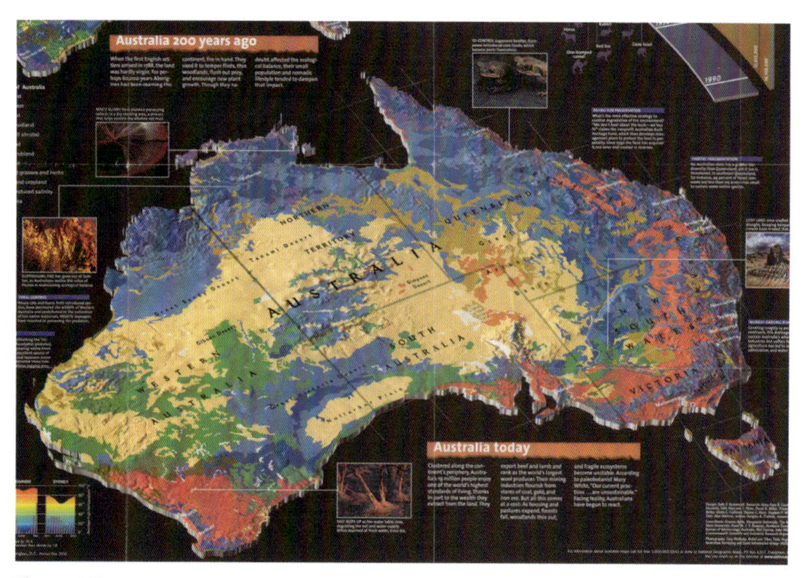

Figure 4.8
"Australia today" illustrates the present state of Australia's land cover.

be spills of industrial chemicals—yellow, blue, green, and red (much of this additionally stippled in red), the brilliant image draws the eye, yet gives the eye nowhere to rest.

Where it roams is up to "Australia 200 years ago." This is a smaller jigsaw piece. It has no name sprawled across it, the states are not delineated, and no physical features are named (though the words "Rain Forest," in white, are attached to a couple of purple blotches). Seven cities are named, but they are accompanied by tags: "founded 1829," "founded 1835." This map, too, is blotched with what could be industrial chemicals, but the absence of the state borders, the nomenclature, and the red produces a comparative calm.

Comparison of the maps—the eyes flitting back and forth—leads the reader to discover the legend (figure 4.9), between them but off to the left: "Vegetation of Australia." Confronted with its twelve categories—Rain forest, Open forest, Woodland, Open woodland, Scrub (tall shrubs), Shrubland, Open shrubland, and so on—the eyes (and mind) glaze over, but the notable difference in red between the two maps singles out its category for attention: pasture and cropland. This suggests that what all this is about is the *growth* in pasture and cropland—and in human-induced salinity, which is what the red stippling turns out to stand for—that's taken place over the past 200 years, and the corresponding loss in rain forest and other natural habitat. The meaning of the graphic maelstrom becomes clear (if it hadn't *been* clear from the moment the reader had finished unfolding the map, with its alarming title, flame-imagery, and matte black background): *thanks to us, Australia is going to hell in a handcart.*

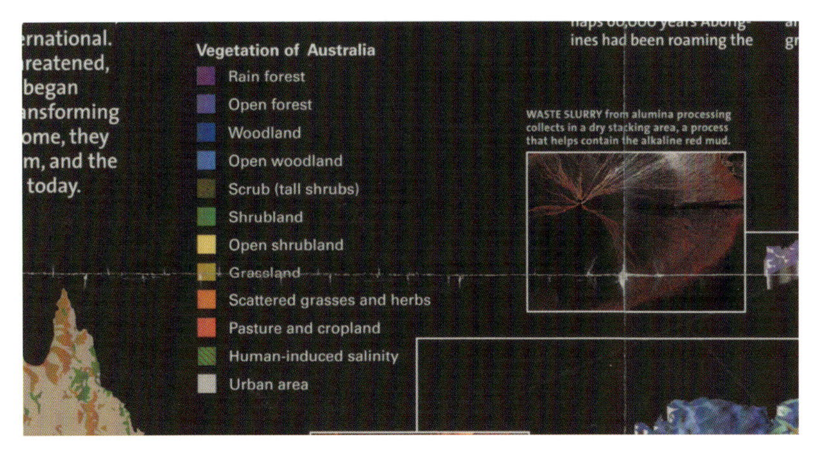

Figure 4.9
The Vegetation legend for "Australia Under Siege" allows us to find meaning in the two maps above (simply put: more pastures, less rain forest).

It doesn't say that.

The *Geographic* would never say anything *like* that. Despite changes at the Society that had been set in motion in the 1970s, the magazine would still never go so far, certainly never get out ahead of the facts that way. To do so would endanger the authoritativeness of the main map *and* violate the essence of the Society's credo, enshrined as the first of the elder Grosvenor's famous guiding principles: "The first principle is absolute accuracy. Nothing must be printed which is not strictly according to fact." It's true that the magazine is no longer under the iron sway of all seven principles, or this map could never have been published in the first place. Certainly the map violates the fifth principle ("Nothing of a partisan or controversial character is printed") but it also violates the spirit of the sixth ("Only what is of a kindly nature is printed about any country or people, everything unpleasant or unduly critical being avoided").[2] If the fifth and sixth principles were in 2000 no longer the watchwords they had been, the first continued to be. How then to make the map's point about the future, without saying the words?

We want you to imagine you are holding this map in your hands, holding it as though you were reading it, as though it were a newspaper. Don't forget that the big Australian jigsaw pieces are perspective views, and that there is a feeling that these pieces are flying toward you out of the map. Recall, too, that the timeline accentuates this feeling. In the upper left, farthest away from you, is "Australia 200 years ago." More or less at chest height, bottom center, right next to you, is "Australia today." Where's "Australia tomorrow"?

Exactly. *You're* Australia tomorrow. Time flies toward you down from the upper left, right off the map's bottom into your lap—and *with* time, all the things changing with it. In graphic terms, these are the red blotches, nonexistent 200 years ago, dramatically present today . . . *overwhelmingly* present tomorrow.

Isn't this the only conclusion? You don't take it in like this, of course, you (1) glance at the today map; (2) glance at the yesterday map; (3) compute that the difference is red; (4) see on the legend—lying between the two maps between which your eyes are flitting—that RED = PASTURE AND CROPLAND; (5) compute that there's going to be still more cropland in Australia's future; and (6) given the "Under Siege" title, the flame imagery, and the matte black background, understand that this *ain't* good.

Why not?

Your curiosity piqued—this whole process might not have taken a minute—you glance at a photo. It's hard to make out, but eerie. The caption reads, "Salt seeps up as the water table rises, degrading the soil and water supply. When deprived of fresh water, trees die." *Oh!* It's a picture of dying trees. A call-out connects the picture and caption to one of the areas stippled in red ("Human-induced salinity"). Another photo of what looks something like Monument Valley is captioned: "Lush land once cradled Lake Mungo, but drought, foraging kangaroos, and human crowds have eroded that memory." A call-out connects this to the Murray-Darling River Basin. An adjacent "purple floater" reads, "Covering roughly 14 percent of the continent, this drainage basin helps sustain Australia's wheat and wool industries but suffers for it: intensive agriculture has led to severe erosion, salinization, and water pollution."

Maybe *now* you read the introductory text. After a few words about Australia's isolation and its biological diversity, it concludes: "This endowment, though, is threatened, largely due to Europeans who began arriving in the 18th century. Transforming an alien landscape to be like home, they upset the ecological equilibrium, and the consequences still reverberate today." Directly below this are the two small maps given up by the second fold: "The Physical Environment" (figure 4.10). The upper one is of soils ("Soil infertility is partly a result of the long absence of volcanism and glaciation—geologic processes that would have helped replenish the soil"); the lower one is of climate ("Prevailing winds from the southeast quickly release their moisture along the coast of Australia, leaving much of the continent arid or semiarid"). And these are about . . .?

The logic is less far from that of "Living Fossils" than the complications here would let you believe. If the main map is factual—and again, who could doubt it?—so is "Australia today." This jigsaw piece has precisely the relationship to the factual that the portrait map of "Living Fossils" had. Recall that "Australia today" retains from the main map (which is simply "Australia") the country's name, the state borders and names, eighteen cities, and many of the physical features ("Tanami Desert," "Nullarbor Plain"). The implication is that the main map couldn't possibly show all the things it does already *plus* vegetation, so the vegetation is on *this* side for no more than *technical* reasons of legibility.[3]

Reframing this in cognitive linguistic terms we would say that the main map opened a base space. Recall from the introduction that meaning construction begins in a base space, which establishes the initial focus and viewpoint. Our base space contains *elements* from the main map (the water/land distinction, shapes, selected cities and other features) *internally structured* by the relationships these elements have on the map *and* by the conceptual frame, Australia. We take this conceptual frame to be an integration in long-term memory of a complicated stew of *experiences* (for example, watching movies about or set in Australia, reading about Australia, looking at maps of Australia, going there), *spatial concepts* (for example, distances, directions), *geographic concepts* (for example, continent, nation, desert), and *metaphorical constructions* (for example, "seeing is knowing," "a nation is a person"). Meaning is constructed in the base space as a person brings to bear

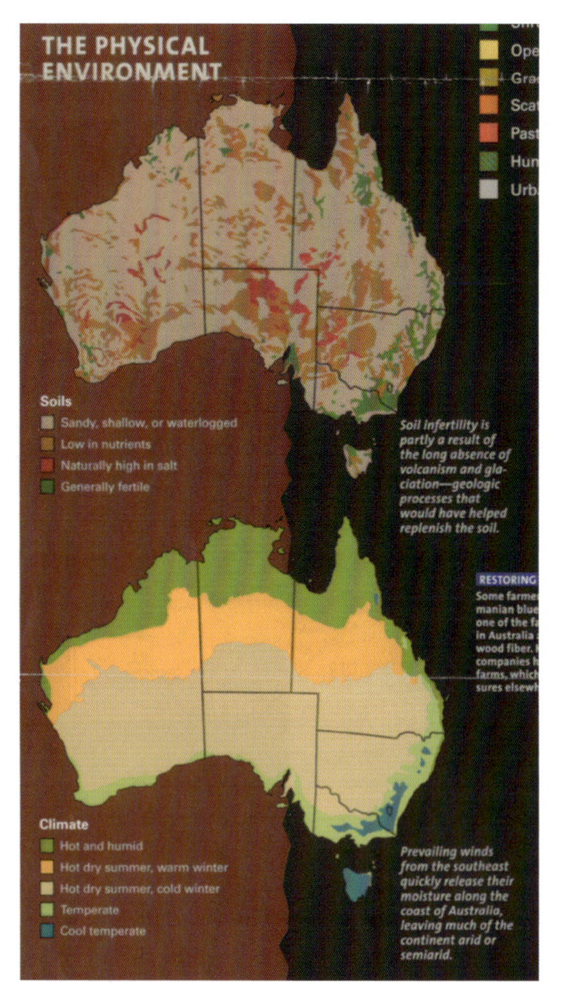

Figure 4.10
Like the backstory to a plot in a movie, "The Physical Environment" maps give us the soil and climate backgrounds to the crisis unfolding in Australia.

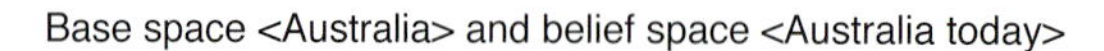

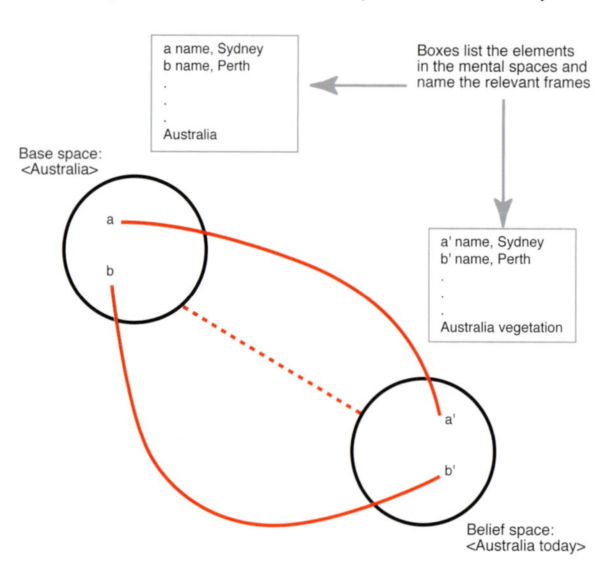

Figure 4.11
Cognitive structure of "Australia Under Siege": base space and belief space. Solid lines map elements from mental space to mental space; dashed lines indicate the transfer of structure, through optimization. The space to which structure is actively being added is said to be in focus.

on her inspection of the main map her rich background knowledge about Australia, and this construction is held in working memory.

Flipping the map back over to "Australia today" causes base space <Australia> to launch a child space, a *belief space* <Australia today>. The base's structure—the whole caboodle—is transferred to this child space by optimization; that transfer-by-default from parent space to child space we introduced in our analysis of "Living Fossils." Keep in mind that *optimization* refers to a linkage that one neuronal assembly is making to another. Through this downward transfer—remember in the space diagrams "base" is at the top—all relevant structure not explicitly contradicted is inherited within the child space, even though the actual *map*, "Australia today," only inherits *selected* features from the main map, "Australia." To this inherited structure, "Australia today" adds additional elements (vegetation and other land-use characterizations) and further structure (including the metaphor "red is dangerous"), so that the meaning is that constructed as a result of looking at the main map *plus* absorbing the <vegetation> mental space content, which is all "Australia today" actually adds. In a diagram, this looks something like figure 4.11.

With "Australia today" now in focus (the mental space to which structure is actively being added is said to be in focus), <Australia 200 years ago> is launched as the map reader's eyes, under pressure from the graphic structure, drift up to "Australia 200 years ago." Recall that optimization transfers only structure *not explicitly contradicted* in the child space. The new space is a *past space*, and because "Australia 200 years ago" forbids the transfer of elements not present 200 years ago, <Australia 200 years ago> inherits from base <Australia> (via <Australia today>) only the shape of the continent, indications of topography, and selected cities (but with the dates of their founding added). From <Australia today> it inherits the vegetation . . . *except for what's red* (which is pasture and cropland and human-induced salinity), at the same time that it adds vegetation present in the past but absent today (e.g., rainforest). We might say that the meaning constructed here is close to that of <Australia today> but with everything developed in the past 200 years erased. It's not the "real" world of the main map, but it's not some wild fantasy (as of the future) either. It's the world of the main map *plus* vegetation *minus* everything that the past 200 years has brought.[4] The network of spaces being held in working memory now looks like figure 4.12.

Cognitive linguists have uncovered a number of ways humans construct meaning in situations like this, where the frames and elements in mental spaces are both similar and different, and that involve our awareness of identity, sameness, and difference. The most powerful of these is what Fauconnier and Turner call a "blend."[5] In blending, the mind constructs a *conceptual integration network* consisting of input spaces (<Australia today> and <Australia 200 years ago>), a new "generic space," and a "blended space," usually shortened to "blend." First a partial cross-space mapping connects counterparts in the two input mental spaces. This cross-space mapping opens the generic space, which maps onto each of the input spaces and captures the structure the inputs share. Now these open the blended space where emergent structure present in neither input is developed. At this point, the network of spaces being held in working memory looks like figure 4.13.

Don't forget that the spaces are actually neuronal assemblies. With that in mind, try to visualize the lines we've been drawing between the spaces as synaptic connections between the neuronal assemblies, that is, as coactivations of these assemblies. In the blend network, four

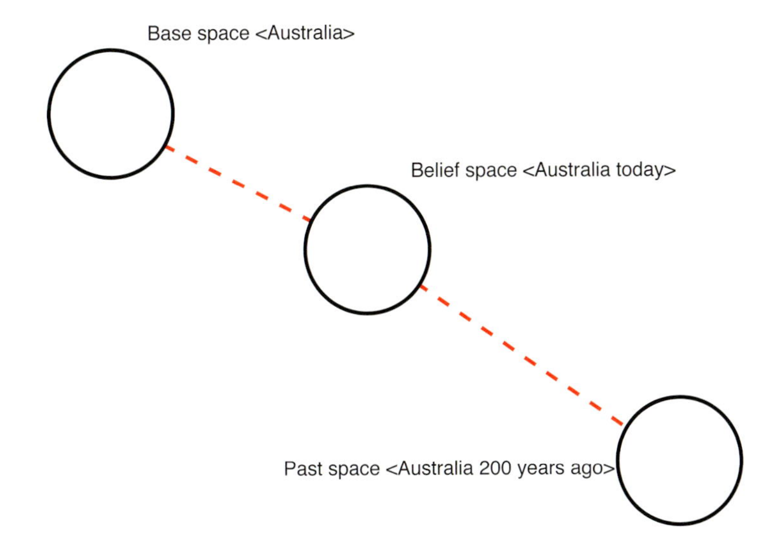

Figure 4.12
Cognitive structure of "Australia Under Siege," adding a second belief space. Since this belief space concerns Australia's past, we're calling it a *past space*. We're showing only the gross network. We've suppressed indications of the frames and the elements, and here the dashed lines represent both the transfer of structure by optimization as well as the projection of elements. In fact, as we'll see in the next diagram, the dashed line between <Australia today> and <Australia 200 years ago> actually represents the conceptual integration network of a blend.

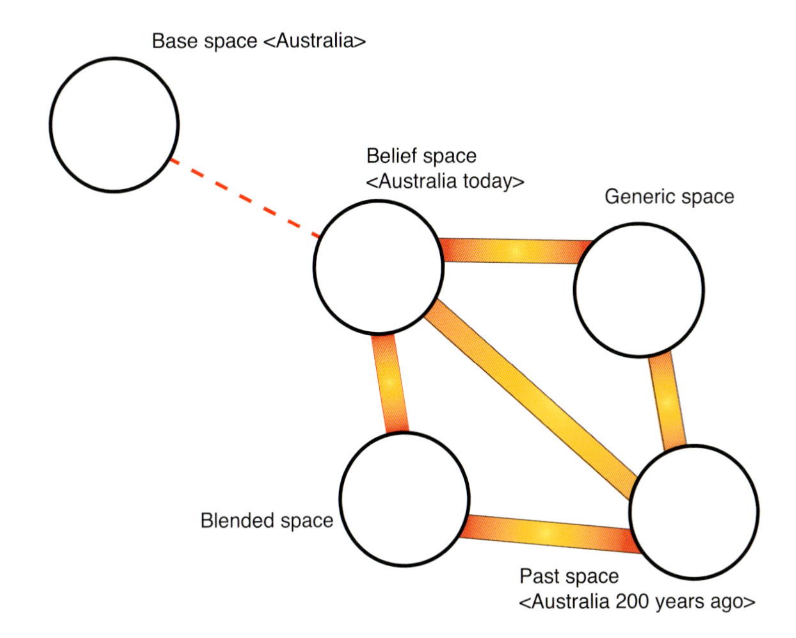

Figure 4.13
The addition of blended and generic spaces. Here the dashed line once again represents the transfer of structure by optimization and the projection of elements from base to belief space <Australia today>. The solid lines are intended to suggest the complicated linkages established between belief space <Australia today> and its child space <Australia 200 years ago>. These open a generic space, which maps onto each of the input spaces to capture and hold the structures they share. This opens the *blended space* in which emergent structure present in neither input space is developed. Drawing on the differences between <Australia today> and <Australia 200 years ago>, this emergent structure includes a future Australia not given in the maps.

such neuronal assemblies have been coactivated. This blend network is dynamic. The matches between the input spaces are not achieved independent of the blend materializing in the blended space; projections to the blend are selective; and the blend is "run" just as a computer simulation is "run." The frame for this "simulation" is recruited to the blend from the cues at hand, which in our case include the map's title, design and color, chart, graphs, and other maps and text, a frame we might call "Alien Invasion." Thinking the maps through this frame encourages the interpretation of the nineteen million Europeans in "Australia today" not present in "Australia 200 years ago" as just such a horde, responsible for the absence in "Australia today" of the rainforests present in "Australia 200 years ago." That is, the rainforest—and so nature—has fallen victim to a human plague. As it continues to run, the simulation churns out a future as well, driven by the graphic structure of the two maps (which drops the future in your lap), by our *general* understanding of how human numbers increase, *and* by the continous growth of Australia's population figured on the chart hovering directly over "Australia today." In this future there are still more humans and still less nature. No map of this future is needed: the meaning (and the fear and anxiety) emergent in the blend is perfectly clear.[6]

This is to say that, in *Geographic*-talk, Past + Present = Future. As the maps make clear, the way the past differed from the present was that it was "more natural." That is, Past - Present = Nature, which is to say in the movement from the past to the present that things have gotten less natural. Almost by way of arithmetic proof, if Past + Present = Future; and Past - Present = Nature; then Future ≠ Nature. Or, as it says under the "Australia today" title block, "As housing and pastures expand, forests fall, woodlands thin out, and fragile ecosystems become unstable. According to paleobotanist Mary White, 'Our current practices . . . are unsustainable.'"

Again, not a single remark about the future, but that we are *fast* going to hell in a handcart cannot be doubted.[7]

In this context, the soils and climate maps play the role of background information. That is, the map doesn't trust us to have the background information that Australia is dry and infertile—in other words, it is fragile.[8] The background spaces—we'll call them Bg_1 and Bg_2—that these maps open up can be launched from <Australia today>, <Australia 200 years ago>, or the blend, depending on which is in focus. Their content can "float" all the way up to <Australia> via <Australia today>; so, this fragile continent (established by these small inset maps), humanized by Aborigines for 60,000 years, but whose numbers and lifestyle minimized their impact (from the caption to and the essential subject of "Australia 200 years ago"), was, in 1788 (it is this date that sets the scope of the past at 200 years), assaulted by Englishmen and other Europeans who besieged nature (from the caption to and the essential subject of "Australia today" and its comparison with "Australia 200 years ago").

Their siege engines? The five maps falling down the right-hand side of the page: "Felling the forests," "A feral infestation," "Fire hazards," "Overgrazing," and "Mining matters," all of which open new spaces launched either from <Australia today> or from the blend. Read top to bottom, these maps comprise an extended essay in land despoliation (LD_1 through LD_5 in figure 4.14), one that builds on and draws out the implications of the larger maps to their left. "Felling the forests," for example, amounts to an explicit gloss on, maybe even a key to, the blend achieved by "Australia today" and "Australia 200 years ago." On "Felling the forests" it is the vegetation *changes* that have been mapped, the red here showing not "Pasture and cropland" but "Woodland or forest loss" (i.e., Pasture and Cropland = Woodland or Forest Loss[9]). The yellow similarly shows "Grassland reduction," and the blue "Shrubland reduction." Guided by this image, which after all amounts to Present-Past, it is possible to revisit the blend, with the blend now constituting an input space to a further blend. The following is the text for "Felling the forests" (authors' comments in brackets):

> Like a balding man's head, Australia lacks much cover in the middle but enjoys lusher growth at the fringes [because the interior is dry, as made clear by the climate map, Bg_2[10]]—yet even that's disappearing. More than 35 percent of Australia's woodlands have been cleared or severely altered, and three-quarters of the rain forest has been destroyed [and hence the special attention drawn to the rain forests on "Australia 200 years ago"]. All this felling hastens soil erosion [see soils map, Bg_1] and has let the water table rise, resulting in an alarming spike in soil salinity [see red stippling on "Australia today," see photo of dying trees with caption "Salt seeps up"]. Queensland has one new region of bright green growth [keyed in a bright green, "Gain in tree cover"], but it is hardly a blessing. An introduced tree—the prickly acacia—is turning native pastures into shrubland [see timeline, see photo of cane toads, see the next map, "A feral infestation"].

Australia under siege!

It is this constant checking and rechecking, the confirmation and reconfirmation—the eye moving from one map to the next, to a photo, to a text—that makes this page such a potent lesson. The reader is not simply told that things are about to get bad, is not simply asked to draw this conclusion, but is actively recruited as a data analyst. Since the endless chunks of information lack cross-references, the reader is encouraged to make his or her own, in the process thinking, "Wow! Look at this! It's a real confirmation of that!" and in this way positively *building* the argument out of facts (apparently) that seem to be just lying around.[11] Though in reality these facts—the maps, charts, graphs, and text—function as pressures and constraints on the kinds of arguments the map reader is able to construct, the sense of accomplishment is merited, for any construction is the result of powerful acts of integrating imagination. In the case at hand, each map, chart, graph, and block of type prompts the opening of yet another mental space, many launch blends, many of these blends act as inputs to yet further blends, structure is projected, and it is recruited from background knowledge. Our diagram of the full network (figure 4.14), which includes only mental spaces opened by maps per se and suppresses the blends as well as the elements and frames—the merest freeze-frame of an elaborate process further involving the deactivation of connections, the reframing of spaces, and other actions—is nonetheless daunting. Yet, in essence, it could be reproduced in the effort to account for the construction of meaning in even *small* portions of the network.

For example, "A feral infestation" helps the reader *pull together the threads* about introduced species, from the "upset the ecological balance" of the introductory text, through the photo of the cane toads (headlined "To control sugarcane beetles"), the purple floater entitled "Feral control," and the timeline (contrasting "Introduced Species" with "Extinct Animals" against a chart of human population growth), to this map of "Introduced game animals," with its martial imagery of a "Fox invasion," a "Rabbit invasion," and the pall of battle smoke hanging over northern Australia. "Pull together the threads" means moving through the network of spaces, continually shifting focus and viewpoint, to build a blend with multiple inputs structured by a variant of the "Alien Invasion" frame that structures the poster as a whole. Indicatively, in the text of "A feral infestation," rabbits are actually called "an alien invader."[12]

Australia under siege!

All the texts can be interwoven (which is to say that conceptual integration networks can be constructed to do so). Thus, "Fire hazards" opens with praise for Aborigine fire management practices (see the text to "Australia 200 years ago"; see the timeline) and immediately connects to "Felling the forests" and the climate map. "Overgrazing" with its text ("Water troughs built for livestock also refresh and sustain wandering feral herds [see "A feral infestation"]. And as pasture expands, natural habitat for native species disappears [see timeline]") all but provides an *explanation* for the red on "Felling the forest," that is, for the woodland and the forest loss, and therefore, as well, for the red on "Australia today" (thanks to a blend with multiple inputs). Everything is pulled together in the timeline, which, with its accompanying chart of human population growth, advances population growth as a first cause (its text starts, "'A plague of Humans'—that's how one overheated 1990 news headline referred to the recent growth of Australia's population, which has contributed to a mammalian extinction rate higher than anywhere else in the world"), but it also manages to get in global warming as a secondary cause of the transformation, which the reader has (by now) concluded is (indeed) threatening Australia.

Australia under siege!

Mining ("Digging massive open-pit mines and building the infrastructure to exploit them inevitably affects animal and plant habitats"), pressure from grazing, forest and brush fires (intensified thanks to fire suppression), introduced species, forest loss, population growth —what does it all remind you of?

The U.S.

Of course: *if Australia is under siege, so is the U.S.!* This is the real text of the map: If Americans don't get their act together they're going to hell in a handcart too. Ditto the Brazilians. The Chinese. The Malaysians. *Everyone.*

Which is why the map wants to be *taught.* "Living Fossils" didn't have to be taught. It *couldn't* have been taught (there was nothing to teach). "Living Fossils" was taken in at a glance, it was swallowed whole: Australia = Koala (= Nature = Soft/Cornucopia). In the then teddy-bear universe of the *Geographic,* Australia and the koala, nature and its cuddlability, might simply have fused: *Australiakoalanaturesoftcornucopia,* celebrated together because equivalently tamed and subdued, because equivalently impotent, because, if not always equivalently warm and soft, at least always warm and wacky. Thanks to the reciprocal effects of optimization and floating, the koala acquires the authoritativeness of the main map, and the main map the fur and feathers of the fauna, in the light of which the main map's white comes off as benign and the green set aside for "nature" as ample indeed.

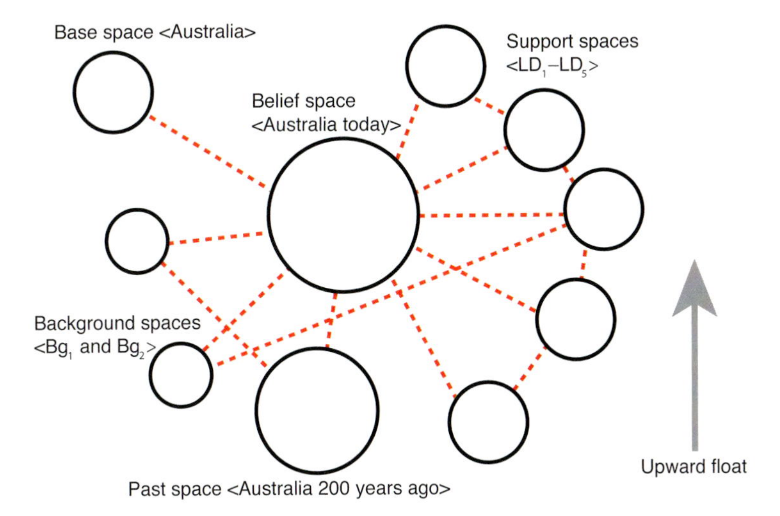

Figure 4.14
Cognitive structure of "Australia Under Siege," full network. We're again showing only the gross network and as before have suppressed indications of the frames and the elements. The dashed lines here represent not only the transfer of structure and the projection of elements, but a number of blends. Most of the work being done by the full network has been cued by the main map "Australia" (base space <Australia>), "Australia today" (belief space <Australia today>), and "Australia 200 years ago" (past space <Australia 200 years ago>), but as pressured and constrained by the two physical maps of soils and climate (background spaces <Bg$_1$ and Bg$_2$>) and the five support maps of "Felling the Forests," "Feral Infestations," and so on (Support spaces <LD$_1$-LD$_5$>). Other aspects of the map—its design and color, chart, graphs, and texts—contribute to the frames that guide the building of structure in each of the spaces of the network. "Alien Invasion," the frame recruited to drive the blend called up by <Australia today> and <Australia 200 years ago>, together with the future that the blend generated, float up and suffuse the rest of the network so that everything is construed in their light.

In comparison, "Australia Under Siege" is in mourning. It's as if someone at the *Geographic* suddenly woke up and realized the party was over, that there had been a dark side to all that white on the main map (which, let it not be forgotten, encodes all the land arrogated by the British through the magic of naming, *not* "returned" or "set aside" for the rare creatures, including aboriginal humans, of this land that time forgot), that this dark side wanted its due, that the bill was going to be enormous, and that it would *have* to be paid. In a way, the map is an effort to explain *why* it will have to be paid: "Why? Because you brought in cane toads to control the sugarcane beetles, and now the cane toads are out of control. That's why."

Consequently what's at stake in "Under Siege" is education, the introduction of a whole new way of thinking about things. It's a way of thinking that might be called ecological, though *ecologistic* is probably closer to the mark. Involved is certainly a sense of the interconnectedness of organisms in their environment—a sense wholly absent from "Living Fossils"—but in ecologistic thinking this comes wrapped in a conservative conviction that the recent (and continuing) growth in human numbers has had a profoundly degrading affect on the environment, and so with a concomitant *anxiety* about resource scarcity, especially that of water and fertile soil. There's a profoundly political dimension to this kind of thinking, maybe even religious overtones,[13] a big part of which derives from the overweening conviction that nature is exactly what "Living Fossils" made it out to be: furry and feathered, bright, warm, and soft; something we can cradle in our arms; something that we can care for; something, in fact, that we *have* to care for . . . because it's weak.

The ecological root of ecologism *has* forced attention away from the birds.[14] Birds remain important in the "Under Siege" world view, but there is an uneasiness about the kind of unabashed adoration characteristic of the earlier *Geographic*, and a greater sense of their importance as parts of the great web of life (and as indicator species). "Under Siege" itself is too anxious—too hysterical—to display any animals at all (except for the introduced cane toad!), but other *Geographic* maps with similar messages have been less reticent. "Amazonia: A World Resource at Risk," with its ocelot-like margay on the cover fold, tosses birds around with abandon, and indeed 29 percent of the animals displayed are birds (and this as recently as August, 1992!)—"With imperious gaze, an endangered Guiana crested eagle searches for prey"—but even here it has to be acknowledged that the animals are absorbed into a "tapestry of life" (they are not portraits), that there's plenty of (unlabeled) vegetation, that 15 percent of the animals are reptiles (!), that two bats are pictured (!), and a tarantula along with the butterflies. "Africa Threatened," of December, 1990, displays portraits of eleven mammals, two birds, and a tortoise. The same tortoise—*Psammobates geometricus*—was the only reptile to make the cut in the tapestry decorating "Africa's Natural Realms" of July 2001, with its text that warns that, "Human beings, once a minor part of all this variety [again almost exclusively mammals and birds], now threaten to destroy it."

We're going to destroy it—that's the point that *Under Siege* hammers home more dramatically than many a *Geographic* map—but this follows only because of a conception of nature that, despite all the extraordinary differences in tone, absolutely unites "Living Fossils" and "Under Siege." "The world is changing," Gilbert M. Grosvenor said in 1977, justifying his publication of articles about Harlem, Cuba, and South Africa,[15] but evidently nature wasn't. Despite the *Geographic's* discovery of ecology, plate tectonics, global warming, vanishing habitat, and microbiology, its idea of nature as soft—and of the human world as not *of* nature—has not only persisted but by, as it were, slipping into the underground (disappearing into the militant hysteria of "Under Siege"), become ever more authoritative (*now* nature's threatened).[16]

But then, nature's *always* been in the underground. That's the problem. "Living Fossils" never presented itself as a systematic description of nature. It was simply a map of Australia. For the *National Geographic*, nature was ever the elephant in the room, danced around, but never directly addressed. Our suggestion that the ceaseless repetition of fur and feather imagery on map after map constituted something more than diverting decoration—constituted an ideology of "nature as soft and cuddly"—would have been greeted by the maps' designers as an astonishing suggestion: "Everyone knows there's more to nature than birds and mammals!" Of course, this is exactly how American universities responded when it was pointed out that their failure to sponsor women's intercollegiate sports amounted to an ideology of "sport as male": "What? We never said that!" True enough, and the ideology of "sport as male" was the more powerful *for that very reason*. Similarly, the refusal of *National Geographic* maps to declare "nature soft and cuddly" also made its assertion of this ideology the more powerful, as map after map after map was bedecked with fur and feathers. Here is "Florida"—from the *Geographic's* "Close-Up:

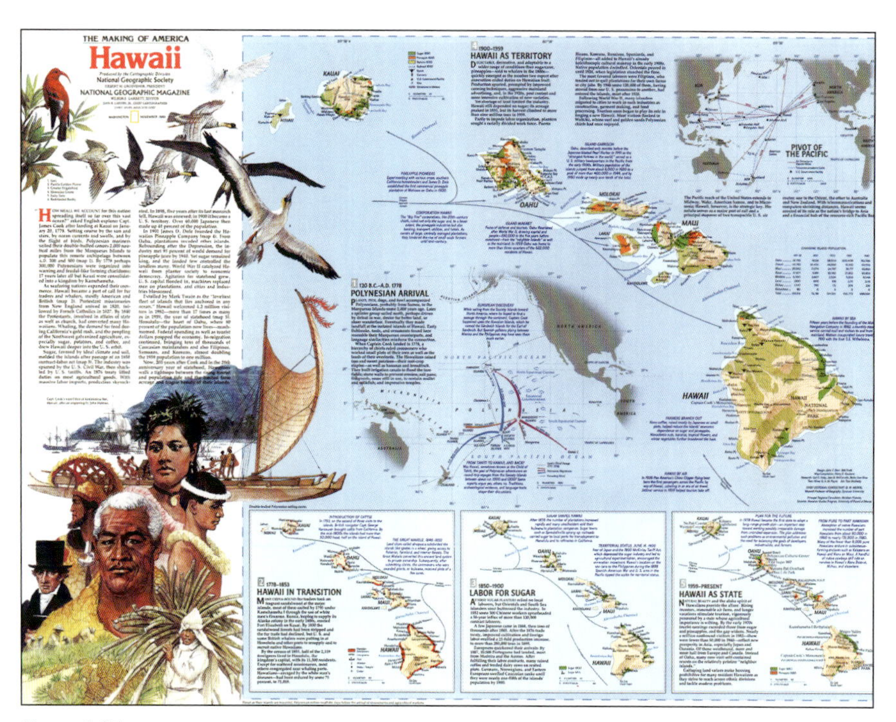

Figure 4.15
"Hawaii" (from "Making of America" series), supplement to *National Geographic*, November 1983. Here, as on so many *Geographic* maps, nature is the elephant in the room, danced around, but never directly addressed. Map dimension (WxH): 27in. x 21in.

Courtesy of Tom McNeely/National Geographic Image Collection.

U.S.A." series —inserted into the November 1973 issue. On the cover fold, looking back at us over its shoulder, a flamingo; in the background, four others in flight, three warping around the fold. Opened, a pelican; opened again, white ibises, great blue herons; completely unfolded, a barred owl, a bald eagle, *thirty-eight images of birds*: "From Everglades trails, along farmland back roads, even by busy highways or at city parks, visitors can view free-roaming specimens of Florida's rich birdlife." "Hawaii," from the "Making of America" series, and inserted into the November 1983 issue (figure 4.15), has six birds, and no other animals, on the cover fold. The second fold reveals a number of *native* people: "Proud as their islands are beautiful, Polynesian nobles recall the days before the arrival of missionaries and agricultural workers," that is, back in the past, back when things were natural.[17] "Indonesia," inserted into the February 1996 issue, does have a large komodo dragon on its cover fold, along with four mammals, two birds, an orchid, and a palm. But there are only three other reptiles on the map, which displays thirteen other mammals (including a mother orangutan and her baby), and six birds (in a country where "timber production is running far above sustainable levels"). Alaska's "Unmatched wildlife," as depicted on the "Alaska" insert to the May 1994 issue, is mostly birds, bears, killer whales, and sea lions. Indeed, except for two sockeye salmon, Alaska's "Unmatched wildlife" is *all* birds and mammals. And so, warm and bright and soft; and so, something we have to care for because it's weak. It's weak and we're killing it, the bad ones of us, anyway; or that bad part of us all that can't stop buying and breeding and building. This is what's threatening nature. It's greed. (It's evil.) And this (essentially) religious message is slipped in with these maps as a fact of geography.

You know, it's hard to say reading one of these maps, handling its smooth paper, scanning its informative and beautifully printed surfaces, where the science—so quietly, and therefore so authoritatively asserted in the signs of science on the map on the map side—slips over into the politics, into the religion of the poster side: "Common sense," it says on the poster side of "Biodiversity" (slipped into the February 1999 issue), "ethics, economic practicality, and a sense of wonder all tell us that we stand to benefit if we protect what protects us." And we may well agree, but this is *not* science speaking. Yet because it is hard to say where the transition occurs, it is hard to say *that* it occurs, and so it doesn't occur, there is no slip, it is all geography, it is all science.

And this particular quotation is more than usually candid; it acknowledges that under the mask of a concern for nature, for biodiversity, for forests, for our furry and feathered friends, beats a rather naked self-interest. Usually, this is harder to make out through the mist of numbers, the chanting charts, the maps with their pall of smoke over the forests of Indonesia and the Amazon: *just counting disappearing species*.

And by no means is it just the *Geographic*. As many maps as the *Geographic* has published, the numbers pale next to those of state highway departments whose ceaseless depiction of state birds and animals work to precisely the same end. Indeed, can anything be more powerfully confirming than turning from the self-consciously educational, the often preachy, *Geographic*, to an avowedly practical state highway map and finding the very same views of nature, the very same ideas about design, the very same structure of the map. It is this unacknowledged web woven by *Geographic* maps, by state highway maps, by the maps of Canon camera advertisements and nongovernmental organizations (the Audubon Society, the World Wildlife Fund, Conservation International [*Biodiversity at Risk!*]), that make of this view of nature and our relationship to it the hegemonic monster that it is. Something horrible is happening to something called nature that used to be a cornucopia. Exactly what is hard to say—the maps are too abstract (or so terrifying we're reduced to gibbering), the photos and paintings too anecdotal—but it's sure something we need to be anxious about.

Threatened, beleagured nature—better catch it with your Canon while you can.

Notes

1. A couple of years earlier, in February 1998, the *Geographic* had run a photo (pp. 36–37) of a map spread out across the floor of the University of Maryland's Cole Field House, a map that had been assembled from 720 smaller images. Even though the map hadn't been composed from *Geographic* maps, the photo illustrated the idea that the world can be disassembled into individual sheets—very like your *Geographic* maps—which therefore are, in effect, just pieces of the world. The photo appeared in John Noble Wilford's "Revolution in Mapping," 6–39, in the "exploration" issue of the magazine's "millennium series." Articles like these, which have become a *Geographic* staple, are another part of the armament defending the facticity of the main maps. Because the stakes are not trivial, the *Geographic* has trundled out a heavy gun here. Wilford is the science editor for the *New York Times*, and author of a history of cartography. (For our take, see Wood's review of Wilford's *Mapmakers* [*Cartographica* 19 (Autumn & Winter 1982): 127–31].)

2. It was, of course, these principles that produced the infamous December 1928 article, "Renascent Germany," with its notorious caption, "Green as goslings now, but practice makes the goose step perfect," to its photo of Hitler youth; and the even less savory February 1937, "Changing Berlin," which managed not to mention Nazi anti-Semitism; but it was also these principles that mandated the endless articles about birds, and that got the younger Grosvenor in such hot water when he ventured even vaguely factual reporting about Cuba, Harlem, and South Africa (because the facts were controversial—as they usually are!). We love the younger Grosvenor's response to his attackers: "The Magazine is not changing, the world is changing," and this in 1977! (For the principles, see Bryan, *National Geographic Society*, 90; for the articles on Germany, 172–3; and for the younger Grosvenor's riposte, 390.)

3. Recall also that the borders of the portrait map stopped at land's end and so less framed a piece of the world than brought a world into being (the land of living fossils). Similarly, this jigsaw piece stops at land's end and so, too, less frames a piece of the world than brings a world into being (Australia today). The qualifier "today" marks the difference between this Australia and the real, unqualified Australia on the main map. The real world exists in the aorist, it just is, without tense, outside past, present, and future. We might think about this as the authoritative, as the reference mode.

4. Having noted in the previous footnote that "today" qualified Australia as "a world brought into being" by the map, it must be noted that "past" really ups the ante in this regard; for though "today" marks the qualified map as other than the real, authoritative, reference Australia, it is at least an Australia that can, in principle, be pointed to. In principle you could go to Australia and find, due north of Adelaide, pasture and cropland with human-induced salinity. You could point to it. This is not possible with a past Australia, which cannot, even in principle, ever be more than a possible reconstruction. In Fauconnier's terms, "Australia 200 years ago" opens a past space, subordinate to "Australia today," itself subordinate to the base space, "Australia." Standing outside the map it is easy to see that all three maps are constructions, including the "real" Australia spoken in the aorist, in the reference mode.

5. Their book, *Way We Think*, is exclusively concerned with blends.

6. Fauconnier and Turner, *Way We Think*, 66, discuss another case in which fear and anxiety, not associated with either input, emerges in the blend.

7. Alternatively, we could construe this as a predictive conditional, a route we'll take in the next chapter in our reading of "Natural Hazards of North America." Here we want to note only that blends and predictive conditionals both involve the construction of conceptual integration networks and to wonder how different the neuronal assemblies and linkages actually are. For predictive conditionals, see Dancygier and Sweetser, *Mental Spaces*.

8. One way of understanding these maps is as providing the "environment" for an environmental determinist reading, and there may be some of that lurking here. However, we see them above all else as establishing the continent's fragility in the same way that the map by Tom Van Sant discussed in *Power of Maps* (Wood, 68–69) labored to establish the planet's fragility: "Don't drop it. It will break." As we'll see, we feel the same sort of millenarianism in "Under Siege" as we found in Van Sant.

9. In the age of "Living Fossils," the magazine would have construed this as Pasture and Cropland = Human Gain. The change at the *Geographic* can be measured in this flip from Gain to Loss.

10. In cognitive lingusitc terms, this would imply that viewpoint had shifted to the space opened by "Felling the forests," from which the space opened—sometime—by Bg_2 was being accessed.

11. This is another form of the participation that comes with being a member of the National Geographic Society. Readers support research through their membership, which results in maps like this, from which readers as data analysts tease still further knowledge!

12. And, incidentally, the text here suggests that the dingo was brought in by Asian seafarers 3,500 years ago, whereas "Australia: Land of Living Fossils" had it possibly arriving "from the Malay Archipelago with early human migrants about 8,000 years ago." This is a substantive, but unremarked, difference.

13. We have benefited here from reading Anna Bramwell's incisive *Ecology in the 20th Century: A History* (New Haven, Conn.: Yale, 1989). Bramwell argues that in the late 1970s the fusion of ecologism's two main roots, the biological and economic, gave it new authority and popularity, especially in the aftermath of the publication of Garrett Hardin's "Tragedy of the Commons" in 1969; the oil crisis of the early 1970s; the founding of the Club of Rome in 1972; Barbara Ward and Rene Dubos's "Only One Earth" report to the United Nations World Conference on the Human Environment in 1972; E. F. Schumacher's *Small Is Beautiful* in 1973; et cetera. See her "Greens, Reds and Pagans," pp. 211–36. We note that it was also in the late 1970s that the *Geographic* began its (very slow) transformation from chief celebrant to chief alarmist. See the map supplement to the July 2002 issue, *A World Transformed*. Its intro text concludes: "Our large brains got us to this point. The question now is, can they get us out?" The titles to the seven maps and charts that surround the central map are "Oceans at Risk," "The Air We Breathe," "Alien Invasion," "Earth's Vulnerable Soils," "Missing the Forests for the Trees," "Energy Binge," and "A Warmer World." It's all very bad news.

14. Bramwell locates ecologism's roots in ecology (biology) and energetics (economics). Both are probably responsible for the greater complicatedness of recent *Geographic* maps, as though the interconnectedness of organisms in their environment required a graphic style of interconnected images. Certainly the many small maps on contemporary *Geographic* maps has something to do with this shift in perspective. *A World Transformed*, for example (see previous note), has nine maps of the world as well as six regional inserts, and these are linked through argument and call-outs.

15. Again, the Grosvenor quote is Bryan, *National Geographic Society*, 390.

16. Indeed, on the cover of the February 1999 issue that contained "Biodiversity" were the words "Biodiversity: The Fragile Web," with "The Fragile Web" in type the size of the "National" in *National Geographic*. Yet, the magazine carried a whole article devoted to the history of extinctions, and the map as a major feature. The lead paragraph in the feature's text ran, "An estimated 99 percent of all species that ever lived are extinct. Yet more kinds of creatures may be alive today than at any other single time in the planet's 4.6-billion-year history." An accompanying chart shows an increase in diversity following each of the five earlier major extinction events. Far from being fragile, these data describe an extraordinarily resilient web, one capable of enduring and coming back from catastrophes vastly greater than anything we can throw at it. The problem is not the web. The *Geographic* is not concerned with the web. It's concerned with us: "The recovery period [from the ongoing human-induced extinction event], if there is one, has little meaning on the timescale of Homo sapiens." That's the problem: we won't be around to experience the recovery. Our lives will be diminished. This would be so easy to say. Why do the *Geographic* and so many others find it so hard?

17. We won't mention "Hawaii," from the Close-Up: U.S.A. series, inserted into the April 1976 issue—"Beguiling isles born of earth's torment"—which despite a text that refers to wind-borne spores and bats (!), displays paintings of only five "forest-flitting honeycreepers," a butterfly, and twenty-two gorgeous blooms, including "plumeria blooms strung into the traditional lei." The diversity of life!!

CIRCIVS.VEL. RESIAS.
SEPTEMTRIO. VEL.APATCIAS.
CAVRVS.CORVS.VEL IAPXSI.VIGESTES.
CIRCVLIVS.ARTICVS.
INSVLE.MALVCHE.
INSVLA.GILOLO.
AEQVINOCTIALIS.
TROPICVS.CAPRICORNI.
TROPICVS.CANCRI.
CIRCVLVS.ANTARTICVS.
FAVONIVS.VEL ZEPHIRVS.
AFRICVS.VEL.LIBVS.
LIBONOTVS.EVRO.AVSTER.
PERV.
MVNDVS.NOVVS.
BRAZIL.
CHINAGVAIRA.
RIO.DE.LA.PLATA.
MAVRITANIA.
LIBIA.INTERIOR.
AEGIPTVS.
CAPO.D
bermuda
quinsai civitas
30
45
60
75
90

Five

Threatening nature

Source: Agnese, Battista. *Portolan Atlas, Oval Map of the World*, ca. 1540.
Reproduced by permission of The Huntington Library, San Marino, California.

Threatening nature

If there is a *threatened nature*, a nature of endangered species and disappearing rainforests, of greenhouse gases and acid rain, of poisoned waters and overdrawn aquifers, strip mines and oil spills, a nature of wholesale exploitation and corporate road kill, then there is also a *threatening nature*, a nature of earthquakes and volcanoes, landslides and tsunamis, of drought and floods and wildfires, of blizzards and ice storms, tornadoes and hurricanes. And, if a threatened nature is threatened because it falls under the dominion of culture, the tables are turned here. Human culture is under the dominion of, at the mercy of, an all embracing and overwhelmingly threatening nature. This is not a nature we can control, much less harm; we can only cower or get out of its way. This is not a nature to be exploited and abused. This is a nature to be feared. We are the victims now, and sometimes there's hell to pay.

Something wicked this way comes

In the southern United States, the summer of 1900 was one of the hottest ever recorded, duly and diligently recorded by the fledgling U.S. Weather Bureau re-formed from the Army Signal Corps Weather Service by an act of Congress just a decade before.[1] In early September of that year, still suffering from the unrelenting heat, the citizens of Galveston, Texas, took to the beaches and stilted bath houses of their Gulf of Mexico seashore seeking some relief in the soup-warm waters.

They didn't have to travel far. The city of Galveston had been constructed on a barrier island at the southern entrance to Galveston Bay and, at the close of the nineteenth century, it was Texas's only seaport—affluent, cosmopolitan for its day, and filled with the civic optimism of more than 37,000 residents over its future as Texas's, even the Gulf Coast's, premier city. (Texas oil had not yet been discovered nor Galveston Bay dredged to make a port of Houston, or "Mud City" as Galvestonians called it.) Galveston had a busy port facility and equally busy mercantile center. It had solid, opulently decorated banks, an opera house, streetlamps, mule-drawn streetcars, an exuberant night life, neighborhoods of finely crafted turn-of-the-century homes, the new southwestern reporting and forecasting center of the U.S. Weather Bureau, and of course, its very own beachfront. Even in an era well before the advent of air conditioning, the city must have appealed as an ideal mix of burgeoning American culture and semitropical (winter free!) nature. Tropical storms? Most Galvestonians had weathered their share of those and just shrugged them off with casual Texas bravado.

That all ended, of course, on September 8, 1900, in the twelve-hour assault of the "Galveston Hurricane"[2] that brought 120-mile-per-hour winds and a 25-foot storm surge to an island city only 8 feet above sea level at its highest point. Egress and communication were quickly cut off, the isolated city torn and hammered to pieces and completely immersed by the sea and, by rough account, more than six thousand of Galveston's inhabitants were killed—drowned in the massive storm surge, crushed under collapsing buildings, or lacerated by the razor-sharp slate tiles whipped off rooftops by the tornado-force winds. Six thousand dead. That's twice the human cost of the 9/11 attacks and, in those terms, the greatest natural disaster in U.S. history.[3] They never saw it coming. Why didn't they?

That was a century ago and weather recording and reporting were in their infancy (never mind weather *forecasting* as we now know it[4]). All observations were local. There was a thermometer to measure air temperature, a barometer to measure air pressure, an anemometer and vane for wind speed and direction, a hygrometer for humidity. These observations were (and still are) postings of localized measurements at fixed times and positions. For any aspiring "forecaster" beginning to grasp the principles of what we'd eventually come to call meteorology, that wasn't much beyond sniffing the wind—the temperature was rising or falling, the barometer inching up or down, the wind coming from this direction or that. Dressed in the instrumentation and enthusiasm of an emerging science, it still required a lot of skill and instinct to make sense of it all. For the recently formed U.S. Weather Bureau to get the "big picture," in other words, to *map weather*, it was necessary to integrate numerous local reports via the communications technology of the day, the telegraph, in its central office in Washington, D.C., where its most skilled analysts could devote themselves to the Bureau's mandated goal of plotting seasonal and regional trends affecting agricultural production.

In the U.S., very little was known about the tropical storm systems affecting the Caribbean and Gulf of Mexico. For that, the Weather Bureau had been reliant on reports and opinions issuing from Cuban "weathermen," who knew from experience much more about those capricious tropical storms than anyone else did. But, following the Spanish-American War of 1898, Cuban weather reporters were supplanted by Cuban-based staff from the U.S. Weather Bureau, who assumed exclusive responsibility for telegraphing tropical reports to head office. Cuban expertise had become suspect and was officially out of the loop.

The Galveston storm was what we now call a "Cape Verde Hurricane," not spawned in the Caribbean but spinning up as a tropical depression off the coast of Africa about a week earlier. It made quick westward progress across the Atlantic and, within a short while, gained enough strength to pummel Cuba with hurricane-force winds and torrential rains. Thence where? American forecasters in Cuba believed that the storm would recurve and eventually track up the eastern seaboard of the U.S. while Cuban forecasters suspected that it would turn westward into the Gulf of Mexico. But it was only the American forecasters' predictions that were conveyed by telegraph to Washington.

After Cuba, the storm did turn north, but across the southern Florida Keys and up that state's Gulf Coast toward the Panhandle. And then it took a sharp turn to the west and did something inconceivable by modern standards. It *disappeared from view* into the Gulf of Mexico.

It literally *went off the map*. No one knew where it was or what it was doing. How was that possible? How could a storm system of that magnitude simply disappear from the maps in the Weather Bureau head office? It had wholly disappeared from the radar screen, so to speak; but, of course, there was no radar screen in 1900. All official weather observations were terrestrially based and no dramatically rising winds or falling barometers signaled its approach among coastal observation stations. Ship-to-shore communications didn't yet exist either, and any ships approaching the storm would surely have been turned back or lost.

Meanwhile, the hurricane had grown to monstrous proportions, feeding on heat from the extraordinarily warm Gulf waters. In the eye wall, wind speeds of 150 miles per hour formed a vortex that sucked sea water two miles into the air and swelled a massive dome of water twenty-five feet above sea level. When the storm reappeared and the eye wall hit Galveston unexpectedly on that afternoon of September 8, it caused the greatest loss of life of any single incident in U.S. history, natural or otherwise. Galveston never saw it coming. They never knew what hit them . . . until it did[5] (figure 5.1).

Figure 5.1
Galveston, Texas, after the hurricane disaster of 1900. Because the storm could not be tracked or mapped, it struck without warning, devastating the town and killing thousands.
NOAA archives.

How could this have happened? Let's try reconstructing an imagined scenario. A big low pressure system rolls out of western Africa and into the Atlantic, where it acquires more energy and quickly becomes what we now call a tropical depression. Some ship at sea encounters this storm and limps back into port with the captain reporting "*Whew!* We just got our pants kicked out there!" And then it's entirely unseen for at least several days (imagine that) until it, presumably the same storm, turns up in the Caribbean and makes a mess of Hispaniola and Cuba. Then, it disappears again and only reappears when it makes landfall at Galveston and destroys everything.

It was a kind of peekaboo game, now you see it, now you don't; and when you do see it again, that's really bad news. The antagonist in our peekaboo game has become the bogey man. There were very few dots to connect, and what the hurricane did in the space in between was highly conjectural. Not so now, of course. We have dots posted every thirty minutes at most, and these things aren't going to take anyone by surprise. But the basic principle of the tracking map remains the same as it has always been, as long as there have been such maps anyway. It's about the construction of space—the space of the hurricane event itself. Individual observations can be assembled into lines—hurricane tracks—and those can, over time, be synthesized into regional concepts like "Hurricane Alley." It's all space building, from point to line to area, and the hurricane is not a single brittle line on a map but, as our meteorologists warn, "effects can be felt as far as 300 miles from the storm's center."

Of course, something like the Galveston disaster couldn't happen now, not quite the way that did. There isn't a speck of land or sea on the planet free from constant satellite surveillance. And of course, after decades of research, we've gained a much greater knowledge of the coastal environment and coastal processes. We know how foolish it is to spend billions building on ephemeral features like barrier islands. Everyone knows they're just (barely) emergent sandbars. Everyone knows that "sand ain't land," right? Sure. Perhaps the knowledge is there, but the allure of the seashore remains undiminished. How many millions in the southeastern and southern U.S. live in coastal cities and towns conspicuously in harm's way? In Virginia Beach, Atlantic Beach, Wrightsville Beach, Wilmington, Myrtle Beach, Charleston, Savannah, Jacksonville, Miami, Fort Lauderdale, and West Palm Beach? In Naples, Tampa–St. Petersburg, Panama City, Pensacola, Mobile, New Orleans, and Corpus Christi? And in between? How many tens of millions altogether? This is not just a high-profit/high-risk investor's crap shoot. We know very well that there are a lot of lives at stake. That's why we have sophisticated weather observation, modeling, and forecasting systems, and up-to-the-minute media reporting of critical storm developments, and federal and state emergency planning agencies and crews, and designated evacuation routes and refuges for evacuees, and a plethora of *hurricane tracking maps*.

Anyone living in the southeastern and Gulf Coast states is familiar with those maps. Every year, with the advent of the June through November hurricane season, a new crop of these appears, issuing mostly from local television and radio stations and variously sponsored by hardware and grocery chains, public utilities, home HVAC manufacturers, and local newspapers. Our small collection includes maps from Raleigh's CBS affiliate WRAL-TV (sponsored over the years by the capital city's *News & Observer* newspaper, regional electrical utility Carolina Power and Light—later, North Carolina's Electrical Cooperatives—Trane heating and air conditioning, Harris Teeter supermarkets, McDonalds, Jiffy Lube, and Chick-Fil-A), Durham's ABC affiliate WTVD (variously sponsored by True Value hardware stores, Lowe's Home Improvement Warehouses, and Winn Dixie supermarkets), The Weather Channel and Home Depot, local radio station WPTF 680, and even a solo effort (2003) from Food Lion supermarkets. And that's just what's to be found close at hand. Along the coast, it's not uncommon in season to find throw-away tracking map placemats tiling diner and restaurant tables.

These are all giveaways, of course, intended to be retrieved from sponsors' franchises ("Would you like fries with your map, sir?") and, not surprisingly, at least as much advertising venues as they are anything else. WRAL-TV's 2004 issue includes fourteen ads: for a furniture outlet, a dental office, a men's store, an eye care office, home repair, fast food, North Carolina produce (with Deputy Barney Fife munching a peach), and no fewer than seven ads for automobile dealerships. WPTF's 2001 map is corralled by nineteen ads: for truck and van accessories, automotive service, pecan products, greeting cards, carpets, mattresses, stair lifts, plumbing and heating services, cellular phone service, car and equipment rentals, generator sales and rentals, television and furniture rentals, debris removal, building foundation repair, a pharmacy chain, more fast food, and three automobile dealerships.

In recent years, these maps have evolved from single folded sheets to booklets of a dozen or more pages; and their titles have swollen with drama as well—from "Hurricane Tracking Map," "Hurricane Tracking Chart," or "Hurricane Tracking Poster" to "Hurricane Preparation Guide," "Hurricane Survival Guide," "Stormtracker," even "Are You Ready?"

The most ambitious among our collection are the recent WRAL-TV issues, which, beginning in 2003, rematerialized as a kind of mini magazine, eight-and-one-half by eleven inches and sixteen pages, and now announcing itself as "Stormtracker: Your Official Hurricane Survival Guide." As such, it proposes to be more than just a tracking map, and it is. We're looking right now at their *2005 Stormtracker*.

At the top of the front cover is a prominent WRAL NEWS banner, then the *2005 Stormtracker* title in bold futuristic italic, and then "Your Official Hurricane Survival Guide" (*Official*, if not to be taken literally, nonetheless affirming its status as the premier publication of its genre still to be found here). Running down the left edge is a list of contents: Tracking Map, 2005 Hurricane Names, Guide to Surviving the Storm Season, Internet Resources, Important Phone Numbers, and Evacuation Plans and Routes. Running down the right edge are the faces of WRAL's team of expert meteorologists: Chief Meteorologist Greg Fishel and his colleagues Mike Maze, Elizabeth Gardner, Chris Thompson, and Mike Moss. At the bottom are the imprimaturs of this year's cosponsors, Harris Teeter supermarkets ("Your Neighborhood Food Market") and Trane Central Air Conditioning and Heating ("It's Hard to Stop A Trane"), and all on a softened background of abstract design incorporating suggestive decorations of hurricane symbols, warning flags, grid lines, and what appear to be strips of punch tape hieroglyphics.

The inside front cover (we'll call that page two) is occupied by a full page advertisement for three Saturn dealerships, and the lessons begin in earnest on page three with two bulleted lists of advice and recommendations, "When A Storm Is Forecast" and "When A Storm Threatens." In the upper right is a list of contents to be assembled in your emergency medical supply kit.

Page four is titled "If You Evacuate" and is occupied by a rather striking map of coastal North Carolina Evacuation Routes (figure 5.2). Coastal counties are tinted overall with an attractive pale green while a darker green paints "Areas that would be flooded by a 5-foot rise in the normal sea level." The evacuation routes are drawn in broad yellow lines with bold arrowheads, and superimposed offshore is a blue-tinted satellite image of an enormous approaching hurricane. This is a forceful and engaging map. Sidling up to the map on the right are a list of radio stations that will simulcast WRAL news in the event of widespread power outages and a table of coastal counties indicating peak tourist populations and hours needed to evacuate—anywhere from nine to twenty-eight hours, although a footnote cautions that "Parts of the Outer Banks can take longer to evacuate, in some cases up to 33 hours." There's also a "Flood Fact": "Flash flood waters move fast enough to roll boulders, tear out trees, and obliterate bridges (a process known as scouring)." The facing page instructs us to *Protect Family And Pets*, with two more bulleted lists for "Keep Seniors Safe" and "Protect Your Pets," and advertisements for Circuit City and Eastcoast Generators.

Lessons continue in the same fashion on the sixth and seventh pages with "When A Storm Hits," supplemented by a diagram of "Hurricane Winds" and an italicized "Wind Fact," as well as another bulleted list for "After The Storm Passes."

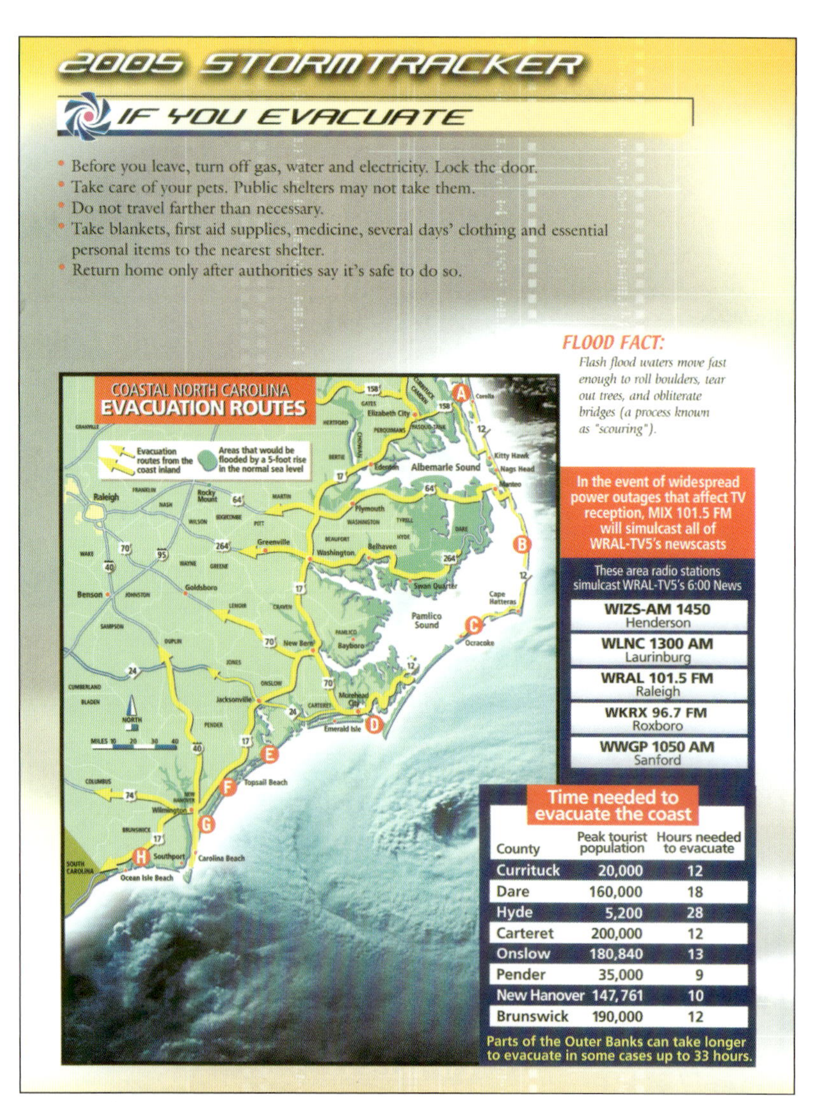

County	Peak tourist population	Hours needed to evacuate
Currituck	20,000	12
Dare	160,000	18
Hyde	5,200	28
Carteret	200,000	12
Onslow	180,840	13
Pender	35,000	9
New Hanover	147,761	10
Brunswick	190,000	12

Figure 5.2
Hurricane evacuation map from WRAL-TV's *2005 Stormtracker* publication. Maps such as this capitalize on the concept that nature can be our adversary and oppressor, and must be met with thorough preparation and constant vigilance.

Courtesy of WRAL-TV5, Capitol Broadcasting Company, Raleigh, North Carolina.

On page seven we find "WRAL-News On Your Cell Phone," which features forecasts (and alerts), static and animated maps (from three Dopplers and regional and national satellites), and hurricane tracking with storm coordinates and wind-speed maps.

The hurricane tracking map itself is spread across pages eight and nine, which, in the stapled 2003 edition, could be easily extracted. The 2004 and 2005 editions are glued instead, so that's not a possibility. We'll flip through the remaining pages and then return to this map.

Page ten provides additional lessons on "Hurricane Dangers," including "Storm Surge," "Winds," "Tornados," and "Inland Flooding," along with a small paragraph of "Safety Tips." There are illustrations for "Storm Surge" and "Tornado" and, while the former is instructive, the latter is simply a cartoon tornado knocking down a couple of trees. "Phone Numbers" are given on the facing page, including the two local energy companies, the Governor's Emergency Hotline, FEMA, and a few you'll need to fill in yourself, along with another "Hurricane Fact" mentioning the African origins of Atlantic tropical storms.

The upper half of page twelve is devoted to "WRAL.com, Your Internet Resource," which features the following:

- Interactive Hurricane Tracker that plots active and past storms
- Live Doppler 5000 radar updated every three minutes
- E-mail alerts when the National Weather Service issues a severe weather alert for your county
- Latest storm coordinates and projected storm path
- More than fifty maps showing national and regional radars and satellite imagery

There's a screen capture of the WRAL.com online tracker, which "allows you to see the paths followed by storms of the past and to track any new storms with up-to-the-minute position and wind speed data. Each point of the storm's track is plotted with a colored dot indicating the storm's intensity at that location."

Page fourteen brings us *The Relentless Season* (figure 5.3), with a brief textual account of the 2004 hurricane season and a corresponding "track chart" from the National Hurricane Center. The text begins with "One of the fiercest, deadliest hurricane seasons on record . . ." and concludes with "Our whole state is now considered part of hurricane alley," and that message is corroborated by the fifteen tracks swarming across the chart (all those red ones were of official hurricane status). With grim irony, however, one notes that the fifteenth and final of the 2004 season was a "letter O" storm, Otto. As we write this in mid-September 2005, just over halfway though this hurricane season, the North Carolina coast is being pounded by Ophelia.[6]

The *2005 Stormtracker* concludes with another full page ad on the inside rear cover and, on the outside, superimposed over an airborne hurricane photo of a big one that bleeds right off the edges, a few final words from (about) the WRAL Weather Center: "Over 100 years of forecasting experience. Enough said." Enough indeed.

Alright, that "100 years" tallies the combined experience of our current WRAL meteorological staff and is not to suggest that they've been forecasting since the Great Galveston Storm. Nonetheless, this is not your father's hurricane tracking map. It's a pretty sophisticated enterprise with a variety of illuminating lessons on a variety of hurricane related topics—ten major topics in all—and an array of supporting graphics—three bona fide maps and another four illustrative

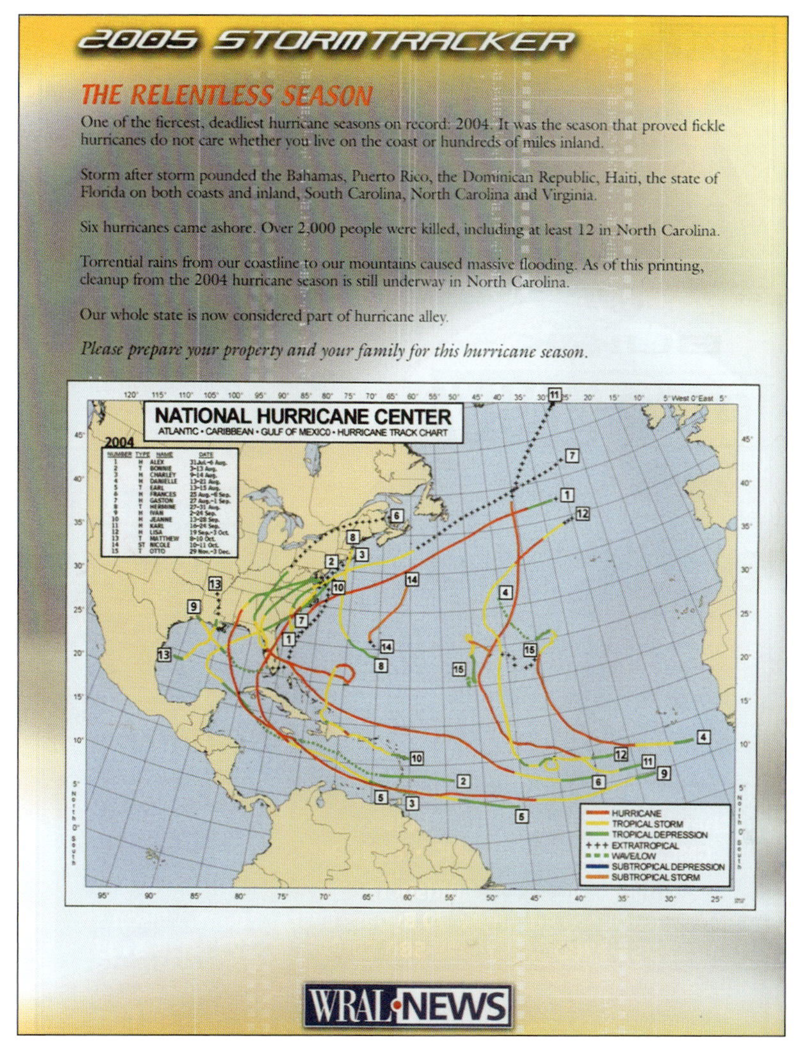

Figure 5.3
2005 Stormtracker map showing the paths of 2004 ("The Relentless Season") Atlantic storms.
Courtesy of WRAL-TV5, Capitol Broadcasting Company, Raleigh, North Carolina.

diagrams. Clearly, this document functions at the level of *myth*, in the same sense that we've employed that term in our previous works,[7] and it does that quite well. Individual maps, charts, and diagrams, each independent and sometimes complex sign constructs in their own right, are appropriated as signifiers of a much larger signified: the broad and deep expertise of WRAL's meteorological team, their state-of-the-art technological resources, their cumulative experience and unflinching objectivity, their capability to offer good advice before, during, and after weather crises.[8] Our observations are not meant to be sarcastic in any way; that's all very much to their credit. But, if anyone missed the point somehow, this should cement the deal. In a hotly competitive broadcasting business, having the best and biggest annual hurricane publication means having the biggest *cojones*, at least here in hurricane country.

And what sponsor or advertiser wouldn't want to get on board with that? Since this process of mythical appropriation consumes signs whole and without discrimination, every advertisement, logo, and slogan rises along with the maps, diagrams, and bulleted tutorials that provide the document's mythical buoyancy. Once formed, this is all of one piece. That's not to say that the WRAL staff heat and cool their

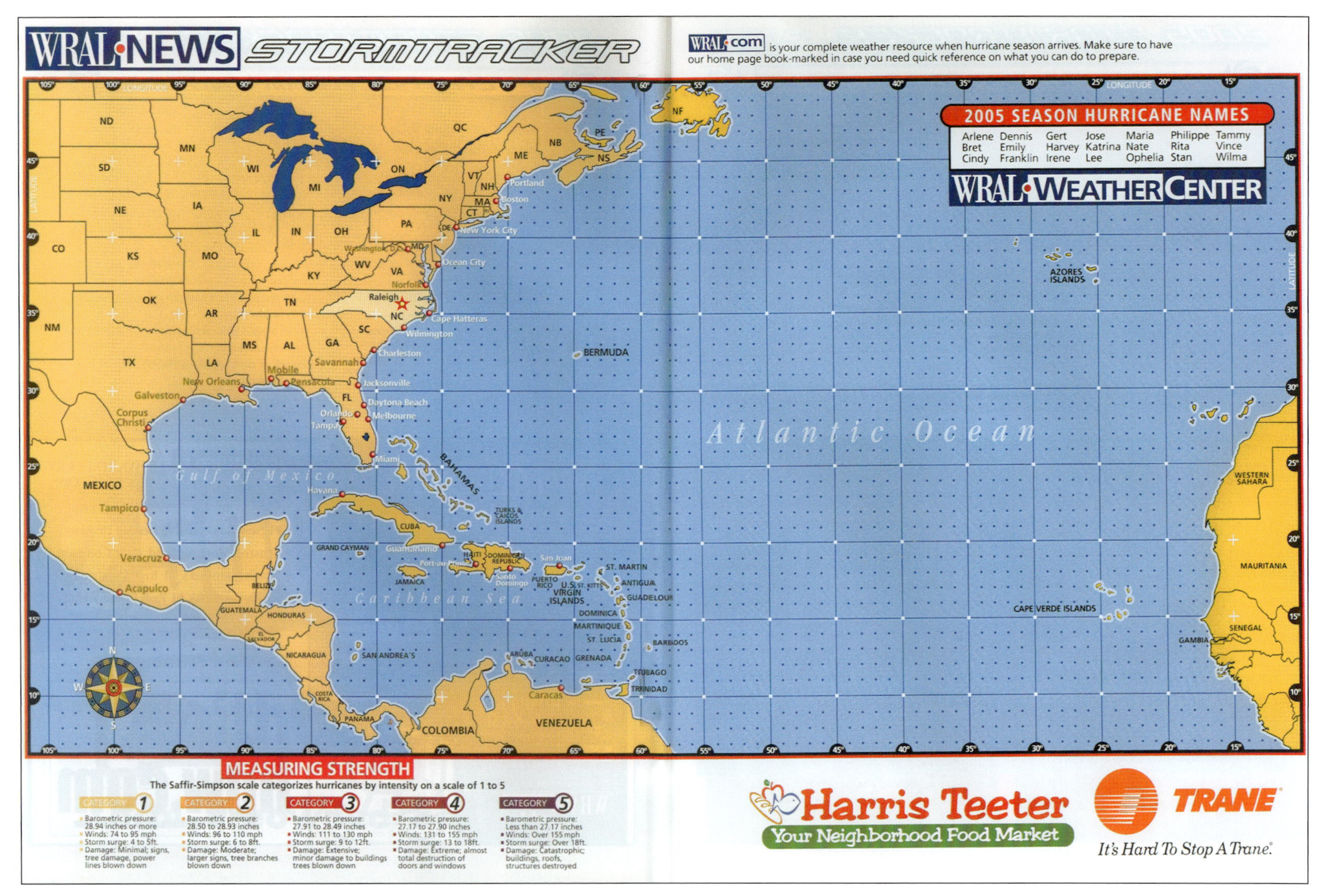

Figure 5.4
Hurricane tracking map from WRAL-TV's *2005 Stormtracker*. Presumably, this is meant to be drawn upon, gradually plotting the path of the approaching storm. But how many people really do that anymore?
Courtesy of WRAL-TV5, Capitol Broadcasting Company, Raleigh, North Carolina.

homes with Trane products or buy their groceries from Harris Teeter or their cars from Doug Henry or Sir Walter; it wouldn't matter at all whether or not they did. In the chaos following a major hurricane hit, does anyone really care where they buy food and water and ice—they just get these wherever and whenever they can—or at just which hardware megastore they're going to get that chainsaw and generator they wish they'd bought last week along with the gasoline to run them, or that stockpile of batteries or canned goods or anything else our Hurricane Survival Guide warned them to provision themselves with? Any hurricane tracking map is of no use after the fact, when the only lines that matter are the ones forming at the supermarkets and hardware stores and gas pumps. The thing's come and gone, for better or worse, and *gone* will do well enough for now, at least until the next time. So what's really the point of these maps in the first place if not to form the nuclei of supersigns claiming and demarcating territories of credibility among competing broadcasting enterprises?

But surely the tracking map itself has some pragmatic intent. We flip back, now, to our *2005 Stormtracker* center spread (figure 5.4), where the map sprawls handsomely across both pages. The Atlantic, Gulf, and Caribbean are awash in a soft blue and the landmasses, warmed with a buff orange in contrast to last year's cool green, stretch from Newfoundland to Venezuela, from Mexico to the Azores and Cape Verde Islands and the northwest coast of Africa. The gridded graticule is now subdivided with dots into one-degree increments and coastal cities, looking like the heads of cherry-red map pins, bead the shorelines from Portland to Corpus Christi and beyond to Veracruz, Acapulco, Havana, Guantanamo, Port-au-Prince, Santo Domingo, San Juan, and Caracas. In the upper right is the customary list of "2005 Season Hurricane Names," from Arlene to Wilma (as always no Q, U, X, Y, or Z), and below left is an outline description of the Saffir-Simpson scale categories—barometric pressure, winds, storm surge, damage. There's a rather fetching compass rose too, replacing last year's awkward harpoon north point and dysfunctional scale bar.

Can we actually plot hurricane tracks on this map? It is undeniably handsome and it wouldn't be an easy decision for map junkies like us to deface it with marks no matter how grave the occasion (maybe if we had a dozen or more copies . . .?). Still, the pregnant possibility of this or any other hurricane tracking map is *exactly* that it be marked up, incrementally posting the progress of the approaching storm in connect-the-dots fashion as it moves ominously closer. It's hardly the same just watching things play out on television or the Internet. This is, at least on the face of it, some kind of *popular participatory cartography*, a cartographic practice in which everyone is a potential cartographer of a sort, and they're participating in a real-world, real-time suspense story, the map connecting the double-wide or the beach house to the impending menace. Maybe *that's* really the essence of the hurricane tracking map, not mapping the space of the hurricane itself but mapping the ever diminishing space *between* the hurricane and ourselves, the space slowly and agonizingly eaten away as the *this/there* of *it* crawls closer to the *this/there* of *us* with each new posting until both become one and the same and the once imagined nightmare becomes reality and engulfs our lives (figure 5.5). And if all of that sounds gratuitously overdramatized, try living through it yourself some time.

Dramatic, yes, and exciting, and genuinely frightening, but this particular tracking map really *doesn't* want to be marked up. It's printed on a lightweight coated stock, a pretty slippery surface affording little traction for drafting pencil, and one more likely to be torn than to accept a trace of graphite. And it won't quite lie flat, at least not near the center "spine" where the two pages join, and it can't, as we've already observed, be extracted from the other pages. Maybe we could plot locations with dots of ink from a fiber-tip pen, one that won't smear or bleed, but then . . . how to connect them? Best, perhaps, to just leave this map alone.

Not all hurricane tracking maps are like this. Some really are intended as substrates for the plotting of hurricane tracks, like our 1999–2002 WPTF maps or a no-nonsense 2001 WTVD map or a 2001 Weather Channel issue, with its pastel-tinted and relief-shaded landmasses and even a set of instructions—"Track Like the Pros." Anyone who wants to plot hurricane positions certainly can. Coordinates and barometric pressure and wind speed data, collected by "hurricane hunter" aircraft and issued by the National Hurricane Center, are still routinely reported on broadcast and cable TV, and radio, and plotting these data on a map doesn't require the skills of a mariner or professional cartographer. But does anyone really track hurricanes themselves anymore, or has this become a semioccult practice like tying trout flies or building miniature ships in bottles? Certainly no one really *needs* to.

In addition to diligent surveillance by the hurricane hunters, armed with drop sondes and an array of instrumentation, and some deployed to forward bases such as Antigua in advance of approaching storms, hurricanes are nearly constantly monitored by a host of weather observation satellites (most notably those of NOAA's GOES system)[9] and, as they approach and make landfall, a plethora of ground-based radars as well. Weather data, in turn, are crunched by supercomputers using a variety of forecasting models (at least fourteen of them) to predict paths and strengths several days in advance. And all this is readily available on TV and the Internet (figure 5.6). Indeed, it's practically inescapable. (The Weather Channel, which began cable broadcasts in 1982, now has more than fifteen million viewers daily.)

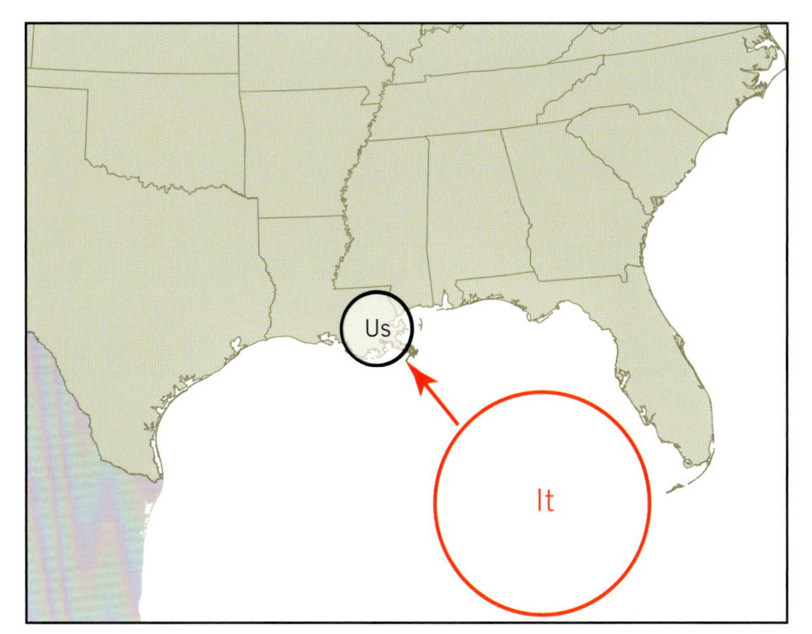

Figure 5.5
A hurricane approaches and makes landfall. The *there* of a major hurricane ("it") is on a collision course with the *there* of a major city ("us"). If its *there* consumes ours, the damage will be almost unimaginable.

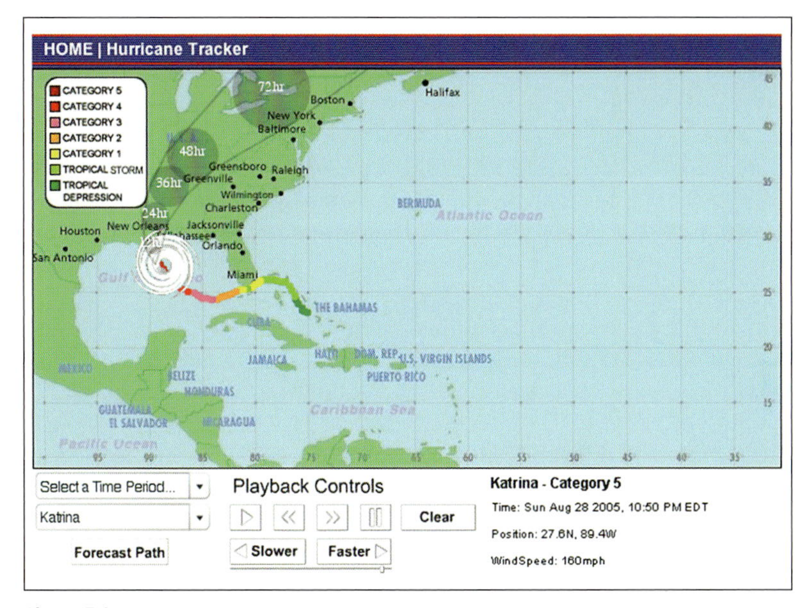

Figure 5.6
WRAL-TV.com's Interactive Hurricane Tracker. "The tracker allows you to see the paths followed by storms of the past and to track any new storms with up-to-the-minute position and wind speed data. Each point of the storm's track is plotted with a colored dot indicating the storm's intensity at that location." This Web site also provides "Doppler 5000 radar updated every three minutes, E-mail alerts when the National Weather Service issues a severe weather alert for your county," and "More than 50 maps showing national and regional radars and satellite imagery."
Courtesy of WRAL-TV5, Capitol Broadcasting Company, Raleigh, North Carolina.

There are no whale oil streetlamps lining our streets now, or mule-drawn streetcars clopping up and down them; and no one runs door to door shouting, "There's a hurricane coming!" We just turn on the television or computer and watch . . . and wait, as the grand drama unfolds in ready-made cartographic form, slowly playing out while we reprovision ourselves with stocks of candles, batteries, canned foods,

Figure 5.7
False-color infrared image of Hurricane Katrina—August 28, 2005.
Courtesy of NOAA, www.NOAA.gov.

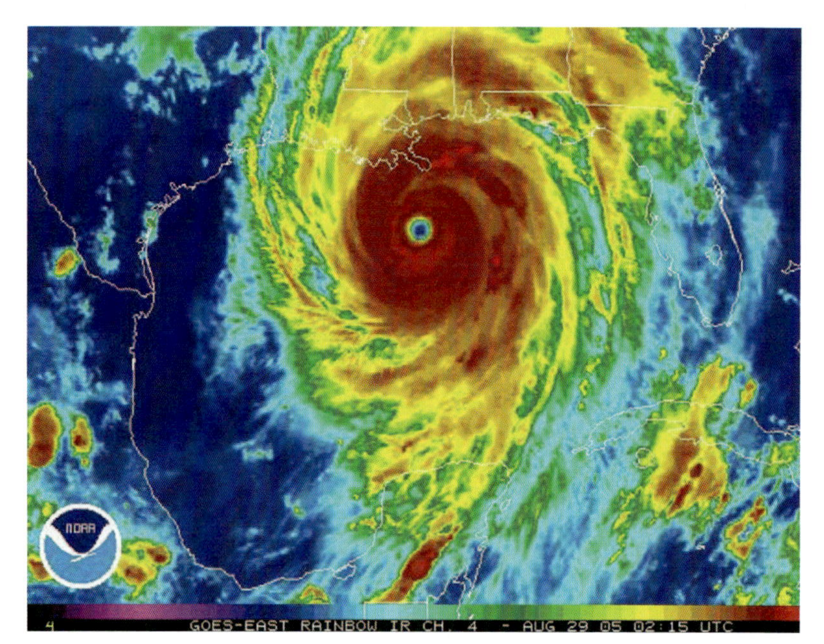

Figure 5.8
Rainbow infrared image of Hurricane Katrina—August 28, 2005. Given images such as these, it's impossible to escape the irony that something so magnificent can also be so terribly devastating.
Courtesy of NOAA, www.NOAA.gov.

and jugged water, and if probability begins to look like near certainty, gas up the family car and secure a stash of cash from the ATM. (See . . . we *were* listening after all.) It would be hard to imagine a form of maps about nature more intimately embedded in culture than these potentially are, but maybe the printed maps themselves are now just the splendid (variously, but the more splendid the better) centerpieces of a table otherwise set full of precautions, warnings, and advice. No one wants to hear unsolicited advice but, if that advice bookends an attractive (and free) map, it is made all the more palatable. And there is some latent geographic education lurking in the map as well. How many Americans really have a mental fix on where the Bahamas are, or Jamaica, or Haiti? If it's a matter of "them today, you tomorrow" though, then that's a very good reason to take some interest. Still, if no one actually *uses* these hurricane tracking maps for their presumed purpose anymore (and we wonder to what extent they ever actually did), why are they still being published? If their pragmatic function—the mapping of the collapsing space separating the monster storm from the minions, the shrinking distance between *there* and *here*, between *it* and *us*—has effectively been usurped by television and the Internet, then what function remains? What, of course, but the mythic.

Nowhere to run, nowhere to hide

For nature at its worst, though, check out the map that accompanied the July 1998 issue of the *National Geographic*. There the words "Natural Hazards" (in red) and "of North America" (in black) are bannered above a storm cloud that unfolds to reveal lightning forked against a lurid sky (figure 5.9). This, in turn, unfolds to a map of North American . . . *population? On a map of hazards?* Quickly the rest of the map is opened (figures 5.10 and 5.11) to reveal four small maps of North American earthquakes, hurricanes, volcanoes, and tornadoes with, in a final unfolding, hail and drought. These are complicated maps. Each is accompanied by a chart and there's a lot of text. In search of some sort of overarching structure, the map is flipped. The reverse gives us not the usual *National Geographic* map smothered in type but nonetheless a single large map of North America (with an unusually complicated-looking legend—this has to be the main map) that carries, in red, the title, "Great Disasters," followed by, in black, "Nature in Full Force" (figure 5.12).

Great disasters! It's like a *Guinness Book of World Records* map of disasters, that were selected, the text tells us, "for their power, their scope, or their effect on human lives." And here are Hurricanes Hazel, Diane, Camille, and Hugo, among others (unnamed Galveston, Andrew), the blizzard of '88, 1993's Storm of the Century, Mount St. Helens, the San Francisco earthquake (the big one, 1906), the Northridge earthquake, the burning of Yellowstone in 1988, the Paricutín and El Chichón volcanoes, the flooding of the Ohio River in 1913, the Great Tri-State Outbreak of tornadoes of 1925 (219 miles ripped apart in less than three and a half hours). The anecdotage is amplified in the surrounding text blocks, one per disaster, each adorned by a map, diagram, photograph, or painting, of Paricutín erupting, of the Mississippi breaching a levee (figure 5.13), of the destruction left by an arsonist's fire in Laguna Beach.

What to make of this? The map is flipped back to the front, where it can now be seen that each of the small maps plots the distribution of a single hazard (volcanoes, hail, drought). *Disasters* on the flip side, *hazards* on this. And in this side's lower left, a map of population (figure 5.14).

Population?

What's this map saying, that hazards plus population equals disasters? Could that be right? And indeed in the text block associated with the population map ("Population: Moving toward Trouble") we read that, "A hazard only becomes a disaster when it occurs where people live." Now we can take in the hazard maps, glancing back and

Figure 5.9
Cover of "Natural Hazards of North America," supplement to the July 1998 issue of *National Geographic*.

Figures 5.9–5.15 courtesy of Wood Ronsaville Harlin Inc./National Geographic Image Collection.

Figure 5.10
As "Natural Hazards" is unfolded, more reasons for anxiety are revealed.

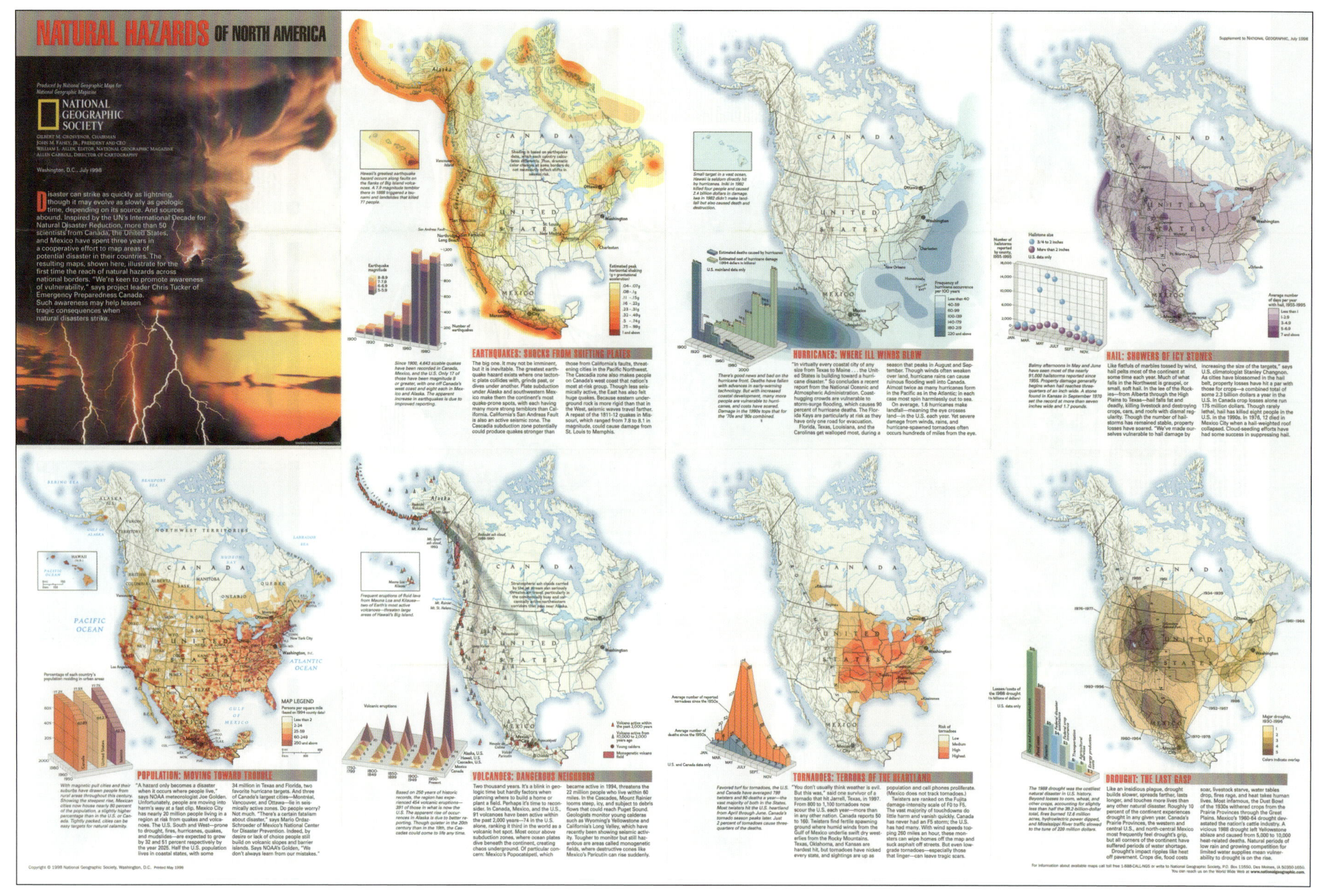

Figure 5.11
"Natural Hazards of North America," entire map. Map dimensions (WxH): 31in. x 20in.

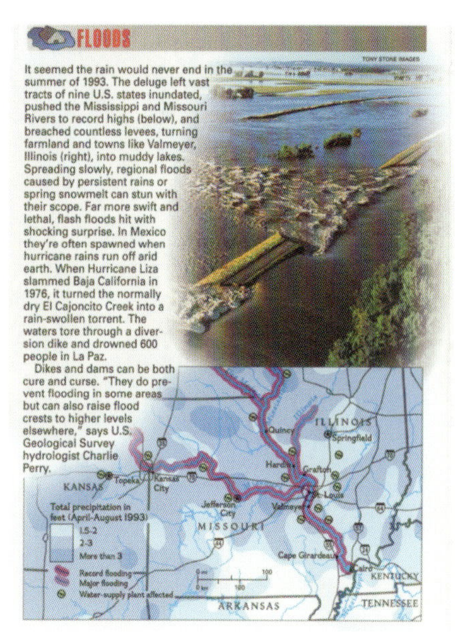

Figure 5.13
"Great Disasters"—detail of "Floods" vignette describing the devastation of various deluges, including the 1993 inundation of Valmeyer, Illinois, (pictured).

Figure 5.12
"Great Disasters: Nature in Full Force" (verso of "Natural Hazards of North America").

Hailstorms photo courtesy of Jim Wells, *Calgary Sun*; Hailstorms art, Earthquake art, Hurricane art, Landslides and Avalances art, Tornadoes art, Volcanoes art, and Wildfires art by Rob Wood, courtesy of National Geographic Image Collection; Earthquakes photo courtesy of Alan J. Oxley, ComPIX; Floods photo courtesy of Tony Stone Images; Hurricanes photo courtesy of John L. Lopinot, *Palm Beach Post*; Landslides and Avalanches photo courtesy of Provincial Archives of Alberta; Tornadoes photo courtesy of Howard B. (Howie "Cb") Bluestein; Volcanoes photo courtesy of Tad Nichols; Wildfires photo by Robert A. Eplett, courtesy of California Governor's Office of Emergency Services; Winter storms photo courtesy of CP Photo Assignment Services (J. Boissinot); Lightning storm over Tucson photo by Warren Faidley, courtesy of Weatherstock.

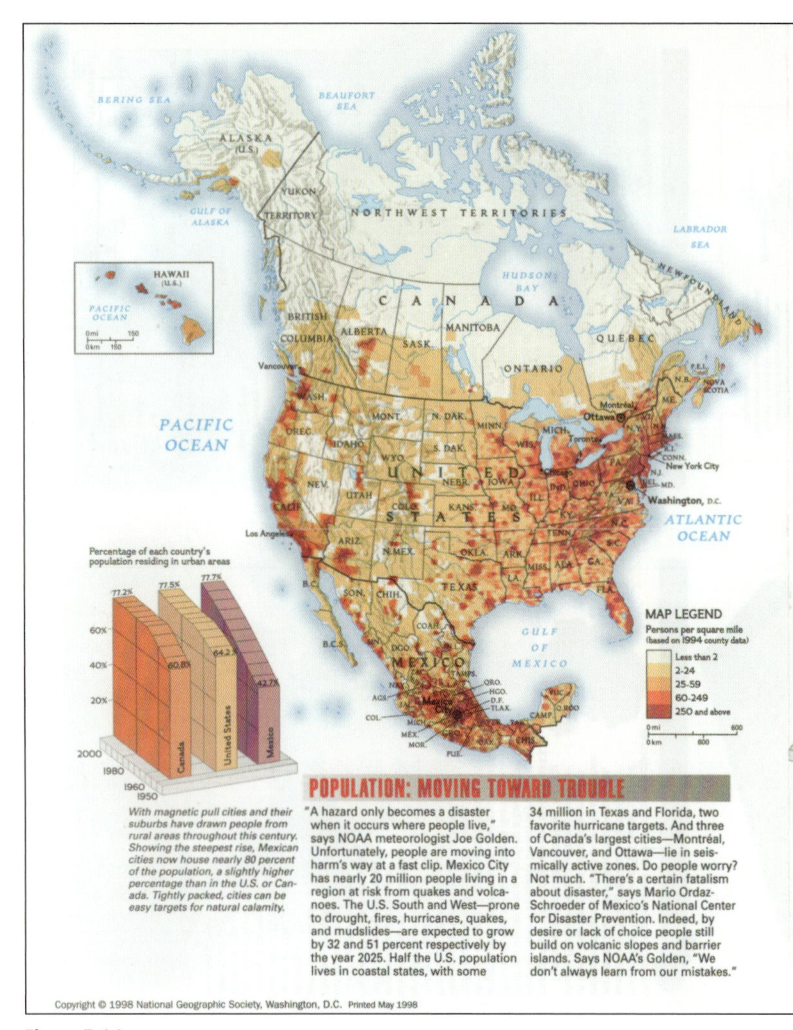

Figure 5.14
Detail of the population map in "Natural Hazards."

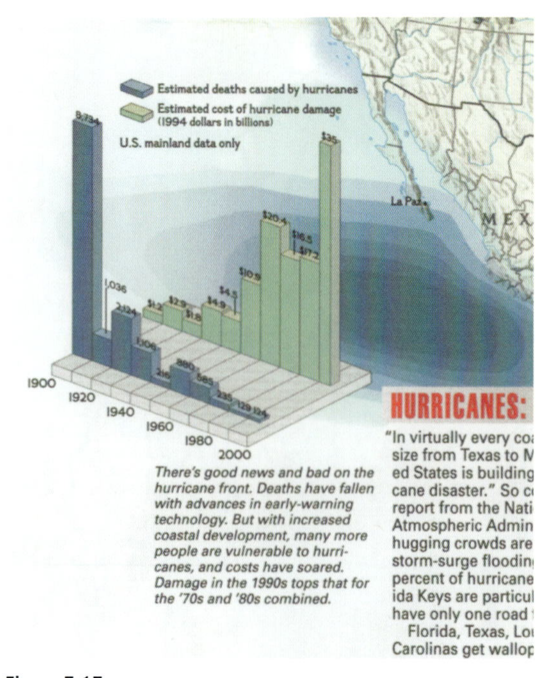

Figure 5.15
Close-up of hurricane chart in "Natural Hazards."

forth between them and the map of population. Finally we can flip the map back over. Look! That's where the great disasters have happened, where people live! Back on the front side we can look at the population map again. "People are moving into harm's way at a fast clip," the text tells us, detailing the growth of cities and population in the West and South, concluding, with a quote from NOAA meteorologist Joe Golden, that "we don't always learn from our mistakes."

Don't always learn from our mistakes. What? And so we continue to move into harm's way? Is that what this map is doing, under the guise of a Cook's tour of great disasters, predicting greater numbers of disasters to come? "The U.S. South and West—prone to drought, fires, hurricanes, quakes, and mudslides—are expected to grow by 32 and 51 percent, respectively, by the year 2025." What could this possibly mean but greater numbers of disasters?

The *Geographic* wouldn't say this. It reeks too much of divination.[10] Instead it says, "Do people worry? Not much," and quotes an expert who says, "We don't always learn from our mistakes." But what else could a reader conclude from a map of great disasters tiled with small maps of hazards and population?

In trying to reconstruct the reading of this map from a cognitive linguistics perspective, we need to acknowledge that while the title alone probably served to open a base space internally structured by the conceptual frame (North America), that meaning construction wasn't really begun until the perusal of "Great Disasters" on the flip side. The front side's seven small maps with their demanding charts—the maps in different colors and the charts in different forms and colors—is initially off-putting (figure 5.15). It looks like too much information to take in and so, we believe, it is initially passed over.

The flip side proves to be another story. Although it, too, is crammed with facts, with their accompanying narratives of tragic death, they are all readily organized under the concept "Great Disasters" (tiny text on the map tells us that El Chichón killed 1,800 people, El Cajoncito Creek flood killed 600, the Great Tri-State tornado outbreak killed 695). Despite its apparent elaborateness, the map turns out to reiterate a single story: *disaster strikes!*[11]

Back on the front side, the base space (<Great Disasters>) rapidly launches six child spaces, or belief spaces$_{2–6}$ (<Earthquakes: Shocks from Shifting Plates>; <Hurricanes: Where Ill Winds Blow>; <Hail: Showers of Icy Stones>; <Volcanoes: Dangerous Neighbors>; <Tornadoes: Terrors of the Heartland>; and <Drought: The Last Gasp>), but the opening of belief space$_7$ (<Population: Moving toward Trouble>) operates as a space-builder to set up what amounts to a kind of *if-then* argument, a predictive conditional. Here the predictive conditional is that *if* people continue moving into the paths of natural hazards, *then* there will be an increase in the number of great disasters; and the *if* calls for an evaluation of the alternative, *if-not*: if people stop moving into the paths of natural disasters, there will *not* be an increase in the number of great disasters:

> The essence of conditional predictiveness is the correlation . . . which allows conditional prediction of one event based on knowledge about the other. Normally, speakers and hearers [or in our case, map readers] assume a causal structure behind such a correlation: that is, two events which are

correlated strongly enough to permit prediction . . . are typically understood as correlated because of some causal relationship . . . . So a conditional prediction normally invites a hearer to imagine what models of the world would lead the speaker to believe in the correlation underlying that prediction. The reader could then draw on this causal model to explain the correlation . . . . In this sense, prediction brings along a causal model into the speaker's and hearer's mental-space structure.[12]

Here, this causal model is our understanding, articulated in the text attached to the population map, that "a hazard only becomes a disaster when it occurs where people live." Coupled with the assertion of population growth in hazard-rich regions, it is this text that sets up the predictive conditional.

As we saw in our discussion of "Australia Under Siege," the *National Geographic* has an aversion to speaking in the future tense. "Natural Hazards of North America" is no exception. Except for the statement that "the U.S. South and West are expected to grow by 32 and 51 percent by the year 2025," the map is set . . . well, it's not entirely easy to say when it's set. The temporal frame of "Great Disasters" is particularly hard to nail down. Earthquakes are reported since 1700, volcanoes since 1750, tsunamis since 1780, landslides since 1841, floods and winter storms since 1850, hurricanes and tornadoes since 1900, hailstorms since 1970. The hazard maps are more determinate: earthquakes and hurricanes between 1900 and 1998, droughts between 1930 and 1996, and hail from 1955 to 1995. But the volcano map plots activity during the past ten thousand years, and tornadoes are reported "since the 1950s." Doubtless some of this reflects vagaries of record-keeping,[13] but there's a sense of wallowing in the past that begs the question, "What time is this place?"

It's not the aorist of the standard *Geographic* political map, and it's certainly not the future, but then it's no simple past either. The use of "since" on Great Disasters implies all the time since 1700 (though only for earthquakes), since 1750 (for volcanoes), since 1780 (for tsunamis), and so on. This creates a *continuing* past of complexly varying depth whose density, however, increases toward the present (that is, more hazards are reported since 1970 than since any earlier time). This pressure toward a present sense is increased, on the hazard side, by the mapping of hurricanes and hail as frequencies, and tornadoes as "risk," and by the population map, which is presented as contemporary. *When is this place?* Given the strong probabilistic flavor, we want to say that it's—more or less—the present.[14]

The question then is, can you make a prediction in the present tense? Actually, there is ample precedent in English. In fact, not only does the future *will* ordinarily not occur in predictive protases (the *if* part of a predictive conditional), but such a "backshifted present" is actually a mental space builder characteristic of background clauses in predictive constructions. An example is, "If she doesn't stop, Miss Minchin will hear her." Of course, in our case there is no *if* and no future in the apodosis either (the *then* part of a predictive conditional) but "in spoken English and in less formal written registers, it is easy to find examples of present form with future reference in predictive apodoses as well."[15] Such uses *convey* without explicitly expressing

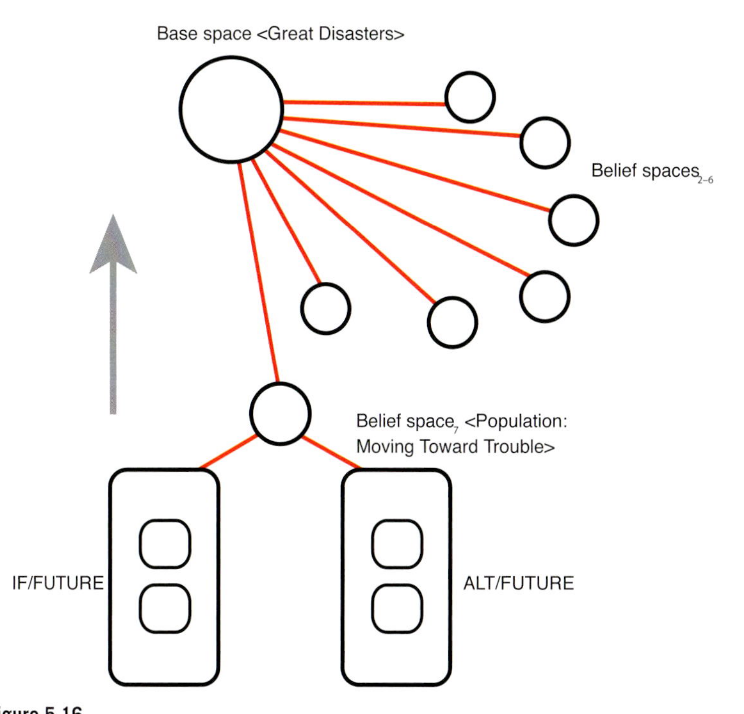

Figure 5.16
Cognitive structure of "Natural Hazards/Great Disasters." Once again, this is a diagram of only the gross network, with the frames and elements suppressed and the lines indicating both the transfer of structure and the projection of elements. "Great Disasters" opens a base space (<Great Disasters>), which rapidly launches six child spaces, belief spaces$_{2-6}$ (<Earthquakes: Shocks from Shifting Plates>, <Hurricanes: Where Ill Winds Blow>, and so on). However, the launching of belief space$_7$ (<Population: Moving toward Trouble>) operates as a space-builder that sets up a predictive conditional network. This network recruits a causal model from the text to "Population: Moving toward Trouble," and is "run" as a blend network is, generating outcomes for the if-then and the alternative if not-then conditions, and comparing them to generate a dire future given in none of the inputs.

predictions, especially in "warnings about negative eventualities"—for example, "If you keep this up, you're toast."[16] English can even dispense with the *if*: "You keep this up, you're toast." The effect, in fact, is to intensify the sense of threat. In this example, the predictive conditional relationship emerges from the construction itself, rather than from *if-then*, or present-future tense pairs.

This is precisely what we're proposing happens on "Natural Hazards of North America." The causal model is spelled out in the text to "Population: Moving toward Trouble"—as well as in the epimap[17]—but it is also constructed into the layout of the piece, with the hazards and population maps on the front followed by the disasters map on the flip side. The present tense of the hazards, and especially of the population map, prompt predictive thinking (with its comparison of alternatives), and the negative outcome is displayed on the map of great disasters, which thus turns out to be a map of what we can look forward to in the future (figure 5.16). In this way, the *Geographic* manages to convey an image of a very scary future while actually presenting nothing but facts about the past. More subtle than "Australia Under Siege," "Natural Hazards of North America" is a rhetorical tour de force.

If that's a reasonable sketch of the maps' cognitive structure, then what about their spatial logic? It could hardly be simpler, if arrived at somewhat indirectly via the maps themselves (but given more explicitly in their accompanying text): natural *disasters* are what we call the outcome of *intersections* between an ever expanding human population and the ever present (if sporadic and unpredictable) potential

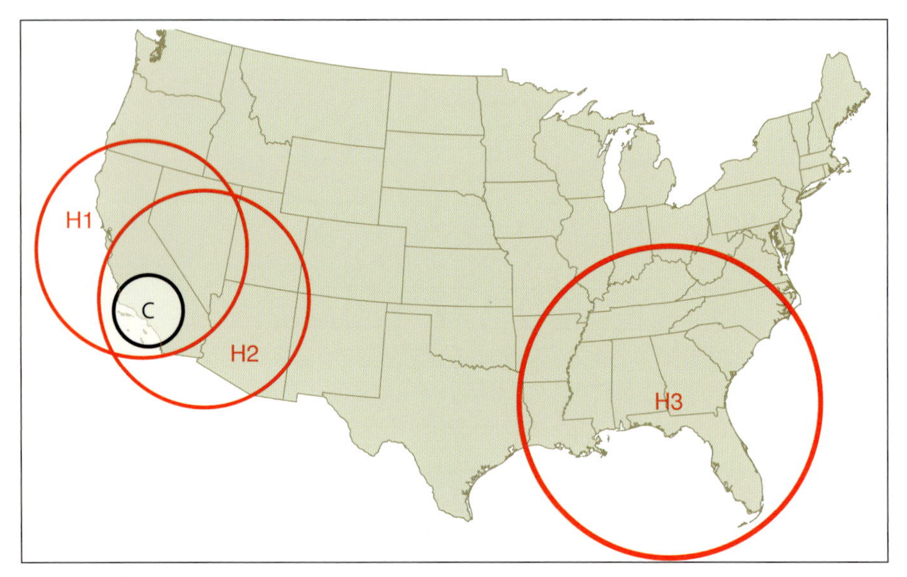

Figure 5.17
The intersection of culture and nature. In this diagram, the *there* of a major city such as Los Angeles (C) is subsumed by the *theres* of two regional hazards—earthquakes (H_1) and wildfires (H_2). The Atlantic and Gulf coasts of the U.S. are subsumed by another regional hazard—hurricanes (H_3).

for destructive natural events. With humanity dug in as widely and deeply as it is, disasters of many kinds are indigenous and inevitable. Just in the past few years, an entirely unexpected Indian Ocean tsunami has taken nearly a quarter-million lives and displaced almost ten times that many; an earthquake in Pakistan has killed at least 87,000 and injured another 69,000 and left 4 million homeless; Hurricane Mitch has killed 7,000 (with at least another 12,000 missing) in Honduras and Nicaragua, and nine years of nearly continuous volcanic eruptions have evacuated and virtually destroyed most of the island of Martinique, to mention only a few of the many natural disasters outside of the U.S. that make any of ours pale in comparisons of human cost.

The intersection of culture and nature—it's almost that simple. And that's largely what maps do—they *intersect* things; and they do that with greater efficacy and force than any other propositional form possibly could. This, in essence, is the framework of spatial logic that implements the conditional predictive in the map: "Here is where you propose to build. And here is the one-hundred-year floodplain. If you build there, sooner or later you'll be shoveling mud out of your living room." $Here_1$ and $Here_2$ in this example are, of course, the $This/There_1$ and $This/There_2$ of our propositional logic. The prediction operates in (through) the map because, at some level, both of those *this/theres* are the *same place*, not coextensively of course, but, at least to some degree, coincidentally. Here's another example. Let's say we stack up a map of human population and a few maps of hazard probabilities in McHargian-layer fashion,[18] or what we call in the GIS world "vertical integration." Let's say that the human or cultural "layer" is a map of metropolitan areas in Southern California and the natural hazards layers are maps of earthquakes, wildfires, and hurricanes. Clearly, the earthquake and wildfire maps intersect this population map (and, of course, one another as well),[19] and just as clearly the hurricane map does not. And it should go without saying that, if this were a population map of North Carolina instead, the reverse would be true (mostly). No doubt, this has been done many times in attempts to identify the most or least hazardous places to live in the United States (figure 5.17). (It is reported that one such effort, exact source and method unknown to us, found that the "safest" city in the U.S. was Gallup, New Mexico).

But this all depends on which criteria we choose—which *natures* or aspects of nature we're really worried about—and that inevitably makes any such enterprise a rather tentative one. In our example above, we neglected to include mudslides, didn't we? But we also neglected to include mountain lion attacks; and we'd bet money that that larger study of unspecified origin didn't include mountain lion attacks either, or rattlesnake bites, or . . .*Wait a minute! Those aren't natural disasters.* They're . . . well, just accidents, just . . . carelessness or imprudence on our part.

Like building reservoirs and hospitals above major geologic faults isn't careless and imprudent? Like destroying the delta marshlands that once offered some protection to the city of New Orleans wasn't careless and imprudent, some might even say reckless and irresponsible? Yeah, whatever . . . we'll just keep building the levees higher. "We don't always learn from our mistakes," says the map. No, we don't. [20]

Still, our rant doesn't address the issues of mountain lion attacks, or grizzly bear attacks, or killer bee attacks, or snake bites, or black widow or brown recluse bites, or the million and a half collisions with deer that occur in the U.S. every year. No, these are not natural disasters, not because they have nothing to do with nature in some way or because they do not demonstrate the requisite intersection of culture and nature, but because each is insufficient, in and of itself, to really qualify as an event rather than just an incident. In other words, disasters don't happen to individuals—accidents do, and tragic encounters do—but disasters happen to *populations*. Sure, it's obvious. That's why this *National Geographic* presentation has to include a population map. Without one of those, this wouldn't make much sense at all; it would be puzzlingly incomplete. The underlying algebra of Natural Hazard + Human Population = Natural Disaster would be lacking an essential term.

But there are also temporal limits of scale regarding what does or does not qualify as a natural disaster. A tornado's destruction might take place in minutes or less, and an earthquake's in fewer than ten seconds, while a snake bite or spider bite is virtually instantaneous. But, gruesomely, a wild animal attack can potentially occupy much more time than an earthquake or tornado. There is something missing from this determination and that's the causal agency itself, which in the case of the snake bite or animal attack is obviously *biotic*, while in the case of the earthquake or tornado is just as obviously *abiotic*. Sure, the lightning strike that killed half a dozen tourists huddled under a tree at Lake Apopka, Florida, was of abiotic origin, but it was also below the lower thresholds of spatial and temporal scale allowed a bona fide disaster. And maybe a single tornado, no matter how massive, doesn't fully qualify as a natural disaster either, but how about that 1925 tornado *outbreak*? *National Geographic* clearly thinks so, and we do too. If so, then what about a plague of locusts or the boll weevil or the chestnut blight, or an HIV-AIDS epidemic that kills more than 8 million annually or, for that matter, any of the great epidemiological catastrophes of human history? Maybe let's not go there or this will likely get even messier. Let's just conclude that what we take to be a *natural disaster* must be of abiotic agency and must meet certain minimal criteria of spatial and temporal scale and effect. And we're definitely not forgetting that such terms are cultural constructions in the first place, complicated nucleations of meaning with very fuzzy edges and subject to ongoing shift within their respective semantic fields.

As for the upper threshold of temporal scale . . . well, the Dust Bowl of the 1930s receives mention in "Natural Hazards" (under the heading of "Drought"), but the last continental glaciation, or any previous, is certainly not regarded as a natural disaster, nor apparently is the prospect of global climate change looming on our own temporal horizon. *Our own temporal horizon, our own temporal scale*; that's it really, isn't it? If the cities of Minneapolis and Chicago and Toronto and Montreal were to be ground into glacial till by a massive continental ice sheet two miles thick, that wouldn't exactly catch anyone by surprise. That would have been anticipated by dozens, maybe hundreds, of generations of cultural adjustment and accommodation.[21] A natural disaster has to take you by surprise, at least somewhat, even if we can map the flood swelling its way down the Missouri and Mississippi rivers for more than a month prior to its arrival. It's not that we couldn't see it coming; it's just that *it wasn't supposed to happen like that* . . . but it did. And that's exactly why a map supplement like "Natural Hazards/Great Disasters" has to operate in the realm of the probabilistic and conditional, and with an unavoidably ambiguous temporal position.

So in the end what makes it, or doesn't make it, onto the plane of such maps is a matter of whether or not they fit within certain spatial and temporal bounds predicated by the maps' makers, that they are singular events affecting a large number of people, and that their origins be of an abiotic rather than biotic nature. But, as this particular presentation makes unusually explicit, those potentials and those events of nature must, absolutely must, intersect significantly with human culture to meet the most primary qualification for participation. Could such a spatially complex proposition be convincingly advanced in any other than cartographic form? We very much doubt that it could.

Less firmer, more terror

Here we have another of the National Geographic Society's maps. This one is titled "The Earth's Fractured Surface" (figure 5.18), originally published as a supplement to the April 1995 issue of *National Geographic* magazine and its feature article on "Living with California's Faults."[22] The original supplement, which we have here, also has on its reverse a map of the U.S. West Coast titled "Living on the Edge," and presenting a multitude of regional tectonic features on a background of shaded relief.

"Earth's Fractured Surface" is nearly filled by a vividly colored map of the earth, shown in Winkel Tripel projection on a ground of silt-smooth calcareous gray. The surround is remarkably reticent for a *National Geographic* map. In the upper left is the title in a bold but unassuming Helvetica, followed by the Society logo. Below these are projection and scale statements and below those is the key for the map's hypsometric and bathymetric color schemes—seven hypsometric tints rising from green through yellow to tan and brown and six bathymetric tints sinking from light to dark blue, shown together in a segmented scale running from "More than 23,000" to "More than 16,400" feet.

In the lower left corner is the map's only summary text, a single paragraph that reads:

> Like the pieces of a giant jigsaw puzzle, slabs of rocky crust known as tectonic plates fit together to form the earth's outer shell. The puzzle changes as the plates slide over the hotter, softer rocks beneath them. Moving by mere inches annually, they reshape continents and ocean basins over millions of years by colliding, separating, and scraping past one another with relentless force. These interactions set off earthquakes, fire up volcanoes, and wrinkle the earth's crust into mountains, valleys, and deep-sea trenches.[23]

Immediately below the text is a small map sketching out the earth's sixteen major tectonic plates and, to its right, a legend for the main map presenting six tectonic features: Strike-slip Fault, Thrust Fault, Fracture Zone, Subduction Zone, Spreading Center, and Hot Spot.

In the upper right are keys to the "Notable Earthquakes of the 20th Century" and "Notable Volcanic Eruptions of the 20th Century," indicated in the map by small black circles (earthquakes) or squares (volcanic eruptions) containing the white numerals that place them in their respective lists. There are twenty-six earthquakes listed in chronological order, along with the year and magnitude of each event and the approximate number of resulting deaths. Twenty volcanic eruptions are enumerated, also in chronological order.

In the lower right is another legend, partly replicating the one in the lower left, showing Subduction Zones, Spreading Centers, Transform Faults, Earthquakes (twentieth century only), Volcanoes, and Hot Spots. In this legend, each feature symbolized is accompanied by a brief explanatory text: "When two plates collide, one often dives beneath the other—a process called subduction . . ."

The map itself is dressed in simple but forceful hues—the greens, yellows, and browns of the continents, the rich blues of the oceans —all underlain with a crisp (in most places) digital shaded relief credited to Peter W. Sloss of NOAA's National Geophysical Data Center. ("Anomalies in relief are due to inconsistencies in the availability and scale of the digital data.") This map is *all rock* and, despite its cheery colors, it looks *hard*. There are no countries, no rivers; there is no vegetation, no ice—just lithosphere, *just rock*, an earth stripped bare.

There are forty-six cities inconspicuously posted, and somewhat curiously chosen. There is a Montreal but no Toronto, a Vancouver but no Seattle or Portland, a New York but no Washington, D.C., or Miami, a London, Paris, Rome, Madrid, even Murmansk but no Berlin; there is a Denver but no Dallas, Houston, or St. Louis (despite its proximity to the magnitude-eight New Madrid quakes of 1811–12 that rearranged the course of the Mississippi River and rang church bells as far away as New England).

The major plates keyed out in the lower left-hand corner are announced in the map with bold, black Copperplate type while corresponding continents are modestly labeled in a smaller outline version of that same type face. With very few exceptions, only the

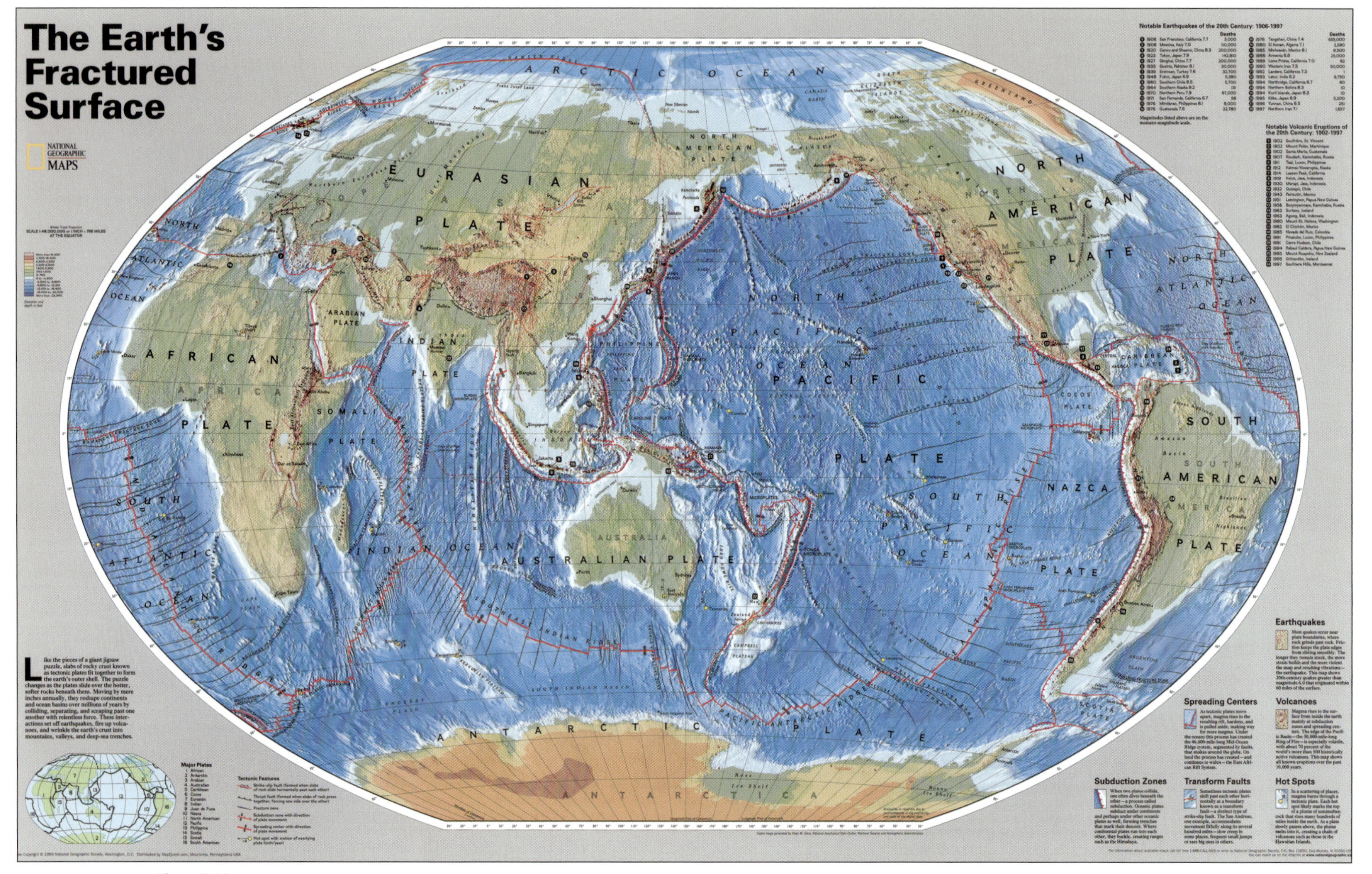

Figure 5.18
"The Earth's Fractured Surface," originally published as a supplement to the April 1995 issue of *National Geographic*, presents a frightening view of a dynamic and unsettled planet, a planet that produces, among other threats, earthquakes, tsunamis, and violent volcanic eruptions. Map dimensions (WxH): 36in. x 22in.

Figures 5.18 and 5.19 courtesy of National Geographic Cartographic Division/National Geographic Image Collection.

largest isles and islands are named: the British Isles, New Zealand, Madagascar, Cuba, Iceland, Greenland, Baffin Island, the Queen Elizabeth Islands, the New Siberian Islands, and a few other remote northern candidates. Labels for the four oceans (but no seas) are nearly as prominent as those for the plates, with which they compete for jurisdiction: the Pacific (North and South), the Atlantic (North and South), the Indian, and the Arctic.

On the terrestrial surface we find mountains and mountain ranges, highlands, plains, and basins. On the submarine surface there are ridges and rises, plains and plateaus, troughs and trenches, basins, sea mounts, escarpments, and continental shelves. Also labeled, but not depicted, are the Ross and Ronne Ice Shelves, and the Canadian Shield. The North and South Magnetic Poles are indicated, and the map is laced with a graticule marked out in ten-degree increments.

But the most salient features of the map are, of course, the *tectonic*. Hot spots are symbolized by yellow circles with projecting arrows indicating the motion of their overriding plates. Most, but not all, of the thirty-one hot spots here are also associated with volcanic eruptions, and a few—Iceland, Galapagos, and Hawaii-Emperor—with crowds of them. But it's the plate boundaries that really scream for attention. Bold, fiery-red lines tear across the blues, greens, yellows, and tans. On the oceans' floors, plate and microplate boundaries are demarcated by thick red lines—spreading centers, transform faults, and the barbed-wire rifts of the subduction zones. The zigzagging spreading centers along the mid-ocean ridges are stitched with hundreds of pointed gray lines—fracture zones—extending perpendicularly to either side. On the land, thinner red lines trace terrestrial faults—mostly strike-slip faults—and, in a couple of places

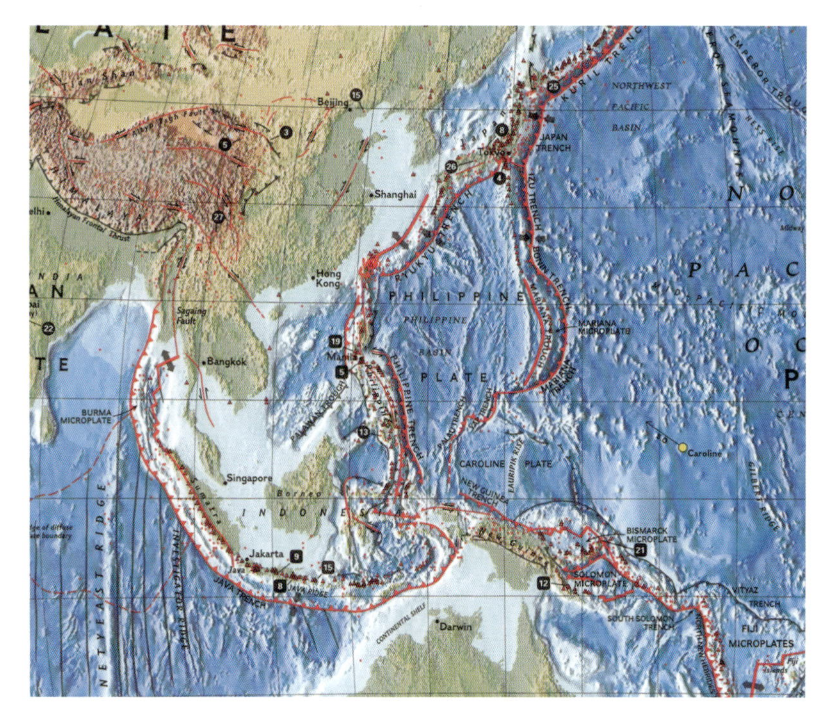

Figure 5.19
Detail of "Earth's Fractured Surface," showing the deep trenches along the western margins of the Pacific Plate.

like the East African and Baikal, these are plentiful enough to constitute a "rift system." Thrust faults are portrayed in a dark gray, and somewhat thinner, version of the barbed-wire line used to symbolize subduction zones on the ocean floors

Since the map's projection is centered on the 150th meridian, the Pacific and Indian Oceans are displayed unbroken. Earthquakes (circular red dots) and volcanoes (triangular red dots) swarm like angry ants around the edges of the plunging Pacific plates (figure 5.19), the "Ring of Fire"—along the Peru-Chile Trench, the Middle America Trench, the Aleutian, the Kuril, Izu, Bonin, Ryukyu, and Philippine, along the Mariana, Yap and Palau, New Guinea, South Solomon, New Hebrides, Tonga, and Kermadec. And there are other concentrations of seismicity—along the Java Trench and the East African Rift, around the Aegean Sea, in the Caucasus Mountains and the Hindu Kush, in the Yellowstone Basin and along the North American West Coast. The fractures in "Earth's Fractured Surface" tear across the map like throbbing scars. It's as if the entire globe had been hurled through an automobile windshield and provisionally stitched back together. It looks like a planetary Frankenstein monster, an uneasy assemblage of parts with all of its slices and sutures disturbingly conspicuous.

This torn and tortured earth—this *Frankenstein planet*—is not the earth we grew up with, not our generation anyway. Going through public school in the 1950s and '60s, there was but the slightest hint of the awesome global processes oozing and grinding far beneath our suburban streets. *Plate tectonics* was definitely not part of our vocabulary.[24]

We knew something about dinosaurs, of course; couldn't get enough of them, romanticized as they were in Conan Doyle's *Lost World* (essential reading for every adolescent male) and, thanks to Ray Harryhausen, terrorizing the local movie theater on Saturday afternoons. We couldn't possibly imagine, or care, that much of what we thought we knew about these *terrible lizards* would be proven nonsense within a few decades. And of course there were those fossils, so abundant in the limestone terrain of Missouri where Fels grew up. Just what *were* those things and how old were they and where did they come from? Our teachers, transgressing the prescribed boundaries of curriculum, told us that they were chrinoids, or trilobites, marine creatures that lived on the ocean floor millions of years ago. OK, but what were they doing in our schoolyard?

The instruction given offered no suggestion—at least none discernable to us—that the earth had ever been very different from what it is today. Of course, it had been differently populated. The chrinoids and the trilobites had come and gone, the dinosaurs had come and gone, and there were once warm shallow seas that nourished each in their day (Aha! Those fossils!). But that wasn't too much of a stretch. On the other hand, the notion that the configuration and global position of the continents themselves—so firmly and deliberately fixed in our young minds—could have been almost inconceivably different in times past? Now *that* would have been a stretch.

But then there was . . . *that map*. Mounted on the wall, rolled up above the chalkboard, were the ubiquitous classroom maps, canvas-backed and tightly wound on spring-loaded spindles—heavy, clumsy contraptions that surrendered their contents reluctantly like obese roller blinds. Grab a metal ring and you could pull down a map of the United States; grab another and you unroll a map of Europe; grab another and you get a map of . . . *The World*. But certainly not *this* map of the world, not the one glaring up at us from the table now.

Those maps were just the usual pedantic classroom cartography of their day (as if that's changed much)—the oceans, the continents, the crudely colored countries, major cities, major rivers, *et cetera, et cetera*—and in Van der Grinten or Winkel Triple projection[25] with North and South America on the left and Europe, Africa, Asia, and Australia on the right and Greenland looming uppermost like an overinflated air balloon. Still, as banal as this map was, it held a kernel of fascination that swelled each time the map was unfurled to reveal . . . what else but *those coastlines*. There in plain view, staring directly at one another over the expanse of Atlantic Ocean, were the African and South American coasts, their striking correspondence unequivocally displayed in the map as a simple and timeless *fact of nature*. There had to be something going on here. This couldn't be mere coincidence. (Can't everyone else see that too? Why does no one ever mention it?) It's so easy to just *will* the two continents together in the mind's eye—to slide them, all the way across the South Atlantic, together and consummate their union to form . . . well, to form *what*? And what happened to the South Atlantic Ocean? Maybe, just maybe, the earth wasn't always the way it is but, if that were possible, there were far more questions here than answers.

Some maps, and a mystery

Of course, such profound ideas aren't born in the minds of curious schoolchildren. They're born on the thresholds of discovery. If, in the 1950s, the ancient bond of the African and South American continents was apparent even to children, then why hadn't it been explained a

Figure 5.20
Cantino Planisphere, anon., 1502. This map portrays all that was known of the South American coast when the map was produced. The bold meridian passing through what is now Brazil separates the territories granted to Spain (west of the line) from those granted to Portugal (east of the line) under the 1494 Treaty of Tordesillas
Courtesy of Biblioteca Estense.

long time ago and incorporated as mundane fact in our textbooks? Because, in fact, *this was not fact,* not yet. The construction of plate tectonic theory as scientific and popular fact was just around the corner, but we didn't know that then. Nor did we know that it would be built largely on cartographic evidence.

The first piece of this evidence, the congruity of those two coastlines, had been compiled nearly five centuries earlier. It required three things: a reasonably accurate charting of the western coast of Africa, an equally credible charting of the eastern coast of South America, and their joint rendering in a map centered on the Atlantic Ocean.

As we also learned in school, credit for the mapping of Africa's Atlantic coast is officially given to the captains, navigators, mariners, and chartmakers of the "great Portuguese age of exploration." In the early 1400s, Portuguese ships visited the Madeira Archipelago, the Canary Islands, and the Azores, and traversed the African coast as far as Cape Bojador, just south of the Canaries, where the continent begins its westward bulge into the Atlantic. The Portuguese were the premier European navigators of their time. They knew how to fix their latitude using Polaris and the sun, and their charts, unlike anyone else's, marked lines of latitude. Cape Bojador was successfully rounded in 1434 and, by 1460, Portuguese exploration had progressed as far south as Sierra Leone, beyond the furthest extent of Africa's westward bulge. In 1473, Lopo Gonççalves's ship was the first European vessel to cross the equator.

Portuguese captains and navigators charted their way ever further down the coast, seeking a sailing route to lucrative trade with the East. Diego Cão found the mouth of the Congo in 1484 and, four years later, Bartolomeu Dias finally attained, and rounded, the southern tip of the continent. Passage to the East was achieved in 1497–99 when Vasco da Gama sailed from the Cape Verde Islands around the Cape of Good Hope, up the east coast of Africa to Malindi, and then across the Indian Ocean to Calicut.[26] Not long thereafter, an African *continent* began to appear in European maps.

What maps? The earliest extant map showing a complete African coastline is probably Juan de la Cosa's *Portolan World Chart* of 1500, which renders a somewhat truncated southern tip of the continent but follows the coast of the Indian Ocean as far as the Persian Gulf. De la Cosa had sailed with Columbus on his voyages of 1492 and 1493, and his map is also the first known to show discoveries in the New World, both those of Columbus and of the Cabots in 1497–98. A more accurate and detailed African coast is found in the *Cantino Planisphere* of 1502 (figure 5.20), drawn in Lisbon by an unknown chartmaker for Alberto Cantino, envoy for the Duke of Ferrara. This colorful parchment is one of the handsomest of its genre and, along with accurate

Figure 5.21
From Battista Agnese's *Portolan Atlas*, 1540. Even though the African coastline appears rather "squared-off" in this map, there is still a strong hint of its congruity with the coastline of South America. The blue line traversing the oceans traces the route of the Magellan expedition's voyage of 1519–22.

Source: Agnese, Battista. *Portolan Atlas, Oval Map of the World*, ca. 1540. Reproduced by permission of The Huntington Library, San Marino, California.

and exhaustively annotated European and African coastlines, it also charts the coasts of Madagascar, India, and Ceylon, the furthest eastward extent of Portuguese exploration at that time. With somewhat less certainty, it shows Newfoundland, Hispaniola, Cuba, a portion of Florida, and the northeastern portion of South America, in particular that part under Portuguese influence as established by the 1494 Treaty of Tordesillas. In this map, the rendering of the South American coast is incomplete and rather tentative, and generalized to a degree that recognizing its fit with Africa would have required considerable imagination, and that would be the case for another two decades. Nicola Caverio's *Planisphere* of 1505 is nearly identical to the *Cantino* and is almost certainly based on it. Martin Waldseemüller's 1507 *Universalis Cosmographia* shows the New World as two separate continents (which are labeled "America" for the first time) but they are made strangely distorted and narrow to conform to an underestimation of the circumference of the earth and an overestimation of the extent of Asia. Francesco Roselli's *World Map* of 1508 and Bernardo Silvano da Eboli's of 1511 also resemble the *Cantino* in that their portrayal of South America is limited to the coasts of the Caribbean and Brazil.

Exploration of the South American coast was, for the most part, a Spanish enterprise. During his third expedition in 1498, Christopher Columbus had made landfall on Trinidad and sailed briefly along the coast of what is now Venezuela, but which European first reached the Atlantic coast of South America is a matter of some debate. Credit is "officially" given to Vincente Yáñez Pinzón in 1500, but Amerigo Vespucci also participated in the exploration of much of coast south of the Caribbean in the first few years of the sixteenth century. He may have sailed as far south as the fiftieth parallel, just short of the continent's tip, and might have encountered the mouth of the Rio de la Plata, although other accounts place that much later, in 1516, and give credit to Juan Diaz de Solis. (Vespucci was well known as an energetic self-promoter.)

For exploration of the Río de la Plata and, of course, attainment of the tip of South America, credit is given to Ferdinand Magellan's voyage of 1519–22. The Magellan expedition's circumnavigation of the globe put to rest any remaining doubts about a spherical earth and, for enlightened Europeans, relegated the Ptolemaic view of the world to their dustbins. It also brought a South American continent into European maps for the first time. Diego Ribiero's *World Map* of 1529 displays—along with Europe, Africa, India, and Indonesia—Central America, the eastern coast of North America, and a nearly complete South America, all shaped much as they are in modern maps. Following Magellan, the opposing shapes of Africa and South America would be regularly fixed in European world maps: in Sebastian Münster's *Typus Orbis Universalis* of 1540; Battista Agnese's *Planisphere, Portolan Atlas* (figure 5.21), *Atlantic Ocean,* and *Magellan's Voyage of the 1530s and '40s*; Giacomo Gastaldi's *Universale Novo* of 1548; Pierre Descelier's *Planisphere* of 1550; Andreas Homen's *Portolan World Map* of 1559; Abraham Ortelius's *Orbis Terrarum, Typus Orbis Terrarum, Theatrum Orbis Terrarum* (figure 5.22), and other works in the latter part of the sixteenth century; Gerard Mercator's *Mappamonde* of 1587; and in many other maps produced by the European cartographic establishment of the sixteenth century.[27]

These European world maps, customarily centered on the Atlantic Ocean, must surely have invited speculation regarding the "fit" of Africa and South America (figure 5.23) and, not surprisingly, it seems that the earliest remarks on the apparent congruity of the two continents were made by a European cartographer. Abraham Ortelius, in the third (1596) edition of his *Thesaurus Geographicus,* wrote the following:

> The Greeks used to have the name Eumelus, giving a translation of the native language of Gadir, as Plato reports in the *Critias* or *Atlantis.* If this work is not a fiction, then Gadir or Gades will be the remaining part of the island Atlantis or America—which was not sunk (as Plato reports in the *Timaeus*) so much as torn away from Europe and Africa, by earthquake and flood—and accordingly will seem to be elongated toward the West . . . the vestiges of the rupture reveal themselves, if someone brings forward a map of the world and considers carefully the coasts of the three aforementioned parts of the earth, where they face each other—I mean the projecting parts of Europe and Africa, of course, along with the recesses of America.[28]

Ortelius's remarks are significant, not only in proposing that America was once conjoined with Europe and Africa, but also that their separation was brought about by geologic (in this case, cataclysmic) events. And he offers as evidence . . . a map. The notion that Europe, Africa, and the Americas were once part of the same landmass would persist, as would the idea that their separation was a sudden event of unimaginable scale and violence, quite possibly related to the Biblical flood or the destruction of Atlantis.

Two hundred and fifty years after Ortelius' obscure remarks, the great features of the earth were still attributed to cataclysmic events—the rifting of continents, the opening (or flooding) of oceans, the upheaval of mountains—events so compressed in time by modern standards that they could be scripted (well, actually have been) into Hollywood disaster films of Biblical scope. The earth was, and always had been, essentially as it was created and, if it had changed, it must have done so catastrophically at the hands of the Creator. Among Renaissance scholars, in fact, the Americas were widely interpreted as Plato's Atlantis, not sunk after all but thrust violently westward in an event that probably coincided with the Noachic Flood.

Drifting continents?

That was the view of nineteenth century geologist Antonio Snider-Pellegrini, who was aware of recent evidence regarding the correspondence of geological formations in Africa and South America. Snider-Pellegrini proposed that a single violent expansion of the earth's crust and massive upwelling of magma along a great north-south fissure drove the Old and New Worlds apart and created the Atlantic Basin.[29] Aside from the "single violent expansion" part, that's not far at all from a plate tectonic explanation. The science of geology was in its infancy in the nineteenth century and, in order to legitimize itself as a science, some explanation of the earth's major features—the existence and configurations of landmasses and oceans, the origin and arrangements of mountain chains—was required, and one not based on classical or Biblical mythology. One line of reasoning invoked an expanding earth, gradually rupturing and rifting its crust with the oceans filling between continents. (Frances Bacon had offered similar speculation two centuries earlier.) But this did little to explain the existence of mountains. And just where did all those ocean filling waters come from? Perhaps, as Bacon had mused, from the interior of the earth. Another line of reasoning, which appeared for a while to hold greater explanatory capacity, invoked a shrinking rather than expanding earth, could seemingly account for the existence of mountains, and did not require the materialization of vast amounts of water. A shrinking earth would shrivel like a desiccating apple, crinkling to form mountain ranges and chains, and splitting the crust to fragment continents and rearrange oceans.

But the increasing pace of geologic mapping was raising more questions than answers. Why was there such a marked difference between terrestrial rock, the granitic *sial* of the continental cratons, and oceanic rock, the basaltic *sima* of the seafloors, and what were marine sediments doing at the tops of some mountains? By the turn of the century, notions of both an expanding earth and a contracting earth had been largely supplanted by that of a rigid earth, one that neither swelled nor shrank as a whole but instead whose surface rose and

Figure 5.22
World Map, from Abraham Ortelius' *Theatrum Orbis Terrarum*, 1595. Although produced some fifty years after Agnese's map, the shape of South America here is noticeably less accurate. The fit of South America and Africa, however, is more credible.

Source: Ortelius, Abraham. *Theatrum Orbus Terrarum*, 1595. Reproduced by permission of the Huntington Library, San Marino, California.

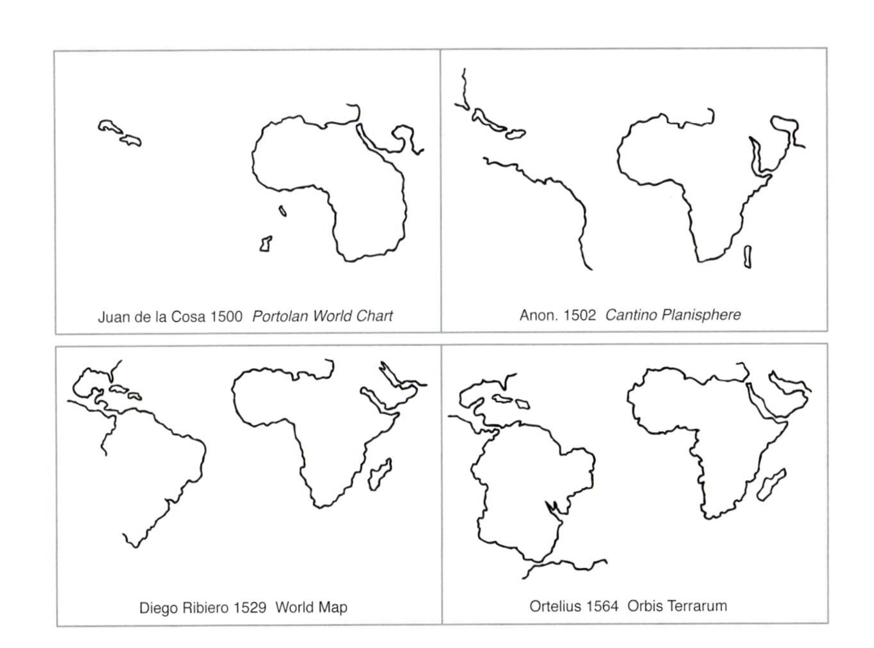

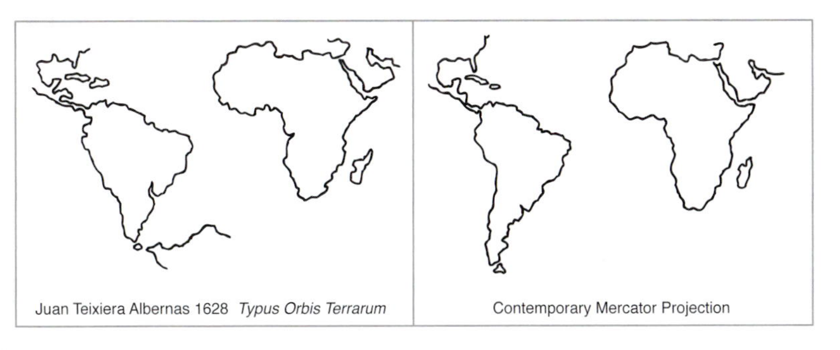

Figure 5.23
The cartographic evolution of the African and South American Atlantic coastlines, redrawn from a selection of maps created between 1500 and 1628.

fell. Mountains were gradually worn down through mass wasting and erosion and all that eroded material eventually found its way into vast geologic depressions called geosynclines. As enormous amounts of material accumulated in the geosynclines, its sheer weight depressed the earth's crust even further, pushing up mountains in other places. In essence, what goes up must come down, and vice versa, in a cycle of isostatic equilibrium. At last a plausible explanation for those enigmatic and pesky mountains. Here's a treatment of that topic from a 1955 geology text:

> . . . the excess volume represented by mountains is offset by deficient density beneath. We infer, then, that great mountain chains are approximately balanced against adjacent lowlands and continental masses against the deep-seafloors . . . that an overloaded part of the crust sinks, that parts lightened by long continued erosion rise, and that the adjustment between them is made by very slow flowage of rock in a deep zone.[30]

and,

> By this compression the rocks of the geosynclinal belt were squeezed into a smaller area. In other words, the crust under the mountain belt was thickened. How much thickening there was depends on the depth to which the compression reached, and this we do not know. Any considerable thickening must have lifted the surface, and we find no record that the sea has ever returned to any large part of the folded belt since the folding occurred more than 200 million years ago. [31]

Huh? Alright, it's really not so simple after all, but there are worse problems. In large part, montane geologic strata do not appear to have been just pushed or gently compressed upward but tilted almost on end, fractured and broken and in places even thrust over one another. And there are entire blocks of strata, sometimes very large ones, that bear no resemblance at all to anything else within thousands of miles. (J. Tuzo Wilson would later coin the term *terranes* for these.) And there are glacial deposits near the equator and coal beds in Antarctica. Maybe geosynclines weren't the answer to every large-scale geological question. Maybe if the earth's crust could move vertically it could move laterally as well.

It was against this prevailing background of rigid earth theory that German meteorologist Alfred Wegener proposed the first comprehensive theory of continental drift in 1912. Similar ideas had been expressed two years earlier by geologist F. B. Taylor,[32] but it was Wegener who constructed and promoted a coherent theory. In *The Origin of Continents and Oceans*, Wegener writes

> The first concept of continental drift first came to me as far back as 1910, when considering the map of the world, under the direct impression produced by the congruence of the coast lines on either side of the Atlantic.

And he adds that

> At first I did not pay attention to the idea because I regarded it as improbable. In the fall of 1911, I came quite accidentally upon a synoptic report in which I learned for the first time of paleontological evidence for a former land bridge between Brazil and Africa (1).

Former land bridges, even sinking continents, were the rigid earth theorists' explanation for the obvious similarities of species now residing oceans apart, but continental drift could also account for that, and more. Wegener advanced his argument on five fronts: geodetical, geophysical, geological, paleontological and biological, and paleoclimatic. His "Geodetic Arguments" cited observed increases in the displacement of Greenland from northern Europe over the previous hundred years or so, measured by astronomical observations and later by radio telegraphy. The longitudinal displacement of Greenland was found to be increasing between nine and thirty-two meters per year. "Geophysical Arguments" pointed out differences in gravity measured on mountains and at sea, suggesting that continental crust is lighter and less dense than seafloor crust and affirming the possibility that continental crust could "float," and therefore move laterally, on the underlying magma. Wegener opened his "Geological Arguments" stating,

> By comparing the geological structure on both sides of the Atlantic, we can provide a very clear-cut test of our theory that this ocean region is an enormously widened rift whose edges were once directly connected (61).

And he went on to present geological maps of South America and Africa and a "Tectonic chart of Gondwanaland" reuniting those two continents, along with Antarctica, Australia, and India[33] (figure 5.24). Geologic similarities of structure, rock series, and fossil content were described, and corroborated by fifteen specific examples drawn from South African geologist Alexander du Toit's *A Geological Comparison of South America with South Africa*.[34]

His geological arguments are probably the most persuasive, but he followed with a chapter's worth of "Paleontological and Biological Arguments" presenting observations from the perspectives of paleontology, zoogeography, and phytogeography, and another of "Paleoclimatic Arguments," including, among others, a map showing indications of carboniferous glaciation in eastern South America, the Falkland Islands, southern Africa, India, and Australia,[35] along with a hypothesized carboniferous equator and south pole. Wegener's reconstruction of the ancient configuration of continents not only reunited South America with Africa but also with Antarctica, Australia, and India to form the ancestral *Gondwanaland*, and ultimately with the northern continental masses of Europe and Asia (*Laurasia*) in an even more ancient *Pangaea* (figure 5.25), an all-embracing union surrounded by a single global ocean called *Panthalassa*.[36]

If Wegener's geodetical and geophysical arguments persuade us to mentally push continents apart, then his geological, biological, and paleoclimatic arguments persuade to pull them back together again, back into their former positions and unions. In terms of the propositional logic of the map, that is straightforward and incontestable. If two *thises*—instances of, say, geological formations or faunal or

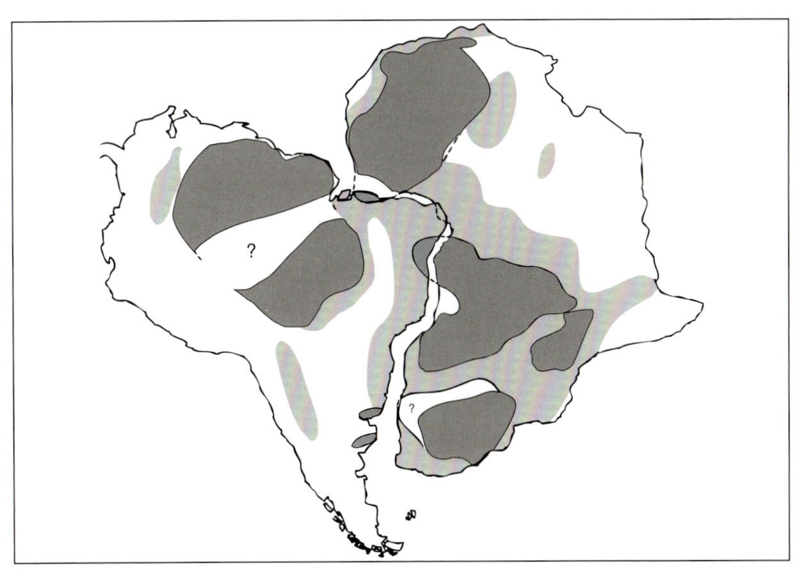

Figure 5.24
Africa and South America reunited, showing geology. Evidence that the two continents were once one is provided by the geologic formations that form continuous units when the continents are brought together. The dark gray areas are ancient cratons, the nuclei of the continents, and the light gray areas depict somewhat younger rocks.

Data source: Hurley, 1973.

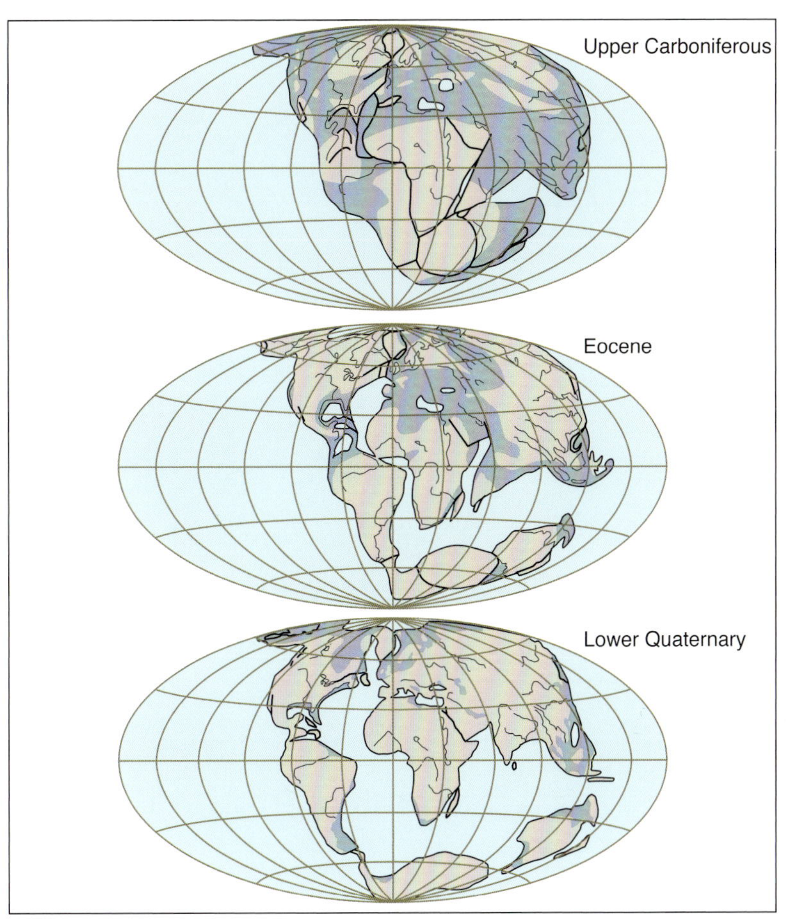

Figure 5.25
Alfred Wegener's reconstruction of the map of the world for three epochs according to continental drift theory. Light blue denotes oceans and dark blue shallow seas. Present-day outlines and rivers are provided for reference and the graticule is arbitrary.

Data source: Wegener, 1966.

floral populations—can be argued to have formerly been one despite their present spatial separation, then their reunion must invoke a spatial transformation of not just the extent but also the location of their corresponding *theres*. But those *this/theres*, those geological formations or biological populations, can't just be cleaved off from their larger encompassing *this/theres*, the African and South American continents, so the pulling of those two instances back together must also drag whole continents along with them, pulling those back together as well (figure 5.26).

Straightforward and incontestable . . . or maybe not. Over two and a half decades, continental drift gained significant support in Europe and South Africa, with du Toit and Scottish geologist Arthur Holmes among its most vocal advocates. In the United States, however, the theory was firmly resisted by most of the geological establishment, Harold Jeffreys in particular, and years of vigorous and often acrimonious argument culminated in a 1928 symposium of the American Association of Petroleum Geologists where the topic was the center of debate. The apparent fit of continental margins was dismissed as inaccurate, geologic similarities either side of the Atlantic as inconclusive, glacial deposits at unlikely latitudes as misidentified, and biological kinships as the result of former land bridges. Wegener himself, a meteorologist (with a doctorate in astronomy), explorer, and accomplished balloonist rather than a certified geologist, was dismissed as an "amateur" and the rigid earth proponents for the most part remained, shall we say, unmoved. What was being threatened was not the well being of humankind but the sanctity of prevailing ideology and the credibility of prestigious academic careers. A theory of continental drift was too ambitious, too unifying; it was a theory in search of evidence instead of the other way around and it offered no convincing causal mechanism that would drive entire continents through oceanic crust like giant snowplows. Wegener died two years later during a Greenland expedition and, in North America, the debate gradually subsided in favor of rigid earth theory (hence those passages from our 1950s textbook). But, in Europe, geologists such as Holmes and Emile Argand and, in the southern hemisphere, du Toit and, later, S. Warren Carey, continued to favor continental drift and to gradually accumulate supporting evidence for the theory.[37]

Plate tectonics . . . finally

Following World War II, new techniques for measuring and mapping began to reveal perplexing new evidence in fields of study as diverse as paleomagnetism, the topography of the ocean floors, and the structure and workings of the earth's interior. In the early 1950s, British physicist P. M. S. Blackett invented the astatic magnetometer, an extremely sensitive instrument for measuring trace magnetism, and it was found that the magnetic polarization of ferric rocks from around the world pointed to very different locations of the earth's magnetic poles. The notion arose that, throughout geologic history, the poles had "wandered" and even periodically reversed, but that alone could not account for such widely varied magnetic alignments in rocks. Not only must the poles have moved but the continents must have moved as well. By 1956, Blackett, Sir Edward Bullard, J. Harpers, and S. K. Runcorn had found that the various directions of geomagnetic alignment could be assembled into a consistent pattern that not

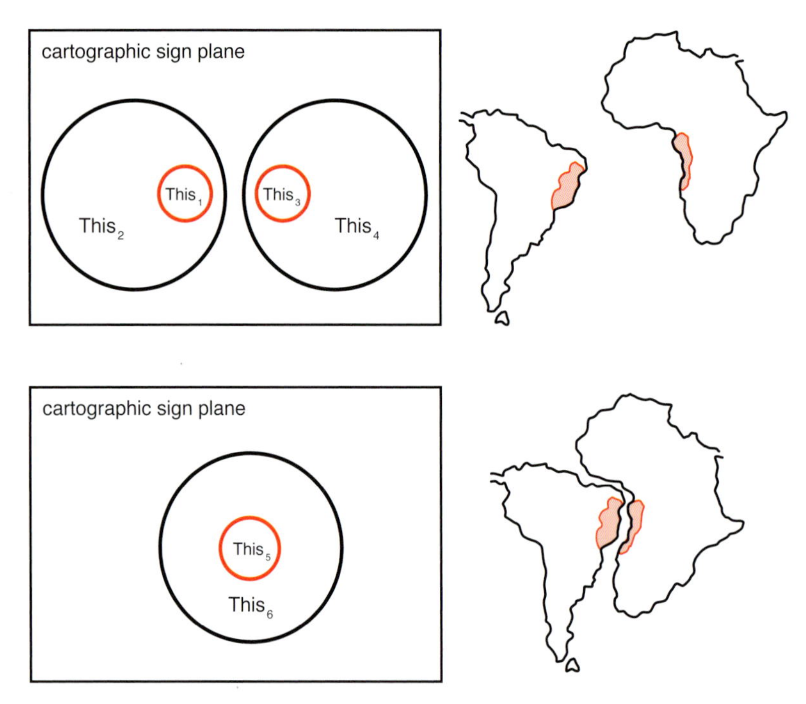

Figure 5.26
Propositional logic diagram of Africa and South America reunited. Here, two geologic formations an ocean apart (*This$_1$* and *This$_3$*) are recognized as having once been part of the same formation (*This$_5$*). Reuniting the two geologic formations inevitably reunites the continents (*This$_2$* and *This$_4$*) in which they are found.

only suggested that continents had moved relative to the poles and to one another but also gave some hint to their positions in times past.

A primitive sonar technology had been employed during the war and, in the 1950s, the U.S. Office of Naval Research secretly sponsored its further development and also an ambitious program to use this technology to map all the world's ocean floors—potential battlegrounds for future submarine warfare. In 1958, the first ocean floor maps were published, the outcome of several years of oceanographic surveys under the direction of Bruce C. Heezen, Marie Tharp, and Maurice Ewing of Columbia University's Lamont (later Lamont-Doherty) Geological Observatory, and they revealed a startling and mysterious feature. Midway down the Atlantic Ocean, from north to south, ran a prominent and nearly continuous ridge jagged with offsetting fractures (those scars and stitches on our Frankenstein planet), and down the center of that ridge was a nearly continuous rift that in places was actively volcanic (figure 5.27). And that feature was not exclusive to the Atlantic but appeared in all oceans, a mid-ocean ridge-rift system totaling some 47,000 miles in length.

In 1962, Harry Hess of Princeton University published "History of Ocean Basins," adopting ideas from British geologist Arthur Holmes, Dutch geophysicist Felix Vening Meinisz, and Australian geologist S. Warren Carey, and proposing that convection currents in the underlying mantle brought new material to the surface at the mid-ocean ridges and conveyed oceanic crust away from the ridges in both directions, creating and expanding the seafloors. Robert S. Dietz of the U.S. Coast and Geodetic Survey had also reached this conclusion at about the same time and called the process "seafloor spreading." Hess's and Dietz's seafloor spreading could potentially move, and even split, continents and, according to this new paradigm, continents did not move independently as continental drift suggested but as "rafts" frozen into and moving with vast sections of ocean floor.

Frederick Vine and Drummond Matthews of Cambridge University proposed an elegant proof of seafloor spreading. If new oceanic crust is being created at the mid-ocean ridges and moving away in opposite directions, and if trace magnetism in the crust recorded periodic reversals of the earth's magnetic poles (by then an accepted notion), then mapping of the magnetic alignment of that crust should reveal a striped or "zebra" pattern lying parallel to the ridges (figure 5.28). Also proposed independently by Canadian geophysicist Lawrence Morley, the Vine-Matthews-Morley hypothesis found evidence waiting in a study of seafloor magnetism off the northwest coast of the U.S., where Ronald Mason and Arthur Raff of the University of California's Scripps Institution of Oceanography had observed exactly such a pattern of alternating normal and reversely magnetized seafloor. In seafloor surveys conducted by James Heirtzler and others at Lamont, similar patterns were soon found in other areas of the Pacific Ocean, and in the Atlantic and Indian oceans as well.

Another proof of seafloor spreading might be found in the sediments covering the ocean floors. If new oceanic crust was being created at the mid-ocean ridges and moving away from those ridges over time, then precipitated sediments accumulated on the crust should be both thicker and older with increasing distance from the ridges. It was recognized that seafloor sediments were thin to nonexistent near the mid-ocean ridges and thicker at greater distances (but nowhere miles thick as would be expected if the oceans were billions of years old), and this was confirmed within a few years by sediment cores collected during the *Glomar Challenger*'s deep-sea drilling expeditions, which demonstrated that, not only were those sediments thicker at increasing distances from the ridges, but also older according to the fossil records they contained.

And improved techniques for the radioactive dating of rocks, the measurement of strontium 87 resulting from the decay of rubidium 87 and of argon 40 from potassium 40, indicated that, not only were seafloor sediments older with increased distance from the mid-ocean ridges, but the underlying oceanic crust was older as well. If the ages of ocean crust at various locations could be correlated with their distances from the mid-ocean ridges, then estimates of the rates of seafloor spreading could be calculated. Edward Bullard and his colleagues at Cambridge determined that rates of seafloor spreading either side of the ridges ranged from as little as one centimeter per year to as much as eight centimeters and that it was fastest in the Pacific basin and slowest in the Atlantic. The rates and directions of seafloor spreading were in general accord with continental drift theory and implied that the opening of the Atlantic Ocean began less than 150 million years ago.

But if the creation of new oceanic crust was not a relatively recent geologic phenomenon, then what had happened to all of that material potentially accrued throughout earth's history? At the time, geologists and geophysicists fell into several broad camps. The primary disagreement was between fixists and mobilists, with the former adhering to the principle of *permanence*—that the configurations of continents and oceans were essentially as they always had been (which did not broadly exclude any general global expansion or contraction)

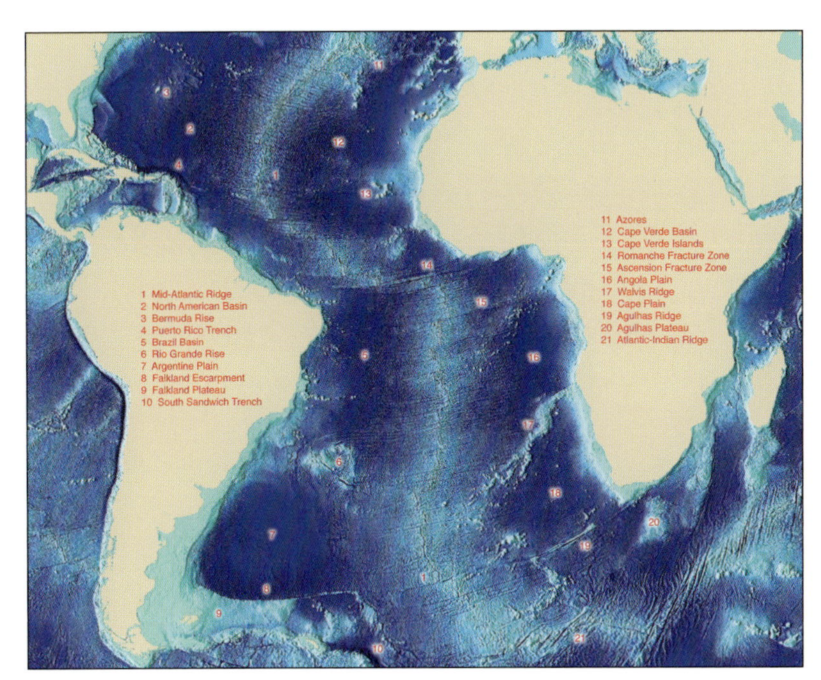

Figure 5.27
A contemporary rendering of the South Atlantic Ocean floor, showing significant features.
Image from ETOPO2 data, NOAA/NGDC; map created by John Fels.

Figure 5.28
Magnetic striping on the ocean floor. This map shows a portion of the Pacific Ocean floor off British Columbia, Washington, and Oregon. Stripes denote areas of anomalously high field strength while straight lines represent faults offsetting the pattern of anomalies. Remnant geomagnetism on the moving ocean floor provides a record of past reversals of earth's magnetic field.
Data source: Raff and Mason, 1961.

—and the latter believing that continents could move and that oceans could open and close. Among the mobilists there were expansionists, who held that the opening of oceans could be explained by a rapid expansion of the earth; drifters, who remained in basic agreement with Wegener's theory; and others who agreed with Carey's and Hess' views that ocean basins expanded from the mid-ocean ridges. The expansionists' view was a difficult one, since the formation of the present oceans would require about a 75 percent increase in the earth's radius over little more than 100 million years. Still, if the earth hadn't expanded and if much of the oceanic crust created over time was no longer with us, then where did it go? Early seismic studies of earthquakes associated with Pacific island arcs and their associated deep sea trenches had found that earthquake foci were concentrated on a roughly forty-five-degree plane extending from the trenches downward as deep as 450 miles into the earth's interior. Seafloor crust was being bent and pushed under the lighter continental crust by tremendous force, much as Hess, who was familiar with those studies, had speculated. *Subduction*. Oceanic crust was a global "conveyor belt" that could indeed convey continents across the earth's surface.

With that understanding, the picture was pretty much complete and the new paradigm of seafloor spreading became the even newer paradigm of plate tectonics. (Tuzo Wilson is generally credited with putting the "plate" into "plate tectonics" and the first integral construction of the theory's principles, along with its corollary "hot spot" theory.) Convection currents in the mantle, caused by radioactive heating from the earth's interior, were forcing up new seafloor crust along the mid-ocean ridge-rifts, *spreading centers*, and that caused the lateral displacement of older crust and the opening and expansion of ocean basins. But rather than expanding the earth's entire surface, crustal materials were eventually thrust back into it in the deep-sea trenches fronting island arcs, the *subduction zones*. The earth's crust was not of one piece—differences in continental and oceanic composition aside—but an assemblage of *plates* (at first, there were believed to be about six, but now at least sixteen major plates are recognized), the largest of which, except for the Pacific, include embedded and entrained continents that move along with them. And where plates are not subducted beneath other plates and melted into the magma of explosive island arc volcanoes, or driven against one another to wrinkle up massive mountain chains, they can grind alongside one another with the friction of their passing generating earthquakes and tsunamis. If this is a dangerous planet, it is so because it is a *dynamic* planet, much more so than a now quiescent Mars, for example. (Not surprisingly, the USGS counterpart to our *National Geographic* piece is titled *This Dynamic Planet*[38]).

Here was the causal mechanism that Wegener's theory lacked, not simple to be sure (and, from an early twentieth century perspective, it would have seemed more than a little fantastic anyway), but it accounted for much and, unlike Wegener's theory, it had the solid stamp of empirical legitimacy. But what about the refitting of continents? Before World War II, French scientist Boris Chambert had constructed a tentative fit along the margins of continental shelves (insofar as they were known prior to Heezen and Tharp's maps) rather than their more transient shorelines. But in a 1964 Royal Society symposium on continental drift, Bullard, J. E. Everett, and A. G. Smith presented a computer-generated matching of southern and northern Atlantic continents fit along the five-hundred fathom bathymetric contour, which they had taken as a surrogate for the shelf margins (figure 5.29). Their continental fittings assumed the principle of Euler's Theorem (from the nineteenth century French mathematician), which stated that the spherical rotation of two bodies around two different axes could be mutually defined as a single rotation around a single third axis. The resulting match of continental margins was undeniable, with less than a one-degree mismatch along most of the boundaries, and not just between South America and Africa but between North America, Greenland, and Europe as well. Of course the eastward bulge of South America could be snuggled into the broad recess of western Africa, and it suddenly seemed rather silly ever to have doubted what had been in plain view—right there *in maps*—for more than four centuries.

In the span of little more than a decade, from about 1955 to 1968, understandings—no, *beliefs*—about planetary geology had been transformed from a speculative theory of continental drift to one of seafloor spreading (the real proofs had lain under the oceans all along) to a globally unifying theory of plate tectonics. Believers in an expanding earth or a shrinking earth or a rigid and forever fixed earth, in permanence or great Atlantic land bridges or dramatically sinking continents were suddenly very hard to find. Time for the wholesale rewriting of textbooks. In the United States, several key meetings sealed the deal: the 1966 annual meeting of the Geological Society of America, a NASA sponsored meeting in New York later that year, and a 1967 meeting of the American Geophysical Union. At the New York meeting, not a single fixist could be flushed out to take stage among the other presenters, who piled up converging evidence like geologic strata, and the AGU meeting five months later cemented what Robert Dietz called "the total and instantaneous conversion of the American community to continental drift."[39]

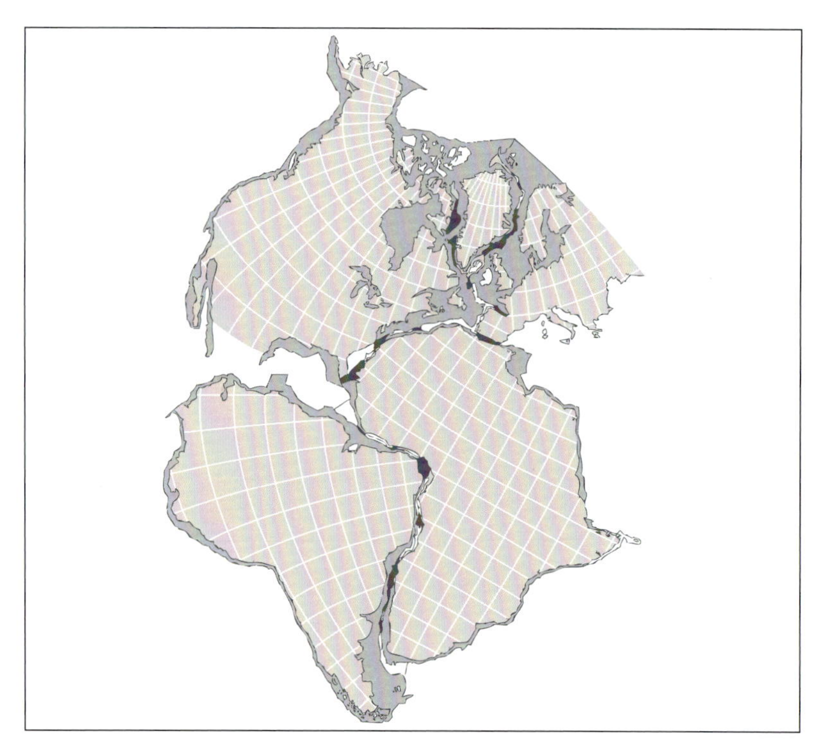

Figure 5.29
Bullard's fit of the Atlantic continents along the margins of their continental shelves. Gaps in the fit are represented by white and overlaps by black. The thin lines represent the present-day graticule. Along most of the fit, the average mismatch is less than one degree.
Data source: Bullard et al, 1965.

The 1970s saw a tremendous expansion of the literature of plate tectonics—one was either on board or simply out of date—and in the following decades the model was broadened, enhanced, and refined on multiple fronts. The oceanographic surveys were stepped up, mainly by Heirtzler's group at Lamont and W. H. Menard's at Scripps. There was increasing focus on seismological studies (now that the causes of earthquakes were known, could they be predicted?) and reflective seismological imaging of subsurface structures led to better understandings of the large-scale workings of the earth's interior, processes of crustal deformation, and interactions between the lithosphere (the crust) and the underlying asthenosphere (the source of new material emerging at the mid-ocean ridges). Gravimetric mapping was also being refined and increasingly employed in charting gravimetric anomalies and, after about the mid 1970s, geologists grew increasingly reliant on Landsat (originally Earth Resources Technology Satellite or ERTS) imagery in mapping major fault systems. New instrumentation and techniques such as very long baseline interferometry (VLBI), global positioning systems (GPS), and laser ranging made it possible to precisely measure slip rates along faults, crustal deformations, even the motions of entire continents, and led to a renaissance of geodesy. The availability of satellite altimetry data, beginning in the mid 1980s with the Defense Mapping Agency's GEOS-3 and the U.S. Navy's Geosat, made it possible to map continental and also seafloor topography in greater detail than ever seen before. Ultimately, every bump, crack, and wrinkle of the earth's fractured surface appeared as if under the lens of a giant space-born microscope.

In the forty years or so since plate tectonics materialized as a coherent theory of the earth, and even extended to other planets, a lot of detail has been sketched in without much rearrangement of its underlying theoretical framework. Plate tectonics has won widespread assent, at least for now and the foreseeable future, and in doing so has gained the authority of *fact*. Just look at any contemporary geological textbook. Still, it's hardly a case-closed theory of anything and everything about the geologic history of the earth, intrinsic force and certainty of established cultural fact notwithstanding, and some aspects of that history—the interior structure of continents, for example—are still ample territory for debate.[40] But that's not the point here at all. The point is that maps aren't just instrumental but *essential* to any cultural construction of a nature this large, and our most basic beliefs about what that is and just how it might do whatever it does. And it's not that we don't have vested interests in these issues either—we have more interests, and more money, invested all the time. So, is this really an angry and raging planet instead of the benign and embracing "blue marble" pictured from space? That really depends on how it's presented.

Which brings us, finally, to this last map, USGS's *Three-Dimensional Anaglyph of the Earth* by Brian Davis, Paul Morin, and Monte Ramstad (figure 5.30). This could hardly be more different from "The Earth's Fractured Surface." Gone are the faults and fracture zones, the spreading centers and subduction zones and hot spots. Well, not gone really, but certainly not *posted* as distinct features in their own right and with their own distinct claims to existence. They're there alright, and in the most up-to-date detail, and even in 3D! (Find a pair of 3D glasses and check it out.) But they're no more or less important than any other salient features of the earth's physiography. There are no hints here of earthquakes or of volcanoes (except on the closest and most informed inspection) or of devastating tsunamis or landslides or lahars, or consuming flows of lava or searing pyroclastic flows or choking clouds of volcanic ash. There is no violence here, manifest or

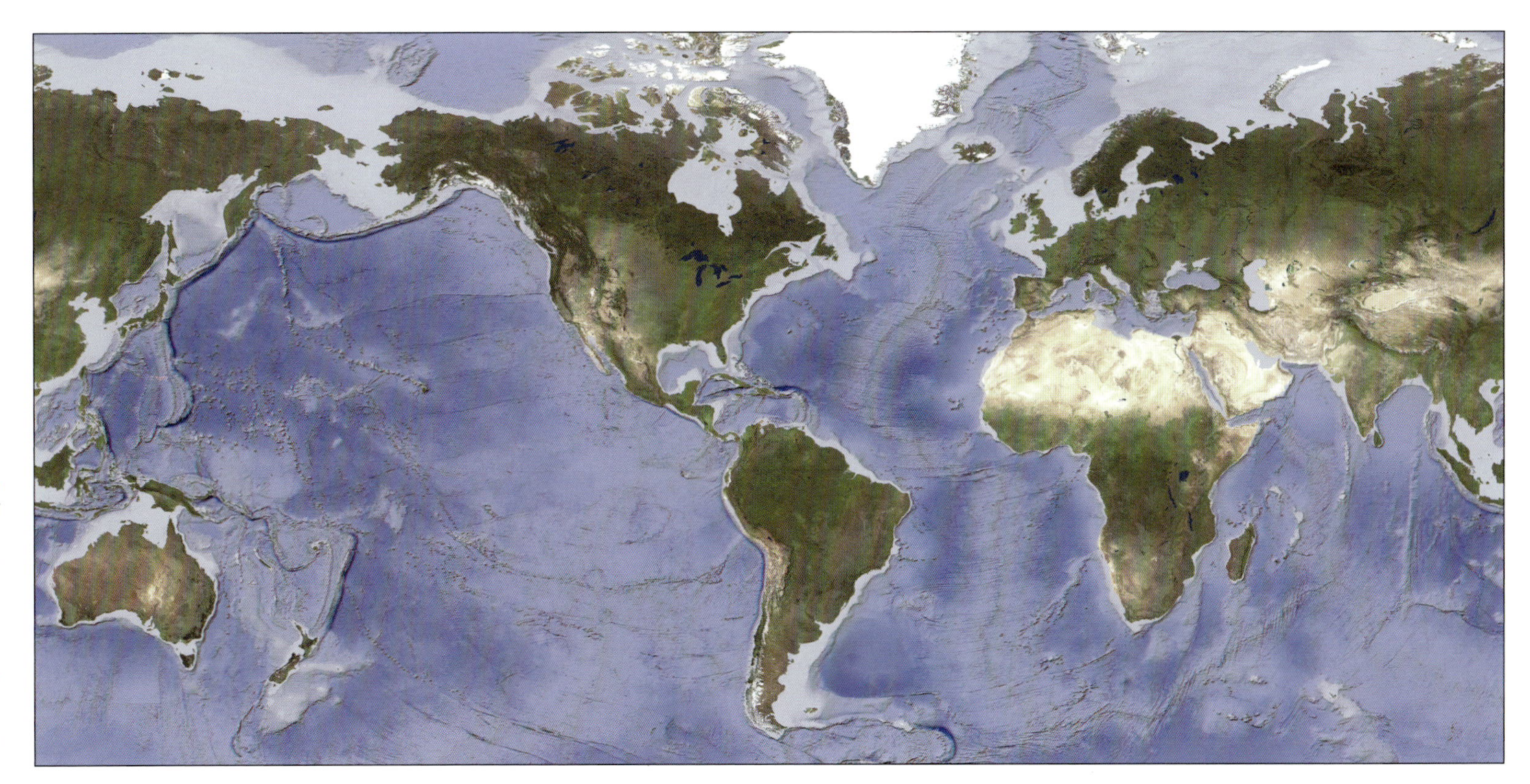

Figure 5.30
Three-Dimensional Anaglyph of the Earth. Here, the earth is portrayed in all of its awesome splendor. Any threat of geologic disaster is veiled by the planet's extraordinary complexity and beauty.

Source: Brian Davis, Paul Morin, and Monte Ramstad. Courtesy of USGS, 2004.

implied . . . none. The planet is presented, seemingly presents itself, as in a state of, if not repose, an elegant equilibrium of grand scale, in a breathtaking global gestalt. And there's little point in any belabored verbal description here . . . just *look* at it.

This planet isn't threatening or frightening at all. It's beautiful, astonishingly beautiful. Drop-dead gorgeous in fact—a whopping piece of cartographic eye candy. Everything is integral to a grand whole, one in which faults are far from faults in any human sense, not flaws in any sense, but just inevitable manifestations of grand forces and processes that we can try to understand but can never possibly hope to influence. It is a coherent synthesis, all of a piece, not wanting to be deconstructed into constituent components. (You can try if you like. Good luck.) This is a nature entirely *apart* from culture, and maybe one that can only be seen through the lens of advanced technologies (and what a helluva big lens that is!). This is not a scientized nature to be dissected and analyzed. It is not a primitive or even classical nature that might be angered or appeased but a completely undeified nature without possibility of cognizance or intent or malice or even indifference—magnificent certainly, and awesomely powerful and beautiful, and maybe physically knowable to some small degree, but ultimately as unreasoning and inscrutable as the face of Melville's whale.

In its consummate dispassion, this map isn't about charting new coastlines in quests for riches and empire, or the allure of imagined former worlds, or a grand theory of the earth. It's not about broken cities or broken lives, or any consequent human risk or response or actuary. (Unlike "Earth's Fractured Surface," the human species—those ants scurrying about on the surface—don't even exist here.) It simply *is what it is,* even if it takes funny glasses to see that.

It's all so undeniably beautiful, not fractured and defaced and angry at all but an elegant and exquisite manifestation of planetary forces of almost unimaginable spatial and temporal scales, and it underwrites any conceivable notion of planet as place. Even ignoring hydrosphere and biosphere and atmosphere, it's still a mysterious and wondrous place in imaginations large enough to catch a fuzzy glimpse of it. But does that make this a less *dangerous* place? Not a chance. We don't *own* this planet and we never will. In the end, we can only stay out of harm's way if we can, or get out of its way when we must . . . if that's even possible.

Notes

1. Transferred to the Department of Commerce, the Weather Bureau's first year of civilian operation was 1891, this according to Monmonier, *Air Apparent: How Meteorologists Learned to Map, Predict, and Dramatize* (Chicago: University of Chicago Press, 1999), 164. The Bureau had originally been created in 1870 under the authority of the War Department. In the 1970s, it was reconstituted as the National Weather Service (NWS).

2. Hurricane records had been kept in the United States since 1851, but tropical storms weren't given personal names until 1953.

3. Total deaths were estimated at around 8,000, with 6,000 in Galveston and another 2,000 in the surrounding area. Some estimates of total fatalities range as high as 12,000. For comparison, the death count from Katrina stands at a little more than 1,300. More than 3,600 homes were destroyed in Galveston alone, which was nearly all of them.

4. In *Air Apparent,* Monmonier recounts, "Numerical forecasting was first proposed in 1922—a half century before even vaguely reliable computer models . . .," (preface, xi) and, "Despite the incorporation of physical principles into complex mathematical equations, weather forecasting in the 1940s and 1950s was still very much a manual-graphic process, in which clerks entered observations into charts by hand and forecasters plotted displaced positions of fronts and zones of precipitation" (87).

5. For a thorough account of the Galveston Hurricane, see Larsen, *Isaac's Storm* (New York: Crown Publishing, 1999). See also Bixel and Turner, *Galveston and the 1900 Storm* (Austin, Tex.: University of Texas Press, 2000).

6. The 2005 Atlantic hurricane season would go on to become the worst ever recorded, with twenty-six named storms before the end of November, stretching for the first time into the Greek alphabet (previous record: twenty-one in 1933), thirteen hurricanes (previous record: twelve in 1969), four major hurricanes hitting the U.S. (previous record: three in 2004), three of the strongest hurricanes on record (Wilma, Rita, and Katrina), and the strongest ever recorded in the Atlantic (Wilma). With more than 1,300 killed, Katrina was the deadliest since the "Florida Keys" hurricane of 1928, and supplanted Andrew as the costliest (estimates now rising to more than 100 billion dollars). Incredibly, the last (twenty-seventh) named storm of the year, tropical storm Zeta, the sixth name drawn from the Greek alphabet, did not form until after Christmas.

7. This idea was essential to our broad interpretation of maps in Wood and Fels, "Designs on Signs," 61–5, and provided the basis for distinction between what we called the intrasignificant and extrasignificant signing functions of the map, which is to say, those operative internally within the map, presumably about its content, and those operating externally in attempting to claim the map's cultural position and status. We borrowed this notion and its indispensable diagrammatic expression, with due acknowledgement of course, from Roland Barthes, who had appropriated it (without acknowledgement or credit) from Roman Jakobson's distinction between speech and language, substituting "sign" for Jakobson's "speech" and "myth" for "language."

8. In the opening paragraph of "Desiring the Weather," Marita Sturken writes the following: "Weather is not what it used to be. It is no longer something one goes outside to register, that one experiences on the ground and in the flesh. It has become, rather, a technological experience, seen from satellites and endlessly monitored on television and the internet. What was once the site of interest for farmers and fishermen has become the source of pleasure and obsessive viewing for urbanites and suburbanites. What was considered to be a boring, uneventful news item has become a primary, if not quintessential, aspect of contemporary cable television. What was understood as a natural phenomenon is now the source of technological fantasy."

9. NOAA employs aircraft primarily for hurricane reconnaissance and tracking and the GOES (Geostationary Operational Environmental Satellites) system for monitoring and surveillance. This is from the GOES Imager Tutorial accessible via NOAA's Web site: "The GOES provide frequent images at five different wavelengths, including a visible wavelength channel and four infrared channels. The satellites scan the continental U.S. every 15 min.; most of the hemisphere, from near the north pole to ~20S latitude, every 30 min.; and scan the entire hemisphere once every three hours in their 'routine' scheduling mode. Optionally, special imaging schedules are available that allow data collection at more rapid time intervals (~7.5-min and 1-min), over reduced areal sectors."

10. The *Geographic* wouldn't say this, but in the issue of the magazine accompanied by our map, there is space for a map of "risk projection" (still not quite prediction) prepared by Risk Management Solutions, "one of many firms that use computer modeling to help insurers assess risk," along with a map of population (*National Geographic*, July 1998, pp. 14–15).

11. Which proves to be the story told in the magazine. The story and map are identically structured, with the photographs in the story magnifying the anecdotes (for example, of foundations from which houses have been ripped, torn, swept, or burned away).

12. Dancygier and Sweetser, *Mental Spaces*, 33–34. Note that their entire book is about conditional constructions. Predictive conditionals are covered in chapter 2, but they come up again and again in the discussion of other kinds of conditionals. Chapter 4's treatment of the use of future and present forms in conditional constructions is especially relevant to the case we're discussing here.

13. But this wouldn't explain why tornadoes on "Great Disasters" are reported since 1900, but on the tornado hazard map "since the 1950s"—whatever that means—to pick one example.

14. Dancygier and Sweetser, *Mental Spaces*, note that "the present is a far less well-defined tense: it ranges from generic statements to specific references to present events and states, to these future-reference uses (see below) in predictive backgrounds and in present-for-future assertions. We would like to propose that all of these are related to each other, as linking the event or state of affairs to a current temporal baseline" (91).

15. The occurrence of simple present-tense in if-clauses with future reference is called tense backshifting (it also occurs in when-clauses and with after, before, and so on). Dancygier and Sweetser, ibid., discuss this on pp. 43–44 (the example's from p. 44). They make the remark about future will in predictive protases on p. 83. Subsequent quotes from pp. 89, 91.

16. Dancygier and Sweetser's example, from Neil Stephenson's *Snow Crash*, is "Hey, if anything happens to that, my ass is grass."

17. "We find a tendency for people to put themselves in harm's way," writes *National Geographic* editor Bill Allen in his "From the Editor" column (p. 1).

18. See McHarg, *Design With Nature* (Garden City, N.Y.: Doubleday, 1969).

19. Simultaneous natural disasters are of course rare, but not unheard of. The 1991 eruption of Mount Pinatubo in the Philippines came at the onset of a major typhoon, whose torrential rains soaked the deeply accumulating volcanic ash, collapsing roofs and even entire buildings.

20. This, again, is from Sturken, "Desiring the Weather," 163: "One of the primary narratives governing the weather is that of revenge. Originally a Christian narrative about the weather as a punishment for sins, this story has evolved in contemporary environmental politics into the weather as nature's anger at humans for all the ills they have perpetrated upon it. The idea of nature's revenge thus provides a contemporary secular theme for the weather: nature's "fury" is aimed at the negligence and indifference of humankind toward the environmental consequences of its actions. Its current manifestation is the compelling argument that the weather upheaval of the late twentieth century is the result of global warming caused by pollution."

21. We're reminded of a *Far Side* cartoon in which two fur-clad Neanderthals are standing beside a crude stone hut and staring upward at a sheer wall of glacial ice towering out of the frame just a few feet away. One muses to the other, "Wall of ice closer today, Grog?"

22. The companion article is by Rick Gore, pp. 2–35, with photographs by Roger H. Ressmeyer. This article contains an additional three cartographic presentations: "The Devils beneath the City of Angels," "California's Push and Pull," and "Anatomy of an Earthquake."

23. And that's a much bigger deal than just weather. Here are a few more remarks from Sturken, "Desiring the Weather," 181: "[Earthquakes] play a particular role in apocalyptic narratives precisely because they remain, despite geological understandings, deeply mysterious in origin. A serious winter storm can cause death and damage, but it is understood as a more intense version of a "normal" weather event. An earthquake, however, is always an anomaly. In other words, because earthquakes are the result of geological phenomena, they are not part of the weather but rather a broader set of physical phenomena that are linked to the earth's status as a planetary body. An earthquake, like astronomical phenomena such as asteroids and solar anomalies, points to general properties of planets and the universe that are at the root of basic human anxieties about the meaning of life."

24. *Tectonic*, which pertains broadly to building or construction, is a term borrowed from the fields of carpentry and architecture. In the *Iliad*, Tecton was a carpenter and the father of Phereclus. A contemporary definition of tectonics, this one from *Webster's Encyclopedic Unabridged Dictionary of the English Language* (1994), reads, "the science or art of assembling, shaping, or ornamenting materials in construction; the constructive arts in general." *The Compact Edition of the Oxford English Dictionary* (1985) cites several early applications of the term to structural geology: by Boyd-Dawkins (1894)—"The relation existing between the tectonic anticlines and synclines in the districts of South Wales, Gloucester, and the West of England"; by Lord Avebury (1902)—"The primary configuration of the country's surface is no doubt due to tectonic causes"; and another from *Athenaeum* (1905)—"Whilst the most powerful and destructive disturbances are of this tectonic character, many other earthquakes are no doubt connected with volcanic phenomena."

25. It surely looked like a Mercator projection, but history indicates that the Mercator had not been employed in standard American classroom maps since the 1920s. As a general gestalt, the Van der Grinten and Winkel Triple look pretty much the same anyway.

26. According to the Eurocentric accounts given in school, the "discovery" of the rest of the world was an exclusively European enterprise achieved over a scant two centuries; but had da Gama reached the Indian Ocean seventy years earlier, he might have encountered the enormous fleet of Chinese junks under the command of Admiral Cheng Ho, who headed Chinese maritime exploration and trade from 1405 to 1433 and visited Southeast Asia, Indonesia, India, Persia, Arabia, Somalia, and Kenya. Also unmentioned were earlier Indian and Korean voyages to the southern tip of Africa, African maritime migrations to Madagascar and around the Cape of Good Hope, Japanese and Arabian voyages to China, Polynesian voyages of more than two-thousand miles to the Marquesas and Society Islands, and the very high probability that both Chinese and West African navigators reached the Americas centuries before Columbus. Maritime mercantile enterprise was hardly the exclusive domain of the Europeans. See Blaut, *The Colonizer's Model of the World* (New York: Guilford, 1993), and Frank, *ReOrient: Global Economy in the Asian Age* (Sacramento, Calif.: University of California, 1998). See also *The Harper Atlas of World History* (New York: Harper Collins, 1992).

27. These and the previous examples are drawn mainly from Portinaro and Knirsch, *The Cartography of North America*, 1500–1800 (New York: Crescent Books, 1987); and Whitfield, *The Image of the World: 20 Centuries of World Maps* (San Francisco: Pomegranate Artbooks, 1994).

28. This translation is drawn from Romm, "A New Forerunner for Continental Drift," *Nature* 367 (1994): 407–8. By "forerunner" Romm means forerunner to Frances Bacon's remarks in *Novum Organum* (1620), which have often been erroneously cited as the first known observations on the congruity of the two continents. On close reading, however, it is apparent that Bacon remarked on the general similarity of South America's western coastline to its African counterpart and not on the congruity of South America's eastern coastline and the west coast of Africa.

29. Romm, 408.

30. Longwell and Flint, *Introduction to Physical Geology* (New York: John Wiley & Sons, Inc., 1995), 411.

31. Ibid., 421.

32. Taylor, "Bearing of the Tertiary Mountain Belt on the Origin of the Earth's Plan," *Geological Society of America Bulletin* 21 (1910): 179–226.

33. The term *Gondwanaland* was originally coined by Austrian geologist Eduard Seuss, after *Gondwana*, a key geological province in east central India. (Seuss, *The Face of the Earth*.)

34. As du Toit's work wasn't published until 1927, these examples appeared in only the last of the four editions of *The Origin of Continents and Oceans*, published between 1915 and 1929.

35. Beds of glacial tillite had been found in India in 1857, in Australia in 1859, in Africa in 1870, and in Brazil in 1888.

36. It was Wegener's view that the southern landmasses of Gondwanaland and the northern landmasses of Laurasia had, at even earlier times, composed the single universal landmass, Pangaea, which consisted of (in some form) all of the present day continents, held within it an ancient Mediterranean Sea, Tethys, and whose shores were washed by the universal ocean, Panthalassa. It is now believed that the earliest known supercontinent, Rodinia, began splitting apart around 750 million years ago and then reunited around 600 million years ago to form another supercontinent, Pannotia. That would also split apart but eventually reunite to form Pangaea around 300 million years ago.

37. See for example du Toit, *Our Wandering Continents: An Hypothesis of Continental Drifting*. (Edinburgh: Oliver and Boyd, 1993), or Holmes, *Principles of Physical Geology* (Sunbury-on-Thames: Thomas Nelson and Sons, 1944).

38. Also of note are Sloss's *Surface of the Earth* (Boulder, Colo: NOAA/National Geophysical Data Center, 1996); Espinosa, Rinehart, and Tharp's *Seismicity of the Earth* (Washington, D.C.: U.S. Navy Office of Naval Research, 1982); and National Geographic's *Millennium in Maps, Physical Earth* (Washington, D.C.: National Geographic Society, 1998).

39. In a such brief account of the complex evolution of plate tectonic theory, generalizations and omissions are inevitable and, of course, entirely our responsibility. Many volumes have been written on the subject including, among the more noteworthy, *Plate Tectonics: An Insider's History* (Oreskes, ed. [Boulder, Colo.: Westview Press, 2001]); Le Pinchon et al., *Plate Tectonics, Developments in Geotectonics* (New York: Elsevier, 1973); and Menard, *The Ocean of Truth* (Princeton: Princeton University Press, 1986), which all provide detailed personal accounts. Seminal articles on various aspects of the topic are gathered together in Wilson, ed., *Continents Adrift and Continents Aground* (San Francisco: W. H. Freeman and Company, 1967). A broad account of "drift theory" as it stood in the mid-1960s is given in Takenichi, et al., *Debate about the Earth* (San Francisco: Freeman and Company, 1976). An account of key discoveries regarding the paleomagnetism of the Pacific Ocean floor is given in Glen, *The Road to Jaramillo* (Standford, Calif.: Stanford University Press, 1982). An exhaustive and up-to-date treatment of plate tectonics is given in Strahler's *Plate Tectonics* (Cambridge: Geo Books, 1998). Powell, in *Mysteries of Terra Firma* (New York: The Free Press, 2001), provides a comprehensive history of theories about planetary geology and evolution, including pre-twentieth century theories.

40. See McPhee, *In Suspect Terrain* (New York: Farrar, Straus, and Giroux, 1982). This is also included in McPhee's compilation, *Annals of the Former World* (New York: Guilford, 2004). See also Wood, *Five Billion Years of Global Change* (New York: Farrar, Straus, and Giroux, 1998).

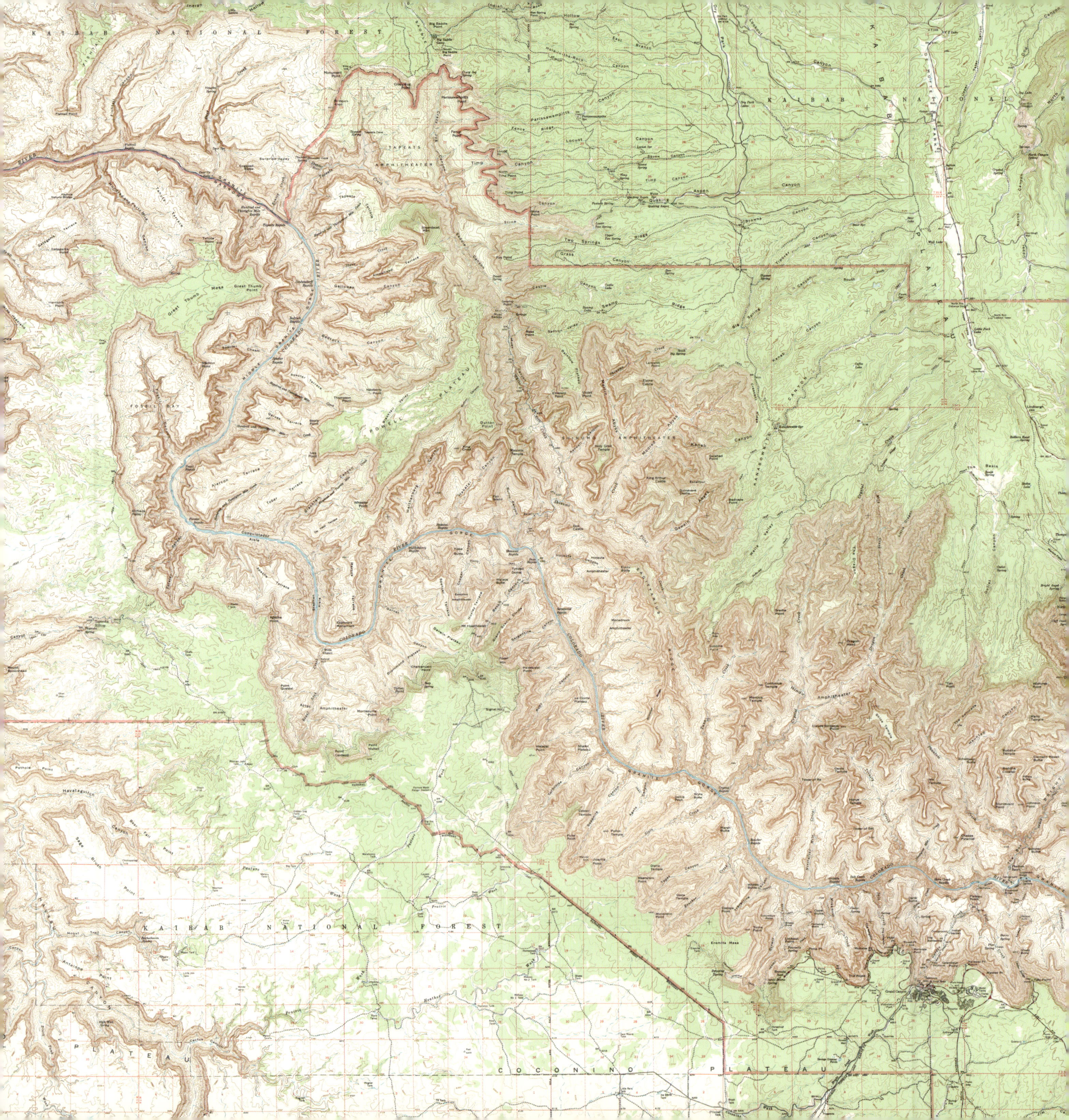

KAIBAB NATIONAL FOREST
KAIBAB NATIONAL
KAIBAB PLATEAU
TAPEATS AMPHITHEATER
POWELL PLATEAU
SHINUMO AMPHITHEATER
Great Thumb Mesa
Great Thumb Point
FOSSIL BAY
Havasupai Point
Point Sublime
KAIBAB NATIONAL FOREST
COCONINO PLATEAU
PLATEAU
Eremita Mesa

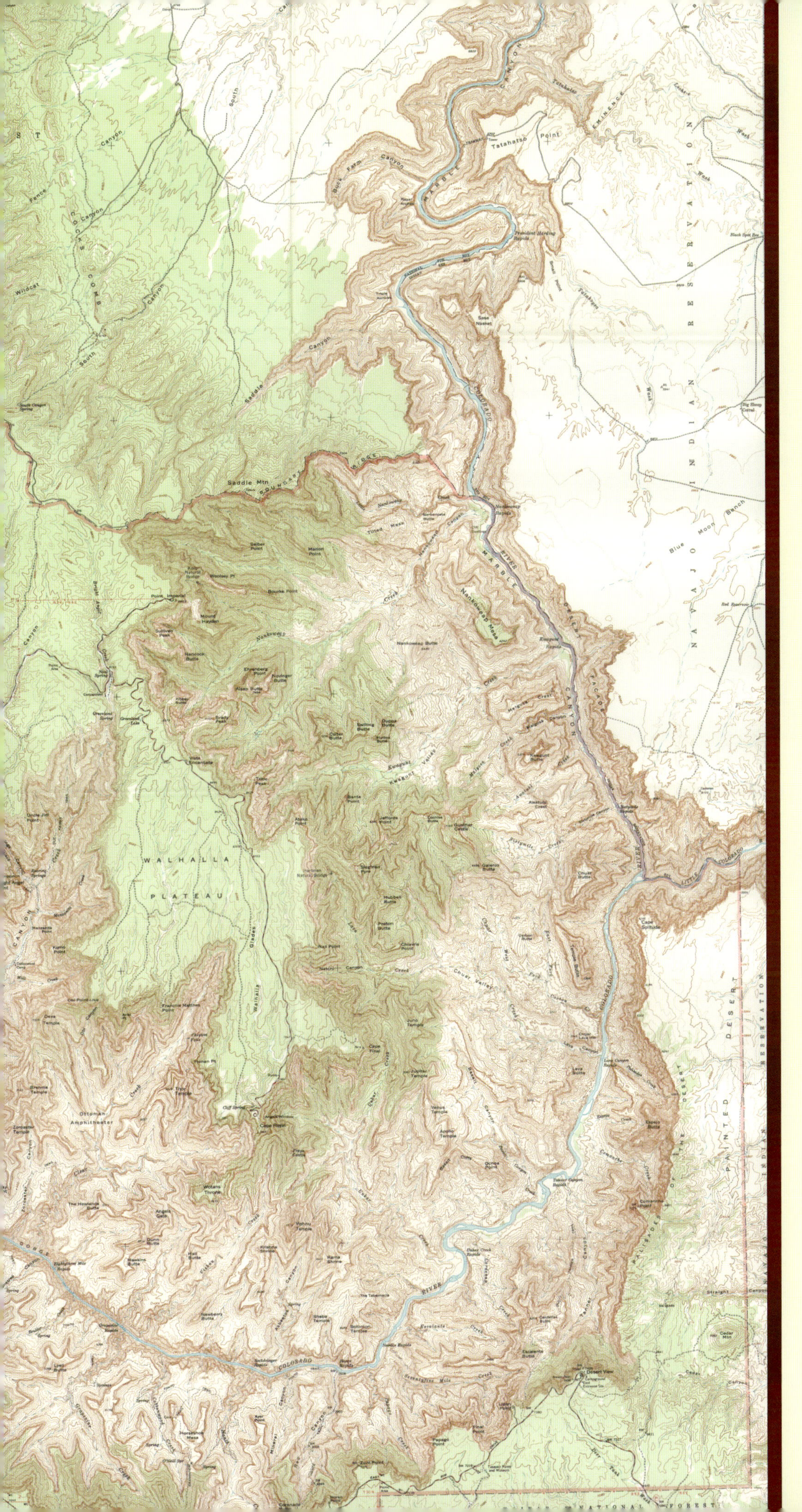

Six

Nature as grandeur

Courtesy of USGS, 1962.

Nature as grandeur

When one thinks of nature, as often as not those thoughts aren't of hurricanes or tornadoes or earthquakes or tsunamis or any of the other natural manifestations that visit destruction and death upon the human race, but of the grand, the majestic, the sublime. After all, natural disasters come and go (even if repair and rebuilding might seem to take forever), but other facets of nature are so distinctive and remarkable that we can easily regard them as the most prominent and permanent features of planet earth: Yellowstone, Yosemite, the Grand Canyon, Mount Everest, the Amazon, the Sahara, and more than a few others. These are forever engraved in our cultural consciousness. Who's never heard of the Grand Canyon (which receives more that five million visitors annually) even if they've never actually seen it? We know that it's there; we know what it is (more or less); and we promise ourselves that if we haven't seen it, we will some day.

These monuments of nature inspired the creation of the United States' first national parks—Yellowstone in 1872; Yosemite, Sequoia, and General Grant in 1890[1]—and ultimately the creation of a national park *system*, first here and later in other countries as well. Today there are more than one hundred nonhistoric national parks and monuments in the United States alone, administered by the U.S. Department of the Interior's National Park Service, and that's not counting various types of national preserves and reserves, national seashores and lakeshores, national rivers and trails and parkways, or any of another nearly two dozen such administrative categories under NPS jurisdiction. And that's also not counting national forests (U.S. Department of Agriculture, Forest Service), BLM lands (U.S. Department of the Interior, Bureau of Land Management), national wilderness areas (variously administered by the Park Service, Forest Service, Fish and Wildlife Service, and Bureau of Land Management), or the seemingly countless state parks, state preserves, and state forests.[2]

To be sure, our widespread awareness of earth's grandest features has more than a little to do with television, where cable offerings like The Learning Channel, Travel Channel, and Discovery Channel regularly serve up top-ten lists of the planet's most noteworthy natural wonders. Here's one such list from the Travel Channel,[3] presented in customary bottom-up order:

10. Ngorongoro Crater
9. Mount Everest
8. Lake Baikal
7. Victoria Falls
6. Uluru
5. The Everglades
4. The Sahara Desert
3. Great Barrier Reef
2. The Amazon
1. The Grand Canyon

And here's another from The Learning Channel:

10. Kilauea
9. Sequoia National Park
8. The Everglades
7. Victoria Falls
6. Yellowstone
5. Great Barrier Reef
4. The Amazon Rainforest
3. The Grand Canyon
2. Mount Everest
1. The Aurora Borealis/Australis

And there are plenty more of those inventories of "nature's longest, highest, largest, and greatest,"[4] often peppered with error and overstatement, and always presented against a backdrop of sternly voiced yet hucksterish narrative. And enumerations and rankings of these *categorillas*[5] abound in atlases and gazetteers as well—the longest rivers, the highest mountains, the tallest waterfalls, and so on. Even the U.S. Postal Service has issued a series of stamps titled "Wonders of America: Land of Superlatives" and including, among other things, The Grand Canyon (largest canyon), Yosemite Falls (tallest waterfall), Lake Superior (largest lake), Crater Lake (deepest lake), the Mississippi and Missouri rivers (longest river system), the Great Basin (largest desert), the Great Sand Dunes (tallest dunes), Rainbow Bridge (largest natural bridge), the Bering Glacier (largest Glacier), Kilauea (most active volcano), Steamboat Geyser (tallest geyser), and Mammoth Cave (longest cave). OK, OK, we get it . . . Wow! North America, and the rest of the world, are brimming over with natural wonders of almost unimaginable scale and majesty. If only one could experience them all in a single lifetime!

Have all of these things been mapped? Of course they have. They've been mapped, sketched, painted, and photographed over and over and over again. Would anyone care to guess how many photographs have been taken of Old Faithful or the Grand Canyon? (It would be nearly impossible to tally just the photographs of the Grand Canyon appearing among the pages of *Arizona Highways* over the years.)

But this book is about maps or, almost as cryptically, what maps do and how they do it. It's hard to imagine a more difficult challenge for the cartographer than to capture the grandeur of the world's greatest natural wonders in the two-dimensional microcosm of the map; but any cartographer worthy of that label couldn't help but leap at such a chance if only given it. Truly great places inspire truly great images, whether those are photographs or paintings or drawings or maps. This is where the craft of cartography has to reach its highest level, whatever other agendas are also in the mix; this is where the very fortunate cartographer gets a shot at creating his or her *magnum opus*. And have no doubt that these are genuine labors of love—we don't—and that's partly what this chapter is about, too.

Figure 6.1
Eadweard Muybridge's photograph *Valley of the Yosemite*, from *Rocky Ford I*, 1872. The kinship between this and figure 6.2 will be readily apparent.

Eadward J. Muybridge (American, 1830–1904). *Valley of the Yosemite*, from *Rocky Ford*, 1872. Albumen print from wet collodion negative; 43.1 cm. x 54.8 cm. Courtesy of the Cleveland Museum of Art, John L. Severance Fund 1990.133.

The Valley

Alright, we didn't mean to say that only maps are worth discussing here; examining images in other media will help broaden the context of our discussion. This photograph by Eadweard J. Muybridge, titled *Valley of the Yosemite*, from *Rocky Ford I* (figure 6.1) is one of the earliest taken of the valley.[6] Born Edward James Muggeridge in Kingston-on-Thames, Surrey, in 1830, he immigrated to California around 1852 and, within a few years had established a book business in San Francisco. Learning the new art of photography some time in the late 1860s, he quickly established himself as one of the premier photographic artists of his day. With his mobile photo wagon, which he dubbed "The Flying Studio," he traveled up and down the North American West Coast taking landscape and portrait photographs. Along with Carleton Watkins, who also worked extensively in Yosemite, he is recognized as one of the finest and most prolific of early western photographers. His large, wet plate photographs of Yosemite (this one about seventeen inches by twenty-two inches) are highly acclaimed but not as well known as his later experimental photographic studies of motion, which he began in the early 1870s and which consumed his attention for the rest of his life.

This photograph was apparently taken from the north side of the Merced River (the viewpoint almost appears to be *in* the river) looking eastward up Yosemite Valley. Beyond the stand of ponderosa pine in the middle ground, the valley is framed by El Capitan on the left and Bridalveil Fall and Cathedral Rocks on the right. Sentinel Rock and Glacier Point, about five miles distant, are obscured in the misty haze of the atmosphere. Indeed, the heavy atmosphere fills the valley, like water in a vessel, with its palpable and weighty presence. Even that not-very-distant stand of trees is softened by all the water in the air, although one can easily imagine that air fragrant with their resinous scent. This is, simply put, a very *wet*-looking image. The air is wet, the trees are wet, the rocks are wet—everything is saturated with water. And the foregrounded river, its surface glossed by the long exposure of the photograph, occupies nearly a full half of the view.

The pervasive presence of the thick, dense atmosphere here is not something we're likely to find in an aerial photograph, or especially in a map. In an aerial photograph or satellite image map, the atmosphere is something to be purged altogether. Clouds are bad; "cloud free" is good. Clouds, and the atmosphere itself, are far too ethereal and ever-changing to merit presence in such images, where all they do is get in the way of what we really want to see—the solid and seemingly fixed surface of the earth. (This topic will be taken up at much greater length in chapter 10.) Unless, of course, the maps are about clouds, like visible satellite or Doppler radar images dressing up the weather forecasts on the evening news. *Weather.* That's it, isn't it? This photograph is filled with weather—weather is a large part of what the photograph is about—and maps aren't supposed to have

Figure 6.2
Albert Bierstadt's *Looking up the Yosemite Valley*, 1863–1875, oil on canvas. Dimension (WxH): 58in. x 36in.

Courtesy of the Haggin Museum, Stockton, California.

weather. The weather, the atmosphere, is rarified out of existence in the map, where only a perfect vacuum can facilitate the map's intrinsic intention of, or pretension to, perfect clarity and objectivity.

Painters as well as photographers explored the wonders of the American West in the latter half of the nineteenth century seeking their, and the nation's, vision of this vast undiscovered country. It was a period of rapidly intensifying national interest in the West, during which a number of ambitious surveys were sponsored by the War Department and the Department of the Interior, and an ever-increasing volume of reports and images from many sources found its way to an ever-more insatiable Eastern audience. The West came to represent, especially following the Civil War, the scale and majesty of the world's greatest nation, a single nation of continental proportions fully in accord with burgeoning notions of Manifest Destiny. Settlement and development in the "Trans-Mississippi Country" intensified rapidly during this period, as the nonnative population west of the Mississippi grew from two million to twenty million between 1850 and 1900. The American West was a magnificent wilderness, the most beautiful on earth, but it was also a vast land of opportunity awaiting unbridled exploitation by anyone with sufficient ambition and a propensity toward taking risks. For landscape painters and photographers, it offered artistic opportunities far surpassing any to be found in the East.

Albert Bierstadt had left the United States for Europe at the age of twenty-three in 1853 to study painting at the prestigious Düsseldorf Academy. There he focused on landscape painting, for which Düsseldorf itself had become widely acclaimed owing to the work of influential painters such as Carl Friedrich Lessing, Andreas Achenbach, and Johann Peter Hasenclever, and a landscape class had been established at the academy by Johann Wilhelm Schirmer in 1839. Throughout three years of study at Düsseldorf, Bierstadt produced a considerable volume of drawings and paintings, mostly of Alpine and Mediterranean landscapes. He returned to New England in 1856 and made his first western trip in 1859, journeying to Wyoming with a railroad survey under the direction of Colonel Frederick W. Lander. Bierstadt became enraptured with the West, making repeated trips there and quickly becoming the most widely traveled and most prolific of early western artists.[7] His paintings found an eager—and wealthy—audience in the East, and he gained widespread recognition with the exhibition of his "The Rocky Mountains" at the New York Metropolitan Fair in 1864. The painting was heralded as a vision of the unrivaled beauty of the nation and a symbol of American nationalism in a country still in the throes of the Civil War and gnawed with doubt over its own future. With that painting, Bierstadt came to be regarded in the company of the most famous large-format landscape painters of the time—Frederick Church and Thomas Moran.

Bierstadt was undoubtedly acquainted with the nascent art of photography. Both of his brothers were photographers, and he had spent the summer of 1860 with them on a photographic expedition to the White Mountains of New Hampshire (although it is reported that he never operated the photographic equipment but served only in the capacity of artistic advisor). And he was certainly familiar with the work of Watkins and Muybridge; in fact, he and Muybridge traveled together to Yosemite in the late summer of 1872. (The dates are indefinite in both cases but it's tempting to speculate whether Muybridge's photograph and Bierstadt's preliminary sketches for the first of his paintings presented here—figure 6.2—might have been created in

Figure 6.3
Albert Bierstadt's the *Domes of Yosemite*, 1867, oil on canvas. Dimension (WxH): 180in. x 116in. Bierstadt's paintings present the grandeur of the American landscape at monumental scales.

Albert Bierstadt, :The Domes of Yosemite," 1867. Oil on canvas, 180 x 116 inches. Courtesy of the Collection of St. Johnsbury Athenaeum, Vermont.

their mutual company.) Photography held some practical import for Bierstadt, and he often painted his finished canvases while referring to not just his own sketches but, when available, photographs of those scenes as well.

Yosemite became the nucleus of Bierstadt's West and, after his initial visit there in 1863, he returned repeatedly throughout his career to paint scenes of a broad variety of subjects and in a broad variety of conditions in all seasons, gradually building a substantial oeuvre and visual documentation of Yosemite.

Bierstadt's canvases ranged in size from quite small to monumental; although, for Yosemite bigger was usually better. The painting reproduced here, *Looking up the Yosemite Valley*, measures fifty-eight by thirty-six inches, which is large enough to accommodate the scale and grandeur of the scene. The vantage of this painting is clearly very near that of Muybridge's photograph, which was probably taken from not far beyond the clump of three trees in the foreground. The framings of the painting and photograph are nearly identical too, with the Merced River and valley floor front and center, and again flanked by El Capitan and Bridalveil Fall and Cathedral Rocks. But the stand of pine that, in Muybridge's photograph, forms a solid barrier between foreground and background, is pushed further back here, almost melding with the near background. That allows the painting to exhibit two well-known traits of the Düsseldorf school: the sweeping horizontal plane of the foreground and the centering of a distinctive tree or small group of trees in the foreground, using vegetation as a focal point rather than framing element. Other hallmarks of the Düsseldorf style are evident here also: broad tonal ranges and often abrupt tonal contrasts, rugged and distinctive rock formations, dynamic skies with clouds obscuring the source of light, and, of course, near photographic detail.

Those same traits are evident in this second Bierstadt painting, *The Domes of Yosemite* (figure 6.3)—an enormous, wall-filling canvas measuring 180 inches by 116 inches. Again, the view is toward the east up the valley, which is again flanked by the rugged canyon walls. This time, though, Half Dome and Quarter Domes can just be made out at right, and to the left we have Washington Column and North Dome (with Mount Watkins above and beyond obscured by clouds). In the far left foreground, though, is something very special—a dynamically plunging waterfall, with the rapids at its base tearing off on their way toward the Merced River. This, of course, is Lower Yosemite Fall, the lower of two falls jointly comprising North America's tallest waterfall. It leaps from far above our view, backlit by reflection off the rock behind it, to come crashing down beyond another stand of pine, and its rapids flash briefly into view before surging away again. One can almost *hear* this painting, as one could almost smell Muybridge's photograph. But this painting is not about that luminous waterfall—that's not even mentioned in the work's title. Those domes are there as promised, but rendered tantalizingly indistinct by the descending clouds. And the sun is there too, not behind the clouds but rising somewhere in the southeast behind Half Dome. And the river is there too, but distant, twisting across the panorama of the valley floor. Clearly, this painting is not *about* any one of those things: it's about all of them at once —it's about Yosemite Valley, the *grandeur* of Yosemite Valley. Could a map, any map, ever be quite *this* grand? We're beginning to have our doubts, but then we haven't looked at any of this chapter's maps yet.

A couple of the attributes found in Bierstadt's paintings—meticulous attention to detail and perhaps the careful rendering of rock formations—can sometimes be found in maps, but the others almost certainly cannot. A map can have more visually foregrounded elements and more backgrounded elements (layers) but seldom a distinct foreground, middle ground, and background in the sense that a painting or photograph does. A shaded relief map can have an apparent source of light (it has to) but no map, except perhaps those labeling themselves something like "3D view of . . .," has a sky and ground plane—maps are *all* ground plane. And a map, of course,

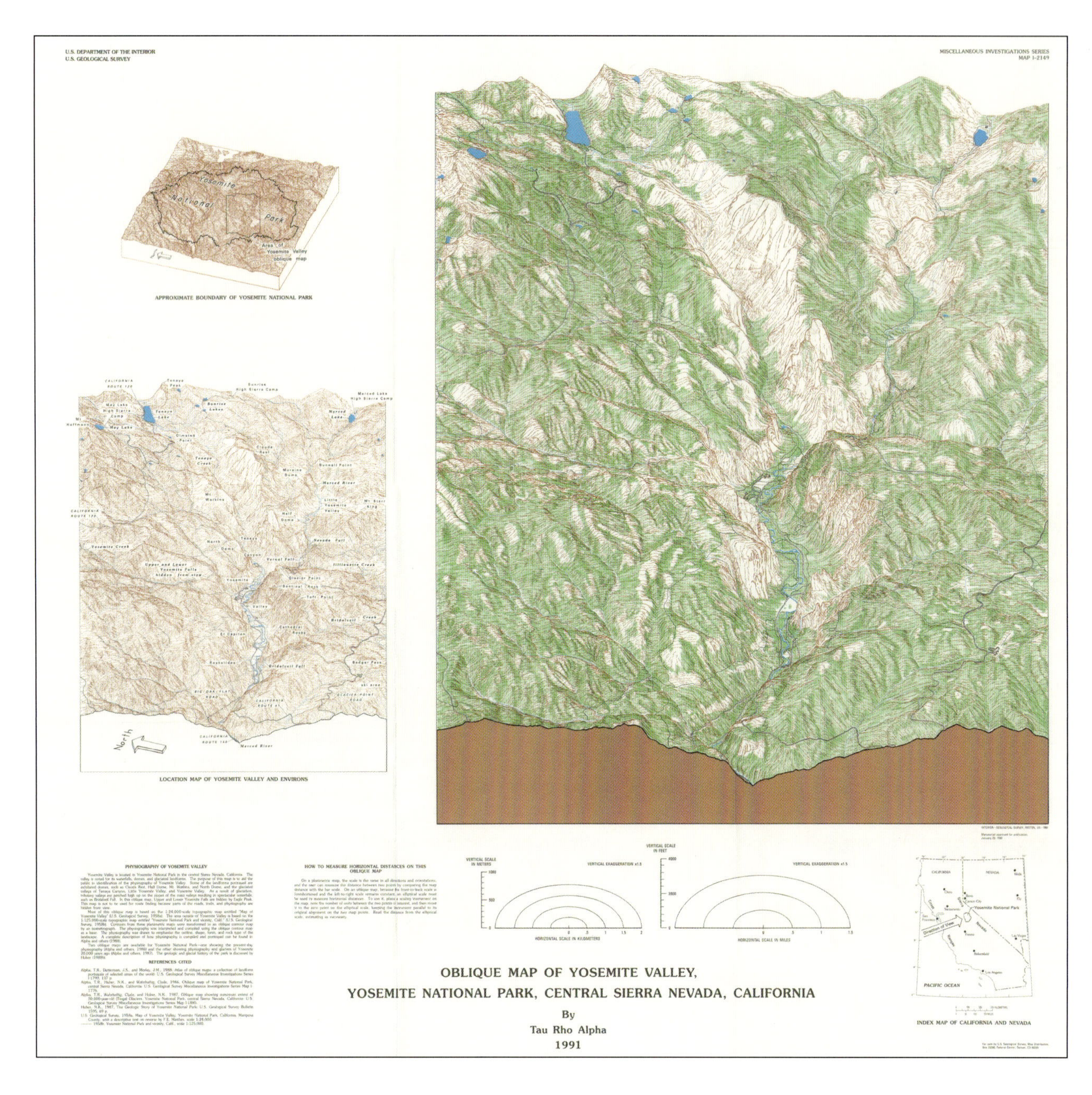

Figure 6.4
Oblique Map of Yosemite Valley. Is this a picture, a drawing, a map, or all of those at once? Map dimensions (WxH): 36in. x 36in.

Figures 6.4 and 6.5 from Tau Rho Alpha, USGS, 1991.

Figure 6.5
Detail of *Oblique Map of Yosemite Valley.*

does not have a singular viewpoint (except sometimes in one of those 3D views); the viewpoint of the map is ubiquitous.

And there's something else in Bierstadt's works we won't find in most maps (the most obvious exception being the NGS portraits we've looked at in earlier chapters)—living things, like individual and uniquely formed trees, or human or animal figures. (Human or animal figures appear with some frequency in his landscapes but rarely in those of Yosemite.) In the previous painting, those trees served their customary purpose of "providing scale," and the figures lent a Western flavor to the overall impression. But maps seldom operate at such a large scale, and besides, that's why they have scale bars anyway. Maybe a map can't capture the grandeur of something like Yosemite Valley in the way that a painting or photograph can, but then a painting or photograph can't *map* the valley either. To accomplish anything approaching a visual mapping would require an ordered sequence or series of images traversing the valley from one end to the other and back again. (Such a technique has actually been employed, although rarely, in speleological surveys.) This is not to say that maps can't be grand—some of the maps in this chapter are especially grand—or that a so-called representational image such as a photograph can't have the utility so often ascribed to maps.[8] (Consider, for example, the photograph of a rock wall taken to assist the climber in planning his or her route. That's not unlike a map is it?) Maps and paintings and photographs all have power. They just have somewhat different kinds of power, and they acquire and realize that power in somewhat different ways.

Can an image simultaneously have the power of both a map and a painting or photograph? After all, isn't there a strong kinship between photography and representational painting, and isn't drawing very much like painting in many respects, and aren't maps (many anyway) still *drawn*, even if on a computer screen rather than on a drafting table or light table? Enter the oblique map or 3D map or panoramic map, a well-represented genre claiming much of the power of both two-dimensional maps and pictorial imagery.

Take a look at USGS's *Oblique Map of Yosemite Valley* by Tau Rho Alpha, 1991 (figure 6.4). (There is also a corresponding map of the entire park by the same author.) Nearly all but the leftmost panels of this folded sheet are occupied by the featured map, with title, index map, and combined horizontal and vertical scale diagrams in meters and feet at the bottom of the sheet. A paragraph of instructions explains how to use the scale diagrams. Vertical exaggeration is stated to be x1.5, which is not a great deal of exaggeration but certainly sufficient for this landscape. To the left is a small oblique map of Yosemite National Park showing the park boundary and the area encompassed by the main map. And below that is the "Location Map of Yosemite Valley and Environs," a half-size replica of the main map with labels for prominent features: Yosemite Valley, Little Yosemite Valley, Tenaya Canyon, Merced River, El Capitan and Half Dome and Glacier Point, Mount Watkins and Clouds Rest, Bridalveil Fall and Vernal Fall and Nevada Fall (with a note conceding, "Upper and Lower Yosemite Falls hidden from view"), and a number of others. Below that is a brief summary of the physiography of Yosemite Valley and a short list of references.

The location map is necessary because the main map contains no annotation whatsoever—not a single word. This map image assumes the guise of a large pen-and-ink drawing, which, in terms of its execution, appears to be just that. Topography is very detailed and very skillfully rendered in multiple pen strokes of brown ink, "field sketch" style. That is overlain with a wash of green ink distinguishing vegetated areas from the untinted bare rock and further modulating the three-dimensional land surface. Turquoise-colored lakes lie here and there in the flatter upper elevations, while turquoise creeks and rivers thread their way among the cliffs and spires of the valley, disappearing from and then reappearing into view. El Portal Road (in black, double-line) follows the Merced River up the valley to Yosemite Village, whose edifices look like little shacks at this scale, while Tioga Road traverses the highlands to the north and Glacier Point Road those to the south. Hiking trails (black, single-line) wind their way upward and outward across the landscape from their trailheads along the main roads. This, as simple as it seems in style and technique, is a highly skilled presentation, not at all like those lurid, shaded and blended, kitschy colored-pencil panoramas for sale in tourist gift shops. This work takes itself very seriously (it is without a doubt one of those aforementioned labors of love) and it admits no gratuitous graphical indulgence outside the bounds of USGS tradition.

But what is this exactly? Is this a map, a picture of some sort, both, neither? And do such distinctions have any meaning here at all? It might be enough just to say, "It looks terrific. Who cares anyway?" but we can't let the matter go at that. This certainly looks more like what we generally think of as a picture than what we generally think of as a map, but it calls itself a map and qualifies that only with the word "oblique." If it genuinely qualifies as a map, then what is the substance of that claim? Framing and selectivity are essential to what this image is and what it has to say, as they are with all maps, but isn't that true of visual images in general? Photographs and paintings and drawings all have to frame their subjects purposefully, and they are all selective in what they choose to include or exclude (photographs a bit less so with respect to the latter, but there are plenty of computer applications that will facilitate that after the fact). Abstraction and generalization are evident here also, but paintings and drawings also abstract and generalize their subjects to a greater or lesser degree.[9] All four of these qualities are essential qualities of human vision itself, which is to say that we choose what we want to see and how we want to see it. And the marks in this map/drawing do symbolize, but less in the manner of, say, topographic map symbols and more in the manner of brush strokes in a painting. These pen strokes, called "form lines," are varied in thickness and density in maps such as this to render, depending on individual technique, steepness of slope and/or light and dark surfaces and/or surface texture.[10]

None of the qualities we've mentioned, not in themselves nor in combination, really substantiate this image's claim to being a map. What does substantiate that claim is something else. This image elevates, or rotates, our point of view from the ground plane upward to an overhead, or in this case oblique, position. But then that's not necessarily true of panoramic maps, which often present their subjects as they might be seen from a distant mountain peak; and a similar view might be photographed from a helicopter or through an airplane window. Similar, yes—the same, not quite—because this image has no perspective in the strict sense, no convergence of lines, no vanishing

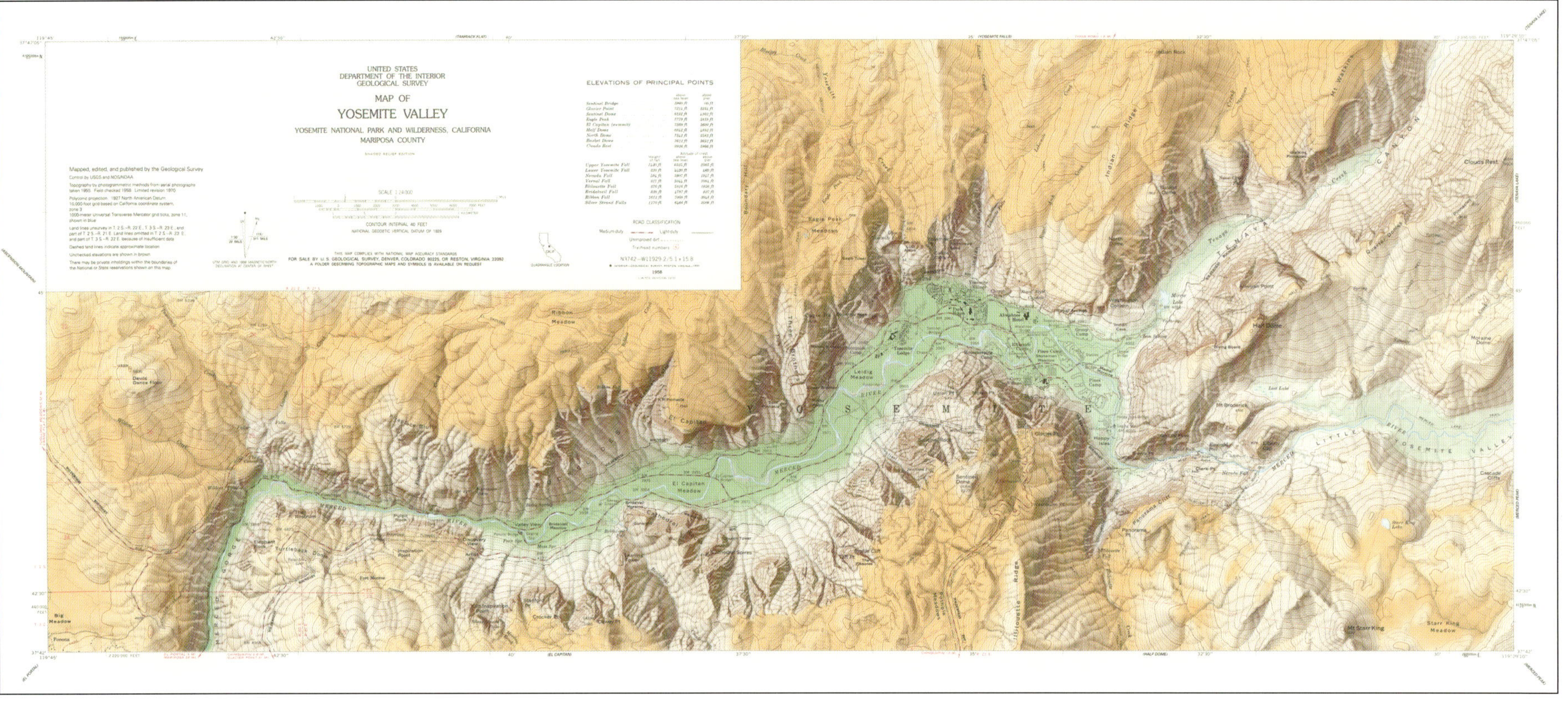

Figure 6.6
Map of Yosemite Valley (shaded relief edition). This is far more beautiful than the utility of a topographic map demands. Map dimensions (WxH): 42in. x 19in.
Source: Courtesy of USGS, 1958.

point or points on a distant horizon. Parallel lines (if there were any here) would remain parallel in this image, and what that means more importantly is that the image has uniform scale—not the same scale in both the horizontal and vertical dimensions (a consequence of foreshortening) but the same across each of those dimensions. And that means that we can measure distance in this map. (Yes, it's a map!) The procedure is not quite as simple as in a planimetric map, since foreshortening makes scale in the vertical dimension about one third the scale in the horizontal dimension. And because of that, the more elevated portions of the land surface might intervene between the features just beyond them and our imaginary viewpoint—as Upper and Lower Yosemite Falls are hidden here by Eagle Peak, and as Half Dome would have been hidden by El Capitan, even on a perfectly clear day, in our first two images. The idea that a map shows everything of interest within its purview simply doesn't hold here.

Here is USGS's topographic *Map of Yosemite Valley* (shaded relief edition), published in 1958 and revised in 1970 (figure 6.6). There is also another edition of this map in more conventional style and sans shaded relief, but, in appearance, it's no match for this one. In the upper left quarter of this elongated forty-two-by-nineteen-inch sheet is the usual documentation: title, credits, scale, key map, magnetic declination diagram, metadata, a listing of "Elevations of Principal Points" such as Glacier Point, Eagle Peak, El Capitan, Half Dome, Clouds Rest, and the elevation and height of eight of the valley's waterfalls, and a minimal legend showing road classifications and trailhead numbers.

On the reverse, there's quite a bit more—"The Story of Yosemite Valley," written by F. E. Matthes in 1922 (and revised in 1929 and 1938). That consists of five subtexts: "A Birdseye View of the Valley" (with a carefully rendered oblique drawing and associated list of twenty-four annotated features), "A Scene of World-wide Fame," "A Story that Began Millions of Years Ago" (with a photograph of Yosemite Falls), "Origin of Hanging Valleys and Cascades" (with a large wide-angle photograph of the valley and a "Generalized Diagram of Part of Tilted Sierra Block"), and "Transformation Wrought by the Ice" (including three sequential cross sections of the valley and a photograph of El Capitan).

The map itself is quite beautiful, with its interplay of contour lines and expertly crafted shaded relief (both in brown), green-tinted valley floor, and orange-washed uplands. The map doesn't say, but it appears that this orange tint is meant to represent areas with at least some vegetation, since it is absent from the bare rock of the valley's cliffs and domes, as well as the uppermost elevations to the east. But, as beautiful as this map is, it is more than just eye candy. It is indeed a fully functional map, marking out streams and roads and trails and buildings with unquestionable cartographic certainty, and naming, with equal certainty, the valley's myriad physical features. In achieving its aesthetic goals, which it does convincingly and with vigor, this map sacrifices none of its utility. On the contrary, the gorgeous colors and vivid three-dimensionality of this map only serve all the better to contextualize—to anchor—the natural and man-made features of this remarkable place. And all of those features come seamlessly together here in an exquisite visual synthesis. This map isn't just about Yosemite Valley—it *is* the valley . . . all of it, at once. There are good maps and

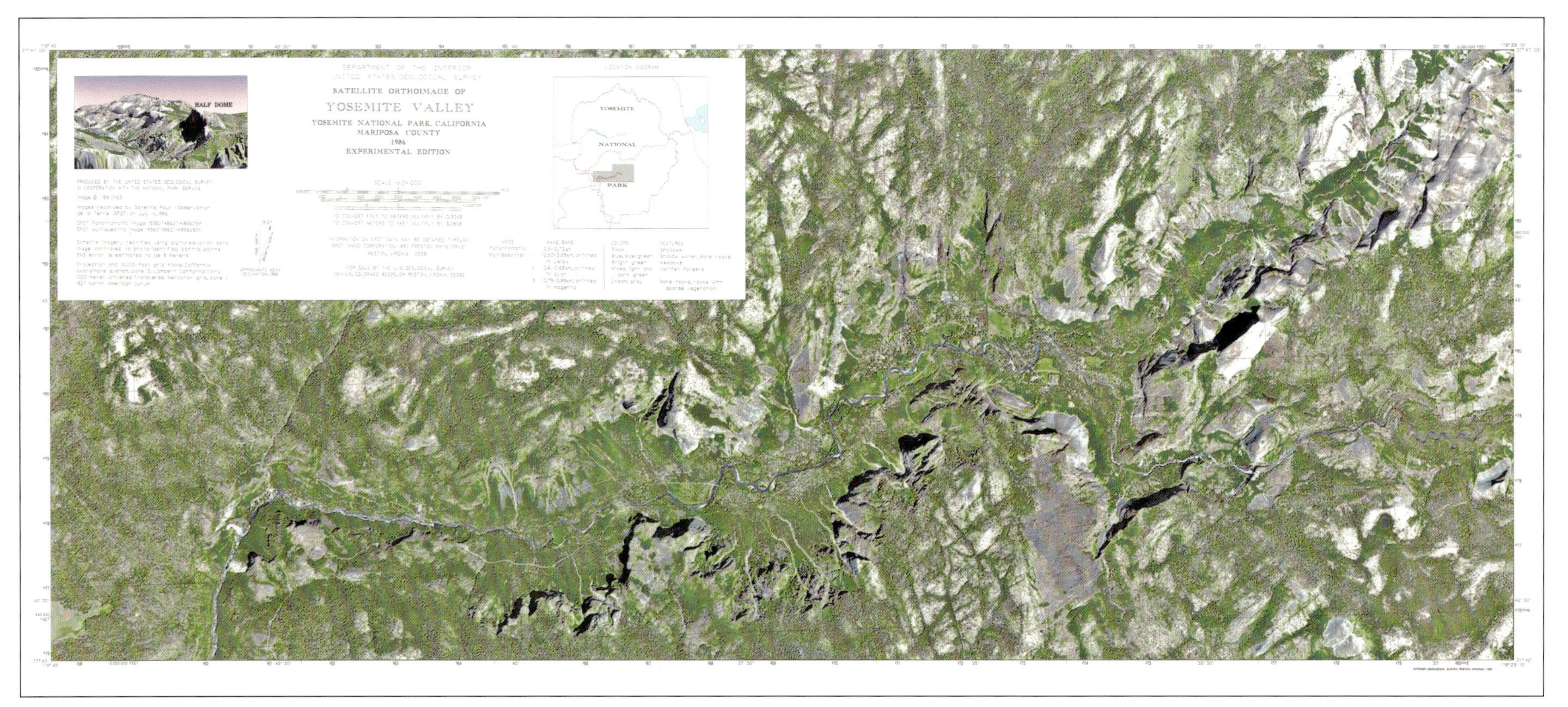

Figure 6.7
Satellite Orthoimage of Yosemite Valley. This is a map by any reasonable standards but marginalizes itself as such. Map dimensions (WxH): 42in. x 19in.
Source: Courtesy of USGS, 1986.

Figure 6.8
Inset detail of Satellite Orthoimage of Yosemite Valley.
Source: Courtesy of USGS, 1986.

there are not-so-good maps and there are truly awful maps (plenty of them), and then, once in a while, there are really great maps. And this, without argument, is clearly one of the latter. Truly great places can and do inspire truly great maps. Any doubts?

Our last vision of Yosemite is USGS's *Satellite Orthoimage of Yosemite Valley*, 1986 (figure 6.7). At a scale of 1:24,000 this covers exactly the same territory as the corresponding topographic maps, from Big Meadow and Merced Gorge at the lower (west) end of the valley to Mount Watkins and Clouds Rest at the upper (east) end, an elevation difference of more than a vertical mile. The upper left quarter of the sheet is occupied by a block of the customary ancillary material: title, scale, magnetic declination diagram, and credit to the U.S. Geological Survey and the National Park Service. On the right side of that block is a "location diagram" for the area covered, and on the left side is a small but quite dramatic three-dimensional visualization of Half Dome and Clouds Rest beyond, draping part of that same orthoimage on a digital elevation model. Below the visualization of Half Dome is a statement indicating that the data sources for this were SPOT[11] panchromatic and multispectral imagery captured on July 14, 1986; that the satellite imagery was rectified using digital elevation data; and that the projection employed (and the ten-thousand-foot grid ticks) conform with the California coordinate system, zone three. Below the location diagram is an accounting of the spectral wave bands employed here: panchromatic (0.51–0.73 μm, printed in black), multispectral 1 (0.50–0.59 μm, printed in yellow), multispectral 2 (0.61–0.68 μm, printed in cyan), and multispectral 3 (0.79–0.99 μm, printed in magenta). We are also informed how those different bands appear in color in this composite image and what they are likely to signify here: black (shadows), blue, blue green, and bright green (shallow water, bare rocks, and meadows), mixed light and dark green (conifer forests), and cream or gray (bare rocks and rocks with sparse vegetation).

Armed with that information, we can begin to decode this image. The bright, light, and dark greens of the valley floor contrast with the cream and pale gray unvegetated and sparsely vegetated rock of the uplands, including much of the area around and above Eagle Peak and North Dome as well as Clouds Rest, Quarter Domes, and Half Dome immediately east and south of Tenaya Canyon. Big, El Capitan, and Leidig meadows are easily distinguishable from their forested surroundings, and thin, near-white traces of intermittent tributary streams etch their way down the valley walls. Even thinner traces fix the paths of paved roads, and close inspection affirms the presence of Yosemite Village below Indian Canyon. The Merced River is conspicuous as a thread of rich blue. Hard shadows are cast by features of strong relief like El Capitan, Cathedral Rocks, Sentinel Rock, Panorama Cliff, Half Dome, and Quarter Domes.

For those unfamiliar with Yosemite Valley, these features can be identified, not because they are annotated here—there are no labels at all in this image—but by reference to the amply annotated topographic maps of the same area. It is as if this image would deny any pretense to even being a map at all. It's not called a map but a "satellite orthoimage," and even the location map is not titled as such but as a "location diagram." Still, this has most if not all of the qualities we habitually associate with maps, like its overhead ubiquitous view, two-dimensional projection, and uniform scale. And the colors employed here are symbolic even if, in concert, they give more the impression of an all encompassing *picture.* But, obvious pictorial qualities aside, this is clearly much more like USGS's topographic maps than it is like Muybridge's photographs or Bierstadt's paintings.

Those paintings and photographs easily demonstrate—are imbued with—the sheer drama of this landscape. It would be difficult for them to escape that if they tried. But here, much of that drama seems to have just drained away somehow, replaced by a dispassionate general inventory of land cover. The dramatic vertical scale and relief of the landscape, so vivid in the small three-dimensional image at upper left, appears curiously compressed here, evident only in the forceful shadows cast by north-facing canyon walls. And that is a significant part of the problem. Illumination appears to emanate from below (south of) rather than above (north of) the land surface—as it must, of course, in the northern hemisphere—and this has the effect of visually flattening the image. Simply turning this sheet upside down, light appears to emanate from above, as we expect, and much of the three-dimensionality of the landscape is restored. One has to wonder whether the sanctity of orienting the map with north up was truly worth the cost.

But, to be fair, the breathtaking three-dimensionality of Yosemite Valley was never meant to be the *raison d'être* of this image—of that we get only the smallest taste. What was very much the essence of Muybridge's photograph and Bierstadt's paintings and Alpha's oblique map and USGS's topographic map is simply not of great importance here. This is just what it appears to be and, no doubt, what it was intended to be—a land-cover inventory, presented with snapshot factuality.

If we'd been forced to choose only one of these six visions to present the grand and the sublime qualities of Yosemite Valley, to convey the sheer emotive power of the place, which one would it have been? Would it have been that moody and mysterious photograph, one of those dynamic and epically scaled paintings, that engaging and handsomely crafted oblique map, that beautifully rendered topographic map, this precise and revealing satellite orthoimage? If we could have just one, it would probably be Bierstadt's *The Domes of Yosemite*; but that's to take nothing away from the maps here—that topographic map especially. So maybe no map can ever be quite as dramatic as an enormous Bierstadt canvas, but these are, each in their own way, quite remarkable achievements.

The Canyon

As grand as Yosemite Valley is, it is certainly not as grand as the Grand Canyon; indeed, many would argue that no feature on earth can match the grandeur of this canyon carved 277 miles long through and more than a mile deep into the high plateaus of northern Arizona. It covers two thousand square miles and is as wide as eighteen miles; but, strictly speaking, it might not be the world's largest canyon. That kudo may well go to Mexico's Las Barrancas del Cobre (Copper Canyon), which is purported to be both longer and deeper than the Grand Canyon of the Colorado River. (And there are a number of other six-thousand-foot-plus deep canyons gouged out in that remarkable twenty-five-thousand-square mile region of the Sierra Madre del Norte.) But the faithful will argue that Las Barrancas del Cobre is actually an enormous *canyon complex* rather than a single great canyon, and point out that its heavy vegetation precludes anything like the vast panoramas of the Grand Canyon, which lies nakedly open in the high desert sun, presenting seemingly endless vistas of bare vertical rock throughout an almost unimaginable volume of space.

A great deal has been written about the history and exploration of the Grand Canyon, beginning with the John Wesley Powell expedition's 1869 traverse of the Green and Colorado rivers[12] and continuing to the present day; and a great deal more has been written about its geology and geomorphic origins.[13] But those exhaustively treated subjects are of little concern here. What is our concern is whether or not anything of such astonishingly complex form and gigantean scale can possibly be presented in something like a map . . . and, if so, how? We already know that a lifetime (many lifetimes) of photographs or paintings could not begin to do that completely; and we know that the canyon's extreme relief would make impossible any synoptic presentation of its features in an oblique map, and that satellite imagery could never hope to capture the canyon's immense depth and volume. So that leaves maps . . . but what kind of maps?

The U.S. Geological Survey, in 1962, issued its 1:62,500–scale map of *Grand Canyon National Park and Vicinity* (figure 6.9), covering much of the area of eleven of their fifteen-minute topographic maps. This was hardly the USGS's first published map of the Grand Canyon, but it was the first in quite a while (a minuscule note here says, "Supercedes map dated 1923") and clearly the Survey's first this ambitious. Of course, it's a very large map, measuring sixty by thirty-eight inches overall, but, apart from that, very much in USGS house style.

This map is printed in the standard four colors of a USGS topographic map: contour lines in brown, hydrography in blue (cyan), vegetated areas in USGS's signature light green, roads and trails and structures and all feature annotation in black, national forest boundary in a patterned black line and national park boundary in patterned black line with a band of red tint, and section lines and labels in red. Surrounding the map are the usual requisite elements: title, date of publication, credits, identification of projection and coordinate system, an accounting of source data, an index of its eleven constituent fifteen-minute maps, ratio scale and scale bars (in miles, feet, and kilometers), magnetic declination diagram, location map, and a fourteen-item legend of a sort (called simply "Explanation").

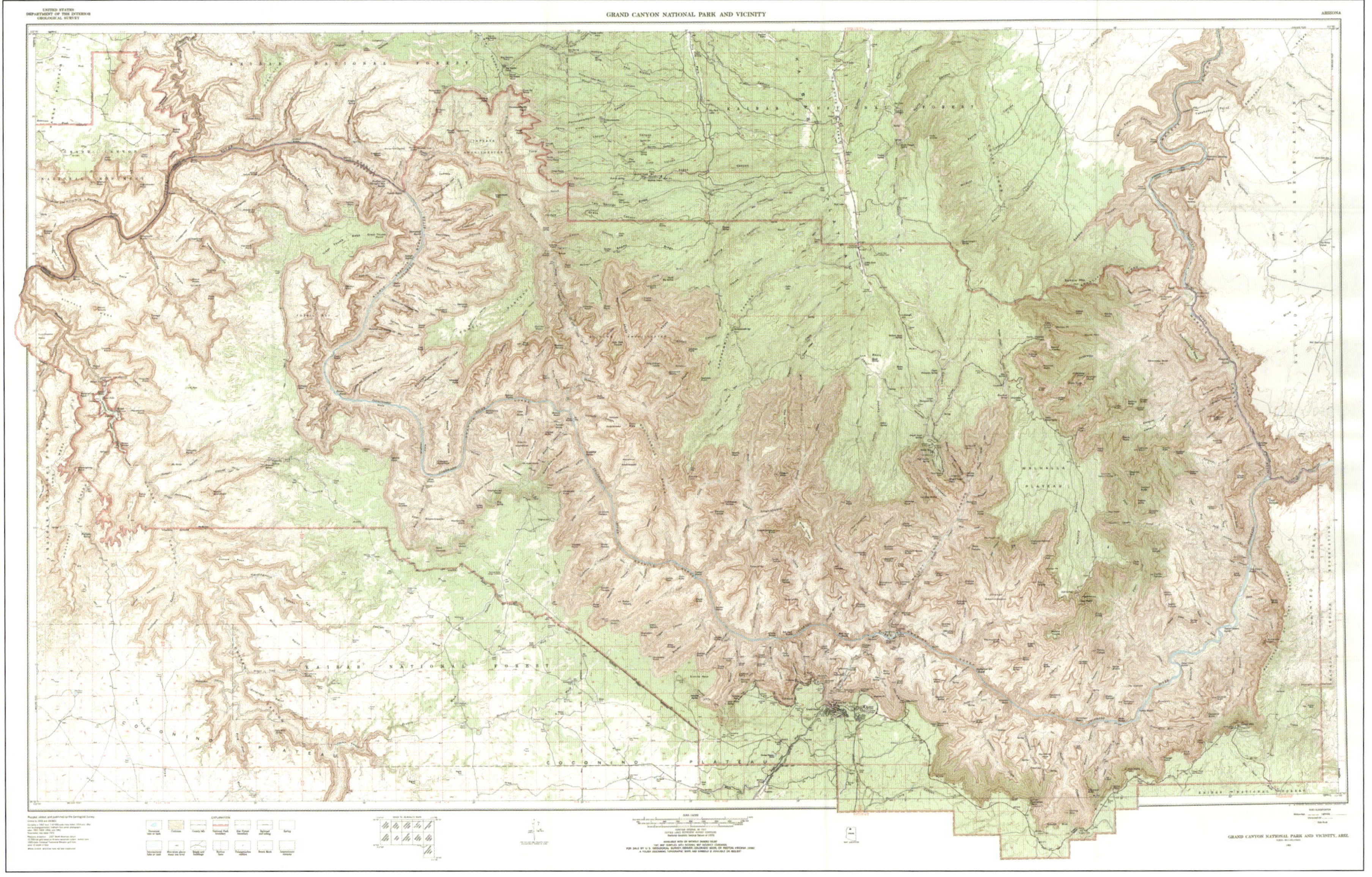

Figure 6.9
Grand Canyon National Park and Vicinity. Can one imagine a more challenging or rewarding subject for a topographic map? Map dimensions (WxH): 60in. x 38in.
Source: Courtesy of USGS, 1962.

Can a stripped-down standardized symbology like the one employed here even begin to describe a place like the Grand Canyon? As it turns out it can, and more than a little bit convincingly. Viewed from a distance of a couple or a few yards, this canyon carves its way right through its presentational plane and into the wall or floor behind it. The endlessly varying browns of the canyon's aggregated contours, from Havasu Canyon in the west to Marble Canyon in the east, contrast starkly with the gentle greens of the Coconino Plateau to the south and the Kaibab and Walhalla plateaus to the north, and the nearly blank white of the Painted Desert and Navajo Lands to the east.

There is definitely something going on . . . something very, very, big. The immediate transition from gently sloping plateau and desert to this astonishing colossal gash in the earth is nearly as jarring here as for the neophyte, digital camera–toting tourist approaching the canyon rim for the first time ("HOLY . . . !"). And all those—*all those*—endlessly convoluted contour lines collude in an infinitely varied systemic dance—here spreading apart to blend a light brown on a (relatively) shallow slope, there crowding closer together on a steeper slope where they collectively paint a medium brown, and there nearly touching to form bold, dark brown bands that leap up from the visual plane and drop us precipitously off cliffs and escarpments. (This is making us dizzy. Who on earth actually *drew* all these contours anyway? They weren't drawn by a computer.)

Approaching the map more closely (figure 6.10), the contours' systemic gestalt begins to resolve itself into delineations of the canyon's countless individual features—the myriad tributary canyons, the points and peaks and plateaus, the towers and buttes and palisades, the fancifully and mythically named (many by Powell) temples and castles. This is the stuff of photographs and paintings—this is the *scenery* of the place.

Figure 6.10
Detail of *Grand Canyon National Park and Vicinity.*
Source: Courtesy of USGS, 1962.

Approaching very closely now, the names of all those features become legible and, with the aid of a magnifying lens, even individual contours can be distinguished and the vertical extent of their features estimated. The elaborate semiotic *system* of the contours resolves itself into its *elemental* components.

And, stepping further back again, the *synthesis* of this map's semiotic systems becomes readily apparent—the conversation between the twisting blue line of the progenitive river and its confining walls in the inner canyon, the tracing of the precipitous edge of the south rim by the tourist road from Hermits Rest to Yaki Point, the commingling of brown lines and light green wash that renders a distinct olive color in the few places where vegetation descends into the canyon, as it does along a small portion of the South Rim from Shoshone to Grandview points and around Moran and Zuni points, and along the expanse of steep slopes below the east rim of the Walhalla Plateau.

For Pete's sake, just how many maps do we really have here anyway? At least three to be sure, but maybe as many as we like, that is, as many as we choose to cognitively construct. It's unfortunate that, in so many semiological and even deconstructionist analyses, the role of the reader/viewer has been glossed over or just ignored altogether, as if all we had to do was regard an image and its meanings would automatically and immediately evoke our response or take up residence in our consciousness. (Umberto Eco calls such stimuli *signals*, not signs.[14] We're looking at signs here.) Maps don't make meaning; *people make meaning*, and it would be much more realistic to say that the visual stimuli of the map trigger the production of meaning on our part, and to recognize that our role in that regard is every bit as motivated as that of the cartographer who created the map or the agency or organization that sponsored it.

But the USGS couldn't just let the Grand Canyon go with their 1962 treatment. Ten years later they published another version of the map, in shaded relief (figure 6.11). The statement "Shaded Relief Edition of 1972" does not even appear in the map's title but only in small brown type in the lower right corner of the margin, below the statement that still reads "Grand Canyon National Park and Vicinity, Ariz. 1962." Alright, it would be difficult to fool the blind with that. This isn't just a different map; it's obviously a very different map, the same perhaps in nearly every respect but one, and that one makes all the difference. The light green vegetation is gone now, replaced by a richly detailed and exquisitely rendered shaded relief; and where that green had emphasized the abrupt transition from the adjoining plateaus to the plunging depths of the canyon itself, that purpose is served here by a subtle wash of orangish brown over the plateaus and desert. The overall impression of the orangish brown wash, brown relief shading, and brown contours is of a very dry place . . . very dry and very hot. The sun beats down on this landscape like a sculptor's hammer, exposing the fine details of canyon and arroyo, cliff and escarpment, peak and palisade. Even geologic faults reveal themselves for scrutiny.

And we thought that all those contours were impressive. Combine this shaded relief with those and it's almost enough to give a cartophile a heart attack. This is as real as it gets, or so it would seem, unless perhaps the park ranger's attention could be diverted and we could crawl across the plate glass over their huge three-dimensional model of the canyon. Of course, there are those vacuum-formed three-dimensional Hubbard maps[15] as well, but they're just not this . . . this *big*. This map is big in every respect—big in size, big in content, big in the scale of the landscape, big in visual depth and sheer visual impact, big in ambition. No, big isn't a big enough word—it's . . . grand. And if the Grand Canyon has often been likened to a mountain range turned upside down, that's exactly the impression we're getting here.

But just how *real* is this anyway? Is this what the Grand Canyon really looks like, really feels like? The stripping away of all the vegetation allows us to see the landscape through a geomorphologist's eyes, right down to the thin veneer of soil and the chiseled face of bare rock; but it all seems rather unearthly—rather, well, Martian perhaps. It would not have been impossible to show both shaded relief and vegetation—that's just what National Geographic's "The Heart of the Grand Canyon"[16] does (another impressive map that, regrettably, we were not able to reproduce here). But the point of this map is not to look real anyway, as unflinchingly realistic as it seems. The point is to tell us something specific, something special, about the physical and material form of the Grand Canyon—about its physiography, its geomorphology in the classic sense. And the point is to take us by surprise, to make our heads snap back from its sheer visual impact, to make us stare open-mouthed at the incomparable scale and complexity of the place, and, most of all, to reflect that raw and majestic beauty in and on the map itself. The point, put simply, is to be Grand.

Before leaving the Grand Canyon, though, there is one more map we need to consider, and that's National Geographic's *Grand Canyon National Park* (figure 6.12). This is one in their Trails Illustrated series, which includes an impressive array of ". . . more than 60 maps of areas within the national park system and more than 100 maps covering national forests and recreational areas across the United States."

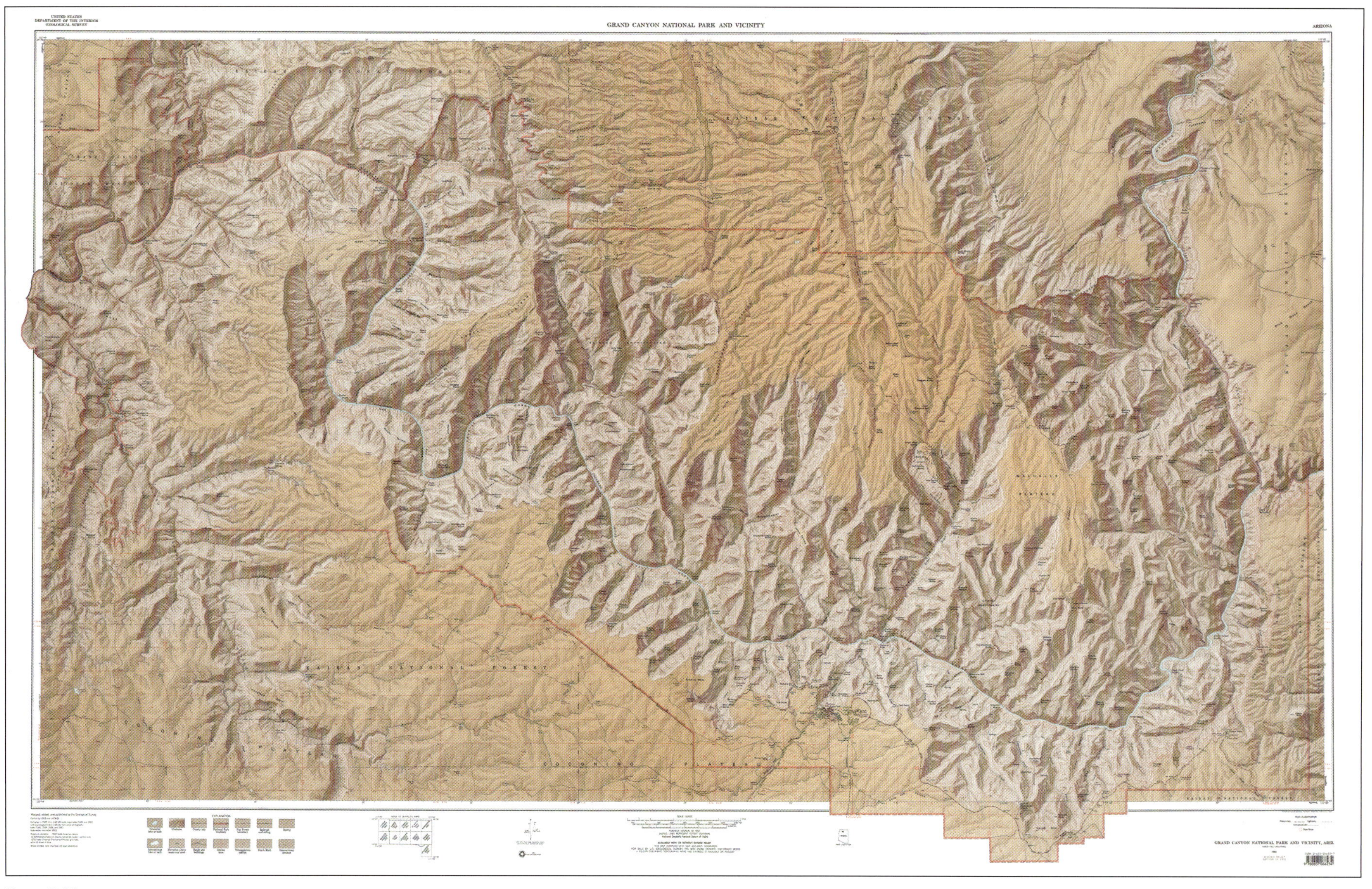

Figure 6.11
Shaded relief edition of *Grand Canyon National Park and Vicinity.* This image is almost as breathtaking as the canyon itself.
Courtesy of USGS, 1972

(There are an additional eleven such maps in the Grand Canyon region alone.) This one is typical of the series, a twenty-five-by-thirty-seven-inch, twenty-seven-panel map that folds down to a backpack or (rather large) pocket size of nine and a quarter by four and a quarter inches (the standard size fitting National Geographic's commercial rack displays). It's printed in smear-proof ink on a waterproof material (we're guessing Tyvek, but National Geographic just calls it plastic) and is ostensibly intended for the hiker or backpacker, who would undoubtedly make short work of a less durable and less substantial product.

The cover panel of the folded map is of the standard design for the series, with a block of National Geographic yellow at the top containing the Society logo and map number (207) and simple title, *Grand Canyon National Park, Arizona, USA.* Running up the left edge of the panel to the bottom of the yellow block is a longer and narrower green block or band with larger type that identifies this as a "Trails Illustrated Map." Wedged between these is a colorful collage of a green location map of the lower forty-eight states superimposed on a small-scale snippet of the map to be found inside, fading down to a wide-angle photograph of Bright Angel Canyon. At the very bottom it says, "Topographic Map /Revised Regularly/Waterproof/Tear-resistant."

On the flip side of the folded map is an index map of "Regional Trails Illustrated Titles," a legend consisting of forty-three map symbols and codes (we'll return to the more important of those), scale bars and ratio scale, statements for contour interval (100 feet) and map projection, and a cryptically cautionary note that reads, "Distances within canyon are typically longer than they appear due to drastic elevation changes and switchbacks" (figure 6.13).

Opening the fold reveals the third cover panel (figure 6.14), which begins with a paragraph telling us that "National Geographic Trails Illustrated Maps are based on USGS information, modified

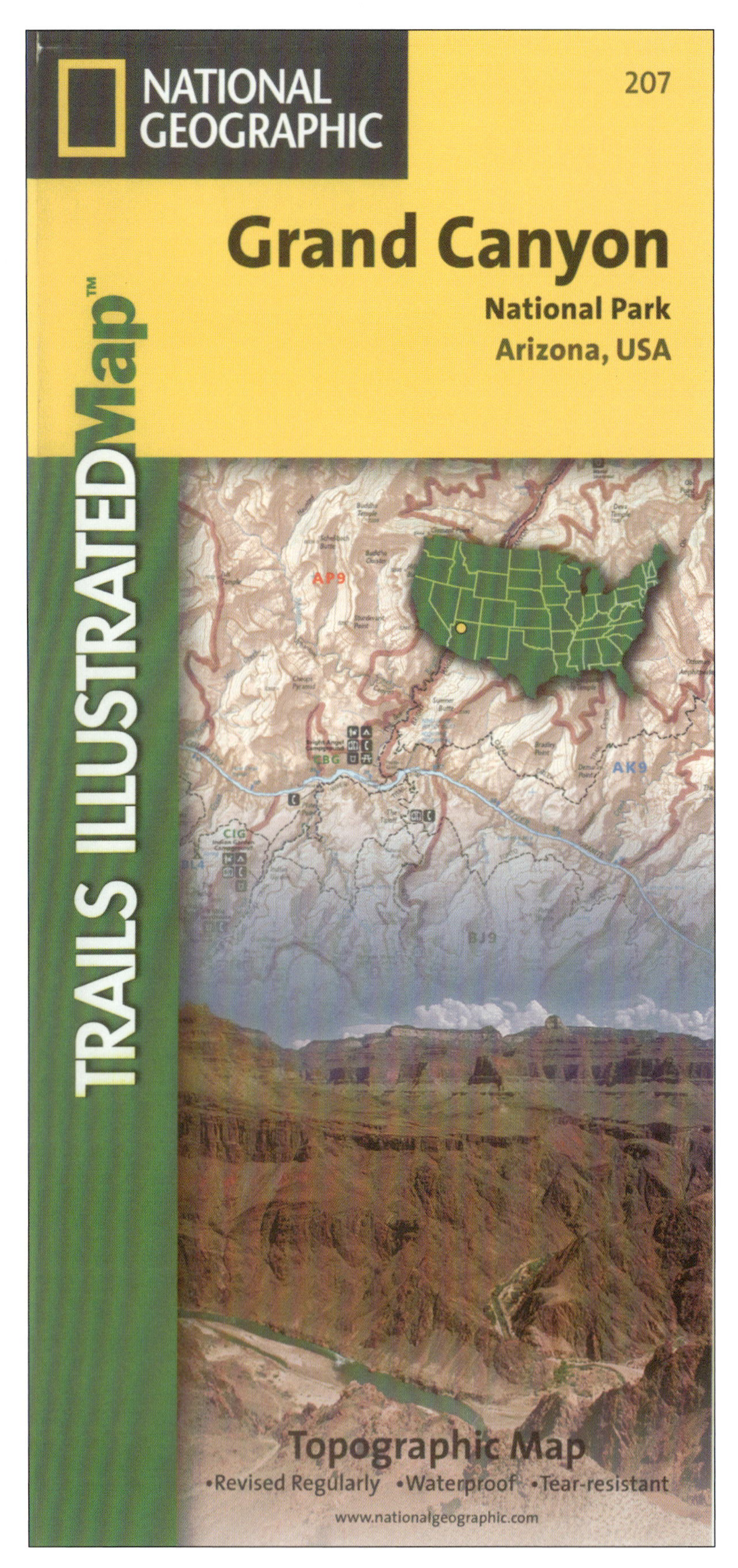

Figure 6.12
The title cover panel of *Grand Canyon National Park* from National Geographic's Trails Illustrated series.

Figures 6.12–6.15 courtesy of Trails Illustrated/National Geographic Image Collection.

and revised by National Geographic Maps in cooperation with the National Park Service, Bureau of Land Management, and USDA Forest Service," and "We welcome your input if you discover conditions other than as shown on the map." A second paragraph assures us that "National Geographic Maps are printed on a 100% plastic material, which is waterproof and tear-resistant with normal use," but also warns to "Keep your plastic map away from petroleum products, such as solvents, stove fuel and sunscreens, which can potentially smear the map's ink." An additional three paragraphs forbid unauthorized reproduction and exempt the publishers from responsibility for personal injury or property damage, remind us of the Society's goals and research and education programs, and provide general and contact information for National Geographic Maps. Finally, there is a brief statement about and example of the UTM grid and GPS waypoints provided in the map.

Having prepared ourselves, we can now pull the thing open and see what's inside. There we find the three cover panels laid out across the top of the sheet above a largish, medium-scale ("approximately" 1:63,360 or one inch equals one mile), pastel-hued map of Grand Canyon National Park, East Side (figure 6.15). In the upper right of that is a listing of "Suggested Hikes and Backcountry Trips," consisting of ten "Suggested Day Hikes" originating from the south or north rims, five "Suggested Backpacking Itineraries" from the two rims (one or two days each), and an additional three "Experienced Canyon Hiker Itineraries." In the lower right is another listing, of "Closed or Restricted Areas," with twelve closed to camping but open to day use and two others closed to all use—the Hopi Salt Mines, which are closed except by permission of the Hopi Tribe, and the Furnace Flats archaeological site. In the lower left and stretching across much of the bottom of the sheet is a listing and classification of "Backcountry Use Areas" along with their respective alphanumeric codes that link it to the indexical list, where it is named and categorized in degree of difficulty and danger. These are presented in four categories: Corridor (coded in green), Threshold (in blue), Primitive (in black), and Wild (in red), and that is echoed in a sort of bar diagram consisting of four colored bands and labeled "Less Experience Required" at one end and "High Degree of Experience Required" at the other. There are only four corridor use areas, which, like the Bright Angel area (CBG), are "recommended for hikers without Grand Canyon experience" and which might furnish such amenities as ranger stations, permanent structures, sanitation, purified water, signs, bridges, trails, and emergency phones. Threshold use areas are "recommended for experienced canyon hikers only" and, like Deer Creek (AX7), Point Sublime (NH1), and the nineteen others in this category, may have nonpermanent structures, sanitation, and signs but are also unmaintained trails with "scarce or no water." The twenty-nine primitive use areas, like Swamp Ridge (NJ0) or Cape Solitude (SA9), and the thirteen wild areas, like Phantom Creek (AP9) or Thompson Canyon (NB9), are "recommended for highly experienced canyon hikers only" and, while both of those categories offer scarce or no water, the primitive use areas have unmaintained trails and routes while the wild areas have "indistinct to nonexistent routes requiring advanced route finding ability." These are not places where one would want to get lost.

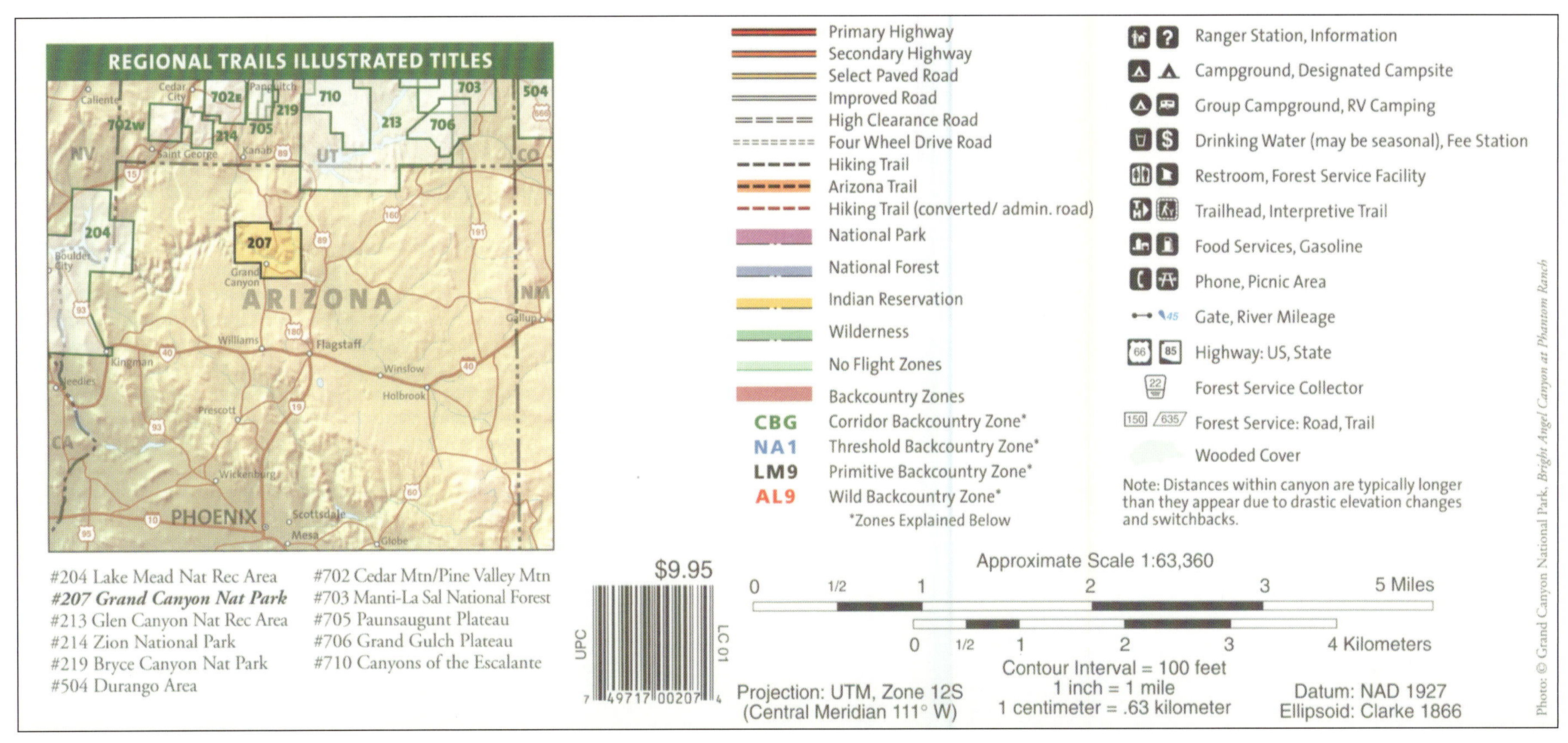

Figure 6.13
The reverse cover panel of Trails Illustrated map of *Grand Canyon National Park.*

On the reverse is a similar map of the West Side of the park (figure 6.14). In the upper right is a brief lecture on safety, with two paragraphs devoted to emergencies (one admonishing that "helicopter evacuations are used in emergency situations only" and that they are "very expensive"). Below that is a bulleted list under the heading "Hike Smart" that advises the hiker to stay hydrated, eat often, rest often (sit in the shade), and to "choose any season *except the summer* to hike in the canyon." We are also informed that cellular phones do not work reliably in the canyon, and given the warning (in red type), "DO NOT attempt to hike down to the river and back in one day." (This warning is also provided on the East Side map.) Also in the upper right are six paragraphs of instructions under the title "Leave No Trace." We are advised to "Plan Ahead and Prepare," to "Travel and Camp on Durable Surfaces" (at least two hundred feet from streams), to "Dispose of Human Waste Properly" (urinate at least two hundred feet from creeks, trails, and campsites; bury feces in a six-inch deep "cathole" at least two hundred feet distant; and pack out rather than burn or bury toilet paper), to "Leave What You Find" (and to not build structures or furniture, dig trenches, or burn campfires), to "Respect Wildlife" (observe wildlife from a distance and never feed animals), and to "Be Considerate of Other Visitors" (among other things, avoid using loud voices or making unnecessary noises).

Along the left edge of the sheet, we find some notes on "Information and Permits," including the Backcountry Information Center and the North Rim Backcountry Office, the Grand Canyon Association, hotels and car campgrounds, Phantom Ranch, and Havasupai and Navajo Tribal Lands.

Most of the lower three panels of the sheet are occupied by a 1:500,000–scale map of the Grand Canyon region from Lees Ferry to Pearce Ferry, including much more designated national park land than is shown in the two trails maps, the Grand Canyon Parashant National Monument, Vermilion Cliffs National Monument, Kaibab National Forest, Grand Wash cliffs, Mount Logan, Mount Trumbull, Kanab Creek, Saddle Mountain, and Paria Canyon-Vermilion Cliffs Wildernesses, and the Hualapai, Havasupai, Navajo, and Kaibab-Paiute reservations.

Finally, in the lower right, is a list of no fewer than seventeen "Backcountry Regulations": a backcountry use permit is required; stay on the trail; carry out your trash; wood or charcoal fires are prohibited; writing or scratching on natural features is prohibited; all weapons, traps, and nets are prohibited; dogs and other pets are not allowed below the rim; and so on. Transgression of any of these regulations "may result in a citation, a fine and/or an appearance before a United States magistrate."

Not atypically for a National Geographic map, this one is absolutely bristling with perimap content that, among other things, explicates the map's symbology, fixes the map's geographic and geodetic positions and boundaries, fixes its temporal position, positions the map in the context of its cartographic and corporate cultures, cites its sources of authority (U.S. Geological Survey, National Park Service, Bureau of Land Management, USDA Forest Service), establishes claims of authorship and ownership, delimits the map's legal responsibility and liability, and establishes links to the map's epimap (the Society's mission and programs, reference to their other publications,

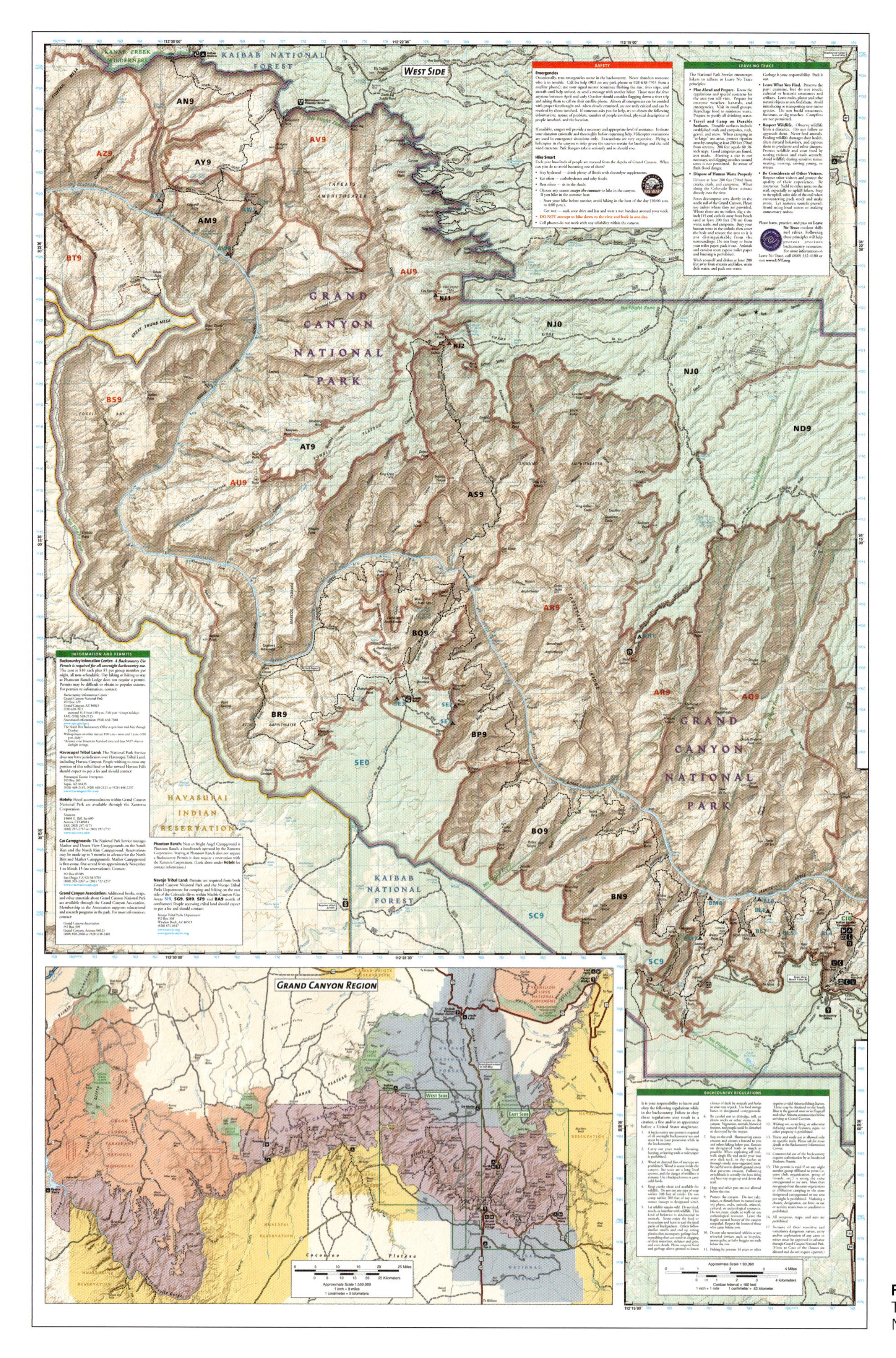

Figure 6.14
Trails Illustrated map of Grand Canyon National Park, West Side.

Figure 6.15
Trails Illustrated map of *Grand Canyon National Park*, East Side.

an address, two phone numbers, a Web site). And all that is from just the three exposed panels of the folded map. On the inside there was more, providing suggestions for the map's use as a backcountry guide, laying down more cautions and restrictions for the hiker/camper, enumerating and codifying regulations for backcountry conduct, identifying sources of information and requisite permits, providing a thorough lecture on hiking safety, and an even more thorough lecture on minimizing one's impact on the canyon's environments and ecosystems. Alright. This is a far cry from the minimal offerings dressing the margins of those USGS maps. Here the perimap is barely peri- at all, penetrating and overwriting the map image itself with its recommendations, rules and regulations, cautions and warnings. But what of the map image, the map's core text?

The overall impression is of another richly detailed map. Again, there is that dense convolution of contour lines (here in a more reddish brown) and a beautifully rendered, reddish brown shaded relief that is subtly different from USGS's. But there is also a light green tint for "Wooded Cover" that is pervasive except in the desert lands to the east and northeast. That is interrupted patchily by exposures of bare substrate on the adjacent plateaus, but hardly to a degree that gives them much visual presence (they appear to be mainly correlated with steeper slopes). The light green tint even appears in many places within the canyon itself—mostly on the more northerly, less exposed sides of features, opposite their more southerly and more exposed (but shaded here) faces. This is quite a different scenario from that found in USGS's 1962 topographic map, where their interpretation of "Woods" appears to be much more conservative than National Geographic's. But just how wooded qualifies as wooded or woods anyway, and what degree of cover qualifies as cover? These are, after all, *a priori* conceptual constructs, and their boundaries are very elastic.

In addition, there are purple lines for national park boundaries, blue lines for national forest, green lines for wilderness, and deep yellow lines for Indian reservation. Light green lines delimit no-flight zones. The thin ribbon of the Colorado River appears again in a light blue. Primary and secondary highways and select paved roads appear as red, orange, and yellow lines with thin black casings, and three classes of unpaved road, from improved to high clearance to four wheel drive, are shown in broken or unbroken black double-line. Hiking trails appear as broken black or red lines. The various point symbols, from ranger station and information to phone and picnic area, appear as abstract white figures on round-cornered black squares, Park Service style.

This is quite a handsome map. And, while it doesn't have the mass and solidity of USGS's shaded relief version, it does have an airy and voluminous feel that is quite appealing. It would, anyway, if it weren't for all those thick, dark red lines that delimit the sixty-seven aforementioned backcountry use areas—called in this legend Backcountry Zones.

This is indeed a very different map from USGS's. Those dark red boundaries of all those variously sized and shaped wilderness zones are second in visual prominence only to the shaded relief, and they plat the canyon like a subdivision. They carve the canyon like a Thanksgiving turkey and serve it up on a plate . . . here's some white meat—a Corridor Zone—and there's some darker meat—a Threshold Zone—and there's some very dark meat—a Primitive Zone or a Wild Zone. While the USGS maps seem to regard the Grand Canyon with almost religious reverence, this map seizes the canyon as property . . . not just an astonishing *this is* but *this is there*, and *that* is another *there* and another *that* is another *there* still. This Grand Canyon is not to be gazed upon in wide-eyed wonder, not through this map anyway; it's meant to be experienced, physically as well as visually, with hiking boots planted firmly (one hopes) on the ground. It's meant to be *used*.[17] It seems that not even the grandest and most sublime of earth's natural wonders can be free of our claim and conquest and, ultimately, possession—all designs served very well by maps. But that's true of more than just the Grand Canyon.

The Mountain

If there's any doubt as to whether the Grand Canyon of the Colorado is the largest and deepest on earth, there is none regarding which mountain peak is the highest on earth. That title, as we all learned in grade school, belongs to Mount Everest. (A challenging claim for K2 was soundly rebuffed several decades ago.) Forming a dramatic border between Nepal and China (Tibet), Mount Everest—"earth's third pole"—is the crown jewel of the High Himalaya, rising to a mind boggling 29,028 feet above sea level. But this mountain is only one of a string of astonishingly high peaks along the spine of the Himalaya: Lhotse (27,890 feet), Lohtse Shar (27,513), Pethangtse (22,106), Khumbutse (21,867), Lingtren (22,142), Pumori (23,507), Nuptse (25,791), and others all rise to thin air within view of Everest's summit. (There are fourteen 26,000-foot peaks on earth, and they are all part of the Himalayan Range.[18]) In fact, it is difficult to say just where one mountain ends and another begins along this 1,500-mile "rooftop of the world," formed by the folding and thrusting of crustal rock resulting from the collision of the Indian and Eurasian tectonic plates.

That remarkable chunk of our planet is the subject of National Geographic's 1988 "Mount Everest" (figure 6.16), an unusually beautiful topographic map of the mountain and its immediate surroundings.[19] This map was not a small project, but an international effort on a scale worthy of the mountain itself. Across the top of the map sheet there are three, not one, titles: in the center in English, to the left in Nepali (Sagarmatha), and to the right in Chinese ideogram (Qomolangma). In the lower left corner of the margin, there are scale bars, a ratio scale (1:50,000), and a scale statement, along with mention of the contour interval (40 meters). There is also a sixteen-item legend, with entries ranging from stone walls, shelters, and houses to lake, river, and crevasse, to rock and scree, to low forest and bushes. Below that is a glossary of Chinese, Tibetan, and Nepali terms and, occupying the next column of text, there is a listing of place names, which ". . . were provided by linguistic experts in consultation with the Chinese and Nepalese mapping authorities." Next, there is a four-part alphabetical listing of contributors, including institutions (sixteen of these from the United States, Nepal, Great Britain, Sweden, Switzerland, and West Germany), individuals (seven of those), special assistance (another twenty-seven individuals and their host institutions, including cartogra-

Figure 6.16
"Mount Everest," supplement to the November 1988 issue of *National Geographic*. This is how the Society saw the mountain in the late 1980s. Map dimensions (WxH): 22in. x 36in.

Figures 6.16–6.18 courtesy Ng maps/National Geographic Image Collection.

Figure 6.17
Detail of *National Geographic's* 1988 "Mount Everest."

phers, photogrammetrists, photographers, printers, pilots and copilots, and a meteorologist), and lastly two consultants. There is also a small, but not-so-simple reporting of technical data telling us, among other things, that "Principal points of control have been determined by aerotriangulation, using vertical photography from a West German metric camera aboard U.S. space shuttle *Columbia*, December 1983, as well as early maps of British, Chinese, and Austrian surveyors." In the lower right corner, there is a reiteration of the map's three titles, along with date of publication and authorship credits for *National Geographic* magazine and The National Geographic Society Committee for Research and Exploration, joint researchers The Henry S. Hall, Jr., Everest Fund of Boston's Museum of Science, and supporters His Majesty's Government of Nepal and the National Bureau of Surveying and Mapping of The People's Republic of China. And that, finally, constitutes the perimap bracketing the map image on this side of the sheet.

But wait . . . there's more. On the reverse side, there is "High Himalaya: A Computer-Generated Landscape Portrait," a sweeping oblique view of the Himalayan Range, complete with key map and a ten-item color key. There are also an additional four columns of text describing the image itself, and the region's geology, climate, ecology, and demography. And there are two more columns about "Hi-Tech Land-Cover Mapping" informing us that "To create this perspective view, a team of computer experts combined satellite pictures with conventional topographic maps. Using elevations from these maps, they took otherwise flat pictures and created the illusion of a three-dimensional landscape."

We're not going to discuss that side, though—let's just say for now that it's a foretelling of things to come—so we flip back to the title side. This map is nothing short of astonishing in its clarity and degree of detail. The myriad ridges are rock solid yet razor sharp, the glaciers —the entire map in fact—icy cold, and the thin air as transparent as a vacuum. The exquisite topographic rendering is a combination of contours, relief shading (in both grayscale and ice blue), and "rock drawing" a lá Eduard Imhof.[20] Come to think of it, this doesn't look like a National Geographic map at all; instead, it has the distinctive look of a Swiss topographic map. A bit unsettled, we scan the accompanying text again and, sure enough, under "Technical Data" there is a tiny credit saying "Cartographic work and reproduction by the Swiss Federal Office of Topography in Wabern, 1986–1988." Now that makes sense. Typical of the signature style of Swiss topographic maps, this doesn't just reduce to ink on very close viewing. No, closer and closer inspection just reveals more and more detail, right down to individual glacier surfaces, crevasses, and bare rock faces. Did we mention that the maps in this chapter were rather, well . . . humbling? This map certainly is and, aside from the few trails and huts and tiny patches of vegetation (if you can find them), it's very hard to escape the feeling that one is staring directly into the face of a raw, eternal, and inscrutable nature, a sort of geologic Moby Dick. And knowing how and why it is what it is doesn't help much at all.

And did we mention that this mountain *kills people,* at least those with enough bravado to try to claim its summit? Nearly two hundred have lost their lives there since George Leigh Mallory's first attempt in 1921, falling from cliffs or into crevasses, buried by avalanches, or perishing from exhaustion or exposure or hypothermia or pulmonary or cerebral edema in the mountain's "kill zone" above twenty-five thousand feet.[21] Even the famed Sherpas, who now contract their services as climbing guides for westerners, never challenged the spirits of their revered mountain before the advent of Western interests.

This map, and the panorama on the reverse, give us an Everest apart from and pretty much untouched by the human species; but we know that is not really the case. Until the 1980s, only one party was allowed to occupy the Nepalese base camp each climbing season. In 1983, that was increased to five parties, each on a different climbing route, and, in 1987, Nepal began allowing multiple permits on routes. The base camp population swelled from about a hundred to five or six hundred per season, and the base camp itself became a sort of international mini-city. And that's not to mention the thousand or so Sherpas constantly coming and going, portering climbers' baggage and provisions into and out of camp, or the tens of thousands of trekkers who come for a possible glimpse of the mountain or maybe even a famous mountaineer. At base camp, would-be climbers can show up with their $65,000 permits, but otherwise nearly empty handed, and purchase climbing equipment, food, coffee, beer, chang (the local alcoholic beverage), and just about anything else they might need on site. The camp has degenerated into something between a human circus and trash dump, with tents and gear and burning piles of rubbish all over the place. It's also become notorious as a hotbed of lechery and debauchery, with adrenaline-filled climbers waiting impatiently for the skies to clear enough to afford a frenzied dash for the top. The Sherpas do all of the work and there's very little for the climbers to do from day to day but play cards and drink and tryst . . . and fight, a lot. Petty thievery has become a problem as well, and many tents now have padlocks on their zippers. And a good night's sleep is not easy to come by. Base camp is a very noisy place, filled with a near round-the-clock cacophony of crunching boots, coughing, vomiting, diarrhea, satellite phones and radios, portable stereos, and even electric guitars. The juniper forests of the surrounding lowlands have been almost entirely stripped and burned, and even the mountain itself is littered with oxygen tanks and assorted other climbing gear abandoned by exhausted climbers over the years.[22] For the Sherpas, Sagarmatha is not just a mountain . . . it's a god, but that's a view rarely if ever shared by the climbers. They haven't come to revere or even admire the mountain (its peak is hidden by clouds most of the time anyway); they've come for a very different reason—they've come to challenge the mountain, maybe even to conquer it . . . or maybe to die trying.

Fifteen years later, National Geographic gave us another vision of Everest (figure 6.18),[23] one that makes their 1988 map look like a cherished antique (which, in fact, it is). Again, there could be a little doubt for some as to whether this is or is not a map, but, again, who really cares? This is yet another spectacular image, and this time the mountain (the western side of it anyway) swells across the full dimensions of the sheet. Atypically, there aren't even any borders or margins; the image just bleeds right off the page. And what an image it is—an eye-popping, three-dimensional portrait map created from literally "millions of measurements" (but "none was made by hand" says National Geographic). The result is a Mount Everest so real that we feel we could reach out and touch it; but there's no need to do that since the thing all but leaps off the page and into our laps. And, also atypically, there's very little perimap here. The map's title occupies the usual position at upper left, along with the Society logo and credits and a promotional paragraph introducing the map. In the lower right is another short paragraph titled "Everest Reborn from a Mountain of Data," which encapsulates the three-step process employed in creating the image. Below that, there is another paragraph titled "Website Exclusive" that begins with, "View a 3-D presentation of Everest, manipulate a 360° view from the summit . . . ," and below that are individual and corporate credits for design, GIS, production, text authorship, research, consulting, and aerial photography and digital terrain data. And that's pretty much it.

Of course, there has to be a lot of perimap material here somewhere (this is a National Geographic presentation after all); that's just all on the reverse, which is titled "Everest 50." There we find, occupying more that half the sheet, a shaded relief topographic map of Sagarmatha National Park, complete with legend, land-use color key, some translations of geographic terminology, more credits, and location maps of both the region and the globe. Associated with this map is a prominent text titled "Land of the Sherpas," but there are also, surrounding the map, lesser texts taking up a variety of topics: weather, glaciers, geology, cold laboratory, regulating herds, sacred mountain, and a half dozen others. To the right are two small texts titled "Tibetan Borderlands" and "Sherpa Homeland," each with its own trademark National Geographic dioramic collage of people, places, animals, and plants. To the left, below this side's title, are two three-dimensional isometrics showing northern approaches and southern approaches, each with its own accompanying paragraph, and jointly mapping out the routes of sixteen of the more famous climbing parties. Between these is a flow-line map showing the paths and volumes of visitors in 1980 and 2000. The increase in traffic over that twenty-year period is well illustrated in this little map and is nothing short of astonishing. The accompanying paragraph here says, "Sixty-two visitors reached Nepal's Khumbu region in 1964, when the

Figure 6.18
"Everest," supplement to the May 2003 issue of *National Geographic*. This high-tech tour de force is how the Society saw the mountain at the beginning of the twenty-first century. Map dimensions (WxH): 31in. x 20in.

airport at Lukla was built. In 2001 more than 21,000 arrived—about six times the resident Sherpa population," but adds that "Relatively few people come to climb Everest."

Flipping the sheet back over, we are reminded why all those visitors come. This is indeed a remarkable place, without doubt one of the most remarkable on earth, and this image makes that point very strongly. There's not much annotation here; there are labels for Everest's summit and for its neighboring peaks, Changtse and Lhotse, and for the Lhotse Face and the North and Southwest Faces and West Ridge of Everest. There are labels for the Rongbuk Glacier, Khumbu Glacier, Khumbu Ice Fall, and Western Cwm (cirque). The locations of the 1953 and present-day base camps are indicated, and the routes of Hillary's 1953 team, the Chinese expedition that was the first to successfully summit Everest from the north side in 1960, and the first successful American expedition in 1963 are traced out. There are also a few small notes, about Shih Chan-Chun's 1960 ascent, James Whittaker's 1963 ascent, and, of course, the 1953 British ascent by Sir Edmund Hillary and Tanzing Norgay.[24]

There is a nearly invisible north point in the lower left, but no expression of scale—just a brief note saying "Scale varies in this perspective. Contour interval is 100 meters." Fine, we get the point anyway—this is a very big mountain. And the superrealism of the image only makes it seem bigger, as Everest rises up to dwarf even enormous adjoining features, like the glaciers and ice falls at its base. The finely detailed textures of all that rock and snow and ice only underscore the overwhelming scale of the mountain, and the sharply

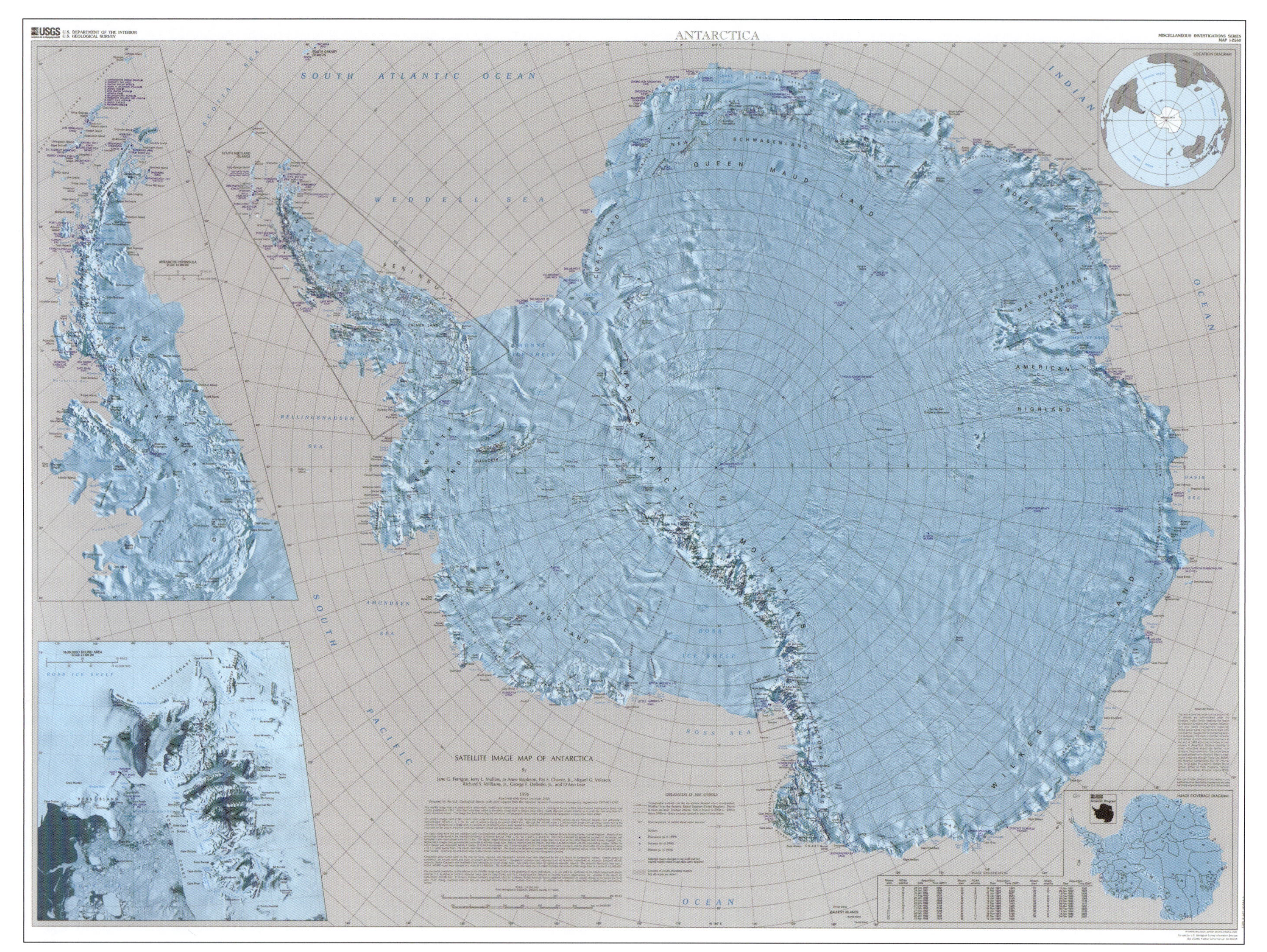

Figure 6.19

Satellite Image Map of Antarctica. This remarkable image is simultaneously forbidding and extraordinarily beautiful. Map dimensions (WxH): 56in. x 40in.

Courtesy of USGS, 1996.

honed ridges and the deep shadows they cast bespeak its immense depth and volume.

Again, wow! Mapping technologies have come a long way in the fifteen short years between the Geographic's two Mount Everest maps, which now seem to stand as icons of an old and of a new cartographic tradition. While the 1988 version is as painstakingly wrought and as skillfully crafted as a fine Swiss watch, and consummately cool, this image is very much in-your-face. Rather than a measured, contemplative overhead gaze, this is a full frontal assault, and it leaves us reeling. Never mind that we can see only half of the mountain here—we're seeing enough. And never mind that we're not even sure this is a map at all. This is a truly awe-inspiring image, and that, of course, is just what it's supposed to be. It tells us, in the most emphatic terms imaginable, that *National Geographic* has arrived in the twenty-first century.

The Continent

Finally, here is the U.S. Geological Survey's *Satellite Image Map of Antarctica* (figure 6.19). It's another big one and, as with those Grand Canyon maps, it demonstrates convincingly that, in this case anyway, size does matter.

There's virtually nothing in the margins of this huge sheet. More or less centered above the main image is a simpler title—just "Antarctica"—and in the upper right corner there is the map's institutional designation, "Miscellaneous Investigations Series, Map I-2560." In the left corner of the upper margin are the usual credits for the U.S. Department of the Interior and the U.S. Geological Survey, and the new USGS logo, now sporting the slogan, "Science for a Changing World." There are actually five maps here. The full map of Antarctica, at a scale of 1:5,000,000, consumes about three-quarters of the sheet. In the upper right is a "Location Diagram" (for those who don't know where Antarctica is—don't laugh, they're out there) with the South Pole dead center in a polar projection and with meridians radiating out to its edge in all directions. In the lower right there is an "Image Coverage Diagram" divided into thirty-eight mosaic areas and accompanied by a table associating those areas with specific NOAA satellite numbers and dates and times of image acquisition. At left are two enlarged portions of the main map—one of the Antarctic Peninsula, or Palmer Land, at a scale of 1:3,000,000 and, below that, another of the McMurdo Sound area at 1:1,000,000. At the lower center of the sheet is the map's full title, "Satellite Image Map of Antarctica, by Jane G. Ferrigno, Jerry L. Mullins, Jo Anne Stapleton, Pat S. Chavez, Jr., Miguel G. Valasco, Richard S. Williams, Jr., George F. Delinski, Jr., and D'Ann Lear, 1996." And there are also the statements, "Reprinted with minor revisions 2000" and "Prepared by the U.S. Geological Survey with joint support from the National Science Foundation Interagency Agreement OPP-9114787." There is a text block of five paragraphs describing the map's creation and geographic annotation, and citing a bevy of individuals and institutions for their assistance. There are, of course, dual scale bars and ratio scale, and a statement that reads, "Polar stereographic projection, standard parallel 71° South." There's another legend that chooses not to call itself a legend but an "Explanation of Map Symbols," showing symbols for three different types of topographic contours (varying in certainty), a small dot for spot heights, and three different types of research stations—Permanent (as of 1999), Summer (as of 1996), and Historic (as of 1996). There is a fine dotted line for "Selected major changes in ice shelf and (or) coastal margin since image data were acquired" and a dashed, diagonal-line pattern for "Locations of clouds obscuring imagery. Not all clouds are shown."

The main map and its enlargements are rendered in an icy blue surrounded by a light gray ocean (three oceans actually: the South Atlantic, South Pacific, and Indian) and indented by the Weddell, Bellingshausen, Amundsen, Ross, and Davis seas. The wrinkled spine of the Transantarctic Mountains runs across the continent from upper left to lower right, passing just below and left of the pole. (Cardinal directions mean little here since every direction from the South Pole is, of course, north.) To the left of the central mountain range are the peninsula and mountains of Palmer Land, the expanses of Ellsworth Land and Marie Byrd Land, and the immense Ronne and Ross Ice Shelves. At the bottom end of the Transantarctic Mountains is Victoria Land and at the upper end is Coats Land, and to the upper right of that are New Schwabenland, Queen Maud Land, Enderby Land, and MacRobertson Land. Just below that, the Amery ice shelf and American Highlands fringe the coast. The entire lower right third of the continent, from the South Pole nearly two thousand miles to the coast between the 90° and 150° east meridians, is the vast and nearly featureless desolation of Wilkes Land—undoubtedly the most remote and extreme place on earth. The map is laced with a one-degree (latitude) by five-degree (longitude) graticule and amply labeled with mountains, mountain ranges, peninsulas, capes, coasts, and bays.

We said this about *National Geographic's* 1988 "Mount Everest" but it's doubly true here. This map is cold—*extremely cold.* Even the rugged mountains don't expose any warm brown rock, at least not in this map. The entire continent is relentlessly ice blue and white, and the sunlight, which appears somehow to come from every direction at once, does nothing at all to warm this place up. It's almost surprising to touch the map and not feel one's fingertips stinging with cold. Once again, great credit is due the cartographers of the U.S. Geological Survey for another fitting and monumental tribute to a truly forbidding and monumental place.

But what of that place? Haven't we claimed, taken possession of, even that? Perhaps for now, fewer have reached the South Pole than have summited Everest, but there are refitted Russian icebreakers plying the surrounding seas, full of shivering ecotourists in search of breaching whales, cavorting penguins, and thunderously calving ice walls. Still, only the most dedicated of polar researchers ever take up brief residence in the continent's interior, which, even for those armed with the most hi-tech equipment, can still, like Everest, mete out death. In principle, all of Antarctica is dedicated to, reserved for, scientific research. It's a vast scientific laboratory of continental scale, yet we still know so very little about it, beset as it is by human-induced climate change, atmospheric ozone depletion, and runaway over fishing of its marine populations. It is international (or should we say multinational?) domain, with research stations established there at various times by the United States, Australia, United Kingdom, Norway, Sweden, Germany, France, Spain, Belgium, Poland, the former U.S.S.R.,

Russia, the Ukraine, Argentina, Chile, South Africa, China, Japan, and India. But would that climate of multinational accord and cooperation hold if significant oil reserves were discovered there? Thus far, subsurface mapping has revealed some interesting things, like a huge subglacial lake of ancient water, but as yet no oil. (We know that most of the continent was once tropical rainforest, so it's far from improbable that there are petroleum deposits down there somewhere.) We'd prefer, nonetheless, to think that this beautiful mapping of Antarctica (like the others in this chapter) was done not to service any potential exploitative designs but for the sheer wonder and love of the place—to borrow from George Leigh Mallory, "because it's there"—and the love of extraordinary cartography. Of course, the last thing we want is to be taken as naïve in this; we go to far too great lengths to be suspicious, perhaps even cynical. Still, this map is a genuine visual treat, and if there's any evil agenda hidden here, it's very well concealed beneath all the snow and ice. We just hope no one ever finds that oil.

NOTES

1. Tiny General Grant National Park (now General Grant Grove) was incorporated into Kings Canyon with that park's creation in 1940.

2. This is not to ignore the accomplishments of our Canadian friends. Canada has more than eight hundred national parks, provincial parks, and other protected areas encompassing nearly 100 million acres of wild lands. Established in 1885, Canada's first national park, Banff, was the world's second.

3. "The Ten Greatest Natural Wonders," Travel Channel, 2001.

4. This language is from "Earth's Greatest Natural Wonders," Travel Channel, 2006.

5. This termed was coined by J. K. Wright on pages 159–60 of "On Medievalism and Watersheds in the History of American Geography," in Wright's *Human Nature in Geography*, 154–67 (Cambridge, Mass.: Harvard University Press, 1956).

6. One of the earliest, but hardly the first. Yosemite's first photographer was Charles Leander Weed, and Carleton Watkins had made thirty large plates and one hundred stereoscopic pictures there in 1861. Muybridge, in fact, worked under Watkins for a while in the late 1860s. The early photographic history of Yosemite is well documented in Nyerges, *In Praise of Nature: Ansel Adams and Photographers of the American West* (Dayton, Ohio: Dayton Art Institute, 1999).

7. See Baigell, *Albert Bierstadt* (New York: Watson-Guptill, 1981) and Carr, *Bierstadt's West* (New York: Gerald Peters Gallery, 1997).

8. We don't believe for a minute—and we've said so on many occasions—that maps are just representational images. Again, see Wood with Fels, *Power of Maps*.

9. See, for example, Obata, Driesbach, and Landauer, *Obata's Yosemite: The Art and Letters of Chiura Obata from His Trip to the High Sierra in 1927* (El Portal, Calif.: Yosemite Association, 1993).

10. Alpha, Detterman, and Morley, *Atlas of Oblique Maps: A Collection of Landform Portrayals of Selected Areas of the World* (Reston, Va.: U.S. Geological Survey, 1988).

11. SPOT Image is a French corporation offering a variety of image products from their fleet of satellites, including medium-resolution, high-resolution, and radar imagery.

12. Powell's official account of his expeditions is given in his *The Exploration of the Colorado River of the West, and Its Tributaries, Explored in 1869, 1870, 1871, and 1872* (Washington, D.C.: U.S. Government Printing Office, 1875). Powell's adventure tale, *Canyons of the Colorado*, which was not published until twenty-six years later, now appears as *Exploration of the Colorado River and its Canyons* (New York: Dover, 1961). See also Dellenbaugh, *Canyon Voyage* (New Haven, Conn.: Yale University Press, 1926). It is possible, however, that Powell may not have been the first through the canyon. There is a controversy that is still, and will probably forever remain, unsettled that ". . . concerns the simple, honest words of a sunburnt and emaciated prospector named James White. Near death, White drifted up to the banks of Callville, Nevada, on a crude log raft, claiming to have spent two weeks floating through the 'Big Cañón' in a desperate attempt to escape hostile Ute Indians. If his incredible tale was true, the 30-year-old White was the first man in history to navigate the Colorado River through the Grand Canyon (albeit not for scientific reasons) two years before Powell's first successful voyage." This brief account is from Annerino, *Canyons of the Southwest* (Tuscon, Ariz.: University of Arizona Press, 1993), 63–66.

13. Among these are Dutton, *Tertiary History of the Grand Canyon District, With Atlas* (Washington, D.C.: U.S. Government Printing Office, 1882); Euler and Jones, *A Sketch of Grand Canyon Prehistory* (Grand Canyon, Ariz.: Grand Canyon National History Association, 1979); and Smith, *The Colorado River* (Provo, Utah: Brigham Young University, 1972).

14. Eco, *A Theory of Semiotics* (Bloomington, Ind.: Indiana University Press, 1976). See in particular pp. 32–36.

15. A series of 1:250,000–scale, three-dimensional topographic maps is produced and marketed by Hubbard in Northbrook, Illinois. Hubbard also publishes plastic vacuum-formed maps of selected national parks.

16. By the National Geographic Society and the Boston Museum of Science. This is the map supplement to the July 1978, issue of *National Geographic* magazine and its companion article "The Grand Canyon: Are We Loving It to Death?" by Associate Editor W. E. Garrett.

17. And use it we do. The following excerpts are from W. E. Garrett, "The Grand Canyon: Are We Loving It to Death?," *National Geographic* magazine July 1978:

For decades the "parks are for the people" policy meant satisfying demands for recreation. This left superintendents to stick their fingers in the environmental dikes to prevent flood damage—usually with hands tied by lack of funds and a public ignorance of ecology.

. . . A 2.8-million-dollar inventory and resource-management plan complains that resource-management funds have been nil in the past and are still inadequate. It admits that years of neglect of the park are apparent. It regrets that pressure groups know more about the park's resources than the staff itself and spotlights staff inadequacy.

Our trails are in poor shape . . . They're a safety hazard to mule trips. I've only got nine full-time people, including myself, to manage almost a million acres. That's just not enough people to do what needs to be done. (Ranger Marvin Jensen, Inner-Canyon Manager)

Mary's (Mary Langdon, in charge of backcountry reservations) office issues permits for 360 backcountry campsites a day. For Easter week, requests had to be submitted between October 1 and 5 to be eligible for a lottery. Only 20 percent of the requests could be filled.

This July as many as 200 people a day will line up to take the [river] trip. More than a quarter of a million of us hiked into the canyon last year, while another quarter of a million toured overhead by plane and helicopter.

Some of the beaches here are as messy and cluttered as a kid's sandbox. (Dr. Roy Johnson, Park Scientist)

Also piling up on the beaches, an esthetic and health problem, is most of the twenty tons of fecal matter produced by river passengers every year.

Clearly, some of these problems are addressed in the current park regulations, but there are now nearly twice as many visitors as in 1978.

18. For expert mountaineers, 8,000 meters (26,247 feet) is the benchmark to be attained. Mount Everest is 8,848 meters.

19. The epimap included four articles on Mount Everest in the same issue.

20. Imhof, *Cartographic Relief Presentation* (Berlin and New York: Walter de Gruyter, 1982).

21. More than twelve hundred men and women from sixty-three nations have now summitted Everest, but many have also perished there. Between 1921 and 2004, 185 died in the attempt, and it is estimated that more than 120 of those still remain on the mountain, the locations of their remains unknown in most cases. The worst years for fatalities were 1982 (eleven dead), 1988 (ten dead), and 1996 (fifteen dead, nine in one snowstorm).

22. Most of these disturbing facts are gleaned from Wetzler, "Base Camp Confidential," *Outside* (April, 2001), which compiles accounts of thirty-six of Everest's most renowned climbers, including Sir Edmund Hillary and his son, Peter; Jim Whittaker; Doug Scott; Reinhold Messner; Kurt Diemberger; Pete Athans; and Alan and Adrian Burgess.

23. The issue in which this map was found included *five* articles on Mount Everest.

24. New Zealander Hillary and Sherpa Tanzig were the second of two assault teams leapfrogging up a series of eight progressively higher camps toward the summit. The first team, consisting of Tom Bourdillon and R. C. Evans, had come within three hundred feet of, but failed to attain, the summit two days earlier. After Bourdillon and Evans's return to camp, Hillary and Tanzig set off up the mountain and established a ninth high-altitude camp, where they endured a strong overnight storm. The next day, May 29, 1953, they reached the summit in clear weather. Hillary and expedition leader Colonel John Hunt were knighted by a recently coronated Elizabeth II and, within three years, five of the world's six next-highest peaks—K2 (Mount Godwin Austen), Kanchenjunga, Lhotse I, Makalu, and Nanga Parbat—were all successfully scaled by others.

POT MARIGOLD
FOXGLOVE
STOCK
BELLFLOWER
FORSYTHIA
GODETIA
CHECKERED LILY
SWEET SCABIOUS
ENGLISH DAISY
PHLOX
WALLFLOWER
FORGET-ME-NOT
COLUMBINE
POLYANTHUS
LUPINE
ROSE
LILY OF THE VALLEY
ASTILBE
PRIMROSE
EUROPE
SPRING
CROCUS
CHINA ASTER
CALIFORNIA POPPY
SHIRLEY
POPPY
PENSTEMON
SNOWDROP
TULIP
GARDENIA
BLANKETFLOWER
HYDRANGEA
PANSY
FLAME AZALEA
MEDITERRANEAN
TURKEY
DAY LILY
OLEANDER
SNAPDRAGON
CROWN IMPERIAL
SWEET PEA
HYACINTH
IRAN
GRAPE HYACINTH
SWEET
ALYSSUM
STAR-OF-BETHLEHEM
CANDYTUFT
GLADIOLUS
CARNATION
SPIDERFLOWER
ORIENTAL POPPY
POKER PLANT
BIRD-OF-PARADISE
FLOWER
AFRICA
EAST INDIAN LOTUS
AFRICAN LILY
SCARLET SAGE
CALLA
GERBERA
VANDA ORCHID
NASTURTIUM
FRINGED
HIBISCUS
FREESIA
CASTOR-OIL PLANT
CAPE MARIGOLD
ROYAL POINCIANA
PLUMBAGO
PETUNIA
FLORISTS' BEGONIA
PELARGONIUM
"GERANIUM"
BUTTERFLY
FLOWER
SOUTH AMERICA
MADAGASCAR
AFRICAN VIOLET
GLOXINIA
NEMESIA
CALCEOLARIA
VERBENA
IMPATIENS
LOBELIA
PORTULACA
CUPFLOWER
VICTORIA WATERLILY

Seven

Nature as cornucopia

Courtesy of Ned Seidler/National Geographic Image Collection.

Nature as cornucopia

And then—from awe to aww—there is the nature we can hold in our arms, nuzzle and kiss, nurture and protect. This is the nature, domestically, of puppies and kittens, of foals and lambs. It is that of babies. It merges seamlessly with that of bunnies and nestlings, fawns and kids, cubs and whelps. It evokes tender feelings in us. Who has not heard the melting "Awwww . . ." of an elementary school classroom looking at pictures of harp seal pups, of tiger cubs, of the whelps of wolves gamboling in the wild? It is a fond sound, an adoring sound. It evokes the maternal, the paternal in adults. It calls for kindness and caring and protection. It asks us to open our arms and *enfold*.

Caring for a nature . . .

At the top of a letter from the National Wildlife Federation (dated February 5, 2004) is a photograph of a "FREE polar bear plush family waiting for you if you act promptly. These awesome but gentle creatures of the North symbolize our fight to save endangered American wildlife." Stuffed polar bears: this is the wildlife adored by kids, by the kids in all of us. "Support from new members like you," the plea continues, "is the key. It is only through help from caring people like you that we are able to continue our work to protect the bears, wolves, birds, prairie dogs, buffalo, lynx, and many other species under attack today." It is a nature under threat, but it is one we care for. That is, it is a nature for which we feel love and affection.

With the offer of the plush polar bear family come five "stunning nature note cards." These frame the National Wildlife Federation's concern for bears, wolves, prairie dogs, buffalo, and lynx through a view from the garden. On the first note card, a bluebird perches on the rim of a birdbath nestled in a riot of poppies, daisies, and roses; on the next, a monarch butterfly alights on a cluster of phlox; the third gives us dragonflies, lilies, and bellflowers; on the fourth a hummingbird hovers outside a window; a flurry of butterflies flutters across the fifth. There is nothing here of the sublime. It is all what Kant called beautiful, a category he understood to include "birds, sea shells, articles of dress and furniture, dwellings, trees, gardens, bird song, a summer day . . ."[1]

. . . of beautiful flowers

"Emblazoned with beauty, this floral map shows the origins of 117 of man's favorite flowers" commences the peritext of "The World of Flowers" (figure 7.1). This is a lovely three-page gatefold that illustrates a *National Geographic* article on the American flower seed industry (May 1968). It's a pale cream, bled to the edges, with the continents pulled out in a light tan. It's covered with flowers. Beguiling bouquets burst from the hearts of the continents; a posy of tulips sprouts in Turkey; a branch of *flamboyán* flowers in Madagascar; in Mexico dahlias bloom, and zinnias, and marigolds. It is the world reduced to a florist's shop, or to the brochure in the florist's shop advertising Flowers by Wire.

Befitting the subject, it's a pretty map, even beautiful. It simultaneously mobilizes the two strategies most commonly exploited to spatialize this adorable nature: at the same time that it *locates individual origins*, it *displays their universality*. To post "the origins of 117 of man's favorite flowers" is to showcase the individual and the particular and allude to a history ("as people began to move from one part of the world to another, they carried plants with them"). On the other hand, to post 117 of them together is to create "the world" of flowers, to abolish history (flowers are everywhere), and to propose a *unitary* nature (the "world" of flowers promiscuously embraces every variation). The trope is notorious, unity in diversity, and as so often it instantiates here what Barthes understood to be the fundamental mystification: "placing Nature at the bottom of History."[2] In "The World of Flowers" history is run backward to reveal where the flowers grew before "people began to move from one part of the world to another" ("to trace these blossoms to their source, Geographic artist Ned Seidler consulted Dr. Mildred E. Mathias, Professor of Botany at the University of California at Los Angeles"): that is, history is run backward to reveal nature.

At the same time, the epimap, "The Flower Seed Growers: Gardening's Color Merchants" (figure 7.2), is about how nature is undone, that is, improved, civilized, under the hand of culture: "Behind the steady improvement of garden flowers lie thousands of test plantings, years of careful selection and crossing of varieties, and a determined search for mutants—plants with new characteristics." With the map taking us back to (a bountiful and beautiful) nature, and the text displaying the hand of man, they together proclaim a "nature improved." It is our friend. We love it.

Appropriately, the map is ambiguous about the nature that has been recovered in consultation with Dr. Mathias. Origins are sited in Asia, South America, and so on, but the flowers are corralled into bouquets, corsages, boutonnieres (these are a source of the map's charm). These reduce the continents to vases, the islands and countries to bodices and jacket lapels. Although the map's primary epimap is devoted to the flower seed industry, and so to photographs of seed, to portraits of seed merchants, in what we might think about as the map's secondary epimap—the rest of the magazine in which the map was bound—we find, in the immediately preceding article (on the "Île de la Cité, Birthplace of Paris"), a photograph of a sidewalk flower stall profuse with pots and buckets of geraniums, tulips, lilies; a photograph of a diplomatic luncheon, bouquets of roses bedecking mantel and tables; a photograph of a Princess Bibesco, a vase of orchids in the foreground; a photograph of the library of a Baron de Redé, a vase of roses on his desk; and a photograph of a midnight ball he's given (on a side table, one of those bouquets large and lavish enough to establish without further comment the resources required

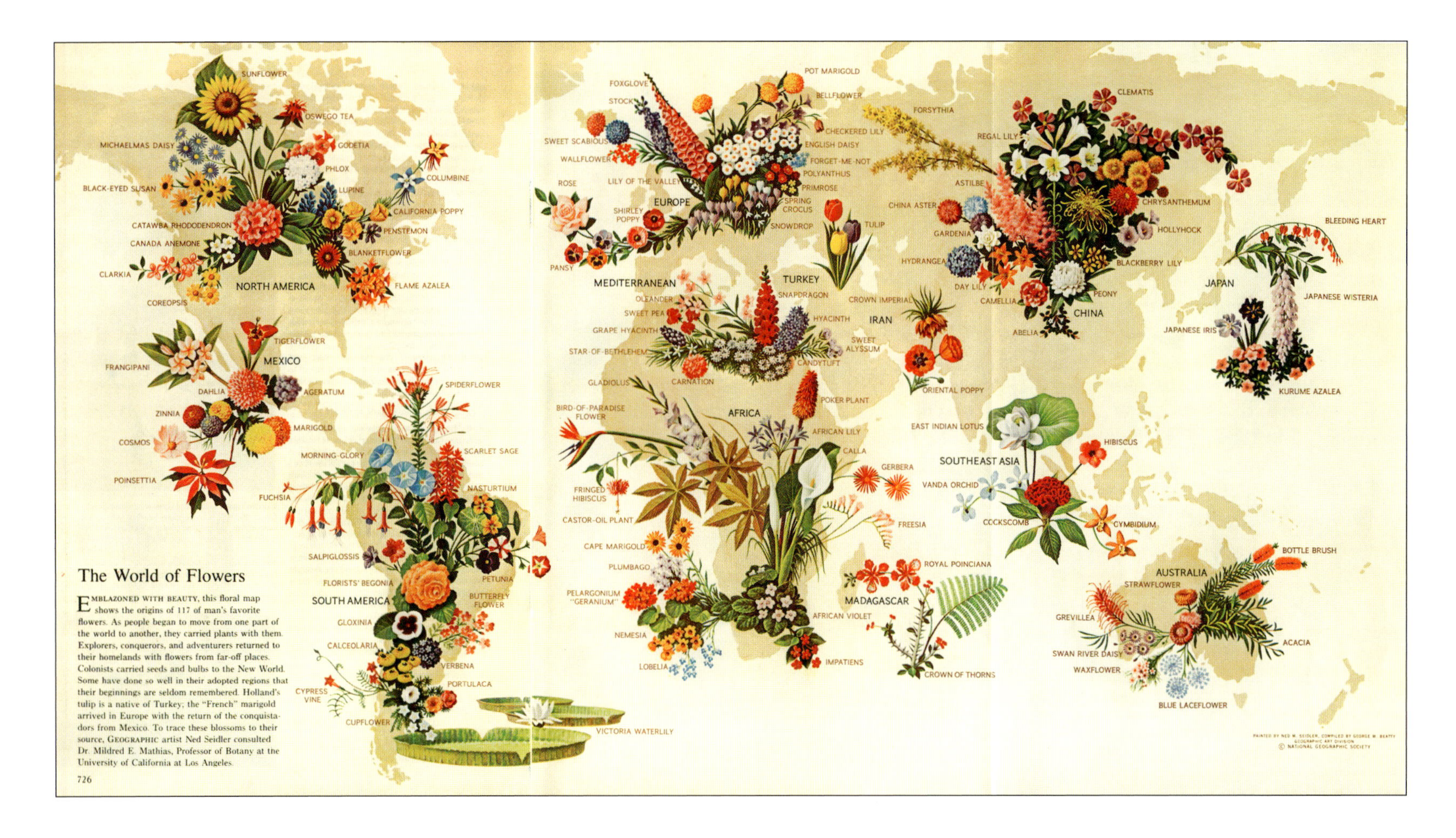

The World of Flowers

EMBLAZONED WITH BEAUTY, this floral map shows the origins of 117 of man's favorite flowers. As people began to move from one part of the world to another, they carried plants with them. Explorers, conquerors, and adventurers returned to their homelands with flowers from far-off places. Colonists carried seeds and bulbs to the New World. Some have done so well in their adopted regions that their beginnings are seldom remembered. Holland's tulip is a native of Turkey; the "French" marigold arrived in Europe with the return of the conquistadors from Mexico. To trace these blossoms to their source, GEOGRAPHIC artist Ned Seidler consulted Dr. Mildred E. Mathias, Professor of Botany at the University of California at Los Angeles.

726

Figure 7.1
"The World of Flowers" map in the May 1968 *National Geographic* article about the flower seed industry reduces the world to a florist's shop. Map dimensions (WxH): 18in. x 10in.

Courtesy of Ned Seidler/National Geographic Image Collection.

Brilliant ranks of ladies and gentlemen march row upon row across a Lompoc Valley field of Bodger Seeds, Limited. Specially bred female marigolds of sparkling yellow alternate with orange males. Bees imported to the field effect more thorough cross-pollenation than is possible by wind. Resulting offspring will reflect the color dominance of the male—orange. Other marigolds must be planted at least a mile away to guard against unplanned crossings.

THIS PAGE FOLDS OUT 729 730

Figure 7.2
"Brilliant ranks of ladies and gentlemen march row upon row across a Lompoc Valley field of Bodger Seeds, Limited," begins the caption to this photograph from Robert de Roos' "The Flower Seed Growers: Gardening's Color Merchants." This *National Geographic* article, in which "The World of Flowers" map appeared, is the map's primary epimap, at once providing a rationale for a map of flower origins and augmenting its content to proclaim a nature improved.

Courtesy of Jack Fields/National Geographic Image Collection.

to command it). Of course, it's not necessary to have noticed *these* bouquets in order to see bouquets when you unfold "The World of Flowers," but to have just done so does make it harder to miss in the world of flowers the implications of culture, and of a certain culture at that (Paris, royalty, decoration, wallpaper).

Because it means so little to cite an origin in Asia or North America, especially with flowers from widely disparate environments clumped together into decorative ensembles, in the end the map less displays the *origins* of flowers than it flaunts floral *variety*, is less concerned about localizing something than it is with showing off a nature of wild extravagance and endless profligacy. So varied a nature is inherently fecund, and ultimately the fecundity and variety come to appear as no more than different ways of naming the same nature, one that is endlessly nurturing because it is inexhaustible. It is a nature we can love because it loves us, that is, it is a nature we can turn into bouquets because it has provided us the wealth and variety to do so.

It's the nature we teach our kids about

This is the first nature we teach to our kids. On its title page, Scholastic's *Atlas of Plants: A First Discovery Book* carries the following note: "Atlas of Plants is a child's very first atlas. Using simplified maps, bright illustrations, and basic information, it introduces young children to the diversity of plant life all over the world." While this *could* pass for a description of "The World of Flowers," in fact the maps in *Atlas of Plants* are vastly simplified, are literally the size of thumbnails, and instead of the plants being on the map—as the flowers are in "The World of Flowers"—here they *surround* them. Otherwise the layout is tricky, employing a kind of peek-a-boo effect that could make reading the book to little kids a lot of fun. (Unfortunately, we were unable to secure permission to reproduce a page from the atlas here.) On each left-hand page, four paintings surround a tiny hemisphere with the continent in question—let's say, North America—centered and highlighted. The paintings in this case are of a sequoia forest, a field of poppies, a cypress swamp, and a range of saguaros—that is, they're of diverse plant communities. On the facing page are close-ups: of the back of a single poppy flower, of the base of a sequoia, of a saguaro trunk, and of cypress knees. The trick is that these are printed on transparent plastic; so when you turn the page, these close-ups of individuals—the front of the poppy flower now—fall *onto* the community pictures, where they fit as images of plants foregrounded in their "natural" settings. On the right-hand page, a specimen poppy plant, a sequoia, a saguaro, and a cypress are revealed. As the parent or child turns the see-through page, now you see the back, now the front of the poppy on the transparent page itself, and on the right-hand page, now the specimen poppy, now you don't. What you take away from this—besides the fun—is an impression of the *diversity* of the world's plants: poppy, sequoia, saguaro, cypress, cacao, lianas, corn plant, tillandsia, date palm, welwitschia, baobab, cotton plant, daisy, chestnut, bulrush, apple, and so on. All of them are useful to humans or other animals ("Pandas live in bamboo forests," "The thick roots of the mangrove make a good breeding ground for fish"). It is a nature of both immense diversity and notable utility.

My First Atlas (figure 7.3) is another atlas for children. Less focused on plants, *My First Atlas* claims to "introduce young children to the countries of the world," but except for the United States and Canada, the approach is more regional than national. The astonishment is the extent to which images of plants and animals dominate the maps. This may be less surprising in a map of Canada, practically a zoo with large pictures of arctic hare, ringed seals, hooded seals, a musk ox, a grizzly bear, a polar bear, an Arctic fox, caribou, beluga whales, a Canada goose, a gannet, a black bear, a beaver, a moose, a salmon, walruses, and porpoises; but the map of the United States, which is shown fully forested east of the Mississippi and peppered with trees throughout the Great Plains, also displays a salmon, a cougar, a coyote, a rattlesnake, a bison, an armadillo, a bald eagle, and an alligator, together with a cattle rancher on horseback, two Kentucky thoroughbreds, giant saguaros, an orange grove, Devil's Tower, Grand Canyon, and wheat harvesting. Africa comes off as little more than an animal theme park, with masked dancers, goatherds, and a camel train providing human color. Only the map of Central Europe, where people are shown making chocolates, sausages, and cars, and where human structures dominate (the Brandenburg Gate, a castle), presents a view of existence in which labor plays any kind of a role. Over all, with its wealth of plant and animal life, the world comes off as a gigantic cornucopia.

Nature is a cornucopia

What better symbol for a nature that we love because it loves us than the cornucopia, that large horn, itself a symbol of power and fertility, gushing, overflowing, with the fruits of the earth? Effortlessly the cornucopia bespeaks what "The World of Flowers" and these children's atlases have bodied forth: a nature at once diverse and abundant, inexhaustible in both kind and quantity. Undoubtedly Edenic, the maps present this nature as a contemporary reality. And so, along with the native fauna, *My First Atlas* displays the flags of 140 nations, and no spread is without its airplanes.

Of course, maps are not necessary for the construction of this nature. The illustrated bestiaries, for example, those twelfth- and thirteenth-century "books of beasts" derived from the second-to-fifth-century *Physiologus*, described a nature of unending variety, and one that grew richer over the years. Forty-nine beasts inhabited the world of the *Physiologus* in its original Greek, but by the twelfth and thirteenth centuries, bestiaries were cataloguing between 110 and 150 different animals: the lion, the tiger, the pard, the panther, the antalops, the unicorn, the lynx, the griffin, the elephant, the beaver, the ibex, and so on, and so on.[3] In the notes to his translation of a twelfth-century bestiary, T. H. White characterizes it as "a serious work of natural history," and observes that it "is one of the bases upon which our own knowledge of natural history is founded."[4] White quotes M. R. James to the effect that "the Bestiary may be reckoned as one of the leading picture-books of the twelfth and thirteenth centuries in [England]. It ranks in this respect with the Psalter and the Apocalypse," in which connection it is worth noting that bestiaries were first and foremost collections of Christian allegories.[5] What *we* want to observe, however,

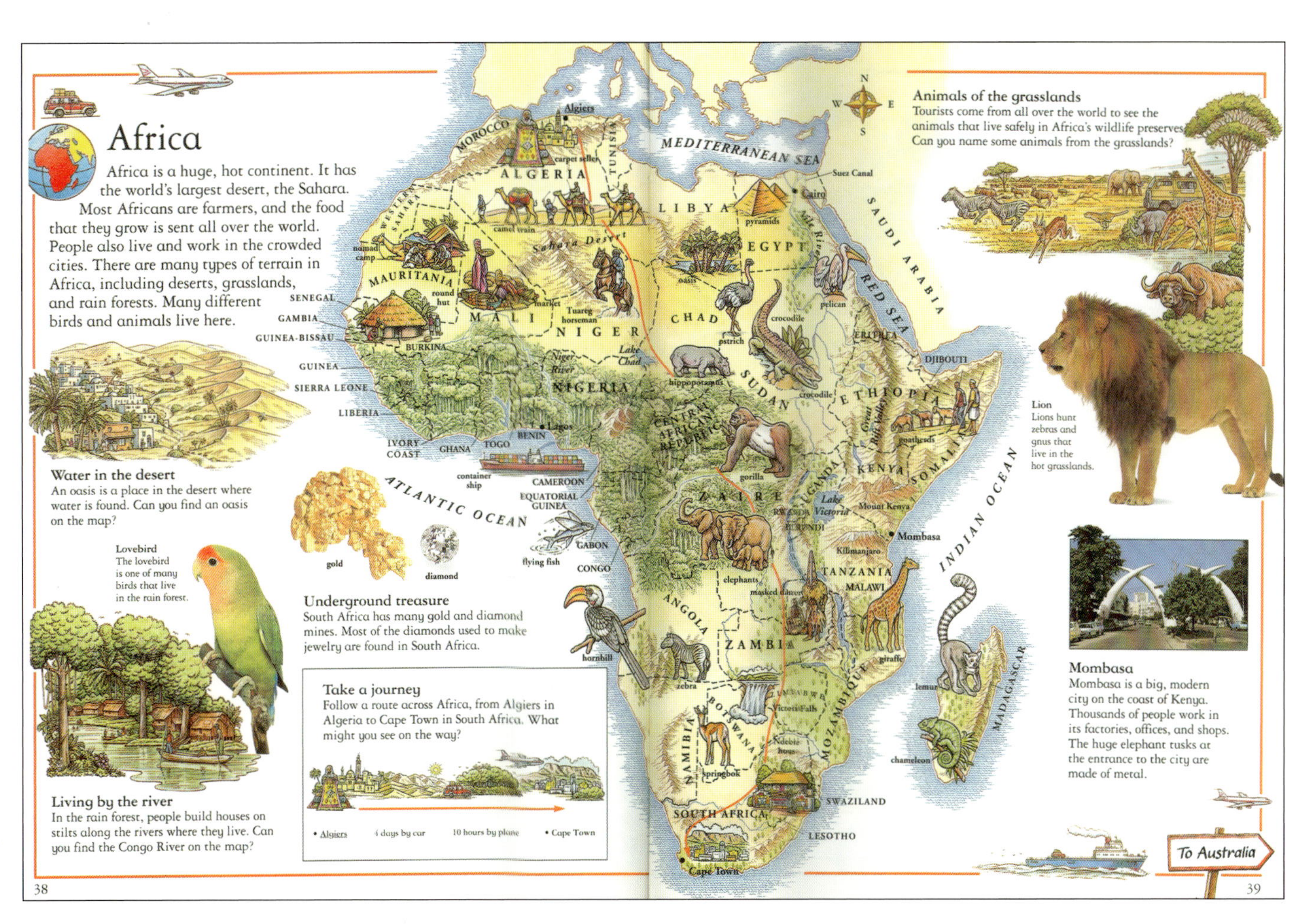

Figure 7.3
"Africa" from *My First Atlas* by Bill Boyle, 1994, a book meant to "introduce young children to the countries of the world," turns Africa into an animal theme park.
Courtesy of Dorling Kindersley Books, London.

is the way the text and pictures constructed a nature, at once diverse and abundant, that, except for its lack of spatialization, is hard to distinguish from that figured in *My First Atlas*.

So what was gained by mapping it?

Posting beasts makes them more real

And the nature of the bestiaries was certainly one that was mapped early on. Wilma George opens her study of the mapping of animals in earnest with the contemporaneous T-O maps[6] of the thirteenth century, the great Ebstorf and Hereford *mappaemundi*. Large, free-standing maps of the world, these *mappaemundi* were works of both art and learning. They were hung on the wall or placed behind the altar where they functioned simultaneously as decoration, as a source of edification and instruction, and as a symbol of the deity.[7] They were *covered* with animals.

Contradicting our traditional dismissal of the animals on early maps as merely decorative (to the invariable accompaniment of Swift's "So Geographers in Afric maps / With savage-Pictures fill their Gaps; / And o're unhabitable Downs / Place Elephants for want of Towns"), what George actually finds is that from the beginning, recognizable drawings of wild animals were usually located precisely where they were thought to be most typical or outstanding. The Ebstorf, for example (figures 7.4 and 7.5), appropriately displays in the Ethiopian region—that is, in the zoogeographical region embracing Arabia Felix and Africa south of the Mediterranean littoral—an elephant, a leopard, a hyena, a mirmicaleon, monkeys, a camelopardalis, a scarp, a deer, a tarandus, many types of snakes, crocodiles, lizard and flying lizard, an ostrich, an ibis, and other birds.[8] The Palearctic region is just as richly and appropriately populated, the Oriental less so.

George regards the Ebstorf and Hereford as zoogeographical maps *strictly speaking*. This is to say that both maps accurately display distinctive faunal associations in all three (then known) regions. Conceding that the maps fail to accurately depict ranges, George insists that they only rarely located animals in the wrong region.[9] She concludes,

> From these early "distribution" maps it could be learnt that, while snakes, dragons, birds, some carnivores and domestic camels were common to all three regions; that elephants, unicorns and crocodiles were shared by the ethiopian and oriental regions; leopards, pelicans and ostriches by the palearctic and the ethiopian; and that, while elk and bears were particularly representative of the palearctic, giraffes were peculiar to the ethiopian, along with the rhinoceros and antelope which, however, occur only once.

What George is saying here is that no matter how "merely decorative" the animals may appear to us today, the logic underwriting their location was genuinely zoological. As we progress through the fourteenth, fifteenth, sixteenth, and enter the seventeenth centuries,

Figure 7.4
Ebstorf World Map, 1235, offers more-or-less reasonable impressions of well-known animals located in the region where they actually lived. Zoologically, the map was comparatively sophisticated.

Figures 7.4–7.5: Map created by Kloster Ebstorf. Concept of this map to Gervasius of Tilbury and dates it between 1241 and 1245. Digital image provided by Martin Warnke.

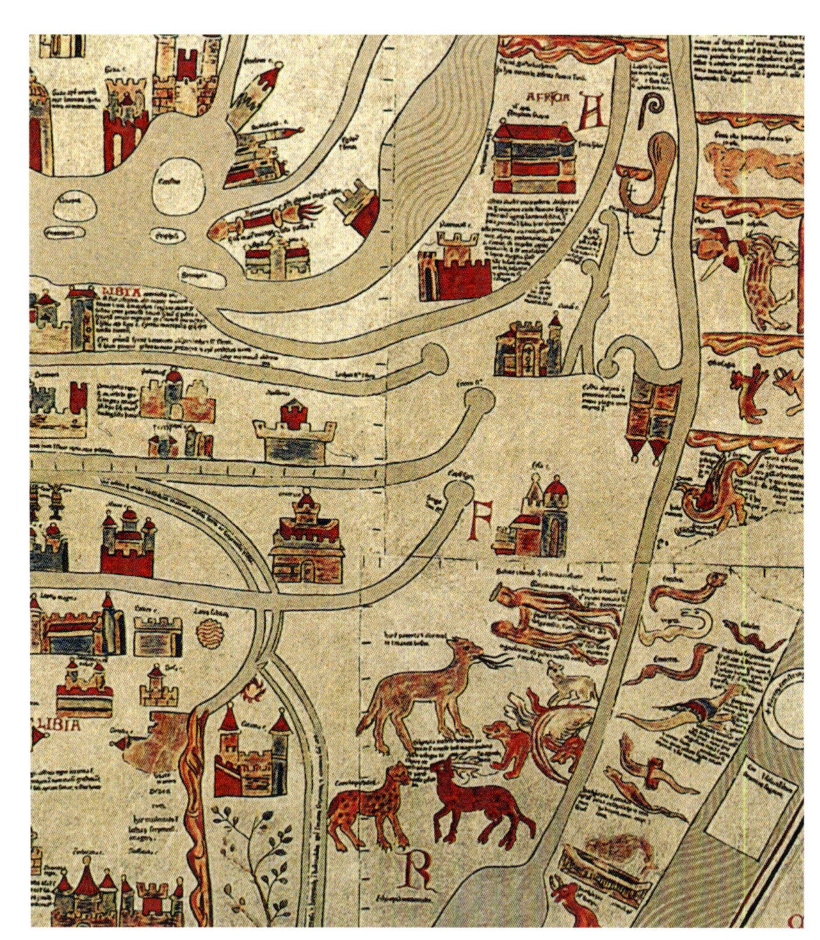

Figure 7.5
The Ethiopian region of the *Ebstorf World Map*.

the maps—and keep in mind that these are "general purpose" maps —only get more and more comprehensive:

> By the end of the seventeenth century, excluding the australian region which was hardly known, an average of twelve mammalian families had been represented in each region [gradually including the nearctic and neotropical] with a scattering of birds and reptiles. This number forms a considerable proportion of the total terrestrial mammalian fauna, if bats are excluded.[10]

In fact, many of the maps were well in advance of contemporary bestiaries.[11]

And why not? After all, the spatial logic of the map imposed demands and restraints on its compilers beyond those imposed on the compilers of the bestiaries (which did happen to be compiled much like maps, bits and pieces added or deleted as the manuscripts were copied to keep them fresh and up-to-date[12]). Beyond knowing that "Leo the Lion, mightiest of beasts, will stand up to anybody" and being able to draw it, the mapmaker had to be able to *post* his drawing to the sign plane of the map. Beyond obligating the mapmaker to assign the animal to a *region*, this posting forced the mapmaker into constructing *faunal associations*. These obligations—imposed by no more than the map's inherent spatial logic—made the mapmaker a compiler *par excellence*. For one thing, not every bestiary situated its subjects in space (White's bestiary, for example, says nothing about where Leo might be found); and when bestiaries did, they didn't necessarily agree. Because the earliest bestiaries predate the Ebstorf and Hereford *mappaemundi*, those bestiaries do *include* all the animals shown on the maps, but as George observes, "*these animals evidently have not been copied directly from any* one *bestiary*." Later, when maps increasingly described territories not treated in the ancient authorities (and not just in the nearctic and neotropics either), high-quality compilation became even more critical, and George suggests that this led the mapmakers to abandon classical sources much more rapidly than did the naturalists. Maps soon enough posted animals found in no bestiaries at all.

The faunal associations called forth by the map's spatial logic *re-sorted* the data of the bestiaries. They pulled the data away from Christian allegory and toward natural history. This was neither unprecedented nor altogether undesired, since the *Physiologus* itself had originated in a related impulse. In the second century,

> Christian teachers, especially those who had a leaning toward Gnostic speculations, took an interest in natural history (e.g., the *historia* of Aristotle's animals, Pliny's *historia*

> *naturalis*), partly because of certain passages of Scripture that they wanted to explain, and partly on account of the divine revelation in the book of nature, of which also it was man's sacred duty to take proper advantage. Both lines of study were readily combined by applying to the interpretation of descriptions of natural objects the allegorical method adopted for the interpretation of Biblical texts.[13]

These second-century motivations are all but identical to those that in the later Middle Ages brought the great *mappaemundi* into being. As David Woodward explains, "The function of the medieval *mappaemundi* was largely exegetic, with symbolism and allegory playing major roles in their conception . . . The central theme is the earth as a stage for a sequence of divinely planned historical events from the creation of the world, through its salvation by Jesus Christ in the Passion, to the Last Judgment."[14] To stick with our example, in the Ebstorf, the body of the Christ is superimposed over the entire map (and so over all its animals) in an all-embracing dying gesture that makes the map as a whole a clear symbol of the salvation of the world, his head at the top in the east next to Paradise, his hands in the north and the south, his feet in the west.[15]

But despite their common religious motivations, and probably their related conceptions of the role of natural history, the differences between the bestiaries and *mappaemundi* in expressive logic, expressly the two-dimensional deictic plane of the map, brought into being in the *mappaemundi* if not a new nature, then a nature—beneficent, abundant, various, cornucopial—newly seen, newly understood, and newly appreciated. On the plane of the map, albeit beneath the conspectus of Christian theology, animals (1) shed their individual allegorical associations, (2) gained locations, and (3) acquired neighbors. That is, in acquiring locations in the map, animals exchanged their allegorical associations for faunal associates. They relinquished their specific relation to Christ to become more wholly animal. *Posting the animals on the plane of the map made them more animal. It made them more real.*

The shifting place of the cornucopial nature

Paradoxically, this shift toward description did not reduce their religious significance. John Pickstone is in the vanguard of those attempting to reconstruct history "by shifting from replacement models in which (Foucault's) 'epistemes' or (Kuhn's) paradigms *follow* each other, towards coexistence models in which paradigms run alongside each other, with different histories and different political relations."[16] This produces what Pickstone thinks about as a *displacement* model of history. Here, in our history of animals and maps, as we move from the thirteenth to the seventeenth century, in the *re*placement model, we would be watching "the enchanted, allegorical, medieval-Renaissance world as a theater of meaning" being wiped out, supplanted by scientific ways of knowing: natural history would supplant allegory, description would trump meaning. But this is not quite what happens. Instead the relationship of and relative weight given to description and meaning shift.

To begin with we have already noticed the mixture of natural history and religion that underlay the *Physiologus* and the later bestiaries. The animals were there as emblems. Thus, "The Lion's second feature is, that when he sleeps, he seems to keep his eyes open. In this very way, Our Lord also, while sleeping in the body, was buried after being crucified—yet his Godhead was awake." When White describes this as natural history, he is reflecting the fact that before Leo can *stand for* something, he has to *be* something: "His first feature is that he loves to saunter on the tops of mountains. Then, if he should happen to be pursued by hunting men, the smell of the hunters reaches up to him, and he disguises his spoor behind him with his tail. Thus the sportsmen cannot track him." Whatever else this and the rest of the pages of observations about tigers, pards, panthers, and so on, is, it is natural history, which in Pickstone's definition is, "the art of describing and classifying which underlie all our more complex forms of 'naturalistic' knowing."[17]

But we have also seen the way putting these emblems on the *mappaemundi* worked to de-emblematize them, even as the animals' presence on the maps helped emblematize *the maps* as valid images of "nature itself, as the Logos of God brought down to earth and made manifest."[18] *Posting* the animals freed them to be *just* animals and worked to construct natural regions and communities. The steady increase on our maps of animals and animal families that took place across the ensuing centuries—on, among other maps, Dulcert 1339, the Genoese world map of 1457, Ribeiro 1529, Desceliers 1550, Ludolfi 1683—as simultaneously across the cataclysm of the Reformation was, in part, a response to the sixteenth-century Italian revival and transformation of the ancient *historia naturalis* into an early modern discipline in the hands of Pier Andrea Mattioli, Ulisee Aldrovandi, and others[19]—and in part to support a Protestant drive to glorify God as a free creator:

> The intermediaries between God and man—the angles and saints and the demons—were stripped out by Protestant theologians as part of their attack on the old Church, and "man" was left to face his maker alone, as part of the religious-politics that proclaimed "the priesthood of all believers." All good Protestants were assumed capable of interpreting the Bible on their own; and just as the Bible could now be read by anyone, so also the "Book of Nature" was open to all. In this way nature became a means of demonstrating the attributes of God through the study of his creation, and religion became a major reason for studying natural history.[20]

In this way, the animals that in the twelfth and thirteenth centuries had been props for Christian allegory had become, by the sixteenth and seventeenth centuries, legitimate, even important topics of interest in (almost) their own right.

Although natural history *was* part of material culture because in a sense you were studying God when you studied nature, natural history was also moral; and "in as much as it was about hierarchy, it also bolstered the social pyramid. For example, in the years after the Restoration of the English monarchy in 1660, taxonomic hierarchy became a common theme in natural history and through the cultivation of heraldry."[21] In this way, even as the study of animals was changing in thoroughly decisive ways that are the subjects of our next two

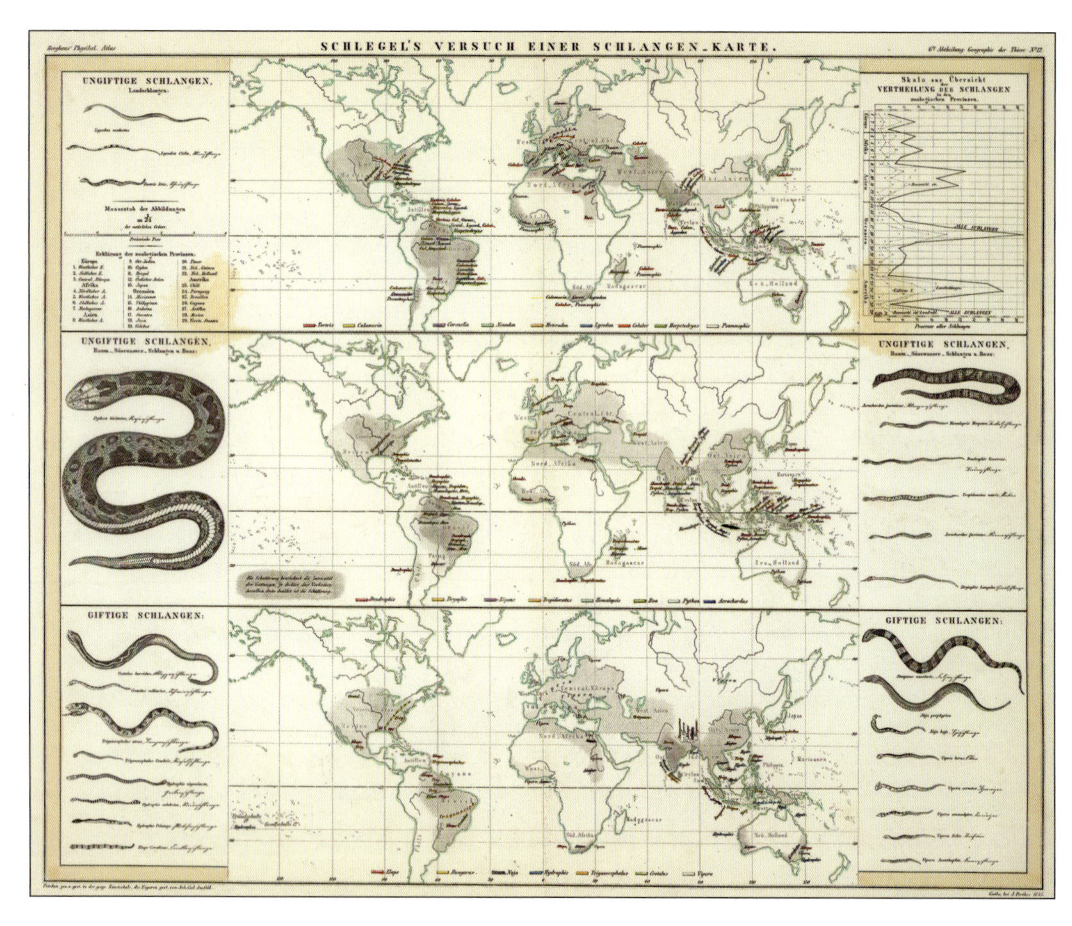

Figure 7.6
"Schegel's Versuch Elner Schlangen-Karte" (The Distribution of Snakes) from *Heinrich Berghaus' World Atlas* of 1845.
Source: David Rumsey Collection.

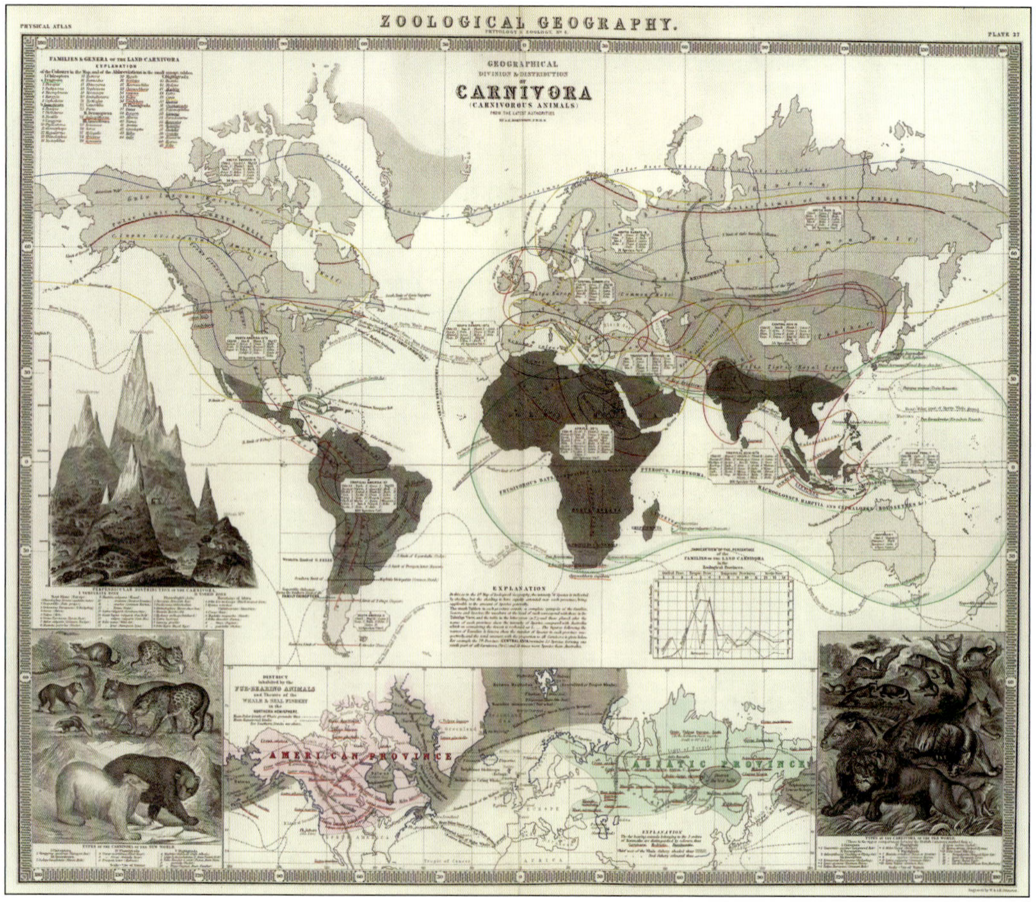

Figure 7.7
"Geographical Divisions and Distribution of Carnivora," from *A. K. Johnston's Physical Atlas* of 1856.
Source: David Rumsey Collection.

chapters (on the collectible and the scientific natures), at least one motivation that had been crucial in the second century, namely that it was man's sacred duty to study the divine revelation in the book of nature, continued to play a crucial role, the new "science" less wiping out the older approach than investing and reshaping it.

As a consequence of an accelerating "reshaping" that took place in the eighteenth and nineteenth centuries, the mapping of animals split into two streams. One, which we might call scientific—but which Pickstone would rather have us distinguish from "natural history" as "analytical"—mobilized a new spatialization strategy, that of depicting animals as ranges. Arthur Robinson describes this transition from the natural historical to the analytical in his *Early Thematic Mapping in the History of Cartography*:

> The first maps of animal distribution appear to have been made somewhat earlier than those of vegetation. Although Eckert refers to the 1783 map of quadrupeds by Zimmermann as being the first map of animal geography, Zimmermann included an earlier version in a book published in 1777. Both merely have names scattered over the map along with some limits of distribution. Ritter's *Sechs Karten von Europa* also contains a primitive map of tame and wild mammals with names here and there and with some limits shown. None of these maps qualifies as a good example of thematic cartography, since they do not portray distribution very effectively. Not surprisingly, the clear display of distribution and variation starts with *Berghaus's Physikalischer Atlas*, which included twelve zoological maps mostly dated 1845 (104–105).

In other words, animal *pictures* are abandoned to be replaced, first by *words*, and then rapidly by all the methods we use today: *lines* are drawn around ranges, ranges are variously *stippled* or *colored*, and variable density is shown by *shading* or *contours*. Heinrich Berghaus' "Schegel's Versuch Elner Schlangen-Karte" in his *World Atlas* of 1845 (figure 7.6) nicely encapsulates this transition: the animal pictures are still here but they've moved from the map proper to its margins where they're named. It's these names that now appear on the map itself, but against a background of lightly stippled zoological provinces keyed in the map's upper left. Ten years later, as exemplified in A. K. Johnston's 1856 world map of carnivores, this transition had been all but completed (figure 7.7). Now the animal pictures have been pushed into two corners (all but off the map), while the animals appear in the map as thoroughly modern ranges, against a stippled background indicative of provincial species density. That is, by the middle of the nineteenth century, animals have acquired a completely spatial identity (species = range) and in the process participated in the construction of the nature of science.[22] We will take up this turn toward the end of our next chapter.

In the other stream—the one we'll continue to follow in *this* chapter—pictures continue to be the way animals are depicted. These animals continue to be *located* (say in Asia), but are not *distributed* (where in Asia?). "Distribution of Animals in Zones," from an 1895 *Frye's Grammar School Geography*, is exemplary, uniting a pictorial depiction of the world's animals with the ancient torrid-temperate-frigid zonal conception of the earth (figure 7.8). Though laid out in the accompanying text, the *zones* are less easy to make out here than they are on the contemporary "Plants and Animals," from an 1898 *Maury's Manual of Geography*. Here the earth is literally divided into zones, each populated with its characteristic animals (figure 7.9). By the late nineteenth century, this pictorial stream has become distinctly *popular* (it included maps made for children), but it brought into the present, unbroken and unshifted, the tradition we've been tracing from the *Physiologus* and the bestiaries, through the great medieval *mappaemundi* and the so-called "decorative and printed maps" of the sixteenth and seventeenth centuries. While "scientific" mappers in the eighteenth and nineteenth centuries turned their attention toward the mapping of animal *ranges*—what Robinson thinks about as distribution—these others continue to put pictures of elephants in Africa and pictures of bison in North America, and they do so up to the present. Just as Pickstone insists that the turn toward science—that is, away from meaning toward description—produced a *new* nature,[23] so the continuation in popular cartography of a pictorial tradition preserved an older nature, one in which an element of "meaning" was a powerful, if not predominant component. Comparison of the Africa on the thirteenth century Ebstorf, the Africa of the nineteenth century Frye, and the Africa of *My First Atlas* of 1994 (figure 7.10) reveals powerful resemblances that are more than genetic. The *My First Atlas* Africa is not merely heir to a tradition that runs back to the Ebstorf, but shares with it and with the rest of the lineage a commitment to the depiction of a nature of abundant variety, a nature that from the beginning *spoke* of God's benevolence, but also of his puissant inventiveness; then *displayed* the workmanship of a maker God and his wondrous designs; and today *describes* a nature that is varied but inexhaustible.

These natures *do* change, but powerful continuities are evident too. Compare across seven hundred years the Ebstorf's Ethiopian elephant, leopard, hyena, mirmicaleon, monkeys, camelopardalis, scarp, deer, tarandus, snakes, crocodiles, lizard and flying lizard, ostrich, ibis, and other birds with *My First Atlas's* African elephants, cheetah, lemur, gorilla, giraffe, zebra, springbok, crocodile, chameleon, ostrich, hornbill, pelican, lovebird, lion, camel, and hippopotamus. More than the animals demonstrate this continuity; so do the vignettes of local life, the vegetation, and the relief. The biggest difference is the nation-states that show up on the *My First Atlas* Africa. Whereas peoples, as nations, had, like animals, been mapped from the beginning, these nations turned into *states* during the eighteenth and nineteenth centuries, where the states had a relationship to territory (nations turn into states = territory) akin to that of the ranges of animal species (animals turn into species whose ranges = territory), but ranges did not enter popular mapmaking, while nation-states most energetically did. But we are not claiming that the maps are identical. The twentieth century of *My First Atlas* is not the thirteenth of the Ebstorf. Religion no longer plays the role it did then and is certainly anything but unitary, science has been a cultural force of overwhelming consequence, and nature has become a matter for description far more than for meaning making. Yet still, this beautiful nature of abundance and variety *means*

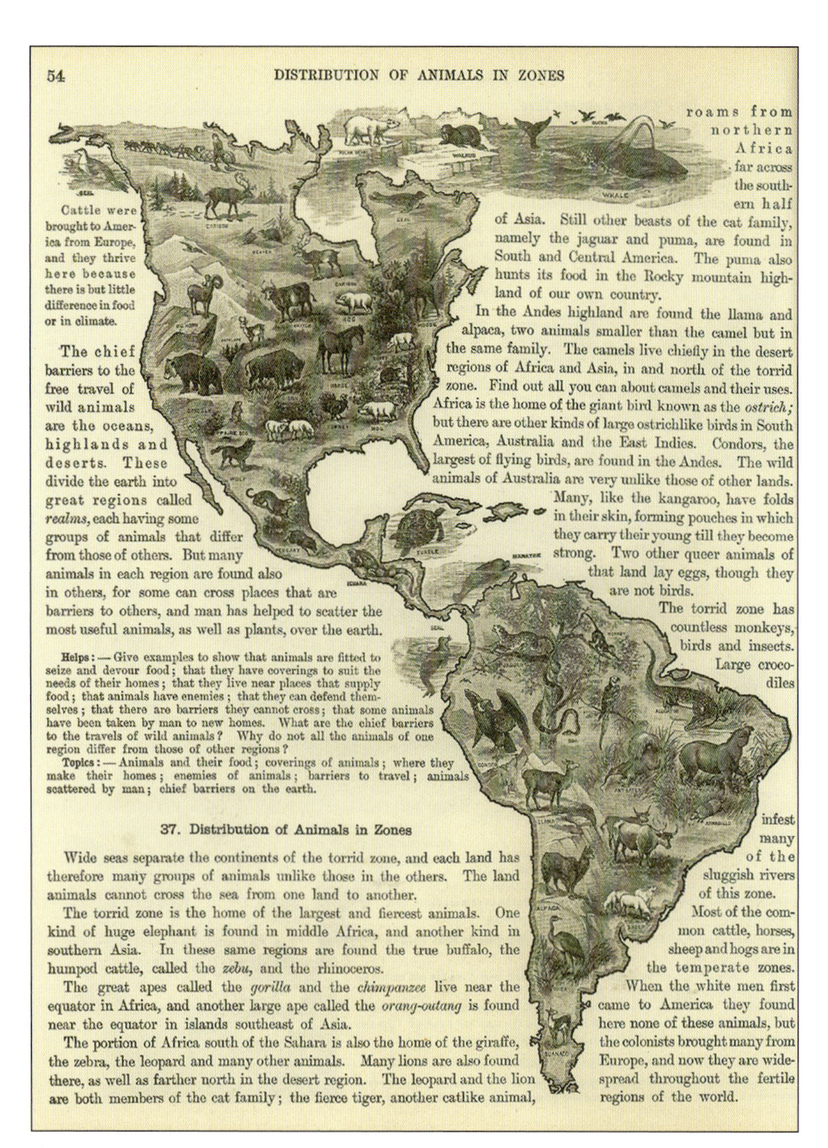
54 DISTRIBUTION OF ANIMALS IN ZONES

Cattle were brought to America from Europe, and they thrive here because there is but little difference in food or in climate.

The chief barriers to the free travel of wild animals are the oceans, highlands and deserts. These divide the earth into great regions called *realms*, each having some groups of animals that differ from those of others. But many animals in each region are found also in others, for some can cross places that are barriers to others, and man has helped to scatter the most useful animals, as well as plants, over the earth.

Helps: — Give examples to show that animals are fitted to seize and devour food; that they have coverings to suit the needs of their homes; that they live near places that supply food; that animals have enemies; that they can defend themselves; that there are barriers they cannot cross; that some animals have been taken by man to new homes. What are the chief barriers to the travels of wild animals? Why do not all the animals of one region differ from those of other regions?

Topics: — Animals and their food; coverings of animals; where they make their homes; enemies of animals; barriers to travel; animals scattered by man; chief barriers on the earth.

37. Distribution of Animals in Zones

Wide seas separate the continents of the torrid zone, and each land has therefore many groups of animals unlike those in the others. The land animals cannot cross the sea from one land to another.

The torrid zone is the home of the largest and fiercest animals. One kind of huge elephant is found in middle Africa, and another kind in southern Asia. In these same regions are found the true buffalo, the humped cattle, called the *zebu*, and the rhinoceros.

The great apes called the *gorilla* and the *chimpanzee* live near the equator in Africa, and another large ape called the *orang-outang* is found near the equator in islands southeast of Asia.

The portion of Africa south of the Sahara is also the home of the giraffe, the zebra, the leopard and many other animals. Many lions are also found there, as well as farther north in the desert region. The leopard and the lion are both members of the cat family; the fierce tiger, another catlike animal, roams from northern Africa far across the southern half of Asia. Still other beasts of the cat family, namely the jaguar and puma, are found in South and Central America. The puma also hunts its food in the Rocky mountain highland of our own country.

In the Andes highland are found the llama and alpaca, two animals smaller than the camel but in the same family. The camels live chiefly in the desert regions of Africa and Asia, in and north of the torrid zone. Find out all you can about camels and their uses. Africa is the home of the giant bird known as the *ostrich;* but there are other kinds of large ostrichlike birds in South America, Australia and the East Indies. Condors, the largest of flying birds, are found in the Andes. The wild animals of Australia are very unlike those of other lands. Many, like the kangaroo, have folds in their skin, forming pouches in which they carry their young till they become strong. Two other queer animals of that land lay eggs, though they are not birds.

The torrid zone has countless monkeys, birds and insects. Large crocodiles infest many of the sluggish rivers of this zone.

Most of the common cattle, horses, sheep and hogs are in the temperate zones. When the white men first came to America they found here none of these animals, but the colonists brought many from Europe, and now they are widespread throughout the fertile regions of the world.

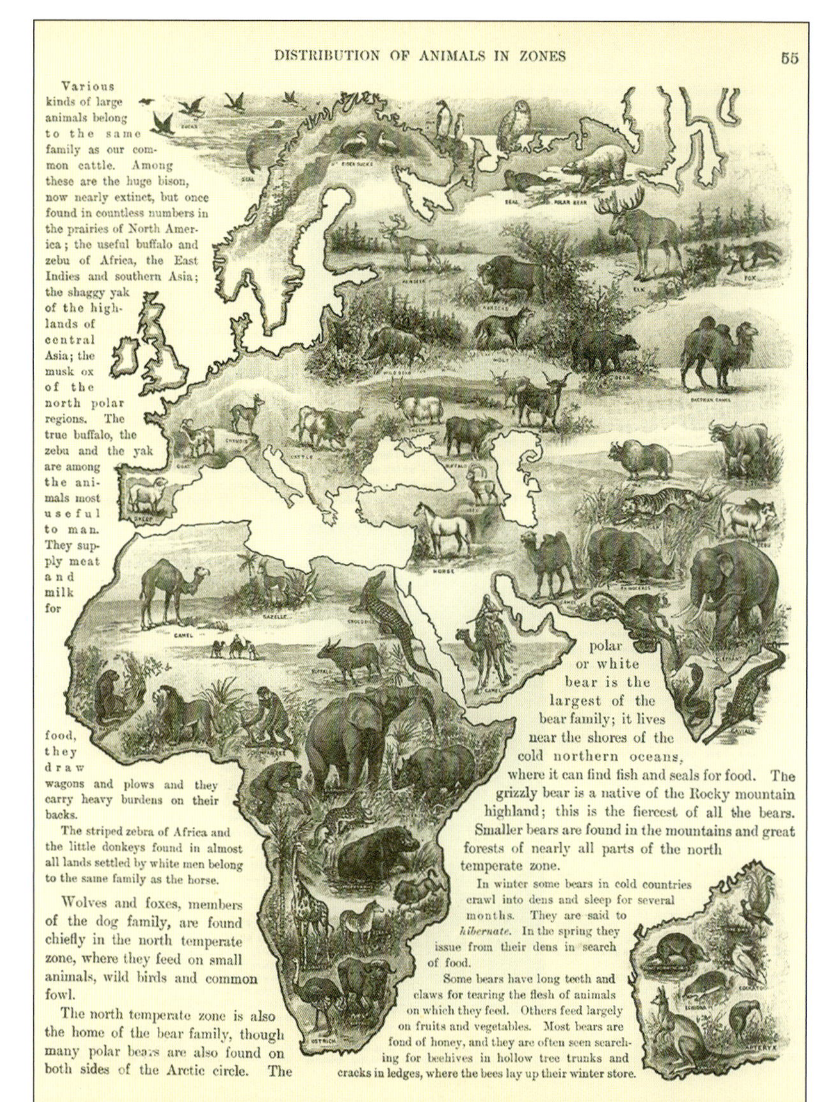
DISTRIBUTION OF ANIMALS IN ZONES 55

Various kinds of large animals belong to the same family as our common cattle. Among these are the huge bison, now nearly extinct, but once found in countless numbers in the prairies of North America; the useful buffalo and zebu of Africa, the East Indies and southern Asia; the shaggy yak of the highlands of central Asia; the musk ox of the north polar regions. The true buffalo, the zebu and the yak are among the animals most useful to man. They supply meat and milk for food, they draw wagons and plows and they carry heavy burdens on their backs.

The striped zebra of Africa and the little donkeys found in almost all lands settled by white men belong to the same family as the horse.

Wolves and foxes, members of the dog family, are found chiefly in the north temperate zone, where they feed on small animals, wild birds and common fowl.

The north temperate zone is also the home of the bear family, though many polar bears are also found on both sides of the Arctic circle. The polar or white bear is the largest of the bear family; it lives near the shores of the cold northern oceans, where it can find fish and seals for food. The grizzly bear is a native of the Rocky mountain highland; this is the fiercest of all the bears. Smaller bears are found in the mountains and great forests of nearly all parts of the north temperate zone.

In winter some bears in cold countries crawl into dens and sleep for several months. They are said to *hibernate*. In the spring they issue from their dens in search of food.

Some bears have long teeth and claws for tearing the flesh of animals on which they feed. Others feed largely on fruits and vegetables. Most bears are fond of honey, and they are often seen searching for beehives in hollow tree trunks and cracks in ledges, where the bees lay up their winter store.

Figure 7.8
"Distribution of Animals in Zones," combines pictorial depictions of animals with the ancient torrid-temperate-frigid zonal conception of the earth.

Source: Alexis Everett Frye, *Grammar School Geography*, Boston Ginn and Co., 1902.

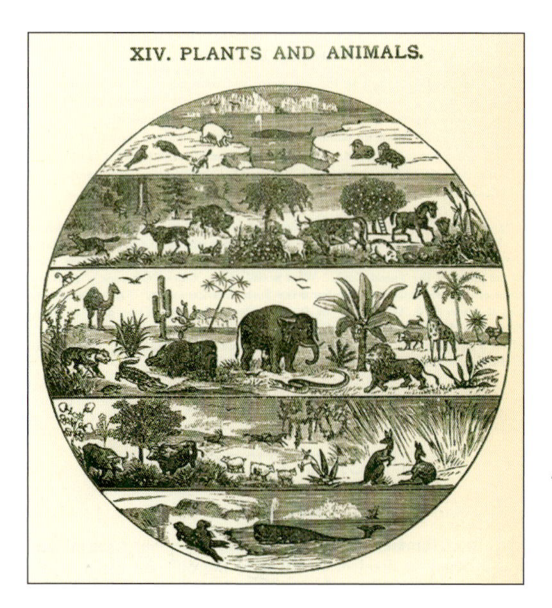

Figure 7.9
"Plants and Animals," also combines pictorial depictions of animals with the ancient torrid-temperate-frigid zonal conception of the earth, but this map makes the zones explicit.

Source: M. F. Maury, "Plants and Animals," *Manual of Geography: A Treatise on Mathematical, Physical and Political Geography,* New York University Publishing, New York and New Orleans, 1898.

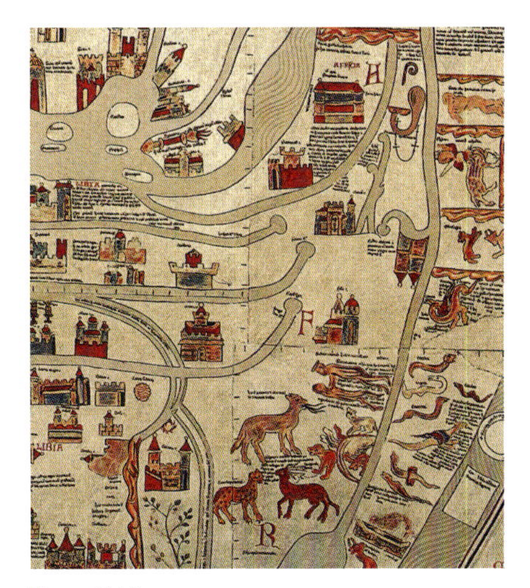

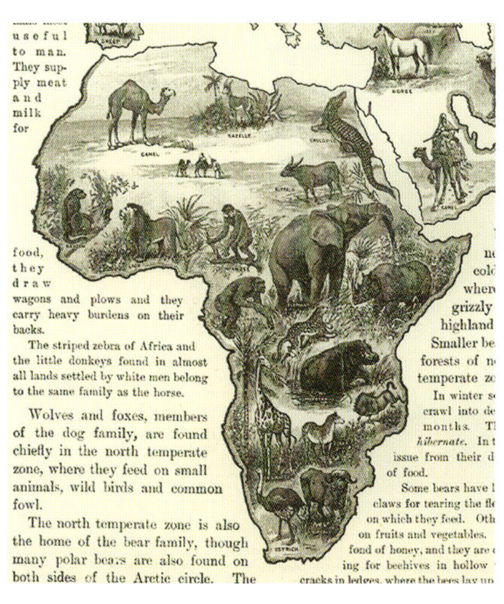

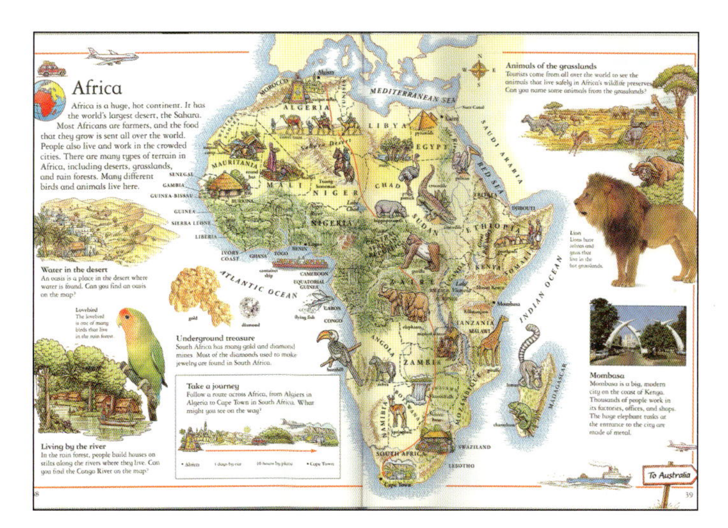

Figure 7.10
Details from the Ebstorf T-O and the Frye map, and the Africa map from *My First Atlas*. Spanning more than seven centuries, these maps make clear the remarkable strength of the pictorial tradition in the mapping of animal locations.

Courtesy of the Trustees of the British Museum London; and Dorling Kindersley Books, London, 1994.

something. It has a straightforward moral force that distinguishes it from the beleaguered, threatening, and awe-ful natures. This nature of plush stuffed polar bears, bluebirds in the backyard, flowers, and children's atlases is a force for good. This nature *is* good.

The source of nature's goodness is diversity in ubiquity

Two elements, then, sign the maps of the cornucopial nature: first, whatever is being mapped will be exhibited in multiple forms, states, or conditions (tulips, roses, lilies; sequoias, cypresses, saguaros; lions, tigers, jaguars); second, it will be everywhere. *Multiplicity* is the pledge of the *collectible* nature we'll be looking at in the next chapter, *distribution* the hallmark of the *scientific* nature we'll be looking at in the chapter after that. It's the commingling of different *and* everywhere, it's the *pervasive variety*, that transforms the variegated and the ubiquitous into the benevolent force that lies at the heart of all that we mean when we say nature.

Here: this from an advertisement for The Cliffs, a collection of private residential communities outside Asheville, North Carolina. The headline says, "Living on the edge of everything." The text adds, "Amid nature's beauty, it is easy to forget the city's hustle and bustle. Hiking on a mountain ridge, traffic and congestion are the furthest things from your mind. Dangling your toes in a clear lake, deadlines and hassles can't even touch you." *How good is that?* Unity in diversity, The Cliffs are "Four Communities. One Wonderful Way of Life." The copy elaborates,

> Why limit your life to one address? At the southernmost edge of the Carolinas' Blue Ridge Mountains are four renowned addresses known collectively as The Cliffs Communities. Each setting is unique. The mountain. The valley. The lake. And now, an idyllic meadow in the mountains of Asheville, North Carolina. But it's what these settings offer together that makes life here beyond compare. Membership in any one of our country clubs gives you privileges at all four communities. And a lifestyle that is simply unparalleled in the nation.

There is even a "resident naturalist" who "teaches everything from fly fishing to herb study to star gazing." Need we add that these are gated communities, that there are golf courses galore, an aromatherapy steam room, and Asheville itself only minutes away?

Or that there's a map? Here the map not only enhances the metaphysical reality of The Cliffs, it renders them physically accessible. Come and see: "Home sites from $150,000, homes from $400,000."

Stripped of its Wordsworthian promise of natural healing, this pitch in its crassest form is the one made by every travel brochure: it's ALL HERE, the mountains *and* the beaches, NASCAR *and* the ballet. You don't need to go anywhere else (all you have to do is change clothes). In one form or another, the claim is made in every popular magazine, in ad after ad, for New Mexico and the British Virgin Islands, Canada and Virginia. In Las Vegas, it's even made for individual casinos, as here for Mandalay Bay: "See it as a fresh and unexpected way of putting your favorite things together. Like surf beaches, lagoons and waterfalls. The Tony award–winning musical *Chicago* and the House of Blues. Aureole and Border Grill. Here, it all works together. Everything goes."

The paradigmatic map for this pitch is the one on the place mat, clotted with clip-art images of golfing in the mountains, surf fishing at the beach, historic sites, shopping. The one in front of us is from The Cottonwood Inn Restaurant and Lounge in Davenport, Washington. Davenport is the "Gateway to Grand Coulee National Recreation Area, Lake Roosevelt, and the North Cascades," and so the map is covered with animals. The placemat/map (figure 7.11) is white with the name of the restaurant and the Columbia River picked out in green. The roads, Indian reservations, and towns are in brown, together with a sheaf of wheat, a bull, a mountain lion, two deer, a bear, a wolf, a pheasant, and a pine tree. It's a simple image—the animals are generalized and silhouetted, the tree is schematic—but it's clearly a descendent of the Ebstorf. This is just its crassest form.

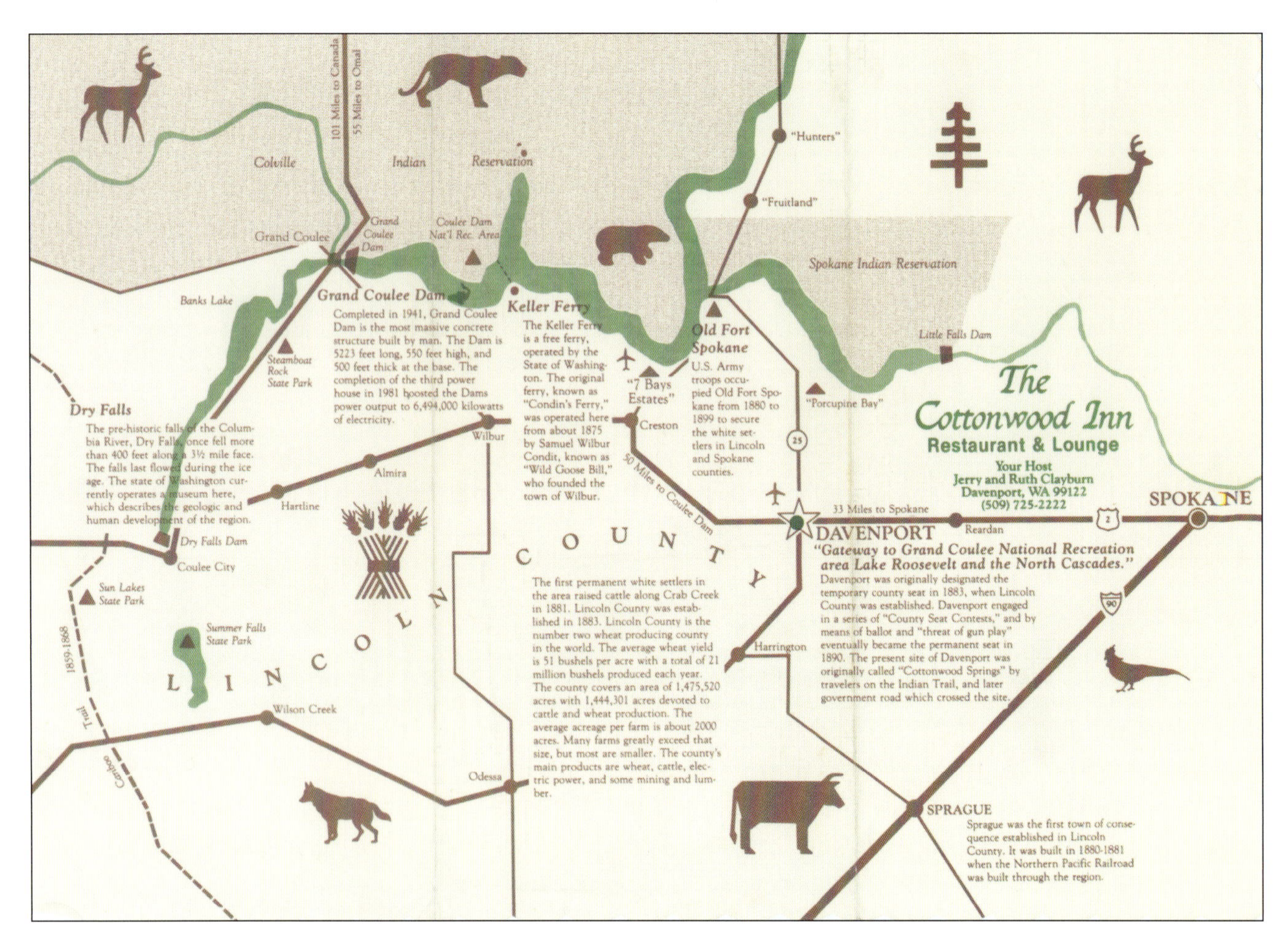

Figure 7.11
This place mat from the Cottonwood Inn in Davenport, Washington, is a vernacular descendant of the Ebstorf mapping of animal locations, stripped to the essentials. Its point? That the area depicted is a veritable cornucopia.

We're now looking at a tourist map—it's older than the place mat, we're guessing it's from the 1950s—issued by the Canadian National Railway (figure 7.12). The *Map of the Arrowhead Country, Minnesota, Quetico Park, New Ontario and Nipigon* is framed by a vignette of an Indian extending a peace pipe and an elaborate north arrow. In one text box it says, "There is a carefree, colorful vacation at Minaki. Golf, Tennis, Boating, Dancing, Swimming, Fishing . . . Balmy, lazy moonlight nights under the glow of northern stars . . . Soft music, Delightful people . . . Everything to crowd your vacation with real happiness. *No hay fever at* Minaki." In another it says, "Not anywhere can be found these vacation retreats where nature is so lavish with all the joys of being alive, except amidst the lake and stream paradise of Ontario and Minnesota." There. It said it for us: *paradise*. Against the background of a thousand lakes are drawings of geese, families of deer, isolated bucks with enormous racks, moose all over the place, ducks, trout leaping on the line, a bear, pheasants. Nature so lavish. Paradise.

Maps of this nature need not be so simple. Consider this map supplement to the August 1979 *National Geographic*: "Bird Migration in the Americas" (figures 7.13 and 7.14). We are acquainted with its form. In fact, it is precisely that of "Australia: Land of Living Fossils," except that here the map folds up instead of down. Over the title on the cover fold looms the pink head of an Andean flamingo. Behind (above) it, there's an Andean flamingo in flight; flying around the fold to the right, a trio of sandpipers, one semipalmated, one white-rumped, one upland; standing on a rock below, its tail, too, stretching around the fold, an American golden plover. Even here, confronted with no more than a sixteenth of the map—a corner!—the theme is already apparent. It is ubiquitous diversity. It is "nature so lavish." It is paradise.

Another fold reveals a partitioning of the field of birds into types, the birds on the cover fold, as these here, being exclusively shore and wading birds. With this is revealed a complication, a conceptual turn toward what Pickstone calls analysis, not *full bore*—not into ranges—and not *replacing* natural history (the bird pictures alone attest to that), but displacing it a nudge. The analytic for Pickstone is first of all about the classification of nature, the living, for instance, into the Linnaean system; or its decomposition, bodies, for example, into organs, tissues, cells. Evidently birds comprise a field susceptible of division and, so, at least this division into shore and wading birds. This second fold adds a whooping crane, the red knot, another American golden plover (this in nonbreeding plumage), a rufous-chested dotterel, and parts of the Hudsonian godwit, the rest of the godwit around the next fold. With this—and we get two folds together now—our first glimpse of a map, of Patagonia and the Antarctic Peninsula.

The Antarctic Peninsula bleeds off the sheet. We are familiar with this trope too: the map's edge less demarcates an entity than frames a piece of the world. Indeed, it is the world, we have merely zoomed in for a closer look, we could pull out, the rest of the continent would reveal itself, it's all real. The variations in plumage—breeding and nonbreeding—have a similar effect: these are but snapshots, we could string them together, they would make a movie, it too would be real.

The next fold brings into view three-quarters of the map, South

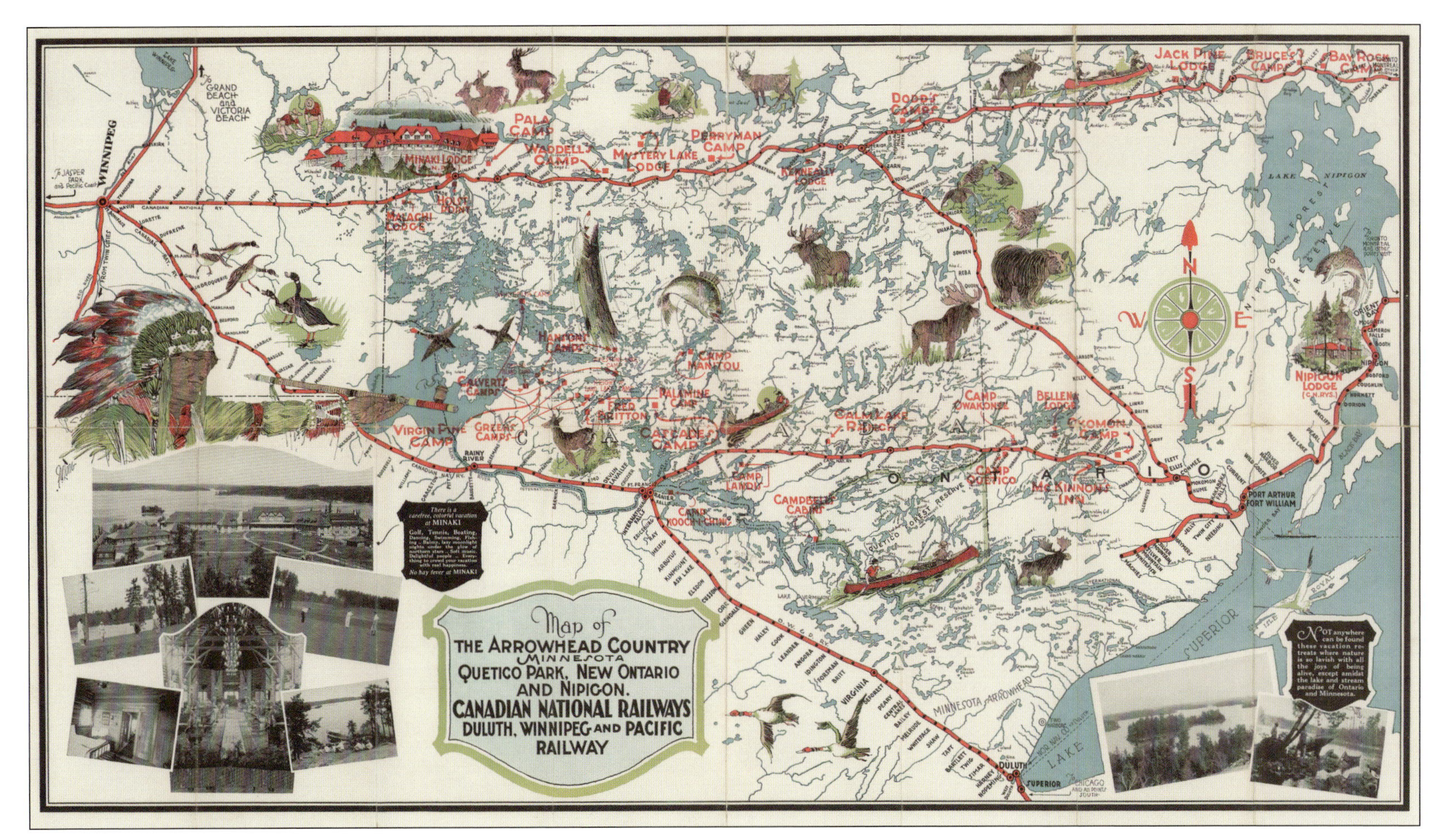

Figure 7.12
A map from a tourist brochure advertising the lake-laden destination as the cornucopial paradise that it (no doubt) resembles. Map dimensions (WxH): 33in. x 19in.

Courtesy of the Canadian National Railway.

Figure 7.13
"Bird Migration in the Americas," map cover and first fold, supplement to the August 1979 issue of *National Geographic*. No more than the cover of this map has to be taken in to understand that nature is a lavish paradise.

Figures 7.13–7.18 courtesy of Ng Maps/National Geographic Image Collection.

Figure 7.14
When "Bird Migration in the Americas" is fully unfolded, the bewildering richness of the natural world is made clear. *And this is just a snapshot of the actual richness*: here are only sixty-seven of the hundreds of kinds of migratory birds that use these flyways. Nature is a cornucopia. Map dimension (WxH): 23in. x 36in.

America, most of North America, and a plethora of birds. The final fold adds Canada, Greenland, and pieces of Siberia and Europe that bleed off the sheet. These too guarantee the rest of the world. Actually, there are only sixty-seven birds pictured, these "of the hundreds of kinds of migrating birds." Again, what we have here is a snapshot, a sample.

Festooning the Americas, seemingly innumerable migratory routes, in four colors, each corresponding to one of the partitions into which this field of birds has been divided: shore and wading birds (in green); seabirds, gulls, and terns (in blue); waterfowl (in brown); and land birds and birds of prey (in orange). The colors are light, like those of the ribbons used to wrap gifts at Easter. Acknowledging the map's complications—which will take study—we complete our tour by flipping the map over. Here, of course, is the "main map"—"The Americas" (figure 7.15)—in the patented house style of the National Geographic Society, its all-but-white continents washed by the palest of blue oceans, Easter-egg tones shadowing the national borders (and those of the U.S. states and the Canadian provinces), the land just *covered* with type. There's not a bird in sight. In the emptiness of the North Atlantic, the title block; in that of the South Pacific, an inset "Physical Map of the Americas." This is one of those vegetationy-landformy maps but, equipped with a full graticule and the tips of Siberia, Europe, and Antarctica, this qualifies less as a portrait than as a perimap's caveat to the pretensions of the main map to be "the Americas." What this "Physical Map of the Americas" says to the main map is, "Uh, you mean, '*Political* Map of the Americas,' don't you?"

As we have seen before, the main map(s)—with their infinitude of *exactly* posted bays, gulfs, rivers, islands, capes, swamps, glaciers, gulfweed, desert, airports, ruins, railroads, canals, states, provinces, nations, and cities, cities, cities, which is to say, with the fullness of their authority—authorize, warrant, the barely constrained profligacy of the other side, the swarming of finches, tanagers, martins, warblers, flycatchers, swallows, parrots, teal, brants, geese, puffins, gulls, penguins, cranes, sandpipers, egrets, and dotterels, with their loopy, lacing, impossibly long ribbons of flight, some of them the length of the earth's circumference, *many* twenty-five hundred miles long—nonstop—at altitudes of up to twenty-one thousand feet. Nature: so lavish!

Unlike the main map side where everything is *certain*—perhaps *contested* like the Falkland Islands/Islas Malvinas, "administered by United Kingdom and claimed by Argentina," but nonetheless *certain*—much on the poster side is not; it's not only barely constrained but guesswork, or mystery, or vague. The routes displayed, for instance, "represent pathways that may be hundreds of miles wide," and the "terminus points shown are general, and the migrants may, in fact, be found at considerable distances in any direction." Texts point out that "Scientists don't know exactly how the migrating birds find their way," and conclude that the process "is a mystery." The routes swerve and cross and many of the (especially) land birds fly common routes, so that it is hard to follow many of them, and the numbers—indicating species—are tiny, and in terminal areas crowded together. You think you're following the route of the red knot north from just south of the River Platte, but somehow you've ended up on the route of the Hudsonian godwit, and it is not easy to find where they crossed. Indeed, it is inconceivable that the map was meant to be used this

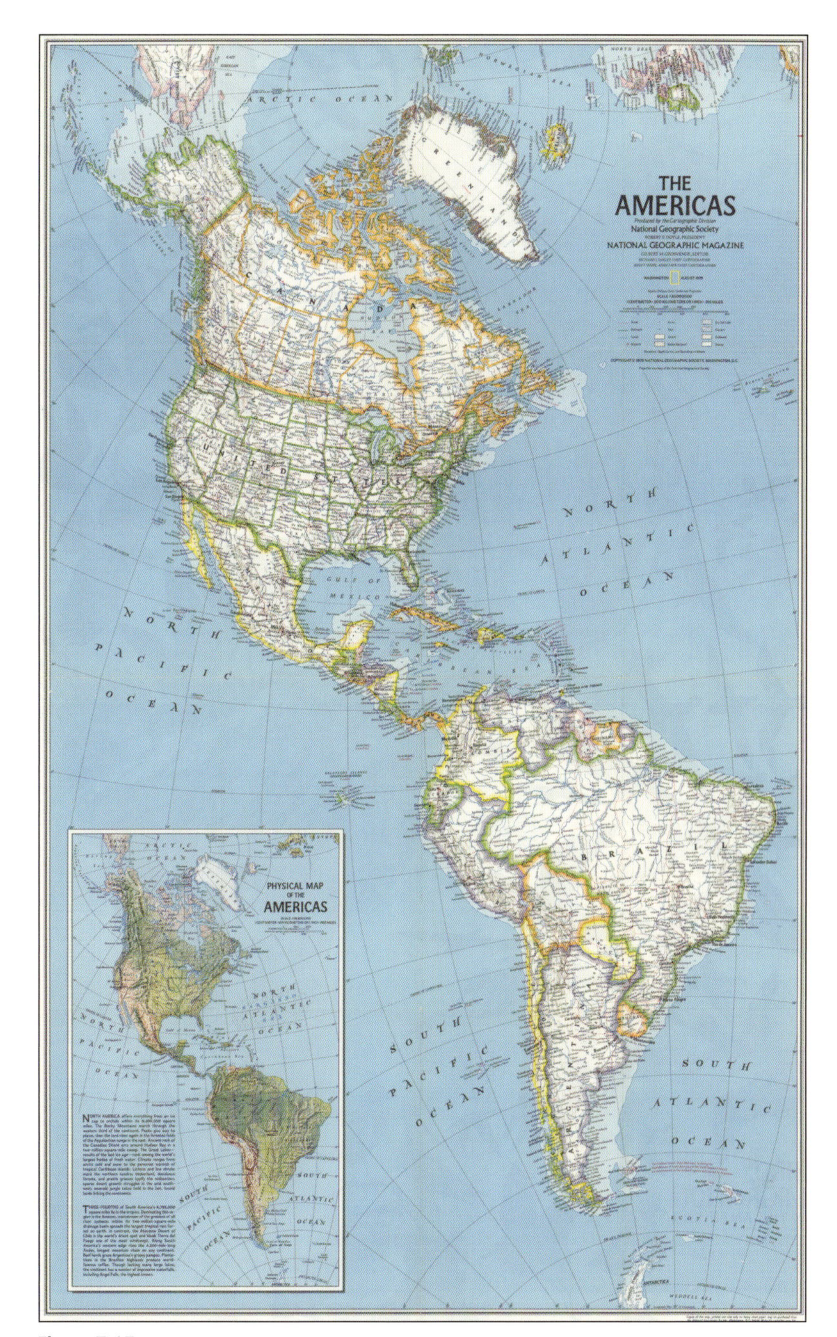

Figure 7.15
"The Americas" (verso of "Bird Migration in the Americas"), in the typical *Geographic* house style for political maps.

way. Again, it's a poster, it's meant to be read from the back of the classroom. It says, "everywhere," "nature," "varied." Yet despite this, the ribbons that indicate the routes do not have fuzzy edges, nor do they fade out or spread at the termini. They are as sharp as any line on a National Geographic map, and their accuracy—given the caveats—has precisely the authority granted them by the undoubted accuracy, indeed, the reality, of the main map on the other side.

As if to gloss our reading, an updated version of this map arrives with the April 2004 issue of the magazine (figure 7.18). On the cover fold, six birds on the wing accompany the title, "Bird Migration (calligraphic, upper and lower case, in black) Eastern Hemisphere (all caps in a condensed typeface in gray)." There are no little numbers tagging the birds, but associated text blocks with the bird's common

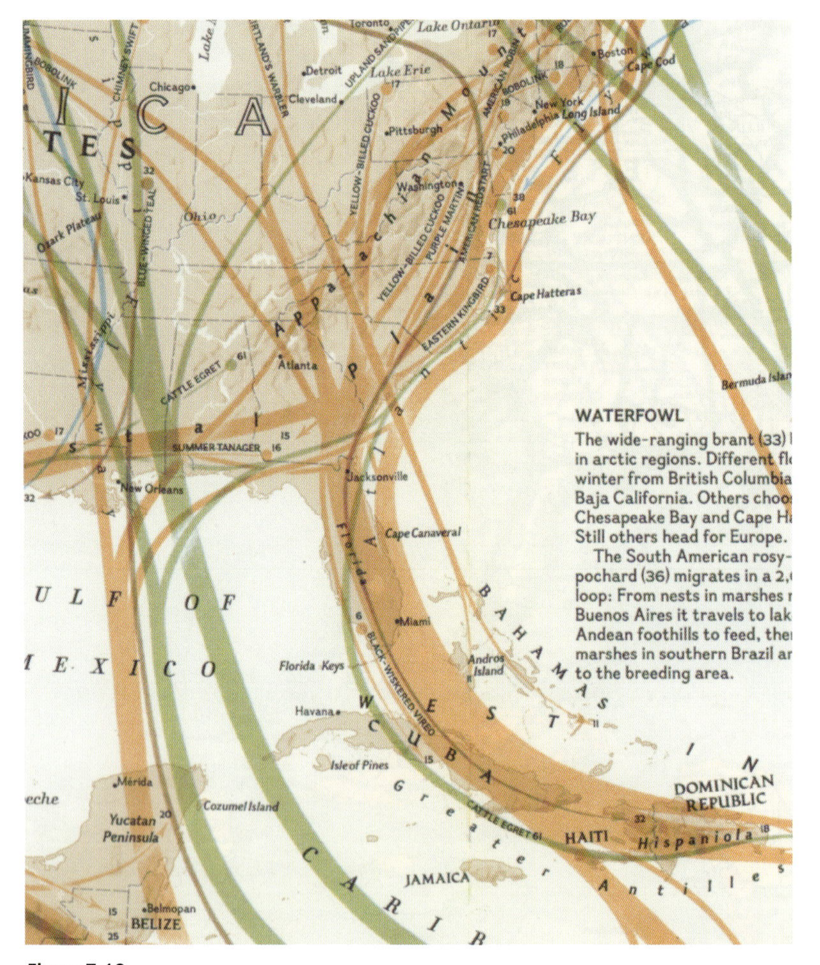

Figure 7.16
Detail of "Bird Migration in the Americas." The definitiveness of the birds' routes, their sharp edges and precise locations, carries precisely the authority granted this side of the map by that of the political map on the flip side.

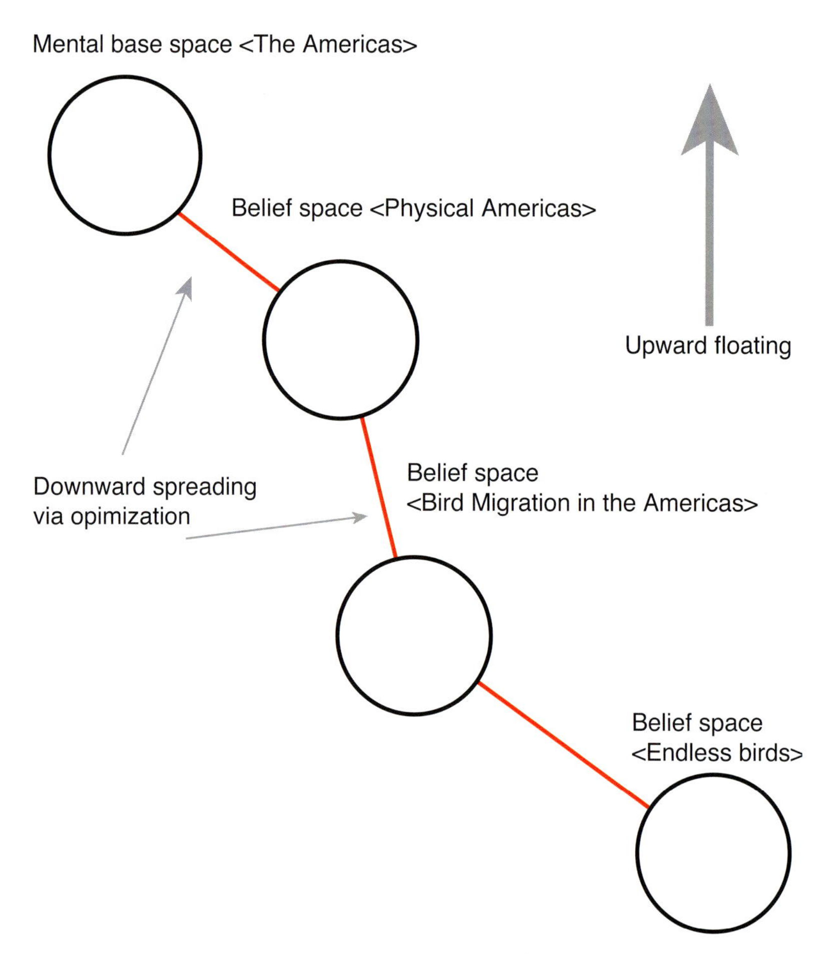

Figure 7.17
The cognitive structure of "Bird Migration in the Americas" resembles that of "Land of Living Fossils." As in that case, the circle at the upper left symbolizes the mental base space opened in the minds of readers who've unfolded and flipped the map over. Here readers construct <the Americas> as a pair of integrated continents—as the "New World"—drawing on their general knowledge about the Americas, and using it to structure their "reading" of the map. The second circle symbolizes the child space propagated by the first when readers drop their eyes to the "Physical Map of the Americas," which modulates the tendency to read the main map in exclusively political terms. The third circle symbolizes the space launched by the <Physical Americas> when readers turn the map back to the poster side. Everything that's been built in the second space gets transferred down to this new space by optimization, while new structure from the bird migration map is being built. The fourth circle toward the bottom symbolizes the child space launched by <Bird migration in the Americas> as the readers' eyes move out to deal with the bird paintings. These birds get Americanized by the structure that optimization builds here in <Endless birds>. The resulting new Americas, with all their birds, then rise back up through the third and second spaces to the first via floating, embedding the animals in the authoritative main map. When they fold the map up and slip it back into the copy of the *National Geographic* it came with, readers will take with them the knowledge that the Americas—and by extension the world—is profligate with birdlife. That is, nature is a cornucopia.

name, its genus and species, breeding ground, wintering ground, length in inches and centimeters, and a color stripe that assigns it to one of *five* classes, our older map's "land birds and birds of prey" being split here into "land birds," and "birds of prey." A bit of map protrudes onto the cover fold with the words, "Guinea-Bissau" and "Senegal" and with what, given our familiarity with the earlier map, we take to be pieces of migratory routes. The map folds down to give us the key, and then up to give us dramatic pictures of another ten birds in flight. The birds on this map seem less posed, are more dynamic, more "natural." The map then accordions out to the right to give us the eastern hemisphere *completely* festooned with migratory routes even paler and more early spring-like in color than those of the older map. These are not of individual species here, but of types, and they *do* fade out and spread at the termini, and mingle and blend in a way that suggests—and maybe it's true—that the entire landmass of Eurasia, and most of Africa and Australia, is flown over or nested in or wintered in by one variety of bird or another; and here they are, flying and fluttering around the edges of the map almost as though we were together with them in an aviary.

To read the other side—"Bird Migration: Western Hemisphere"—we have to rotate the map 90° as we flip it over to accommodate the north–south axis through the Americas. Otherwise there is really no difference in the form and style of the two sides, though this new map of the Americas *does* differ from the old map of the Americas in interesting ways. It's simpler, of course, since the sixty-seven individual routes have been sharply reduced to alternate routes flown by only five types of birds, so the map feels less cluttered. It's certainly less cluttered over Amazonia and the Brazilian Highlands which, formerly dense with flyways, is on this new map all but free of them. Instead, a text block explains that relatively few migrants overwinter there and that Amazonian bird migratory behavior, if any, is poorly understood. A new route has appeared, too, for some land birds and birds of prey

Figure 7.18
"Bird Migration: Eastern Hemisphere," a supplement to the April 2004 issue of *National Geographic*. This expansion and updating of "Bird Migration in the Americas" offers up an even more profligate vision of nature's avian abundance, a vision that is paradoxically more authoritative as it pretends to less precision. Map dimensions (WxH): 31in. x 20in.

that takes them across the Gulf of Mexico via Florida, Cuba, and the Yucatan, reducing their exposure to open water.

But note: no "main map" to authorize these images, which are every bit as "flighty," as profligate, as impressionistic as "Bird Migration in the Americas," as the zoo on the poster side of "Australia: Land of Living Fossils." Taking its place is an apparatus of assurances capable of authorizing these images on its own. Among these is the note just referred to about the Amazonian flyways, an attestation of the limits of the map's knowledge that works to assure us about the reliability of everything else. Operating in this mode as well are the graphic conventions of the fading and spreading of the migratory routes, graphically confessing, as it were, to the limits of what is known and of what can be shown. The listing of the route data sources (for example, "Barn swallow: Rinse Wassenaar, EURING Data Bank" or "Bewick's swan: WWT unpublished data") has a similar effect, undercutting the omniscience of the mapmaker, but at the same time building trust (you can check it out yourself). At the opposite extreme is a display of the routes of individual birds, either tracked by satellite or other means (routes shown in lines of solid green) or derived from banding or ringing recoveries and field observation (shown dashed in brown). These routes of named individuals—"white stork, named Princess, migration Aug. 2002 to Feb. 2003," for instance, or "European honey buzzard, tagged 41504, migration August to October 2003"—annotate the colored bands of the generalized routes, enabling us to compare them and so evaluate the generalized routes ourselves. (It is also a trumpeting of this amazing new knowledge.) In this way, the reader is recruited to validate, and so authorize, the migratory routes, the pervasiveness of bird life, its ubiquitous variety.

It would be fatuous to insist that this new map, with its individual routes, carefully rendered flyways, and exquisite bird portraiture was simply "Bird Migration in the Americas" redux. Indeed it represents, with its radar and satellite tracking, a movement in sensibility beyond even the analytic, toward what Pickstone has characterized as experimentalism.[24] Perhaps here in the map, where despite the classificatory overlay and advanced technology what is recorded remains so deeply natural historical, the sense of the experimental is muted, but in the map's epitext—an article in the magazine on cranes—genuine experiments are described, efforts at helping young, captive-bred cranes raised by crane-costumed people reestablish dormant routes by teaching them to fly behind ultralight planes flown by pilots in crane costumes.[25] Will this work? Only time will tell. It's an experiment.

It would be just as fatuous, however, to deny the family resemblance, the very *powerful* family resemblance, that links this map not only to "Bird Migration in the Americas" and "Australia: Land of Living Fossils," but to "The World of Flowers," the maps of *My First Atlas* and *Atlas of Plants*, and all the rest of the maps we've discussed in this chapter, the railway tourist and place mat maps, the Ludolfi of 1683, the Desceliers of 1550, the 1529 Ribeiro, the Genoese world map of 1457, Dulcert 1339, and the thirteenth century Ebstorf. It's step-by-step, and a fairly big one from the Ebstorf to "Bird Migrations: Eastern Hemisphere," but it is negotiable and this is because the resemblances aren't just superficial, but deep in a shared understanding of nature: that it is ubiquitous and unendingly varied.

Mingled natures

There are maps of this nature that startle in the conviction with which they hew to the straight line of tradition. "Pinnipeds around the World," a *National Geographic* map of April 1987, scatters portraits of fifty seals, sea lions, and walruses around the world, plopping them "where they're found" in *exactly* the same way the makers of the Ebstorf did, and the makers of *My First Atlas* would. Since the animals are only casually located—the Mediterranean monk seal, for instance, sprawls across the Sahara and Arabia Felix—the role of the map is patently less to pinpoint ranges than to provide a stage for the seals to fill: ubiquity and diversity.

Yet other maps may mingle aspects of this nature with aspects of other natures. The December 1976 "Whales of the World" brings aspects of this nature together with aspects of a beleaguered nature and the nature of science. Certainly the title side of "Whales of the World" is a straightforward image of diversity, over two dozen whales of seventeen species cavorting together in a nondeictic space of pure display. But the flip side is another story. "The Great Whales: Migration and Range" may have pictures of six species sprawled across the five great continents and Greenland—anticipating "Pinnipeds around the World"—but it also charts a complicated pattern of ranges in overlapping stipplings and colors, migratory routes, and, in bright red, the whaling fleets of the Soviets and Japanese. As we saw in chapter 1 with the example of "Wildlife as Canon Sees It," and will see in example after example in the next chapter, there is no difficulty about displaying an animal's range without implying its ubiquity. By displaying the ranges of six species, and scattering their portraits around the world, the map in hand exhibits all the stigmata of the benevolent nature of ubiquitous diversity. But by linking the species to ranges, the map displays the essential hallmark of the nature of science, and by including pictures—in bright red—of Japanese and Soviet "floating factories," it indulges as well in the alarmist polemics of a nature under attack.

Why not? This *is* the good nature, the one whose loss we will mourn when it's gone, that we strive today, with all the resources of science, to protect. Let's face it, there's a reason the National Wildlife Federation decorates its solicitations envelope with butterflies, a stuffed polar bear family, and the highlighted, "Plus: Wildflower Seed Mix!" Fur and feathers, butterflies and wildflowers, these constitute the beautiful nature that brings out the protectionist in all of us. The sublime may be *essential*, but it's not something we can put our arms around. Hurricanes and wildfires? Not even with all the science in the world explaining how important for everything these are can we altogether accept them. As for cockroaches, slugs, and mosquitoes . . . *don't you have to draw the line somewhere?*

There is a certain irony that "Bird Migrations: Eastern Hemisphere" so emphatically presents a nature of inexhaustible diversity when the technology exploited to make the map—and we're thinking here especially of the radar and satellite tracking—has been mobilized by the threat of the extinction of, among others, the Siberian crane. Yet the map as a visual artifact is unequivocal in its claim that birds are varied—the portraits attest to that—*and* everywhere—as the migratory routes so clearly demonstrate. This nature threatened? *How?* Yet the peritext is more mixed. Phrases like, "endangered Siberian cranes" speak of threats to diversity, and sentences that read, "The cranes' future depends on wetland protection" suggest that "everywhere" may be more than a bit hyperbolistic. It is science, with its spatialization of the animal as such, that has taught us the relationship between species diversity and habitat. Raising species diversity and habitat together, as this map's peritext does, raises the specter of a nature under attack, and about this the epitext is absolutely explicit. In part, the quandary is rhetorical and motivational. Do the National Geographic Society and the National Wildlife Federation get more bang for their buck sounding the alarm (as in "Australia Under Siege," as in "The Great Whales: Migration and Range") or extolling the world's wonders, as in "Bird Migrations: Eastern Hemisphere" and those stunning nature note cards?

At one level this is a question of strategy, but at another it dives to the very idea of nature we're exploring. In one way or another ubiquity and variety lie at the very core of *any* idea of nature, threatened or threatening, sublime or beautiful, collectible or scientific. As our canvassing of these natures has demonstrated, there is nothing *inherently* benevolent about universal variety. To cull from experience a benevolent nature, we have to *edit* what we imagine gushing from cornucopia's horn (no cockroaches, slugs, or mosquitoes), and then forget that we edited it, replacing the whole with a *beneficent* flood of fruit and flowers, birdsong and butterflies. Science has taken upon itself the effort of understanding how the edited and unedited flows imply and require each other (there would be no birds without the bugs), but after a nearly two centuries of widely supported effort to make this case, *in the popular imagination* it is good things alone that gush from the cornucopia. It's as if there had been a collective shudder:

"If cockroaches, slugs, and mosquitoes must be, let them be, but we don't want to know about them."

It's the beautiful we want, which as Kant pointed out means birds and sea shells, trees and gardens, birdsong and a summer day. For seven hundred years we have mapped this nature, seeing in it first attributes of Christ himself, later the puissant creativity of a maker God, today the beneficence of an unfailing nature. It may be a secular nature, but in "Bird Migrations: Eastern Hemisphere" it is impossible to overlook the emblematic meaning of each and every bird: that of our soul and *its* unfailing promise.

NOTES

1. Kant, *Observations.* This summary list is from John Goldthwait's introduction, p. 36. For another perspective, try Bataille, "Language of Flowers," in *Visions of Excess* (Minneapolis, Minn.: University of Minnesota Press, 1985). For example, Bataille draws attention to the fact that "even the most beautfiul flowers are spoiled in their center by hairy sexual organs" (12), and in general connects them to the excremental whence they come and go.

2. Barthes, *Mythologies* (New York: Hill and Wang, 1972), 101.

3. White, *The Bestiary, A Book of Beasts* (New York: Putnam, 1954). This is a wonderful translation of immense affection carried out by the well-known author of *The Once and Future King* (and so many other terrific books). The notes are at once erudite and a pleasure to read. The numbers of animals in the *Physiologus* and other bestiaries comes from p. 234, and the list we give is simply that of the first eleven animals in White's *Bestiary.* On the *Physiologus,* see the full article in the *Britannica, Eleventh Edition,* where the author lists all forty-nine of the animals of the original.

4. Ibid., 231.

5. Ibid., 234. The character of some of these allegories caused the *Physiologus* to be placed on the *Index Prohibitorium* as early as AD 496, and perhaps a century earlier as well. See White, 244–5, and the *Britannica* article, 552d. The protoscience and the allegory were mixed in this way: after a description of a beast, or of some of its habits, the text would turn into, "It was in this way that our Savior (i.e., the Spiritual Lion of the Tribe of Judah) once hid the spoor of his love . . ." and so on, before returning to, "The Lion's second feature is, that when he sleeps, he seems to keep his eyes open" (White, 8).

6. T-O, or T and O, maps are so called because they're wrapped around by the circular world ocean (the O), within which the Don and the Nile form the horizontal stroke of the T and the Mediterranean the vertical (T-O maps were oriented with east up). In the resulting tripartite division, Asia occupies the top half of the map, Europe the lower left, and Africa the lower right. Conventionally Jerusalem was in the center.

7. Harvey, *Medieval Maps* (Toronto: University of Toronto Press, 1991), 25.

8. George, *Animals and Maps* (Berkeley, Calif.: Secker and Warburg, 1969), 28, 30.

9. Ibid., 186, 187. What she says is, "[The maps] are not outstanding for the accuracy with which the animals are distributed. But the errors are usually those of omission: the animals tend not to have a wide enough range to satisfy accuracy. Only rarely do animals occur in the wrong region. Camels are the main offenders but, in the domesticated form, they have been used all over the Old World for centuries and their presence in all three regions is not, therefore, particularly misleading,"

10. Ibid., 204.

11. Ibid., 206.

12. "So little was the collection considered as a literary work with a definite text that everyone assumed a right to abridge or enlarge, to insert ideas of his own, or fresh scriptural quotations; nor were the scribes and translators by any means scrupulous about the names of natural objects, and even the passages from Holy Writ," B*ritannica,* 553a.

13. Ibid., 552b. Quoting as we do simplifies a complicated subject, and whereas the idea of the natural history in question is that of Aristotle's *historia,* of Pliny's *historia naturalis,* in fact the writers of the *Physiologus* were more likely reading commonplace books with their undigested scraps of folklore, travelers' tales, and so on.

14. Woodward, "Medieval *Mappaemundi,*" in *The History of Cartography: Volume One,* ed. J. B. Harley and David Woodward (Chicago: University of Chicago Press, 1987), 334.

15. This is Woodward's reading on p. 290.

16. Pickstone, *Ways of Knowing,* 42.

17. White, *Bestiary,* 7-8; Pickstone, *Ways of Knowing,* 37.

18. Klonsky, *Speaking Pictures: A Gallery of Pictorial Poetry from the Sixteenth Century to the Present* (New York: Harmony, 1975). 10.

19. See Findlen, "Formation of a Scientific Community: Natural History in Sixteenth Century Italy," in *Natural Particulars: Nature and the Disciplines in Renaissance Europe,* ed. Anthony Grafton and Nancy Siraisi (Cambridge, Mass.: MIT Press, 1999), 369–400.

20. Pickstone, *Ways of Knowing,* 45.

21. Ibid., 45.

22. This is pretty much the sequence P. D. A. Harvey described in *History of Topographical Maps* (London: Thames and Hudson, 1980). The progression is the subject of the entire book, but see especially the introduction, pp. 9–26. It is also pretty much the sequence Wood described in "Now and Then," *Prologue: The Journal of the National Archives* 9 (Fall 1977): 151–61. Harvey compares the two models on his p. 26.

23. Pickstone says, "Thus, in the terms of a displacement model, the 'disenchantment of the world,' beloved of many historians, becomes a matter of changing boundaries and relations between different ways of knowing. As we shall see . . . some seventeenth-century 'subcultures', perhaps especially in Protestant countries, began to give priority to 'natural history'—to description rather than meanings. We shall explore some of the causes and consequences of that displacement, but we can note now that it did not produce only a new 'nature'; it also necessarily produced new kinds of literature," *Ways of Knowing,* 43–44.

24. Pickstone, *Ways of Knowing,* 2, 135–61.

25. Ackerman, "Cranes: The Long Road Home," *National Geographic* (April 2004): 38–57.

Eight

Possessable nature

Possessable nature

Here's another kind of map. It's the size of a postage stamp. It's black and white and has a pink blob on it. It's being all but pushed off a page of a book called *Trees* by a block of type that squats under a painting of pin oak leaves and a seasonal vignette (figure 8.1). In this, a hunter with gun and dog works his way along the edge of a yellowing field in the late afternoon. In the distance, there's a purple mountain and a sky edging into pinks and oranges; in the foreground, a flammiferous pin oak. Those are the words that in bold head the text block: **PIN OAK**.

Trees is a field guide (its subtitle is *A Guide to Familiar American Trees*), one of the ubiquitous Golden Nature Guides, and none the most recent.[1] It was first published in 1952, and is one of a handful we pulled from the shelves of a local used book store. The layout with the map and the text and the leaves and the vignette is repeated ninety-three times, each for a different tree. Navajos gather pine nuts beneath a branch of a pinyon. Cows graze in the shade of an American elm. A quaking aspen surmounts a covered bridge in a snow-covered landscape. In the spring, an ailanthus leafs out over a sidewalk on a city street. Except for the ailanthus—which "thrives in city backyards and lots"—each of these has its little outline map of the contiguous fifty states with a colored blob or line pattern on it (sometimes more than one). Altogether in *Trees*, there are 109 of these maps (some trees have maps but not vignettes). The maps are the trees' spatial signatures.

Range maps

The maps found in *Trees* are range maps. In the section "Using This Book" it says, "Range maps show where various trees are likely to occur." What this means seems unproblematic unless you stop to think about it. Then it's not clear that it means anything at all. Certainly the pink is not posting pin oaks. The pink on this map covers Chicago, Detroit, and Cleveland, covers most of Missouri, Illinois, Indiana, Ohio, West Virginia, and much of the Mid-Atlantic states. It's not all pin oaks. Maybe the key is the word "likely," but what exactly does this key unlock? Does it mean we will "probably" find a pin oak in the pink? Or that we will have a "good chance"? Is this about statistical distributions? Is it that most of the pin oaks will be found in the pink? Is the assertion probabilistic? If it is, what kind of probability are we talking about? Furthermore, how was it determined? Is there a map someplace with all the pin oaks on it, and the range is the area inside a line drawn around all of them? Or most of them? Or some density of them? Or are we to understand that pin oaks are unlikely to be found outside of the pink?

Some field guides are more reticent than *Trees* about these maps. All it says about the many, many range maps in *The Audubon Society Nature Guide to Western Forests* is, "A range map or drawing appears in the margin for birds, mammals, reptiles and amphibians, and for many of the trees, shrubs, and wildflowers." The idea of a range map is presented as common knowledge, like the binoculars the text refers to but likewise doesn't bother to explain. Other field guides are more loquacious. *A Guide to Field Identification of Birds of North America* says, "The range maps use North America as a base except for birds of limited range. The winter range of a species is shown in blue; the summer or breeding range in red. Purple shows where the bird occurs all year. Within its range a bird is found only in certain habitats, such as cattail marshes or pine woods." While helpfully particularizing—"occurs" *doesn't* mean throughout the range but only in appropriate habitats within it—this is less clarifying than it seems, for habitat raises nearly as many questions as range. Marginally more helpful is the description in *Complete Field Guide to American Wildlife*: "Range Maps: Each species has it own particular range. Within this range it lives in habitats suitable for its existence. Some migratory species have a different range in winter than in summer. The range map will tell whether or not a species is supposed to be in your area."[2]

Ranges, habitats, and niches

This formulation—the normative "supposed to be" included—begins to approach contemporary ecological definitions of *geographical range*. This comes from the third edition of Robert Ricklefs' *Ecology*: "The distribution of population is its geographical and ecological range, determined primarily by the presence or absence of suitable habitat conditions." We take this to imply that the range is the area in which the population is distributed. Unfortunately, Ricklefs continues,

> Hence the distribution of the sugar maple in the United States and Canada is limited to the east by the Atlantic Ocean, to the west by low precipitation, to the north by cold winters, and to the south by hot summers. Undoubtedly much suitable habitat exists throughout the world, especially in Europe and Asia, where sugar maple does not occur because it has not had the opportunity to disperse to such areas. (280)

What we have here are a number of ideas circling some core idea of range, but like planets around a sun, also attracted to each other. First is what seems to be the straight-forward idea of range as the area in which an organism is distributed. This is not straight-forward, but we'll get to that in a minute. Second is an idea of range as the area in which an organism *could* be distributed because certain of the area's characteristics permit the organism's existence, a sort of ideal or potential range. This not only allows Ricklefs to refer to suitable habitat to which an organism has not yet dispersed, but allows the rest of us to refer to "original range" (as on the map of Osage oranges in *Trees*) or to speak of animals that survive only in zoos as

PIN OAK takes its name from the many short, pinlike twigs that clutter the horizontal or downward-sloping branches. These make identification easy in winter. The leaf has five to seven deep lobes with long teeth; it is dark green above, lighter and smooth below. The gray-brown bark remains smooth for some time, gradually breaking into scaly ridges. Pin Oak is partial to moist soil. It is a hardy tree, widely used in street and ornamental planting. Pin Oak is widely cultivated in Europe. Acorns are small (about ½ in. long), rounded, with a shallow cup.

Height: 70 to 80 ft. Beech family

Figure 8.1
The PIN OAK page from Herbert Zim and Alexander Martin's Trees: A Guide to Familiar American Trees, rev. ed. (New York: Golden Press, 1956). The tree is caught here in numerous guises: as a name in bold capitals, as a graphic depiction, as an herbarium specimen (the spray of leaves), as an icon (autumn, hunting, distant hills: "our regional American nature"), in a verbal description (with a hint of taxonomy), and in a map (the tree is turned into space).

"extinct in their range," which under the first idea would have meant the range no longer existed, had disappeared, shrunk to zero (or at most to that of the zoo or zoos where it continues to exist). Other planets in this conceptual solar system include the ideas of habitat and niche. Ricklefs defines habitat as the "place where an animal or plant normally lives" where what distinguishes habitat from range is the difference between "distributed" and "lives" (it's a question of scale). Ricklefs defines *niche* as "the ecological *role* of a species in the community; the many ranges of conditions and resource qualities within which the organism or species persists," including the moisture and temperature constraints he described when defining the range of the sugar maple. None of these ideas turns out to be very simple—no matter how taken-for-granted the apparently simple maps in the field guides—and since each amounts to a spatialization of nature that arises from mapping and that motivates the extraordinary proliferation of range maps, we'll take a quick glance at each in turn.[3]

Range. The use of "range" to refer to a stretch of ground over which ranging takes place reaches back into the fifteenth century. By the seventeenth century "range" was being used to refer to grazing land, so its reference to land over which animals ranged had been well established when it was adopted by naturalists in the mid-nineteenth century to refer to the land *occupied* by a species, that is, *naturally* ranged over. The earliest attestation to this usage in the *Oxford* is given in "the reindeer, who is even less Arctic in his range than the musk ox" (1856), where already an idea of "natural habit" has been stirred into the idea of "ranging over." This is the less surprising given that, as we saw in the last chapter, maps of animal *distributions*, as differentiated from *locations*, had begun to appear in the late eighteenth century, with E. A. W. Zimmermann's 1777 map of animal distributions. In 1805, Carl Ritter's map of wild trees and shrubs appeared. Both of these are outline maps covered with names—the ranges are little more than implied—but by midcentury the full panoply of cartographic techniques was being deployed to *delimit* ranges—lines, colors, stippling, shading—and the ranges were playing active roles in the emerging theory of evolution.[4] The question is, how was it being done? This turns out to be the same as asking how it is being done.

In his *Areography: Geographical Strategies of Species*, Eduardo Rapoport says, "Ranges are generally established from maps of points in which each point corresponds to a locality where a species or a genus has been found."[5] He elaborates,

> The problem of outlining the limits of an area is similar to the one posed by the capture points of an individual when we want to establish its home range [the area over which an individual ranges]. Therefore, there are many criteria to define an area, especially if we include the procedure of adjusting-by-eye. This criterion is used by biogeographers; in general there is a cloud of points and the biogeographer draws the limits of the species' range. Sometimes in the case of an animal species, he even extrapolates the range to regions where the species has not been found but is suspected to exist, that is, to regions covered by ground vegetation where the animal normally lives. The method, however, is quite ambiguous for obvious reasons, not only with respect to the aspect and dimensions the range will have but also with regard to its compactness. In this case there is no criterion for determining whether the area will have to be compacted or separated into isolated packets. (45)

In addition to the problems posed by the cloud of points is that of the points themselves. What does it mean to say "where a species or genus has been found"? What does "found" mean? Rapoport says, "If I put a mark on a given place of a map, I shall be able to say that the species Z exists there if, and only if, there is the certainty that any individual of the species passed or will pass over this point at any given moment." This opens the question how long the window has to stay open to see if a member of the species passes: "There is a consensus based on common sense that if in 5–10 years a species is not seen in a given place it is because that species does not inhabit that place," but 5–10 years is an arbitrary length of time, too short, for example, to reliably catch species of fungi that fruit only every thirty years, or fruit unpredictably, and that may require windows of fifty years if changes in ranges are to be caught. Certainly no range map can be relied on that does not specify the length of time embraced by the observations on which it is based. Maps showing the range of wolves drawn on observations of the past hundred years radically exaggerate the range of the wolf today.

"Apart from the fact that any definition of species or genera is controversial, the definition of what the geographical area of a taxon is, or how much it measures, or how it 'behaves,' is very imprecise," Rapoport cautions. "The geographical areas of distribution are the Chinese-lantern shadows produced by the different taxa on the continental screen; it is like measuring, weighing, and studying the behavior of ghosts (1)." Rapoport's conclusion is that "the only alternative we have for the moment is to 'trust' that range maps in some way represent an approximate and rough model of what has occurred in recent years. These maps represent neither an instantaneous photograph nor a film (46)."

Past, future, ideal, and potential ranges. Chinese-lantern shadows or not, such geographical ranges are generated by drawing lines—"not so sharply bounded in nature as on the map" as J. F. Schouw observed as long ago as 1823[6]—*around*, or where the density breaks up into galleries or dendrites or honeycombs, *through the edges of* clouds of points supposed to post the locations of sightings. Imagine the effect on these Chinese-lantern shadows of trying to determine *past* ranges, *future* ranges, *ideal* ranges, *potential* ranges, all of which depend on extrapolation from historical data, or from models of niches into habitats and the mapping of the habitats, or both. Peter Boomgaard's *Frontiers of Fear: Tigers and People in the Malay World, 1600–1950*, a history of human/tiger interactions, provides a recent example of the difficulties. Boomgaard's first problem was to tackle the terminology, that is, Rapoport's "any definition of species or genera is controversial." What they called tigers in the past is not what we call tigers today. At various times "tiger" has referred to leopards and clouded leopards—in all the languages of the Malay world, including Dutch and English—as well as to what we call tigers today (the royal tiger, *Panthera tigris*). These are animals of distinctly different habits *and* habitats, crucial considerations when trying to create a model of the animal (for example, the ungulate biomass required to support an adult male, that is, some approximation of carrying capacity). Such a model is essential in any effort to come to terms with historical data of diverse character, provenance, and quality, all of which vary through the 350 years of Boomgaard's study. Also seen to vary, through all this variance, is the behavior of the tiger. Among other things Boomgaard concludes is that, "Man-eating as a specialized activity of decrepit individuals or of those who have no alternative prey available is probably a modern phenomenon. Earlier sources suggest that the tiger used to be an opportunistic predator who made a 'rational' choice between easy and difficult, unarmed and armed, weak and strong." This and other changes in tiger behavior lead Boomgaard to argue that tigers have a *history* as well as a past:

> Modern observers, who formulated the now orthodox view, never knew the tiger before he learned to avoid humans, and conceived of the tiger—or, for that matter, any animal—as an ahistorical being. Tigers, however, can and do learn. They adapt their behavior to changing circumstances, and the tigresses transmit what they have learned to their offspring, which is what history is all about.[7]

These are interesting and potentially seminal conclusions, but they depend on the reliability of Boomgaard's estimates of range and density. Boomgaard would be the first to admit that his effort to establish the range of the tiger was "like measuring, weighing, and studying the behavior of ghosts"—especially given the "ghost tigers" that figure so prominently in the Malay world—but he would also insist that his "represent an approximate and rough model" of past reality. The prominent population ecologist, G. Evelyn Hutchinson, would support this conclusion. Hutchinson argues that this mixture of modeling, data mining, and extrapolation is the only way to go, the pressing problem being "to uncover possibilities, by any kind of theoretical analysis that proves helpful, and then to see how many of these possibilities are indeed realized in nature."[8] We concur, but doing so should only heighten our appreciation of the tentative, essentially *hypothetical* nature, of range maps.

Habitat. Getting a handle on tiger habitat was crucial for Boomgaard, for it turns out that far from inhabiting virgin forest where tiger food—deer and boar—are scarce, tigers—because deer and boar—prefer transitional zones between biotic communities, such as forest fringes, and especially, during the past 350 years, areas recently disturbed by humans. Boar and deer—as Eastern-seaboard suburbanites have recently come to understand about deer—are culture followers. So, then, are tigers. In the Malay world, "Human action creates the typical tiger habitats, and areas undisturbed by people are often undisturbed by tigers."[9] As these uses of the word exemplify, habitats are the places where an organism lives but *as characterized by the features these places have in common.* What this means is that habitats are less *places* than *kinds* of places, they are a level of abstraction up from places, they have a generic aspect. One speaks of stream habitat, forest habitat, desert habitat. Or, to return to Ricklefs' sugar maple:

> Individuals will be found only where the habitat is suitable. Sugar maples do not grow in marshes, on serpentine barrens, on newly formed sand dunes, in recently burned areas, or in a variety of other habitats that are simply outside their range of ecological tolerance. Hence the geographical range of the sugar maple is a patchwork of occupied and unoccupied areas, just as it is for most other species.[10]

With respect to habitat, then, range is the geographical area containing *occupied* habitats, while potential range is the geographical area containing *suitable* habitats. In either case, the patchiness means that range is something like a quantum wave function: it says that the organism in question may be found somewhere within it but neither where nor how commonly. The organism could be widespread but rare, common but concentrated in few places, or present everywhere in great numbers. Range maps appear to be more informative than they actually are.

Niche. Whether or not a habitat is suitable is a function of the niche. The niche is a level of abstraction up from the habitat (see figure 8.2). As such it is a purely ecological concept (not a geographical one) and cannot be posted on a map. Niche refers to the *role* of a species in its community. "By niche is meant," John Tyler Bonner says, "the place in nature of the organism. The important emphasis, and in fact the value of the concept of the niche, is that it pinpoints the function, the activity, of the organism within its environmental community. It designates what the animal or plant does rather than what it looks like."[11] In *Animal Ecology*, the book that established the paradigm for the twentieth century, Charles S. Elton wrote that when an ecologist sees a badger "he should include in his thoughts some definite idea of the animal's place in the community to which it belongs, just as if he had said 'there goes the vicar.'"[12] By "role," however, the ecologist means something more than simply being a carnivore or an herbivore (or frugivore, or nectarivore or, for that matter, vicar); by "role" is understood "the many ranges of conditions and resource qualities within which the organism or species persists, often conceived as a multidimensional space,"[13] a hyperspace, where, to repeat, this space is not physical but abstract. Niche space coordinates are defined by the values of continuously varying resource attributes that may include temperature, insolation, acidity, humidity, salinity, soil particle size, branch density, food size, prey size, defenses, nutritional value, soil moisture, light intensity—anything one can measure about the ecology of a species.[14] In order, however, for these niches to be occupied by actual organisms, the niches—*this* temperature range, *that* degree of salinity, *that* light intensity—have to be afforded, expressed, exhibited by or in habitats.

The resultant confusion arising from the expression in physical space of a location in an abstract space has been apparent from the beginning.[15] The word, after all, is a figurative application of the word for "a shallow ornamental recess or hollow formed in a wall of a building, usually for the purpose of containing a statue or other decorative object" (*Oxford*). Although its first ecological use dates only to 1910,[16] the word had been used to describe the places of animals far earlier, as in this *Oxford* citation from 1725: "The way to destroy the Niches of Spiders in our Gardens;" or this citation from 1750: "When the animal returns into its nich [sic], the proboscis sinks into itself." Indeed in its first use in ecology, R. H. Johnson's "one expects different species in a region to occupy different niches in the environment," Hutchinson actually hears little more than "an ordinary example of the figurative use of this architectural term."[17] Joseph Grinnell is the first to unambiguously employ the word in its modern sense, but even he has a hard time keeping his footing, at one point defining the niche as the smallest unit of the habitat, the "ultimate unit . . . occupied by just one species or subspecies." Hutchinson's gloss that "though Grinnell is thinking geometrically all the time, the space occupied by 'just one species' is an abstract space that cannot be a subdivision of ordinary habitat space," only underscores, for us, the slipperiness of these terms in "concept space."[18] Certainly it is no surprise to find Bennett and Humphries referring to "habitat niche" ("Its habitat niche can be thought of as

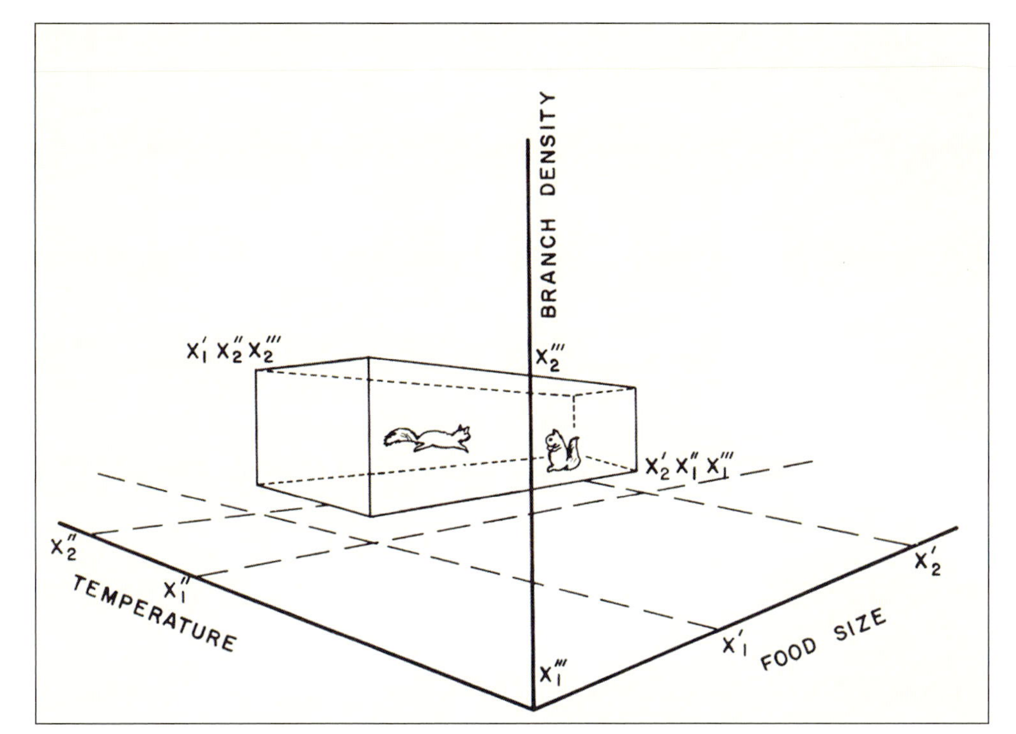

Figure 8.2
A three-dimensional orthogonal fundamental niche. If habitat is a level of abstraction up from place, niche is a level of abstraction up from habitat. It is a purely ecological concept and cannot be posted on a map. The niche pinpoints the *role* of an organism, where role refers to the conditions and resource qualities within which a species persists. It is often conceived as a multidimensional space, as here, where three dimensions have been illustrated: temperature, food size, and branch density. Significant values along these dimensions describe a volume—the niche—within which the species persists and which may, in fact, be thought of as a spatial analogue of the species in niche-space.

Source: Evelyn Hutchinson, *An Introduction to Population Ecology*, 1978. Courtesy of Yale University Press.

the sum total of its many effective environments throughout life") or subsequently complaining that "the term 'niche' on its own is much misused in ecological writing to mean food-niche, habitat-niche, habitat, or microhabitat."[19]

Putting the range to work

But then this problem already existed with respect to the idea of habitat, which as we saw was simultaneously type *and* place. The duality is beautifully caught in the *Oxford* definition of habitat, which not only slips from locality to type, but en route sweeps up the ideas of niche and range as well. It's worth giving in full:

> 1. a. *Nat. Hist.* The locality in which a plant or animal naturally grows or lives; habitation. Sometimes applied to the *geographical area* over which it extends, or the special locality to which it is confined; sometimes restricted to the particular *station* or spot in which a specimen is found; but chiefly used to indicate the kind of locality, as the sea-shore, rocky cliffs, chalk hills, or the like.[20]

The general imprecision surrounding these terms, especially the second you step away from the specialist literature, is perfectly in keeping with the general give and play in the field guides: "At least 50, perhaps 75, species [of oak] occur in this country, mainly in the East," says *Trees*. "Pin oak is partial to moist soil," is the full description of its habitat, unless "a hardy tree, widely used in street and ornamental planting" counts.

And yet, isn't it precisely the *specialist literature* that the range map evokes? Not much else on the pin oak page alludes to the authority that justifies the guide's definitive tone. The "1/2 inch" in "acorns are small (about 1/2 inch long)" carries some air of precision (tempered by the "about"), and the "Beech family" at the end of the text does wave toward a taxonomy, perhaps even a systematics, lurking in the wings. Only the description of the leaves ("five to seven deep lobes with long teeth") vouchsafes much information, though this only verbalizes that offered by the illustration. The soul of the page is the vignette, its evocation of the Midwest in the fall. You can all but smell the stubble in the field and the hunter's jacket tells us that it's cool outside—*crisp* would be the word—as what is sure to be a Saturday draws to its leisurely close. What saves the page from descending irretrievably into "poesy"—or turning into a dinner plate from the Franklin Mint ("The Seasons")—is that crisp, no-nonsense outline of the United States (perhaps, given its size, a little undergeneralized in the Vancouver, Great Lakes, Maritime Provinces, Long Island, and Chesapeake Bay regions (though the visual density *may* manage only to hint at further detail held in reserve)), that crisp, no-nonsense outline and the pink blob, this actually a complicated shape (acknowledging the Appalachians) with two outliers, both smaller than the periods in the text. This overwhelming sense of precision so casually appended to the page, this sense of carefully rendered accumulated fact, sweeps all before it. Less than a square inch, the map manages nonetheless to visually balance the vignette *and* the painting of the leaves that occupy the upper half of the page, its black and white against their vibrant hues, its sharp edges against their blurs, its facts against their fancies.

There is in this "this is the range of the pin oak just the facts ma'am" presentation not even a whisper of an acknowledgment that the range posted could be anything like the Chinese-lantern shadow of a ghost, no indication of when the data were collected from which the range was drawn, not even—anywhere in the book—the data's sources. Yet *this* is not the source. Everything about *Trees* speaks of books standing *behind it*, of monographs, encyclopedias, atlases, of field trips and research and universities.[21] The senior author, "PhD, ScD," is an "outstanding authority," and "well known in professional circles." The second author, the specialist on trees, a "PhD" and "Former Senior Biologist, U.S. Fish and Wildlife Service" is also "an authority." The illustrators have worked on the *Encyclopedia Americana*. But *Trees* wears its erudition lightly. Its massive authority is condensed into a slim, colorful, welcoming volume that encourages you to "slip this book into your pocket" (it's that small), to "use it in the park or along the street" (it's that easy to use), and to "thumb through it in your spare time" (it's that undemanding).

And with this, an amazing thing: effortlessly, without drawing attention to itself, a tree, a species, has turned into a piece of space. There is about the pin oak page an apparent simplicity that belies its complications. The pin oak, the tree species, which—we cannot say this too often—*is a concept*, is mated to four distinct signifiers, in fact four *classes* of signifiers: it appears to us as a text in English, as a painting of leaves, as a vignette, and as a map. The text is largely, almost rigidly denotative (that is, it operates at the level of language), but the leaf painting and vignette are variously connotative (that is, they operate at the level of myth, the vignette exclusively so). The map trades simultaneously in both modalities.

The ensemble has almost the character of a calculus. What we mean by this is that, precisely as in a computation, the pin oak is propagated across media until it appears as the map. In this analogy the pin oak, as a concept, as a signified, is the "statement" of the problem. As it is propagated across media (through the painting and the vignette) it is "re-signed" until it turns into the map, which is the "solution." As Herbert Simon pointed out, "solving a problem simply means representing it so as to make the solution transparent."[22] For instance, in solving an equation, nothing appears in the solution that wasn't present in the initial statement. With each "step" the contents of the previous step are simply rearranged. Cognitively this is precisely what happens on the pin oak page. Cognitive linguistics might suggest that the bold heading PIN OAK is base, and that this spreads, via optimization, first to the flammiferous pin oak in the vignette (which thereby gets "named"), and thence to the painting of the leaves (which is strictly synecdochic), the vignette as a whole (which is perfectly metonymic), the rest of the text (where it is "identified"), and finally the map (where it is spatialized), which then spreads bottom to top via floating, thereby spatializing *everything*, including, ultimately, PIN OAK. Circulated *with* the idea of the pin oak is the idea of nature, which in the end has been spatialized too. Each page in the guide performs this calculation again and again: practice makes perfect!

PIN OAK is nearly as denotative as you can get (okay, not entirely, but it's close), but the spray of oak leaves is unavoidably connotative (alluding simultaneously to Zeus and Druidic practices), though nothing like the vignette, which is almost wholly mythic. Here the myth recruits the pin oak as the signifier of a certain way of life in the Heartland (= Midwestern), as on earlier pages of the field guide the white oak, the post oak, the bur oak, the overcup oak, the chestnut oak, and so on, had been recruited as signifiers of a certain way of life in New England (stone walls, autumn foliage), the Antebellum South (paddlewheel steam boat), the Mississippi Valley (farmland in summer), the Deep South (Spanish moss), the Blue Ridge (touring car on the parkway, purple mountains), and so on; the oaks (as "nature") regionalizing the United States (as "culture"), inescapably conflating nature and the American Way. In this work, the maps play essential roles, establishing these regions in *unmistakable cartographic terms*. In a way, the tree in the vignette (the flammiferous pin oak), the vignette as a whole, and the map look at each other—spread down to or float up to each other—through the medium of the text (figure 8.3). Together the name PIN OAK, with the picture of the tree, the painting of leaves, the vignette ("way of life"), and the pink blob on the map of the United States connote "our regional American nature."[23]

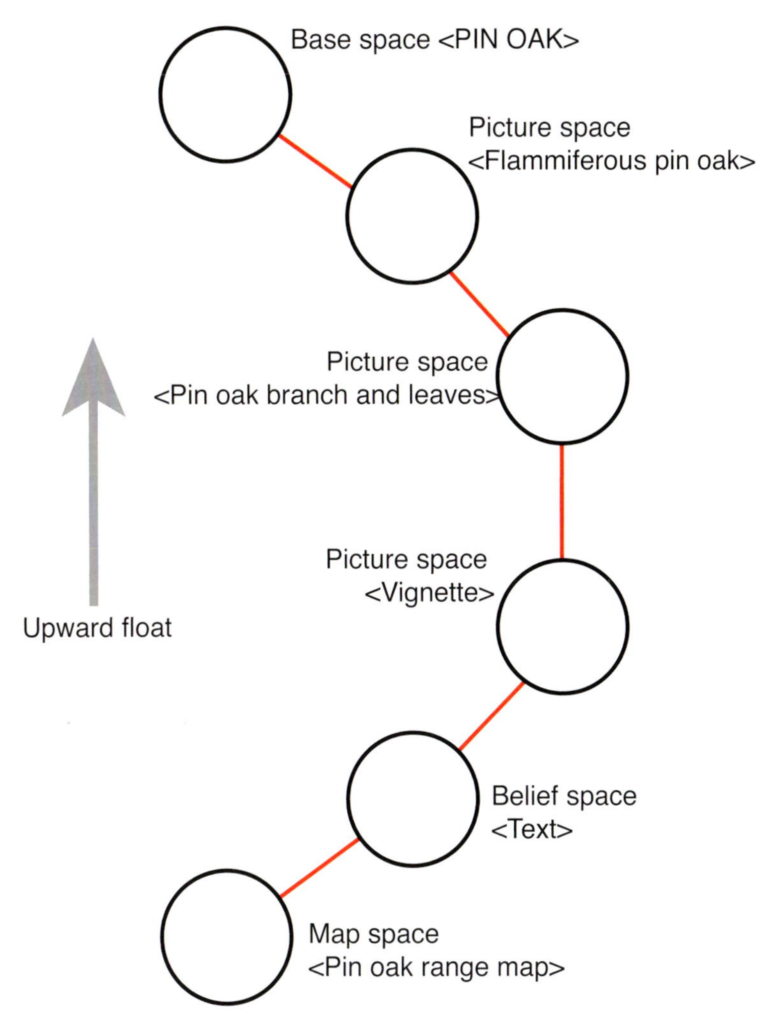

Figure 8.3
The cognitive structure of the pin oak page. The concept name PIN OAK is base and spreads, via optimization to the flammiferous pin oak, the branches and leaves, the vignette, the text and the map, which spatializes everything, by floating

Possessable nature

What kind of nature is this? Simply put, it is ours. It has been claimed for the United States through a thousand individual acts of tree identification.[24] What else is all this identifying but a claiming? "The inventory," Barthes reminds us in his discussion of the plates of Diderot's *Encyclopedia*, "is never a neutral idea; to catalog is never merely to ascertain, as it appears at first glance, but to appropriate."[25] While *Trees* cautions that "collections are not so important in studying trees as you might believe," it does allow that "a collection of leaves, twigs, or fruits may be of real value," before going on to detail when to collect them, how to preserve them, and how to mount and display them. Censuses, it advises, "can be interesting and valuable to both the census taker and to the community," and "The recognition of different woods by their pores, rays, and grain is an interesting hobby, especially if you like wood and use it." In general, "Trees always will be one of the important resources of our country. Their timber, other wood products, turpentine, and resins are of great value. Trees also have value beyond reckoning in holding the soil, preventing floods, and perhaps influencing the climate. In addition, the beauty of trees, the majesty of forests, and the quiet of woodlands are everyone's to enjoy."

In *Trees*, all this utility is ultimately caught up in, reduced to, a name, for example, PIN OAK ("widely used in street and ornamental plantings"), and its analogues, including the map.

Barthes has pointed out that the objects portrayed in Diderot's *Encyclopedia* seem friendly to man, locating this effect in the fact that "it is on each occasion *signed* by man; the image is the privileged means of this human presence, for it permits discretely locating a permanent man on the object's horizon." Almost as if he had *Trees* open in front of him Barthes continues, "You can imagine the most naturally solitary, 'savage' object; you can be sure that man will nonetheless appear in the corner of the image," hunting, as on the pin oak page, or riding a horse, or canoeing, or touring; or there will be no person, but in his or her place a bridge, or a bucket attached to a maple tree, or shocks of corn, or a cottage, fence, or road:

> Take the Giant's Causeway, that mass of terrifying basalt composed by nature at Antrim, in Ireland; [in the plate in the *Encyclopedia*] this inhuman landscape is, one might say, stuffed with humanity; gentlemen in tricornes, lovely ladies contemplate the horrible landscape, chatting familiarly; farther on, men are fishing, scientists are weighing the mineral substance: analyzed into functions (spectacle, fishing, science) the basalt is *reduced*, tamed, familiarized because it is divided: what is especially striking in the entire *Encyclopedia* (and especially in its images) is that it proposes *a world without fear*.[26]

This is precisely because it has been known, that is, divided, named, mastered, that is, because it is ours. And if the *Encyclopedia* has this effect, how much more so is it established by the field guides, small, friendly, and unassuming? The world can be slipped into a field guide, and the field guide slipped into a pocket!

But this possession is by proxy

Yet this possession is of a special kind. Despite the collection of leaves, twigs, or fruit encouraged by *Trees*, the physical possession of nature is precisely what these field guides are *not* about. The guides are in fact *substitutes* for possession, especially when read at home. Even in the field, though, "special equipment for collecting specimens is not necessary," as *Western Forests* points out. "Even many butterflies can be identified without recourse to capturing them—much less mounting them," where it is unequivocal that identification is a proxy for physical control. Naming again replaces taking in: "Similarly, it is generally unnecessary to uproot or otherwise damage plants in order to learn their names," and "Where such extreme measures are required, they are best left to biologists with legitimate research needs. For the rest of us to damage or even kill a plant or animal merely to learn its name would be to lose sight of what is truly important in our relationship to the natural world. Better to let a butterfly ride the winds unnamed."[27] For all of these reasons, "A notebook is the single most important piece of equipment a naturalist takes into the field." This is a remarkable admonition for an activity almost reflexively associated with the butterfly net and killing bottle, if not with the shotgun.

Natural history, at least in its modern phase, that is, from the sixteenth century on, had been all *about* collecting.[28] In contradistinction to the form natural history had taken in the hands of Aristotle, Pliny, and others, from the Renaissance forward the field was increasingly constructed around the collection and display of objects of nature. Paula Findlen convincingly argues that collecting was *central* to the reinvention of nature as an object of humanist inquiry, and ultimately, thereby, to its disenchantment.[29] Central to the growing culture of collecting were early museums with their complicated ties to systems of patrician patronage. Collections grew rapidly. By 1570, Ulisee Aldrovandi had 2,000 illustrations of plants and 14,500 actual specimens in his herbarium.

By the mid-seventeenth century, "collecting had become an accepted practice for Aristotelian naturalists, Jesuit experimentalists, and self-professed Baconians and Galileans, all of whom claimed it as an integral feature of their scientific practice."[30] At stake were not merely the implicit "facts," but the ability provided by museums to *establish* such facts through their display and demonstration to competent observers. By the mid-seventeenth century, authority was "no longer grounded only in canonical text but also in the personal testimony of naturalists who performed experiments," as well as in the testimony of their witnesses, not only the visitors to Athanasius Kircher's museum at the Roman College in Rome, but also the guests of the Royal Society in London, and indeed those of studios, museums, and laboratories springing up throughout Europe.[31] Public display and witnessing—what Findlen sums up as the "culture of demonstration" and the "theatrical culture of science"—rapidly took their place as essential elements in Enlightenment systems of knowledge.

The growing prestige of collecting made collections an increasingly valuable marker of status: "By the eighteenth century, gentlemen considered a modest collection a necessary accoutrement, like a carriage or set of silver. The 'cabinets' of these amateurs generally contained shells, minerals, ancient coins, and books."[32] These collections could become enormous and many became pilgrimage sites for those making the Grand Tour. A popular guide of the late-eighteenth century listed forty-five of these "cabinets" in Paris alone. As the larger collections opened their doors to an ever larger public, fees began to be charged as over the course of the century cabinets of marvels (and the cornucopial culture of wonder) spawned natural history museums (and the possessive culture of systematic collecting).

Collecting became a business that engaged the interest of growing numbers. With the advent of the modern percussion cap and breech-loading shotgun, collecting even began to make itself felt in plant and animal populations: "Naturalists regularly sought the nests of rare birds with the express intention of taking all of them for collections, that is, with the goal of doing so before the species became extinct. Audubon, and other artist-naturalists, thought nothing of shooting one hundred specimens in a day—the rarer the species, the better."[33] By the century's close, the passenger pigeon, which had once darkened the sky in its migration, no longer existed in the wild; and the buffalo had all but disappeared. Alarmed by these losses and inspired by the writings of Henry David Thoreau, John Burroughs, John Muir, and others, organizations concerned with natural history began to promote its *appreciation* instead of its collection. Given that the trade in feathers for women's hats was a particularly egregious offender, and that the absence of the passenger pigeon was a particularly inescapable sign of the threat, it is perhaps no surprise that it was bird *watching*—heavily promoted by the Audubon Society (founded in 1886)—that first gave shape to a socially acceptable alternative to collecting. Making bird watching easier was the invention of prism binoculars and *the evolution of field guides for bird identification.*[34]

The evolution of these field guides into the *field guide culture* that exists today—in addition to field guides to bats, amphibians and reptiles, squirrels, rocks and minerals, wildflowers, mushrooms, fish, coral reefs, ferns, insects, *ad infinitum*, there is even a guide to the splatterings of bird droppings on windshields[35]—took the better part of a century. Roger Tory Peterson has written that "In 1934 when bird-watching was beginning to emerge from the shotgun or specimen-tray era of ornithology, my *Field Guide to the Birds* first saw the light of day. This book was designed so that live birds could be readily identified at a distance by their 'field marks' without resorting to the bird-in-hand characters that the early collectors relied upon. During the half century since my guide appeared the binocular has replaced the shotgun."[36] Modern birders are discouraged from even *disturbing* birds. "WARNING," says a paragraph in a bird book *for children*. "When watching birds, always avoid disturbing them. Be especially careful when watching or photographing parent birds with their young."[37] The tools this guide book recommends for the field include, most prominently, *a notebook*, together with pencils, binoculars, a tripod, and camera; with a magnifying glass, tweezers, and a ruler for studying feathers and pellets. It is *all* about looking, not about taking. It is about possession with the eye, not the hand. It's about *le regard*. It's about the gaze. It's about the very essence of Michel Foucault's power/knowledge.[38]

All the looking, all the identifying, all the naming, is about having. Our *Western Forests* guide may plead that "to damage or even kill a plant or animal merely to learn its name would be to lose sight of what is truly important in our relationship to the natural world," but the butterfly taxonomist, Vladimir Nabokov (better known as the author of *Lolita*), doubtless spoke for many when he wrote, "*I* cannot separate the aesthetic pleasure of seeing a butterfly and the scientific pleasure of knowing what it is."[39] Edgar Anderson is explicit about the possessiveness of a taxonomist in a cloud forest: "He rushes about with a great excess of energy, throwing the plants into presses, searching here and there for something yet uncollected. It would seem as if somehow he is trying to make the cloud forest *part of his professional self*, to get the forest *into his grasp*."[40] Field guides may extol the value of the notebook, but for a birder the life list is what it's all about. This can be as simple as the lists so many collectors carry when on the hunt in flea markets and used bookstores, or it can be as elaborate as the *Bird Lover's Life List and Journal* we described in our opening chapter, that luxurious, hardbound volume for keeping score based on the checklist of the American Ornithological Union:

> Most birders I know keep a Life List, a continuing record of species seen or heard. There are many ways of doing this, but most of us are content to use the lists provided in field guides. The procedure is simple. See a new lifer; put

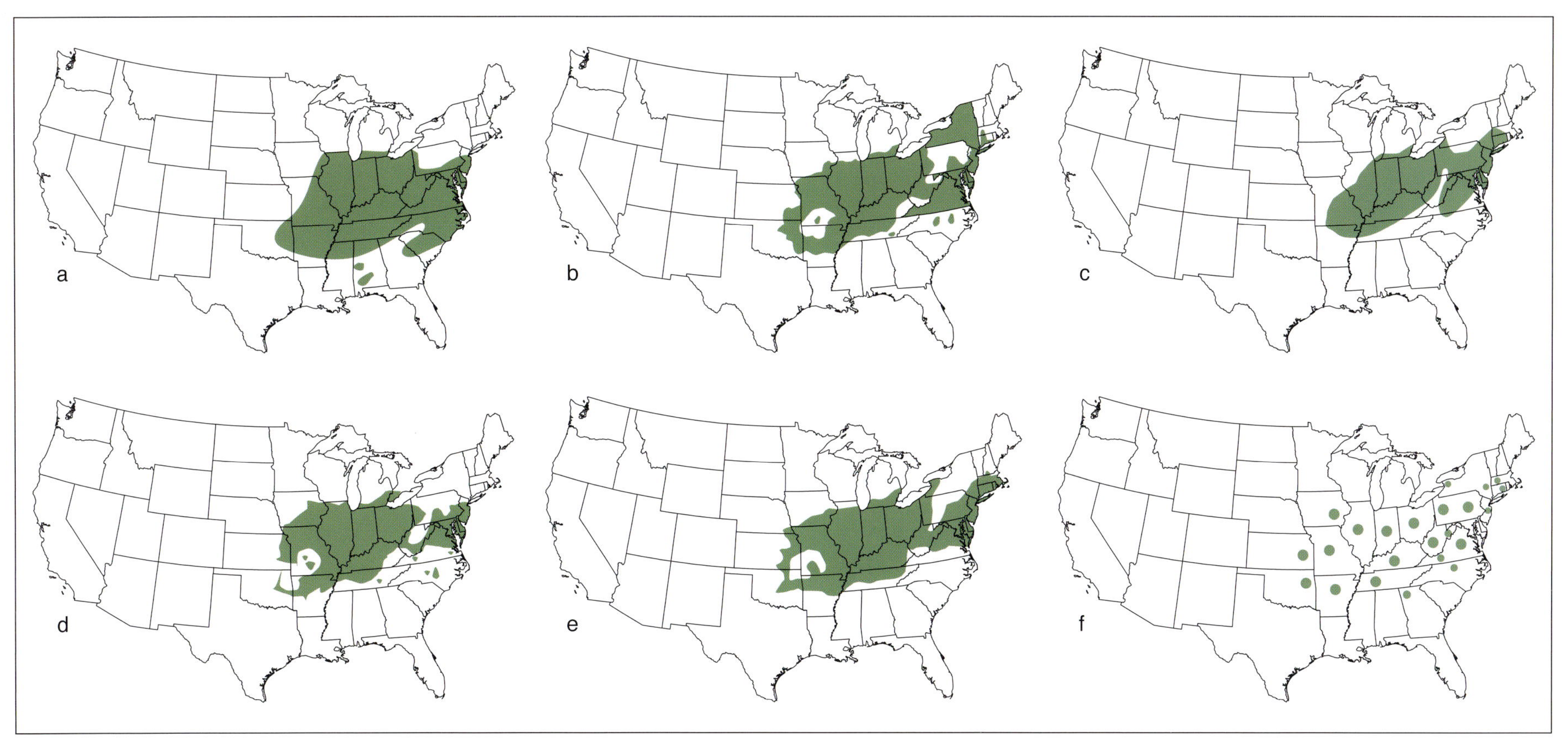

Figure 8.4
Six pin oak range maps, published throughout a span of four decades. The differences between them illustrate the changing face of fact.

> a check mark beside its name. By writing in tiny letters you might be able to squeeze in a date, too, but nothing more. The trouble with this system is that, over time, the list grows longer while memory grows shorter. The concrete details that evoke the cold salt wind of a sighting in Maine, the sweet primordial musk of a mangrove creek along the Gulf of Mexico, the ocher dust that settles on your shoes in an Arizona canyon . . .[41]

In short, the details surrounding the find that enliven every collection, *these* require the space for annotation, the 238 pages, the sewn signatures, the illustrations by John James Audubon. "Noble reasons" for compiling life lists are enumerated—helping scientists track distribution and population trends—but the joy is plainly in the hunt, seeing who can check off the most "lifers" in twenty-four hours, or within a given state or national park, though the hunt be with binoculars instead of guns.

But what exactly do range maps let us possess?

Life lists, checklists, catalogs, inventories, taxonomies—all these surrogates for possession—none *seals* the possession like a range map, which may do this even more effectively than the possession of a physical specimen, for like a *deed* the map seems to let us have them all. More than their certification of scientificity, it is this possessive effect that explains the remarkable proliferation of range maps in field guides (which even without the range maps are manuals for a visual possession): the 8 range maps in the Johnson Nature Series *Squirrels: A Wildlife Handbook*, the 9 range maps in Merlin Tuttle's *America's Neighborhood Bats*, the 32 range maps in the Stokes *A Guide to Amphibians and Reptiles*, the 109 range maps of the Golden Nature Guides *Trees*, the 165 range maps in the United States Department of Agriculture's *Important Forest Trees*, the 266 range maps in the Petrides *A Field Guide to Eastern Trees: United States and Canada*, the 362 range maps in the Peterson *A Field Guide to Birds of Britain and Europe*, the 628 range maps in the Golden Field Guides *A Guide to Field Identification: Birds of North America*, the more than 700 range maps in the National Geographic *Reference Atlas to the Birds of North America*, the more than 800 range maps in the Eyewitness Handbooks *Birds of the World*, and we could go on, but the number of range maps in field guides today is beyond counting.[42]

To be sure, some field guides proffer nonpossessive rationales for the maps, and some of these are more persuasive than others. This, for example, from the Macmillan Field Guides *Trees: A Quick Reference Guide to Trees of North America*, tries hard to justify the map's *utility*: "The

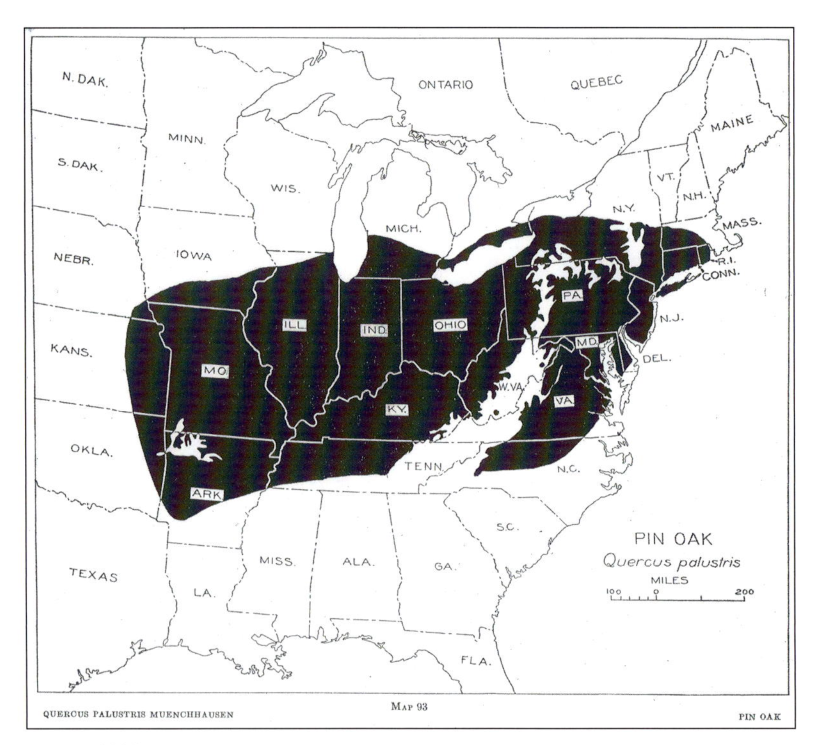

Munns, 1938

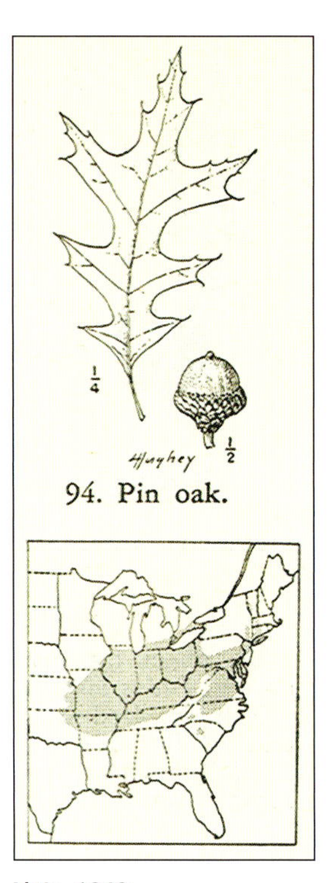

Little, 1949

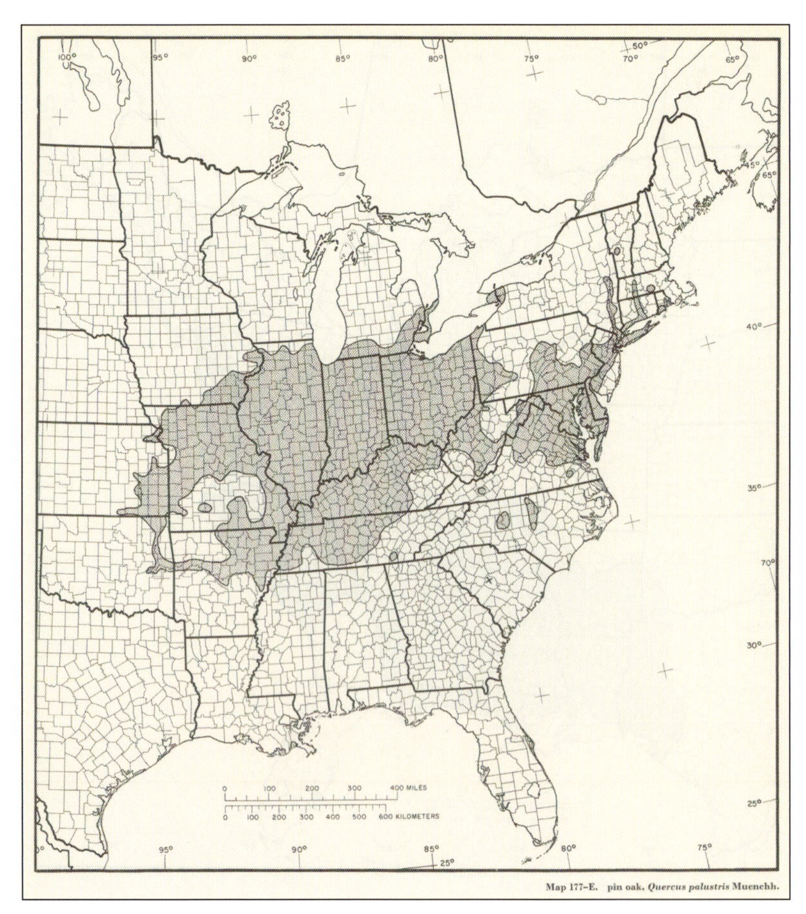

Little, 1971

Figure 8.5
Comparison of pin oak ranges as depicted in USDA maps of 1938, 1949, and 1971.

Courtesy of USDA.

maps show the overall range of each species illustrated. Knowledge of the geographic region(s) in which a tree grows is often of great value in identification. Indeed, once you think you have determined the general kind of tree you are working with—for example a fir of some sort—you should first of all check the maps to see which fir grows where you are," though this help is greatest when there are fewest overlapping species. *Western Forests* merely advises, "To find out whether a bird, mammal, reptile, or amphibian is in your area, check the range map," but *Trees of North America: A Field Guide to Identification* admits that the maps give only a "general picture."[43]

What this "general picture" amounts to becomes clearer once you compare range maps. Here are six pin oak ranges (figure 8.4), all from well-known guides published from 1952 to 1988. Though it's possible to accept these as maps of a common reality—the range of the pin oak—it's just as easy to imagine they're maps of unrelated species. Notice the not-so-subtle differences between the maps, all of which are supposedly showing the same thing. What's going on?

Though few of the guides feel compelled to reveal anything about their sources,[44] there turn out to be really only three for American trees: E. N. Munns' 1938 *Distribution of Important Forest Trees of the United States*, Elbert Little's 1949 "Important Forest Trees of the United States," and Little and others' 1971–1981, six-volume *Atlas of United States Trees*, all of which happen to be genetically related products of the United States Department of Agriculture's Forest Service.[45] Charles Sargeant published the first distribution maps of native species in 1884 in the Tenth Census of the United States with sixteen colored maps prepared by Andrew Robeson; but the range mapping of American trees began in earnest in 1905 with George Sudworth, the Forest Service's first dendrologist.[46] Over the years, Sudworth compiled thousands of locality records—each on a separate numbered card—which he culled from botanical lists, forest surveys, unpublished field notes, herbarium specimens, whatever he could find. These locations were plotted, by number, on large cloth-backed maps, one map per species. As many as two hundred locations might be plotted on these manuscript maps.

Sudworth's goal had been to publish a forest atlas, but only "Part I—Pines" ever appeared, in 1913. However, twenty-five years later, Munns published 170 maps based on Sudworth's manuscript maps. Munns' *Distribution* was criticized on publication for its inaccuracies, but it had no competition and was often reprinted. In 1945, H. S. Betts redrafted these maps at a smaller scale for a series of USDA leaflets. Four years later, Little recompiled 165 of the maps, "hastily, to meet a publication deadline," for *Trees*, the 1949 *Yearbook of Agriculture*. Despite the deadline, Little attempted to rectify errors in Munns with data from an additional two hundred published references and, for some states, with his own personal field work. Little redrafted these maps again in 1965 for H. A. Fowells' *Silvics of Forest Trees of the United States*, expanding his reference base with another hundred publications. When it was decided to launch the 1971–1981 *Atlas of United States Trees*, it was these maps Little had made for Fowells that comprised the heart of the corpus. In fact, Little used "a projector in a dark room . . . to copy exactly from other maps."[47] Of these range maps, we have reproduced those of the pin oak from Munns, Little 1949, and Little 1971 (figure 8.5).

Comparing the USDA maps with those in the field guides makes it obvious that Munns and Little 1949 are the combined source for the original 1952 *Trees: A Guide to Familiar American Trees*; that Little 1949 is the sole source for the revised 1956 edition of the same; and that Little 1971 is the source for the *Trees of North America* of 1979 *and subsequent guidebooks*, Duncan and Duncan included.[48] It's equally clear that none of the maps in the field guides was *traced* from the USDA sources; and that haste, difficulties in copying, diverse notions of generalization and symbolization, and other sources of variation have contributed to the range of propositions, which could easily have been multiplied, about where pin oaks might be found. Certainly it was Little's haste in compiling the portfolio for the 1949 *Yearbook* that spread pin oaks across the Ozarks—and so across the Ozarks in the 1952 and 1956 Golden Guides—and that merely *added* outliers in the Carolinas to the existent range in Munns, thus forcing the outliers south (deep into Georgia in the 1952 Golden Guide), instead of *breaking* Munns' *North* Carolina range into two as Little finally did in 1971, when he also deleted much of Munns' Virginia range.[49] We could continue, but your eyes are as good as ours.

These variations and this history, which are more or less the variations and history of all range maps, only raise with greater force the question what, if anything other than a spatial token of possession, is posted on a map with the posting of a range. Plainly it is not pin oaks that are posted, though this is the sort of automatic response you get from people if you ask them. "It shows you where the pin oaks are," they say. And to be sure, standing behind these range maps is not only George Sudworth's manuscript map with its individual postings of actual tree sightings, but the *idea* of this manuscript map, which is the *ideal* behind all maps: *these are there*.

As if acknowledging this ideal, Little remarks in the introduction to his *Atlas* that "Mapping of the range as a dotted area bounded by dots would perhaps be more natural, as individual trees would appear as dots on a large scale map."[50] But he doesn't do this, he says, because "in spite of problems connecting dots a line serves for emphasis." Undoubtedly this is the case, but the drawing of this line has momentous consequences. For one thing, it changes the subject of the map. No longer is it a map of pin oak *sighting locations*, but a map of pin oak *range*, where the change from plural to singular marks the change from many individual *trees* (pin oaks) to a single *species* (*Quercus palustris* Muenchh). This change, brought *about* by the line, in turn *changes* the line. No longer is the line just a kind of fence corralling the *locations* of the *sightings*, as Little implies with his "a line serves for emphasis." Instead the line has become an articulation of the *ecological limits* beyond which the *species* cannot thrive.[51] As it says in Neil Campbell's introductory college biology text, "A population's geographical range is defined as the geographical limits within which it lives."[52] In effect, the range is the species' spatial analogue. With this twist, from sightings to range, individuals to species, and line to limits, we have moved from what Pickstone asked us to think about as natural history to what he asked us to think about as analysis (figure 8.6).

But analysis of what? *Of nature*. When we said earlier that nature circulated on the back of the "pin oak" concept through the vignette, painting, text, and range map of the pin oak page in the Golden Nature Guide, we had in mind not so much the idea that "trees" are

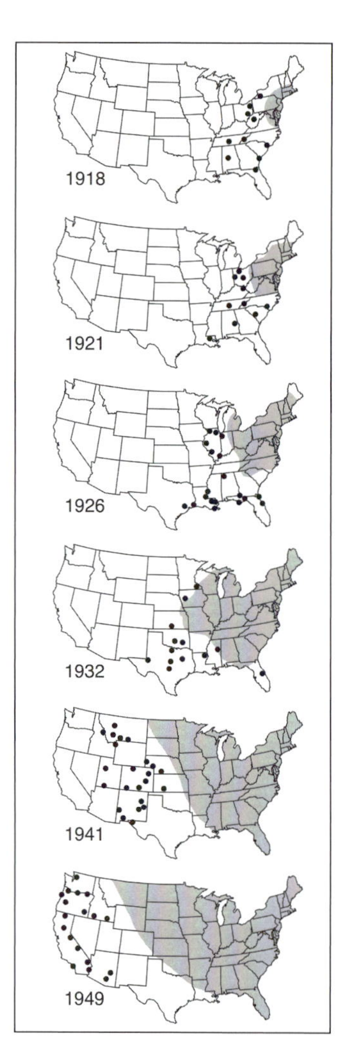

Figure 8.6
The western expansion of the range of the European starling in the United States is often used to illustrate what happens when the limits imposed by "natural" barriers (in this case the Atlantic Ocean) are overcome (in this case by the humans who imported the European starling into the United States). But it also illustrates the fact that ranges have histories, that they are less synchronic spatial facts than dynamic processes.

Robert E. Ricklefs, *Ecology*, 3rd. ed. © 1973, 1979 by Chiron Press, Inc.; 1990, 2000 by W. H. Freeman and Company; adapted from B. Kessel.

"natural" as the fact that range maps post "the *natural* distribution or range of a tree species," as Little puts it in his introduction—that is, "the geographical area where the tree species, including all varieties, is *natural* or *wild*."[53] This is to say that the limits brought into being by the line are natural *limits*, the *limits of nature*. These limits imply an answer to the question, "Where do pin oaks grow?" *They grow where nature lets them.* Implicit here is a kind of environmental determinist argument—a niche-like concept—but written at a geographic scale. As articulated, the idea is either empty (tautological: it reduces to "trees grow where they grow") or false, and this in its most trivial sense, that it can be falsified. Here is Little falsifying it himself: "However, many species have been planted beyond their native ranges, and often have spread by escaping from cultivation. A few, such as Osage-orange, black locust, and northern catalpa, have become widely naturalized."[54]

It's enough to make you dizzy, or would be if it weren't such a common move. Note first the way "native" has replaced "natural" as a modifier of range. "Native" invokes history.[55] It speaks of a time before humans intruded on the scene, at least before European humans intruded on the scene—that is, of a time when the scene was natural. Little is explicit about this temporal dimension: "Where changes have occurred following European settlement, the distribution is intended to be before Columbus, or pre-Columbian."[56] It's important to follow Little's argument: the range is where the species is natural or wild, that is, native, which is to say pre-Columbian. This is to say, pre-Columbian = Nature, the very archetype of every lapsarian myth.[57] We see immediately that through these moves Little has manifested the dichotomy of internal and external spaces that Jacques Derrida faults in the work of Jean-Jacques Rousseau, Ferdinand Saussure, and Claude Lévi-Strauss: that of innocence as opposed to corruption ("beyond their native ranges"), proximity as opposed to alienation ("spread by escaping"), authenticity as opposed to artifice ("become naturalized");[58] and that it is in his effort to stay within the confines of this system of thought that Little has been forced to make his range maps *not* of where the trees *are*, but of where they *used* to be (perhaps, though we have seen the difficulties involved in attempting this). So whereas Little's text acknowledges that the northern catalpa "is naturalized elsewhere in the eastern United States," his range map has it confined to a small central Mississippi and lower Ohio Valley region; whereas his text acknowledges the black locust had been "widely naturalized in eastern half of United States and southern Canada," his range map has it confined to the Appalachians and Ozarks; and whereas his text acknowledges the Osage orange was "naturalized in eastern half of United States except northern border," his range map has it confined to a small region just east of the borders between Louisiana and Texas, and Arkansas and Oklahoma.[59] We can now see that it is this *innocent, proximal,* and *authentic* space—that is, this *natural* space—that is signified in the Golden Nature Guides *Trees* as "original range."

What is less easy to see is that it is this very same innocent, proximal, and authentic space—this *natural* space—that is signified *in every other range map as well,* for these, too, post the pre-Columbian range. The difference is that whereas the range maps of the northern catalpa, black locust, and Osage orange fail to indicate where the species *has spread,* the rest of the range maps fail to reveal where the species *has been decimated,* a word we choose to draw attention to the reality that between 1870 and 1970 Americans managed to clear-cut 500 million acres of virgin forests east of the Great Plains, including, we should emphasize, most of the pin oaks.[60] This depiction of a precolonial America makes of the field guides—all of them—*historic* guides, guides to a vanished America; it renders "Field guides show where various trees are likely to occur" true only of an eidolon of an unspoiled land we cannot get out of our heads. What we suggest with the posting of a range is an image of nature transmuted into space, a historical should be–could be projection of a native species onto a regional screen, identified, named, delimited, mapped, and so, seized by us to be made part of the national patrimony.

Even if long since mostly cut down.

Not even the stars are safe from our grasp

Nor are we content to limit our identifying, naming, delimiting, mapping, and so seizing, to trees and birds, insects and wildflowers, rocks and minerals, reptiles and bats, but we so take the very stars. In May of 1609, Galileo Galilei, then a forty-five-year-old professor of mathematics at the University of Padua, heard rumors of a spyglass.

He immediately threw one together using ordinary spectacle lenses. A few months later, having taught himself to grind and polish the more powerful lenses he could not buy from the spectacle makers, he made a spyglass that magnified eight or nine times. This was a device with obvious military benefit and he promptly donated it to the Republic of Venice. In gratitude, the Republic renewed his contract with the university for life and doubled his salary. Later that year, now in his native Tuscany, Galileo made an instrument that magnified twenty times. He immediately turned this on the moon and, early the next year, on Jupiter. He soon realized that what he had taken to be fixed stars in Jupiter's vicinity were moons, and he promptly dedicated them to the Medicis, the Dukes of Tuscany. Here is Galileo in February, 1610, writing to the ducal secretary:

> As to my new observations, I shall indeed send them as an announcement to all philosophers and mathematicians, but not without the favor of our Most Serene Lord. For since God graced me with being able, through such a singular sign, to reveal to my Lord my devotion and the desire I have that his glorious name live as equal among the stars, and since it is up to me, the first discoverer, to name these new planets, I wish, in imitation of the ancient sages who placed the most excellent heroes of that age among the stars, to inscribe these with the name of the Most Serene Grand Duke. There only remains in me a little indecision whether I should dedicate all four to the Grand Duke only, calling them *Cosmian* after his name [Cosimo de Medici] or, rather, since they are exactly four in number, dedicate them to all four brothers with the name *Medicean*.[61]

The secretary notified Galileo by return mail of the preference for *Medicean Stars*, and an unbound copy of Galileo's *The Sidereal Messenger* was soon on its way to the Tuscan court. The book's full title page is worth quoting:

> SIDEREAL MESSENGER
> unfolding great and very wonderful sights
> and displaying to the gaze of everyone,
> but especially philosophers and astronomers,
> the things that were observed by
> GALILEO GALILEI,
> Florentine patrician
> and public mathematician of the University of Padua,
> with the help of a spyglass lately devised by him,
> about the face of the Moon, countless fixed stars,
> the Milky Way, nebulous stars,
> but especially about
> four planets
> flying around the star of Jupiter at unequal intervals
> and periods with wonderful swiftness;
> which, unknown by anyone until this day,
> the first author detected recently
> and decided to name
> MEDICEAN STARS

Lest there be any doubt about his intentions, Galileo followed this with a fulsome, four-page dedicatory epistle in which, after reviewing the practices of the ancients in naming heavenly bodies after gods and heroes, and Augustus' disappointment in having the comet he named the Julian Star swiftly disappear, Galileo says, "But now, Most Serene Prince, we are able to auger truer and more felicitous things for Your Highness, for scarcely have the immortal graces of your soul begun to shine forth on earth than bright stars offer themselves in the heaven which, like tongues, will speak of and celebrate your most excellent virtues for all time. Behold, therefore, four stars reserved for your illustrious name, and not of the common sort and multitude of the less notable fixed stars, but of the illustrious order of wandering stars." In fact, Galileo goes on, "the Maker of the Stars himself, by clear arguments, admonished me to call these new planets by the illustrious name of Your Highness before all others," for, in the first place, Cosimo is a Jupiter, and in the second, Jupiter graced Cosimo's birth. Finally, "Since under Your auspices, Most Serene Cosimo, I discovered these stars unknown to all previous astronomers, I decided by the highest right to adorn them with the very august name of Your family. For since I first discovered them, who will deny me the right if I also assign them a name and call them the Medicean Stars?"

Who indeed? And who would deny Cosimo the right to name Galileo First and Extraordinary Mathematician at the University of Pisa—this a munificently compensated position with few duties—as well as First Philosopher and Mathematician to the Tuscan court? Galileo soon presented the Grand Duke with as fine a telescope—soon so named by the poet and theologian John Desmiani—as he could make, so that Cosimo could see the *Medicean Stars* with his own eyes. Findlen puts Galileo's dedication into the context of the patrician patronage of natural philosophy:

> Galileo's transformation of the satellites of Jupiter into a form of Medici mythology was not a unique occurrence, although it is one of the most spectacular examples of how scientific culture made its place in the political climate of the courts. The ink had hardly dried on the dedication of his *Sidereal Messenger* when other natural philosophers began to contemplate how to imitate his actions. Peiresc, for example, contemplated reworking the political mythology of the four "moons" to direct its message toward the French monarchy. Accordingly he proposed renaming two of the satellites Catherine and Marie in honor of the two Medici who had married into the French royal family.[62]

Later in the century Johannes Hevelius would dedicate the constellation Scutum Sobiescianum to honor Jan Sobieska III, King of Poland;[63] Augustin Royer would litter his *Cartes du Ciel* with constellations honoring Louis XIV;[64] and Edmund Haley would name a constellation Robur Corinum after the great oak at Boscobel in which Charles II had hidden from Republican soldiers in 1651.[65] Later in the century, Erhard Weigel would recast *all* the constellations into the coats of arms of European royal families.[66]

Though Johann Bayer had created a simple system for naming stars (one still in use today) as early as 1603, identifying, naming, and claiming stars to lay at the feet of potentates remained an active pursuit through the eighteenth century (for example, Uranus, found by William Herschel in 1781, was known as the Georgium Sidus). But then, as Galileo had observed in his dedication, the naming and claiming of stars was an ancient practice. Ptolemy's second century catalog of stars had been based on precedents such as Hipparchus's of four centuries earlier, and although these older catalogs have all been lost, it is not doubted that the tradition is much older still.[67] Ancient Islamic and medieval astronomers had all drawn pictures of individual constellations and some had made globes, and in 1515 Albrecht Dürer published a flat star map covering a large portion of the sky. Star maps like Dürer's multiplied into the nineteenth century—they were the stages on which the acts of political homage were enacted —when, overwhelmed by the growing flood of data and the endless proliferation of constellations, astronomers "reduced the number of recognized constellations [to eighty-eight], rationalized constellation boundaries, making them follow arcs of constant right ascension and declination [essentially tiling the sky], and increasingly omitted the constellation figures."[68] In the middle of the nineteenth century the Bonner Dürchmusterung listed the position of 320,000 stars and identified each with a BD number, for example, BD+5° 1668. The age of modern astronomy had dawned.

It is no coincidence that the Rev. T. W. Webb published the first handbook for amateur astronomers, *Celestial Objects for Common Telescopes*, in 1859. Since from the beginning, looking was the only way to collect stars, it is easy to imagine that the emergence of a skywatching culture wouldn't depend on their endangerment, but the fact is that with the growth of urban industrial culture, and especially the development of street lighting, stars *were* becoming endangered, at least for rapidly increasing numbers of people, a process continuing to accelerate today.[69] Webb went through many editions and remains in print today. In 1907, Martha Evans Martin published her classic *The Friendly Stars: How to Locate and Identify Them*. When Donald Menzel, himself author of the 1963 *Field Guide to the Stars and Planets*, came to revise Martin's book in 1964, he redrew the constellation figures after the ingenious lead of H. A. Rey, author of the Curious George books, whose *Stars* had come out in 1952[70], the year following the Golden Guide's 1951 *Stars* (by Zim and Baker). Field guides to the stars were soon indistinguishable from those to other things of nature, even to a shared language of collecting: "Many amateur astronomers," says *The National Audubon Society Field Guide to the Night Sky*, "work their way through the sky constellation by constellation, or count off the deep-sky objects one by one as they find them," just as birders do with their life lists.[71] Indeed the estimable Robert Burnham has written the following:

> Considered as a collector of rare and precious things, the amateur astronomer has a great advantage over amateurs in all other fields, who must usually content themselves with second and third-rate specimens. For example, only a few of the world's mineralogists could hope to own such a specimen as the Hope diamond, and I have yet to meet the amateur fossil collector who displays a complete tyrannosaurus skeleton in his cabinet. In contrast, the amateur astronomer has access at all times to the original objects of his study; the masterworks of the heavens belong to him as much as to the greatest observatories in the world. And there is no privilege like that of being allowed to stand in the presence of the original.[72]

If this is not sufficient, today you can buy a star: "Buy a star name. An ideal gift for birthdays, Christmas, and Mother's Day. Click here for Star Name kit information and photos," screams one Web site. "Buy stars at StarName from $25.95. Everyone loves it. Be appreciated," shouts another. At Int'l Star Registry it says, "We have over 1,000,000 stars. Don't be fooled by imposters," while at www.space.com it says, "It's not hard to grasp the romantic or otherwise wondrous reasons someone might have for buying a star name, especially as a gift."[73] It is *unquestionably* a possessable nature.

Though the difference between a birder's life list and an amateur astronomer's star catalog may be faint to invisible, the difference between the birder's range maps and the amateur astronomer's star maps (or charts) is substantial. In contradistinction to the empirically derived range, however problematic, of a species, a modern constellation is more like a nation-state (the very paradigm of possessiveness): its boundaries are absolute and arbitrary arcs of constant right ascension and declination; when you leave one constellation, you enter another; there is no unoccupied space; and all the stars inside the boundary are said to "belong" to the constellation, whether they participate in the composition of the figure or not. In a sense, this last has been true for centuries, since Bayer's system depended on it, assigning to a constellation's two dozen brightest stars the letters of the Greek alphabet and so, to pick an example, instead of Rigil Kentaurus (the brightest star in Centaurus), Alpha or α Centauri.

We see, we name, we map!

Taking possession

We have been speaking about these . . . *deeds to nature* . . . as though as they were no more than *tokens* of a possession never actually entered into; and whereas this may be the case for the maps of nature in the field guides, it is worth acknowledging that insofar as such maps are used as tools in the control of nature—in its management—that the possession is literally a matter of fact. For example, we have in front of us a study, one of literally thousands and thousands we could have turned to, by Douglas Johnson and Alan Sargeant of the Interior Department's Fish and Wildlife Service called "Impact of Red Fox Predation on the Sex Ratio of Prairie Mallards" (figure 8.8). Since the 1930s, preponderances of males had been reported, but never explained, in many wild duck populations. In the years leading up to this study (of 1977), hunters had begun to express an interest in "harvesting" these "excess drakes." The issue Johnson and Sargeant addressed was whether this would make good management sense.

Sargeant had begun his career in operational animal damage control programs where he had learned the basics of predator behavior, and then he did pioneering work in the Dakota prairies in the telemetric study of the territorial behavior and daily activity patterns

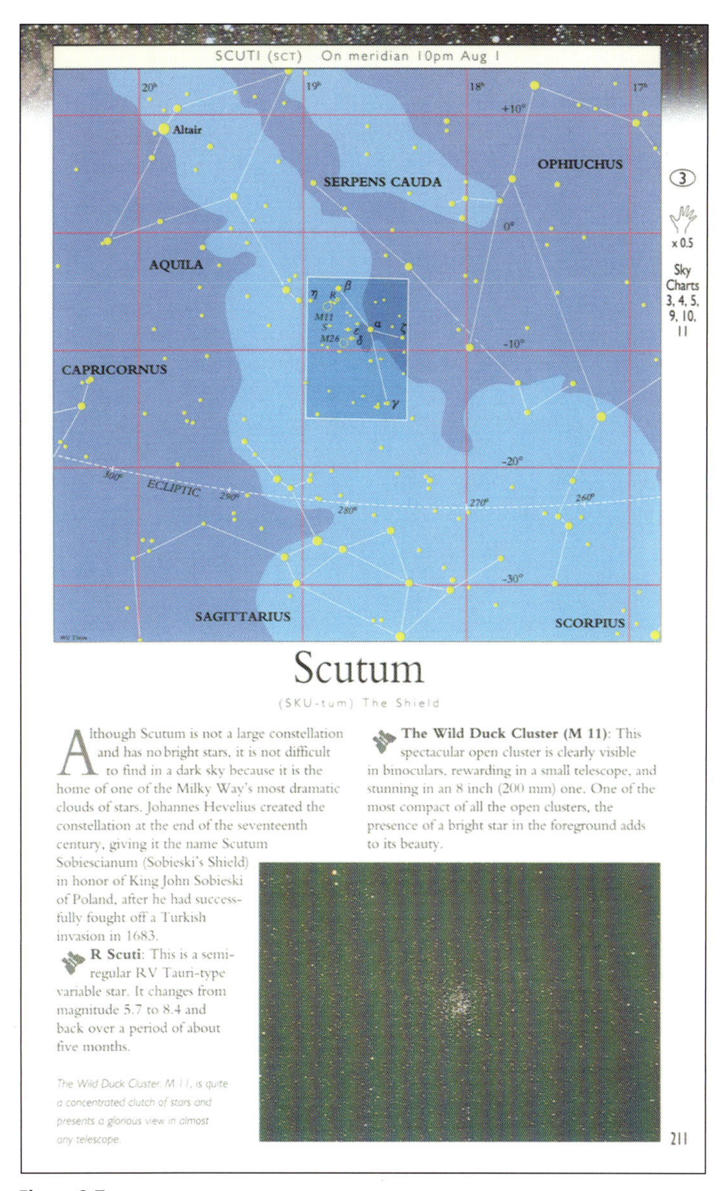
SCUTI (SCT) On meridian 10pm Aug 1

ALTAIR · SERPENS CAUDA · OPHIUCHUS · AQUILA · CAPRICORNUS · ECLIPTIC · SAGITTARIUS · SCORPIUS

Scutum

(SKU-tum) The Shield

Although Scutum is not a large constellation and has no bright stars, it is not difficult to find in a dark sky because it is the home of one of the Milky Way's most dramatic clouds of stars. Johannes Hevelius created the constellation at the end of the seventeenth century, giving it the name Scutum Sobiescianum (Sobieski's Shield) in honor of King John Sobieski of Poland, after he had successfully fought off a Turkish invasion in 1683.

R Scuti: This is a semi-regular RV Tauri-type variable star. It changes from magnitude 5.7 to 8.4 and back over a period of about five months.

The Wild Duck Cluster (M 11): This spectacular open cluster is clearly visible in binoculars, rewarding in a small telescope, and stunning in an 8 inch (200 mm) one. One of the most compact of all the open clusters, the presence of a bright star in the foreground adds to its beauty.

The Wild Duck Cluster, M 11, is quite a concentrated clutch of stars and presents a glorious view in almost any telescope.

211

Figure 8.7
A page from David Levy's 1994 *Skywatching*, a Nature Company Guide about the constellation Scutum. In addition to the text and a photograph, the page includes a contemporary star map. Note that the boundaries of the constellation are absolute and arbitrary arcs of constant right ascension and declination, and that the constellations completely tile the sky. The constellation in question, originally Scutum Sobiescianum, or Sobieski's Shield, was introduced by Johannes Hevelius in 1683 to commemorate the victory of Jan Sobieski III, King of Poland, in the battle of Vienna.

Source: David H. Levey, *Skywatching*, *Time Life*, 1994, 211. Used with permission. © Weldon Owen Inc. 1994.

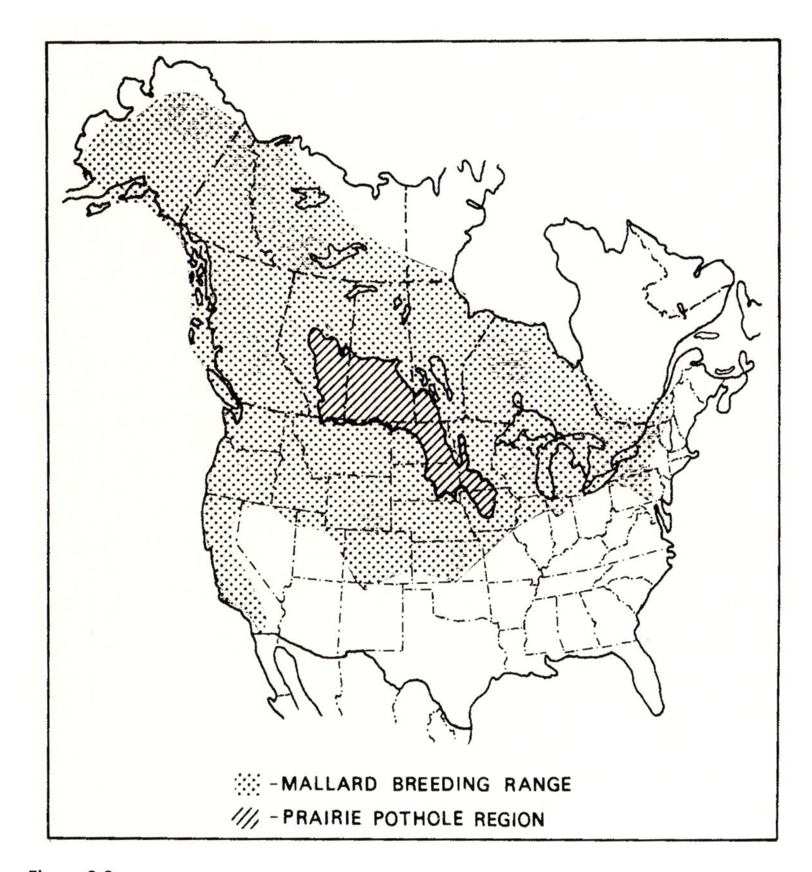

Figure 8.8
North American breeding range of the mallard, from Johnson and Sargeant's 1977 study.

Source: U.S. Department of the Interior, Fish and Wildlife Service.

of red foxes. Information gathered from bird remains at red fox den sites established as fact that a substantial number of nesting female mallards was indeed being taken. At this point Johnson, a mathematical statistician, joined the study to help construct a simulation model. Let us say immediately that the work constituted a significant contribution to studies of the prairie mallard, red fox, predator-prey relations, population dynamics, and the automated radio tracking of animals. Among the numerous graphics illustrating the paper is a map posting both the North American breeding range of the mallard and the Prairie Pothole Region where about half the continental population nests; a map posting twenty-five individual fox litter locations discovered in a 269-square-kilometer area showing the spacing of fox families (figure 8.9); and an incredible map posting the movements of a radio-equipped adult female red fox around her rearing den showing how thoroughly individual foxes cover their home ranges (figure 8.10).

Each of these maps contributes to the authors' arguments articulating the relationship between mallards and foxes. For example, the home-range map establishes the fact that, unlike wolves and coyotes whose much larger ranges prevent them from regularly visiting much of the territory they occupy, the thorough coverage of foxes increases their chances of encountering mallards on nests, and so taking females. Because of this, as wolf and coyote populations have declined and fox populations proportionately risen, nesting females have been put at comparatively greater risk. This is to say that mallard sex ratios are directly affected by canid population dynamics, which have been significantly altered "since pristine times" by selective human predation, among other things by a wolf poisoning campaign "unparalleled in intensity."

In the end, Johnson and Sargeant conclude that "the 'excess' drakes are not truly extra but rather a result of a relatively new additional loss of hens that has developed since pristine times"; and though Johnson and Sargeant suggest that while "it may be reasonable to harvest these drakes, that it might be best to do so with the thought of sparing the hens rather than harvesting a surplus component of the population," the actual policy decisions—whether, for example, to attempt fox population control or to propose changes

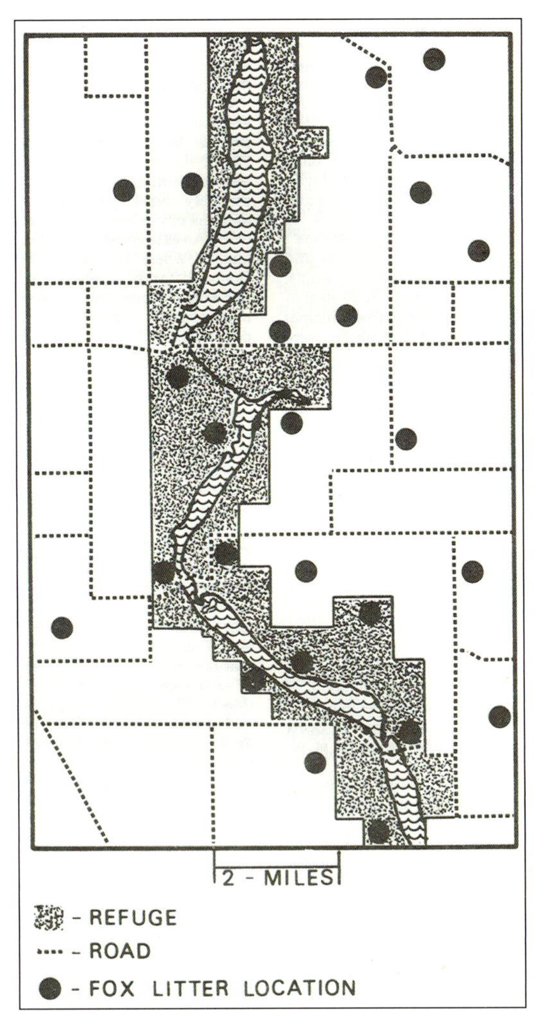

Figure 8.9
The spacing of fox families over a 269-square-kilometer area.
Source: U.S. Department of the Interior, Fish and Wildlife Service.

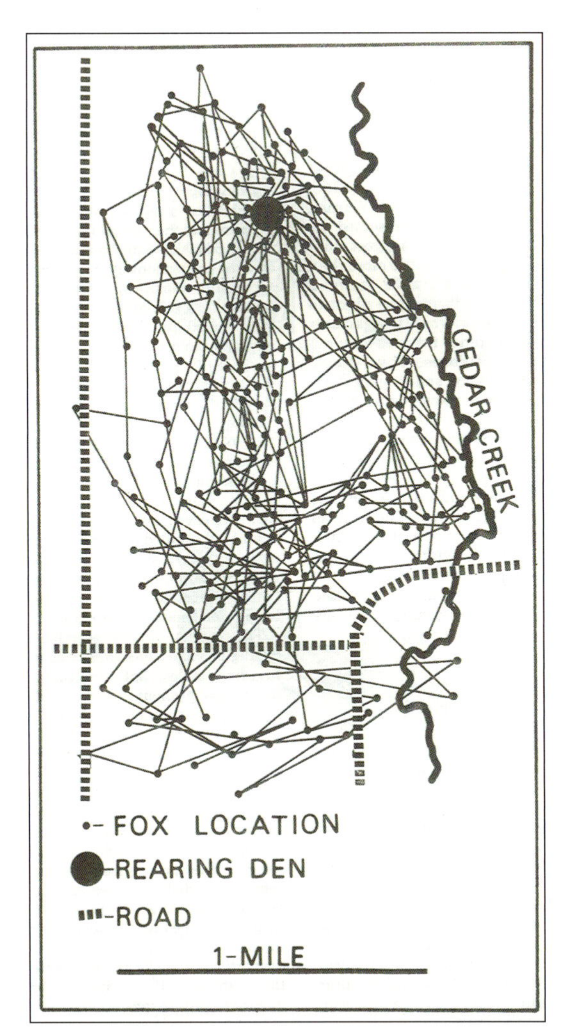

Figure 8.10
Movements of a radio-equipped red fox in Minnesota in 1964.
Source: U.S. Department of the Interior, Fish and Wildlife Service.

in the Central Flyway hunting regulations—would be up to others.[74] What we can't help observing here is the management role played by the range maps—at once inventories and graphic tools used to help think through relationships—of this nature that is here undeniably ours, to protect, eliminate, or harvest as we see fit.[75] In this work the map does not merely clarify what it posts, it participates fully in the establishment of our control.

With this we have passed from the nature we collect to the nature that we study, from tchotchkes to paradigms, from curios to science. In Pickstone's terms we've passed from natural history into analysis and, with the attempt to control fox populations into his third way of knowing, experimenting. This nature of science, of the paradigm, however, belongs to our next chapter.

Notes

1. Zim and Martin, *Trees: A Guide to Familiar American Trees,* rev. ed. (New York: Golden Press, 1956). The pin oak is on p. 101.

2. Whitney, *The Audubon Society Nature Guide to Western Forests* (New York: Knopf, 1985), 8; Robbins, Bruun, and Zim, *A Guide to Field Identification of Birds of North America* (New York: Golden Press, 1966), 15; Collins, *Complete Field Guide to American Wildlife* (New York: Harper and Row, 1959), xii. The beautiful dedication of the last is "To Our Fellow Creatures Whose Names and Pictures Appear In This Book."

3. See Camerini, "Evolution, Biogeography, and Maps," as well as her "Physical Atlas of Heinrich Berghaus," for discussions of the tacit and not so tacit roles played by maps in the articulation of the idea of species and other aspects of the early history of evolutionary thinking. Camerini is particularly insistent about the importance of the "visual thinking" involved, which she elaborated on in a presentation she gave to the History of Science Colloquium at the University of Wisconsin–Madison, "Thinking About Distributions: Biological Maps in the 19th Century," October 12, 1988 (18 page typescript).

4. Augustin de Candolle also published a map of French flora in 1805. This informed Charles Lyell's attempts to distinguish habitation (habitat) and station (niche). See more below. Again, the only general introduction to this history from the cartographic perspective is Robinson's *Early Thematic Mapping in the History of Cartography* (Chicago: University of Chicago Press, 1982), but see also Camerini's detailed discussion of Zimmermann in her dissertation as well as her comments in her 1988 presentation.

5. Areography is the study of the geographical ranges of taxa (subspecies, species, genera, families, etc.).

6. Quoted in Robinson, *Early Thematic Mapping,* 102.

7. *Frontiers of Fear* (New Haven, Conn.: Yale, 2001), 85–86, but Boomgaard argues for the historicity of tigers and other animals throughout. See his opening salvo, "Do Animals Have a History?" (7–9), and his general conclusions, "People and Tigers In the Making of the Malay World" (235–7). Wood reaches similar conclusions on other grounds. See *Five Billion Years*, throughout. For a standard account of the tiger, see the essay "Tiger" in Sunquist and Sunquist's *Wild Cats of the World* (Chicago: University of Chicago Press, 2002), 343–72, with its range maps and rich bibliography. Their discussion of "Conflict with Humans," 364–7, complicates, but does not essentially contradict, Boomgaard's account.

8. Hutchinson, *An Introduction to Population Ecology* (New Haven, Conn.: Yale, 1978), 239. Since Hutchinson is advancing this position against the redoubt of falsification in which he personally is deeply entrenched, this seems to us a perfect example of Andrew Pickering's "mangle of practice." What else, Hutchinson (and Boomgaard) ask, are we to do given the resistances of the world but . . . whatever we can (Pickering's accomodation)? See Pickering's "Objectivity and the Mangle of Practice," *Annals of Scholarship* 8 (1991): 409-25, reprinted in *Rethinking Objectivity*, ed. Allan Megill (Durham: Duke, 1994), 109-25, and incorporated into chapter six of *The Mangle of Practice: Time, Agency, and Science*, by Andrew Pickering (Chicago: University of Chicago Press, 1995).

9. Boomgaard, *Frontiers of Fear*, 24.

10. Ricklefs, *Ecology*, 280–1.

11. Bonner, *The Scales of Nature* (New York: Harper and Row, 1969), 61.

12. Quoted in Hutchinson, *Introduction to Population Ecology*, 157. Elton's *Animal Ecology* was published in 1927 by Sidgwick and Jackson (London). We haven't read Elton, but Hutchinson's chapter, "What Is a Niche?" is indispensable (132–212). All the early work on the idea of the niche was assembled in Whittaker and Levin, *Niche: Theory and Application, Benchmark Papers in Ecology*. Vol. 3 (Stroudsberg, Pa.: Dowden, Hutchinson and Ross, 1975).

13. Ricklefs, *Ecology*, 817.

14. See Ricklefs' chapter, "The Niche Concept in Community Ecology," pp. 728–47 of *Ecology*.

15. For example, Charles Lyell complains that what he was then calling habitation and station, which would more or less evolve into habitat and niche, were being confused when he was writing in 1832! See Lyell, *Principles of Geology*. Vol. 2 (London: John Murray, 1832), 69. We thank Camerini for this reference.

16. See Hutchinson, *Introduction to Population Ecology*, 155; but also Gaffney, "Roots of the Niche Concept," *American Naturalist* 109 (1973): 490.

17. For Johnson, see Hutchinson, *Introduction to Population Ecology*, 155–6.

18. Ibid., 157. See further another example of Grinnell's indeterminate usage, and exculpatory gloss by Hutchinson, on p. 134.

19. Bennett and Humphries, *Introduction to Field Biology*. 2nd ed. (London: Edward Arnold, 1974), 12–13. In Whittaker, Levin, and Root's "Niche, Habitat and Ecotope" (*American Naturalist* 107 [1973]: 321–38), "habitat" too gets claimed for the hyperspace. The authors resolve the "complete hyperspace," which they call an *ecotope*, into a *niche*, which is within a *community*; and a *habitat*, which is a range of gradients along (hyper) *spatial* axes on which different communities could develop.

20. "Station" had been Lyell's word for what would become the idea of niche.

21. Not all field guides are reticent about these. For example, Whitney, in *Western Forests*, lists eight subject experts on his title page; says, "This book represents hundreds of hours of labor by a small army of people"; announces that his guide, "puts information at your fingertips that would otherwise only be accessible through a small library of field guides"; and includes a bibliography.

22. Our argument here might best be described as a figurative exploitation of the argument at the heart of Edwin Hutchins' description of navigation as computation in his *Cognition in the Wild* (Cambridge, Mass.: MIT Press, 1995), from chapter 2 on. Hutchins quotes Herbert Simon's *Sciences of the Artificial*, 2nd ed. (Cambridge, Mass.: MIT Press, 1981), 153. Simon in turn gives credit to Saul Amarel's, "On the Mechanization of Creative Processes" (*IEEE Spectrum* 3 [April 1966]: 112–4).

23. Arthur Krim has suggested that this is likely rooted in the regionalism so broadly characteristic of the American spirit in the Thirties and Forties (personal communication). Certainly it disappears from field guides after the mid-Fifties.

24. For the power of the gaze to claim nature for America, see also Albert Boime's *The Magisterial Gaze* (Washington, D.C.: Smithsonian, 1991), where he sees the "eye" taking "a perspective akin to the devine, represented in the Masonic-influenced Great Seal of the United States," which he relates to Seth Eastman's popular manual for drawing perspectives (all pp. 22–27), and then to Hudson River School painting, among much else. But if we get all this with perspective, how much more from overhead?

25. Barthes, "Plates of the Encyclopedia," in *New Critical Essays* (New York: Hill and Wang, 1980), 26–27.

26. Ibid., 28.

27. Whitney, *Western Forests*, 18–19.

28. For our knowledge of sixteenth- and seventeenth-century collecting we are entirely indebted to Paula Findlen's *Possessing Nature* (Berkeley, Calif.: University of California Press, 1994); for the eighteenth and nineteenth centuries to Paul Lawrence Farber's *Discovering Birds* (Baltimore, Md.: Johns Hopkins University Press, 1997), and to Farber's more general *Finding Order in Nature* (Baltimore, Md.: Johns Hopkins University Press, 2000). See also Pomian, *Collectors and Curiosities* (London: Polity, 1990).

29. Findlen, *Possessing Nature*. Her entire, detailed book could be said to reduce to this point, but she says this in so many words on p. 158.

30. Ibid., 240.

31. The quotation is another from Findlen's p. 240, but the reference to the Royal Society is meant to embrace as well the larger arguments of Shapin and Schaffer's *Leviathan and the Air-Pump*, which is to say, the whole Baconian program for the construction of facts as implemented by Robert Boyle, Henry Oldenburg, and others.

32. Farber, *Finding Order*, 22–23.

33. Ibid., 51.

34. Farber, *Finding Order*, 97–98. Jacob Studer's *Popular Ornithology* (New York: Harrison House, 1977 [1881]) had appeared in 1881, just five years prior to the 1886 founding of the Audubon Society. A year later, in 1887, the first waterfowl regulations were written into the territorial laws of North Dakota, to cite one example, limiting a day's bag to twenty-five birds. It is also no coincidence that the Sierra Club was founded in 1892. We are not interested in suggesting that wanton destruction of wildlife alone drove the evolution of the conservation ethic (which doubtless has numerous, complicated roots in the larger social movements we think about as Romanticism and the age of revolution, the two deeply intertwined, and bearing conservationist fruit as early as 1864 when Yosemite was given to California by the federal government). Frederick Law Olmstead, Thomas Starr King, John Muir, John Burroughs, Joseph LeConte, and so many others made the end of the nineteenth century a uniquely schizophrenic moment, advancing their advocacy even as the

buffalo were being exterminated. It is the conjunction of an appreciation for wilderness and wildlife against the threats to both—e.g., Hetch Hetchy *and* the sixty-one species threatened with extinction by the trade in feathers—that resulted in the twentieth century conversation and preservation movements. In addition to Farber, see Cohen, *The Pathless Way: John Muir and the American Wilderness* (Madison, Wis.: University of Wisconsin Press, 1981), and his *The History of the Sierra Club, 1892–1970* (San Francisco: Sierra Club, 1988), as well as Hays, *Conservation and the Gospel of Efficiency* (Cambridge, Mass.: Harvard University Press, 1959), for the countervailing perspectives (and roles) of Gifford Pinchot and Theodore Roosevelt, among others. The fact is, though, and especially in the long period before World War II, that few Americans could get to Yosemite, much less Yellowstone; what made ardent conservationists of middle class Americans was *bird watching* (as we have seen the National Geographic Society knew so very well), or, to put it another way, bird watching was how the middle class *performed* conservation and preservation. And what *really* got bird watching going was the Peterson guide (see below), which spawned the culture of the guide book with its range maps.

35. A recent piece in the *New York Times* headlined, "So Many Field Guides You Need a Guide" (by James Gorman), observes that an Amazon.com search in response to "field guides to" coughed up 23,593 books, including ones devoted to management, heat and cooling, demons, men, and the American bird (the rude digital gesture), but most devoted to "the natural world." The guide to bird droppings is Peter Hansard and Burton Silver's *What Bird Did That? A Driver's Guide to Some Common Birds of North America* (Berkeley, Calif.: Ten Speed Press, 1991). While parodistic, the guide is also serious (there is an ancient and venerable literature on the subject of animal droppings and other ejecta). Devoting—in standard field guide fashion—one page per species, each includes a photograph of the "splay" or "dejecta" on the windshield, a silhouette of the bird in flight as seen from the driver's seat, a careful description of a typical example ("An extended sklop with an attractive cloud-like appearance. Moist, loose, fragile. The nucleus lacks the ability to retain even small gritty solids, which tend to spread throughout the envelope"), a description of the food the bird eats, its distribution, and the best times and locations for collection. Each example photographed is catalogued by precise location, date, time, weather, and wind speed and direction.

36. From the introduction to *Peterson First Guide to Birds of North America* (Boston: Houghton Mifflin, 1986), 3. Peterson organized birds by appearance, not phylogeny, and his system reduced the time needed to master regional bird identities from half a lifetime to a couple of years. It was a *visual* guide for a *visual* activity (applied with equal efficaciousness to the spotting of enemy airplanes). The excessively chatty biography by John Devlin and Grace Naismith, *The World of Roger Tory Peterson* (New York: Times Books, 1977), provides a useful, if anecdotal, window on the growth of the field guide culture and its significance to the conservation and preservation movements. Just the list of the boards on which Peterson served is instructive (pp. 251–3). Houghton Mifflin, Peterson's longtime publisher, currently lists 125 tiles in its Peterson series.

37. Burnie, *Eyewitness Books: Birds* (New York: Knopf, 1988), 62.

38. It's impossible to think about the possessive power of the gaze without thinking it through Foucault's articulation of ours as a society of surveillance. To think about bird-watching through the section on panopticism in *Discipline and Punish: The Birth of the Prison* (New York: Vintage, 1979), 195–308, is to site bird-watching within an overarching social episteme where it suddenly acquires a new coherence. Foucault writes about the society of surveillance on p. 217. See also his "Eye of Power" essay in *Power/Knowledge*, ed. C. Gordon (New York: Pantheon, 1980), 146–165. Apropos French thinking about the gaze, Martin Jay's overview in *Downcast Eyes: The Denigration of Vision in Twentieth-Century Thought* (Berkeley, Calif.: University of California Press, 1993) provides important balance.

39. As quoted by D. E. Zimmer being quoted by Stephen Jay Gould in the latter's *I Have Landed: The End of a Beginning in Natural History* (New York: Harmony Books, 2002), 51.

40. Anderson, *Plants, Man and Life* (Berkeley, Calif.: University of California Press, 1952), 46. Our emphasis.

41. Boucher, *Bird Lover's Life*, xi.

42. Long, *Squirrels: A Wildlife Handbook* (Boulder, Colo.: Johnson Books, 1995); Tuttle, *America's Neighborhood Bats* (Austin, Tex.: University of Texas Press, 1988); Tyning, *Guide to Amphibians and Reptiles, Strokes Nature Guides* (Boston: Little, Brown, 1990); Zim and Martin, *Trees*; Little, Jr., "Important Forest Trees of the United States" in *Trees: The Yearbook of Agriculture, 1949*, ed. Alfred Stefferud (Washington, D.C.: United States Department of Agriculture, 1949), 763–814; Petrides, *A Field Guide to Eastern Trees, Eastern United States and Canada* (Boston: Houghton Mifflin, 1988); Peterson, Mountfort, and Hollom, *Field Guide*; Robbins, Bruun, and Zim, *Guide to Field Identification*; Baughman, *Reference Atlas to the Birds of North America* (Washington, D.C.: National Geographic Society, 2003); Harrison and Greensmith, *Birds of the World, Eyewitness Handbooks* (New York: Dorling Kindersley Publishing, 1993).

43. Mohlenbrock and Thieret, *Trees: A Quick Guide Reference to Trees of North America* (New York: Macmillan, 1987), x; Whitney, *Western Forests*, 8; Brockman, *Trees of North America*, 3.

44. Though here the Macmillan Guides *Trees*, of Mohlenbrock and Thieret, is an outstanding and laudatory exception.

45. The *monographic* literature devoted to individual species or regions is another matter, though that through 1971 is subsumed by Little, et al.'s *Atlas of United States*. Volumes 1–6 (Washington, D.C.: United States Department of Agriculture, 1971–1981).

46. Little reviews the history of tree distribution maps in "Mapping Ranges of the Trees of the United States," *Rhodora* 53 (1951): 195–203. Little draws on this for his history of tree range maps in the introduction to *Atlas of United States Trees*, Volume 1 (*Conifers and Important Hardwoods*). Shorter versions of this history appear in other volumes of the atlas.

47. This, along with the rest of the information about the *Atlas* comes from Little's introduction to Volume 1. The other major source for the *Atlas* was Critchfield and Little, *Geographic Distribution of Pines of the World* (Washington, D.C: United States Department of Agriculture, 1966). A few maps were created just for the atlas. Others were revised.

48. Recently Little's 1971 maps have been digitized and are now widely available online.

49. And compare Little's 1971 range to that given in Radford et al., *Manual of the Vascular Flora of the Carolinas* (Chapel Hill, N.C.: University of North Carolina Press, 1968)—based almost entirely on herbaria specimens (that is, of locations where specimens have been collected)—pp. 378–9.

Radford and Little 1971 are all but identical. It may not be apparent in our reproduction of Little 1971 that there's an "x" in South Carolina, indicating an "extinct location." *What* could this mean?

50. Arguing this point is a staple of cartography texts. For a pithy but pungent treatment, see p. 35 of Edward Tufte's *Visual Explanations: Images and Quantities, Evidence and Narrative* (Cheshire, Conn.: Graphics Press, 1997).

51. A whole *other* issue has to do with the *filling in* of the line. This has the effect of changing the line from a posting of limits—too dry, too cold—to that of an edge *necessarily brought into being* by the depiction of an area *presumably occupied by the species in question. Can't grow beyond these limits* is a very different statement from *grows in the region indicated*, where *throughout* is a taken-for-granted gloss on *in*. The reality is that for Little the line is clearly both, outer bound *and* ecological limits.

52. Campbell, *Biology*, 1094.

53. Little, *Atlas*, 1:3. Our emphases.

54. Little, "Important Forest Trees," 764.

55. For a really trenchant, and brief, treatment of the history of "native," see Williams, *Keywords: A Vocabulary of Culture and Society* (New York: Oxford University Press, USA, 1976), 180–1.

56. Little, *Atlas*, 1:3.

57. Thomas Berger catches this up in a brilliant expression. Speaking of James Fenimore Cooper's *The Pioneers*, Berger says: "That is, [Cooper] begins fairly close to reality, which in Western culture always implies the loss of something splendid," in his afterword to Cooper's *The Pathfinder* (New York: Signet, 1961), 431.

58. Derrida, *Of Grammatology* (Baltimore, Md.: Johns Hopkins University Press, 1976). Saussure in the first part, "Wring before the Letter," and Rousseau and Lévi-Strauss in part two, "Nature, Culture, Writing."

59. Text and maps from Little, "Important Forest Trees," 774–5 and 781–3.

60. "Decimated" overstates the case, but we use it because "it serves for emphasis." We get the 500 million acre figure from Terborgh, *Diversity and the Tropical Rainforest* (New York: Scientific American, 1992), 188. In his *Where Have All the Birds Gone?* (Princeton: Princeton University Press, 1989), Terborgh breaks this down into six periods, for each of which he gives the forested acreage of six regions, p. 160. For the U.S. as a whole, about 60 percent of what was forested in precolonial times is forested today. The regions with the smallest proportion of original forest, however, are those in which the pin oak lived.

61. This letter, the information in the preceding paragraph, and what follows, is largely drawn from the introduction and conclusion to Albert Van Helden's translation of Galileo Galilei, *Sidereus Nuncius or The Sidereal Messenger* (Chicago: University of Chicago Press, 1989). The letter to the secretary appears on pp. 18–19.

62. Findlen, *Possessing Nature*, 377.

63. Snyder, *Maps of the Heavens* (New York: Abbeville, 1984), 45, where Scutum Sobiescianum is illustrated with another constellation, Gladii Electorales Saxonici, dedicated by Godfried Kirch to the Saxon elector. Kirch's constellation has not survived, whereas Hevelius' has.

64. Ibid., 120–1. One of these constellations, Sceptrum, was replaced by Hevelius a few years later with Lacerta (Lizard), and also by Johann Bode with Frederici Honores to honor Frederick the Great. Lacerta is the name in use today. Another, Lilium, replaced Musca (Fly). Where recognized today, this constellation is once again Musca. See Allen, *Star Names: Their Lore and Meaning* (New York: Dover Books, 1963), for these details.

65. Warner, *The Sky Explored* (New York: Alan R. Liss, 1979), 107. This is a catalog of star charts, and since Warner lists all the non-Ptolemaic constellations for each chart, it also charts the history of honorific naming.

66. Ibid., xii.

67. As old as the Neolithic revolution as far as Giorgio de Santillana and Hertha von Dechend are concerned. See their *Hamlet's Mill: An Essay on Myth and the Frame of Time* (Boston: Gambit, 1969), which reads myth as sky-knowledge in narrative form.

68. Warner, *Sky Explored*, xiii.

69. As a general rule, stars fainter than second or third magnitude can no longer be seen in cities; stars fainter than third cannot be seen in the suburbs; and stars fainter than 4.5 to 5 not even in the distant suburbs. Most field guides indicate the magnitudes of stars on their maps. A good example is David Levy's *Skywatching* (New York: Time-Life Books, 1994), which features a magnitude scale on every sky map. The fundamental division into six degrees of brightness dates to Hipparchus.

70. Rey's innovation was to "show the constellations in a new, graphic way, as shapes that suggest what the names imply." Radical!

71. Charrand, *The National Audubon Society Field Guide to the Night Sky* (New York: Knopf, 1991), 11.

72. Burnham's *Celestial Handbook: An Observer's Guide to the Universe Beyond the Solar System.* Rev. ed. (New York: Dover, 1978), 5.

73. Ingrid Wood brought these wonderful sites to our attention.

74. Johnson and Sargeant, "Impact of Red Fox Predation on the Sex Ratio of Prairie Mallards," *Wildlife Research Report* 6 (Washington, D.C.: Fish and Wildlife Service, United States Department of the Interior, 1977), 50–52.

75. En route we also band it and put radio transmitters on it. However "noble" the intentions behind this, it is inescapably about inventory and control, that is, also like the branding of Africans by slavers. A suite of twenty-eight maps in an appendix to another mallard study, March and Hunt's "Mallard Population and Harvest Dynamics in Wisconsin" (*Technical Bulleting No. 106* [Madison, Wis.: Department of Natural Resources, 1978]), 64–70, locates the sites of "direct recoveries—local females," "indirect recoveries—local males banded west of 90° longitude in Wisconsin," "indirect recoveries —immature females banded preseason east of 90° longitude," and so on. How did that *Guide to Western Forests* put it, "better to let a butterfly ride the winds unnamed"? Yeah, right!

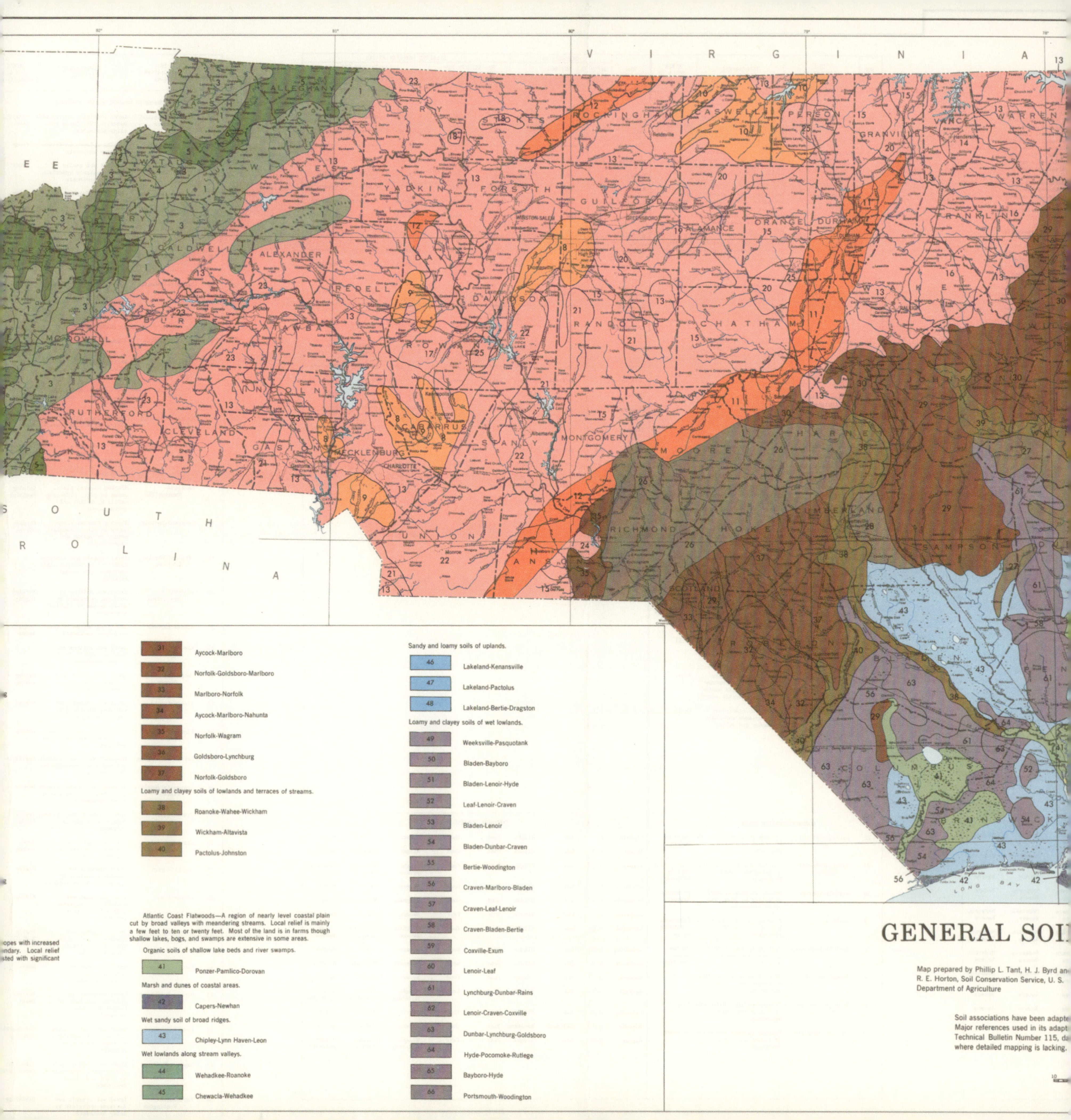

V I R G I N I A
S O U T H
C A R O L I N A
ALLEGHANY
WATAUGA
CALDWELL
ALEXANDER
IREDELL
McDOWELL
RUTHERFORD
CLEVELAND
GASTON
LINCOLN
MECKLENBURG
CHARLOTTE
CABARRUS
STANLY
MONTGOMERY
UNION
ANSON
RICHMOND
SCOTLAND
MOORE
HOKE
HARNETT
CUMBERLAND
SAMPSON
JOHNSTON
CHATHAM
RANDOLPH
DAVIDSON
YADKIN
FORSYTH
WINSTON-SALEM
STOKES
ROCKINGHAM
CASWELL
PERSON
GRANVILLE
VANCE
WARREN
FRANKLIN
GUILFORD
GREENSBORO
ALAMANCE
ORANGE
DURHAM
COLUMBUS
BRUNSWICK
LONG BAY
31 Aycock-Marlboro
32 Norfolk-Goldsboro-Marlboro
33 Marlboro-Norfolk
34 Aycock-Marlboro-Nahunta
35 Norfolk-Wagram
36 Goldsboro-Lynchburg
37 Norfolk-Goldsboro
Loamy and clayey soils of lowlands and terraces of streams.
38 Roanoke-Wahee-Wickham
39 Wickham-Altavista
40 Pactolus-Johnston
Atlantic Coast Flatwoods—A region of nearly level coastal plain cut by broad valleys with meandering streams. Local relief is mainly a few feet to ten or twenty feet. Most of the land is in farms though shallow lakes, bogs, and swamps are extensive in some areas.
Organic soils of shallow lake beds and river swamps.
41 Ponzer-Pamlico-Dorovan
Marsh and dunes of coastal areas.
42 Capers-Newhan
Wet sandy soil of broad ridges.
43 Chipley-Lynn Haven-Leon
Wet lowlands along stream valleys.
44 Wehadkee-Roanoke
45 Chewacla-Wehadkee
Sandy and loamy soils of uplands.
46 Lakeland-Kenansville
47 Lakeland-Pactolus
48 Lakeland-Bertie-Dragston
Loamy and clayey soils of wet lowlands.
49 Weeksville-Pasquotank
50 Bladen-Bayboro
51 Bladen-Lenoir-Hyde
52 Leaf-Lenoir-Craven
53 Bladen-Lenoir
54 Bladen-Dunbar-Craven
55 Bertie-Woodington
56 Craven-Marlboro-Bladen
57 Craven-Leaf-Lenoir
58 Craven-Bladen-Bertie
59 Coxville-Exum
60 Lenoir-Leaf
61 Lynchburg-Dunbar-Rains
62 Lenoir-Craven-Coxville
63 Dunbar-Lynchburg-Goldsboro
64 Hyde-Pocomoke-Rutlege
65 Bayboro-Hyde
66 Portsmouth-Woodington
opes with increased
ndary. Local relief
sted with significant
GENERAL SOI
Map prepared by Phillip L. Tant, H. J. Byrd an
R. E. Horton, Soil Conservation Service, U. S.
Department of Agriculture
Soil associations have been adapte
Major references used in its adapt
Technical Bulletin Number 115, d
where detailed mapping is lacking.

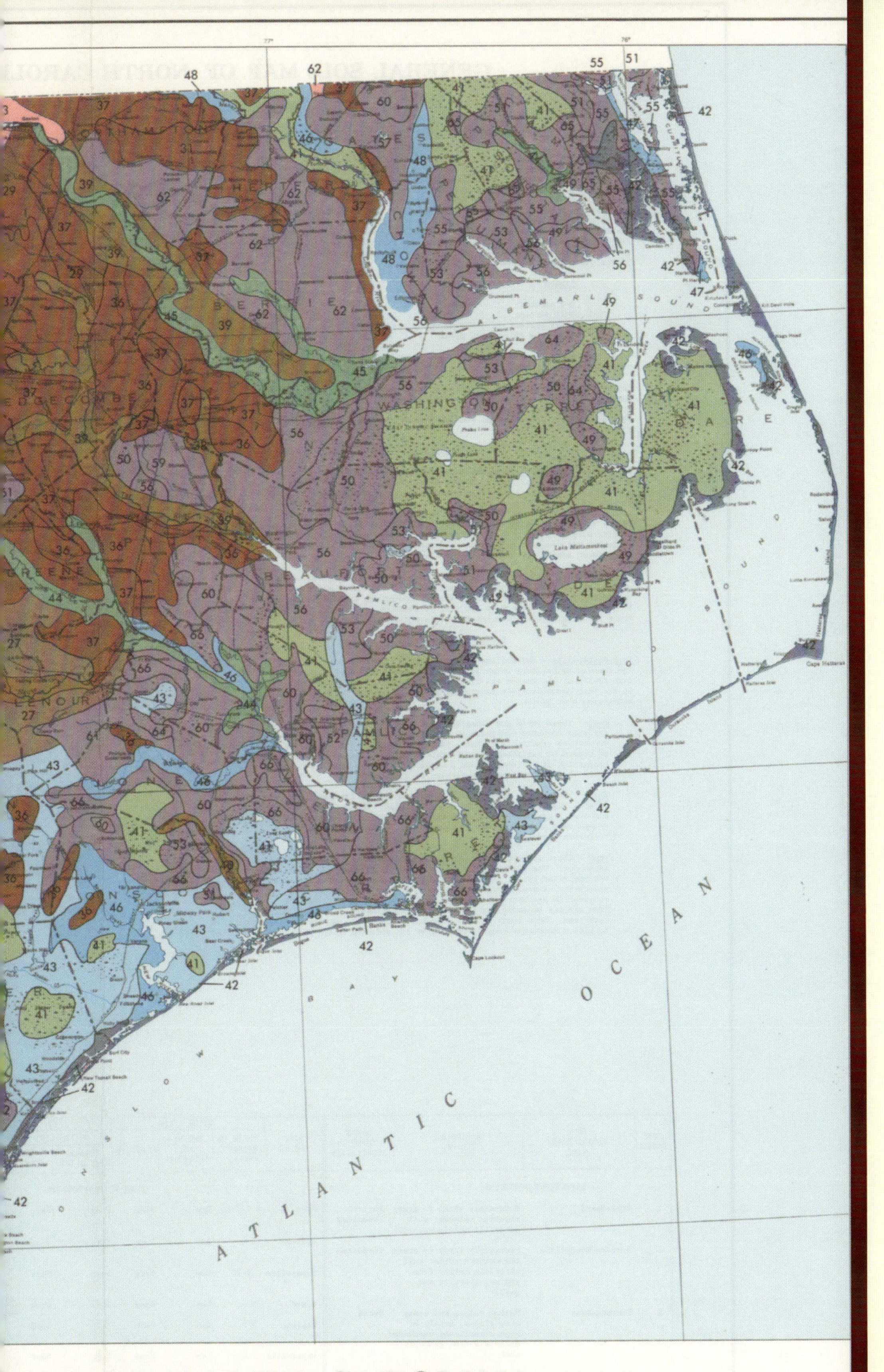

A T L A N T I C   O C E A N
ALBEMARLE SOUND
PAMLICO SOUND
MAP OF NORTH CAROLINA
Published for The Soil Conservation Service by the Engineer Agency for Resources Inventories (EARI), U. S. Army Engineer Topographic Laboratories. Cartopography by Larry B. Staley.
sheets 10 and 11 of a "Soil Map of United States" as associations of phases of subgroups.
ere (a) general soil maps of Counties of North Carolina and (b) The Soils of North Carolina,
cember 1955, by William D. Lee. Soil series names are tentative in portions of the state
Scale 1:1,000,000
0 10 20 30 40 Miles
0 10 20 30 40 50 Kilometers
JANUARY 1990 1005454

Nine

Nature as system

Map courtesy of USDA Soil Conservation Service, 1990.

Nature as system

Can we really *know* nature through the simple (or maybe not-so-simple) enterprises of collecting and cataloging, and constructing maps like range maps? In our birder's life list, we might happily check off the cedar waxwing and, among our manifold bird field guides, read that these gregarious birds "tend to nest late in the summer when there is a good supply of berries" but that "insects are also taken" and that their preferred habitat is "open woodlands, orchards, and residential areas." In the verbal description accompanying the obligatory range map, we might read that the species "breeds from British Columbia and Cape Breton Island south to Georgia, Arkansas, and California" and that it "winters from New England and British Columbia to Panama and the Greater Antilles."[1] But is that actually *knowing nature*, and what could such an audacious and pretentious statement possibly mean anyway? Underneath that idea is the notion that knowing is understanding, like knowing your way around town is essentially understanding what's connected to what . . . and where. This doesn't help much, though, because we're still lost in a fog of metonymy and semantic circularity. If knowing is basically about recognizing what things are and where they're likely to be found, then maybe understanding is knowing (that circularity again) something about how they're likely to behave or how they are related to one another. Clearly, we can recognize (know) the difference between a cedar waxwing and a Cooper's hawk, but it is not until we read about the Cooper's hawk in our field guide that we also understand that it might well have our cedar waxwing for lunch. There is lurking hazily in all this some concept of relationships, beyond just taxonomic constructions—beaver versus badger, pin oak versus post oak, lake versus pond, city versus town (these last two distinctions a bit more fuzzy in our language—de Saussure's *system of differences*).

A *system*. Is that what nature really is? Not simply something to be cared for or feared, admired or cuddled or corralled or cataloged, but something truly apart that has to be taken apart, and then put back together in our own terms, to be truly understood . . . in other words, truly possessed. The epistemological pretense is overwhelming. So nature is a system, at least in our modern populist view, and somehow more than just a collection of everything that we aren't or haven't made. Alright, so what is a system anyway? Here's how our Webster's defines system (from the Late Latin *systema*, which means to bring together, combine):

> a complex unity formed of many often diverse parts, an aggregation or assemblage of objects joined in regular interaction or interdependence, a set of units combined by nature or art to form an integral, organic, or organized whole, an orderly working totality.

So a system has to be more than just an ever more finely dissected classification of the natural world. There has to be some underlying principle or process that furnishes a basis for putting it all back together in some form. And that can be extraordinarily difficult to comprehend. That's why so-called global warming doesn't mean that everywhere just gets warmer and there are more droughts and more hurricanes. Raising the ambient temperature of the planet begins to melt the polar ice sheets, decreasing albedo at the planet's poles and accelerating the warming process there. The northward advance of forests helps warm northern polar lands and, as the permafrost thaws, vast amounts of methane are released into the atmosphere, further accelerating warming worldwide. In the North Atlantic, that means more icebergs and colder water traveling south in the Labrador Current and that means a *colder not warmer* Maritime Canada and, if the Labrador Current becomes cold enough and strong enough to cut off the northbound Gulf Stream, that means a colder Europe as well and perhaps another "Mini-Ice Age" like the last one that began in the thirteenth century and lasted through the Dark Ages. Can we possibly hope to comprehend natural systems, and can maps help at all in that? We'd like to think so . . . like to. After all, we've seen how the mapping of Atlantic coastlines eventually led to a theory of plate tectonics five centuries later. Can deconstructing the natural world and then reconstructing it in cartographic form possibly help us understand that world? Let's look at a few more maps and see what we find.

Down to earth

We'll begin with three maps of the state of North Carolina: a topographic map, a geologic map, and a soils map. Each of these is presented at a different scale, from 1:500,000 to about 1:2,000,000, and each issues from a different government agency: the U.S. Geological Survey, the North Carolina Geological Survey, and the USDA Soil Conservation Service. Each, of course, has its own particular agenda and each addresses its own particular system of interest. At first glance, each looks distinctly different from the others, but closer inspection will reveal more similarities than are initially apparent.

On the surface . . .

First up is the U.S. Geological Survey's map called *State of North Carolina* (figure 9.1). It's quite a large map, large enough that one can easily look *into* the map, at and through its multiple layers of geographic features, pointing to this town or that mountain or tracing the path of a road or river. With its surface convoluted by contours and detailed shaded relief and overlain by layers of rivers and reservoirs (in blue of course), roads and highways (in red), towns posted as black dots of various sizes and larger built up areas as gray polygons with thin black outlines, the matrix of the state's one hundred counties along with federal and Native American lands, and patterns of marsh symbols denoting the swamps of the Coastal Plain, there doesn't seem to

Figure 9.1
State of North Carolina, seems to tell readers *everything* they would want to know about the physical aspects of the land. Map dimensions (WxH): 66in. x 28in.; scale 1:500,000.

Figures 9.1 and 9.2 courtesy of USGS, 1957, revised 1972.

be much that escapes the gaze of this map. This is, after all, what we readily recognize as a *topographic* map (from the Greek *topos*—a place—and *grapheine*—a writing) and the purpose, *duty* in fact, of such a writing of place is to be a sort of *map of everything*—everything physical and tangible at this scale anyway. And this map obviously takes its duty seriously, dressed in the khakis and olive drabs that give it a distinctly military flavor. This isn't a map *about* North Carolina, much less any particular aspect of that, but a map *of* North Carolina; indeed, at some level it inevitably claims to *be* North Carolina, or at least what this state was when the map was made. Of course, there are a lot of things that go unmentioned in such maps, like demographics, economics, vegetation, climate, or any of the myriad other themes that fill the pages of an atlas of the state.

But that's hardly the point. The point is that this is the concrete physicality of the state of North Carolina—the very ground we walk on—and it's all real, very real. What makes it seem so real, of course, is the elaborate substrate of physiography upon which everything else is constructed: the firm shaded relief, subdued contours, and subtle color tints, tones, and shades that write this physiography of place. It's not hard at all to imagine the map's thin veneer of culture entirely stripped away (it doesn't have that much presence here anyway) to purify the terrain's presentation as timeless and eternal fact.

And that clearly constitutes the presence, the existence, of nature in this map. Underpinning the various conceptual constructs of cities and towns, roads and highways, rivers and reservoirs, jurisdictional and political claims, there is always *the land*, the indispensable, rock

Figure 9.2
A detail of *State of North Carolina*, around Asheville. All of the various systems here, natural and cultural, appear thoroughly integrated with one another.

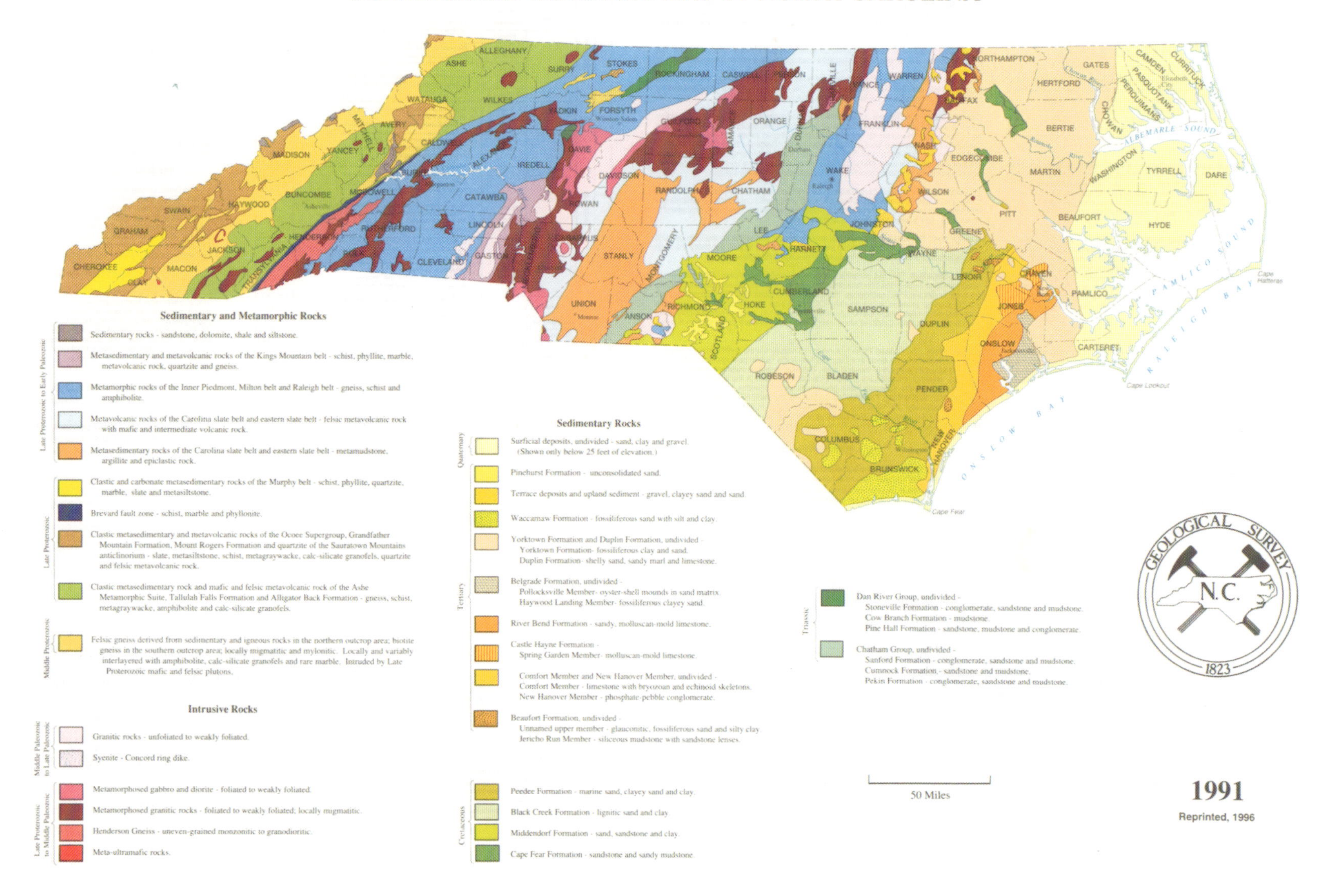

Figure 9.3

Generalized Geologic Map of North Carolina gives us a deeper look at the state's physical surface. Map dimensions (WxH): 17in. x 11in.; scale 1:200,000.

Source: North Carolina Geological Survey, 1991.

solid surface of the earth. Is that all there is to nature? Far from it, very far, but that's what this map presents as nature; and, really, where else could one begin?

Why are there mountains and gorges in the western portion of the state and swamps and estuaries in the eastern portion? Why do the courses of rivers bow toward the southwest? Why, in fact, are the rivers and reservoirs and swamps, or settlements for that matter, even where they are in the first place? On these topics the map is mostly silent (although, one can see why the rivers found the avenues they did in the wrinkled mountainous terrain). This does not have the immediate appearance of a nature deconstructed, in spite of the fact that it most certainly is. Instead, it seems to present itself as one wholly intact in all its physical complexity. This map is not *about* North Carolina; this map *is* North Carolina; and it says so unquestionably and incontrovertibly and with confident cartographic authority. This is not a map of the Something or Other of North Carolina; this is, as the map's title makes quite clear, the State of North Carolina itself—the tangible physical reality of the land. In announcing that *this is there* uncountable times, the map ultimately (and primarily and essentially) proclaims that *this is.*

AND BENEATH . . .

Just *how* this is, however, is the province of another genre, or several genres, of maps. What lies beneath North Carolina's complex physical surface is the theme of the *Generalized Geologic Map of North Carolina* (figure 9.3), produced by the North Carolina Geological Survey. The map's origin is conspicuously apparent because a large N.C. Geological Survey emblem appears here, although credits and homages are reserved for the reverse (more about that shortly). Below the emblem in the lower right corner it says "1991" and, in smaller type, "Reprinted, 1996" and to the left of that is an undivided fifty-mile scale bar. This small but colorful map is adapted from a much larger and more detailed geologic map of the state, unfortunately long out of print and now existing only in rather inelegant digital form on the Internet. Besides geology, there's really not much in this map: the requisite matrix of the state's one hundred counties, a few major cities and major rivers, and the coastline, with labels for major sounds and bays and the coast's three capes, just enough to provide a minimal reference for the geologic formations.

Before looking deeper, though, let's flip the sheet over and read the perimap on the reverse. We find that the upper left of the reverse is occupied by a broad two-column textual description, of the general physiography of the state (its three major provinces of Coastal Plain, Piedmont, and Blue Ridge), of its nine principle "geologic belts" (including remarks on the economic value of mining in each), and a final capsule commentary on the mineral industry in North Carolina, which "has important deposits of many minerals and annually leads the nation in the production of feldspar, lithium minerals, scrap mica, olivine and pyrophyllite. The state ranks second in phosphate rock production and ranks in the top five in clay and crushed granite production." In the lower left immediately below is a simple, variously patterned, black and white map of the nine aforementioned geologic belts which, as it turns out, occupy none of the more recent Coastal Plain geology (about 40 percent of the state). The text continues in the far right of the sheet, with commentaries on "Commercial Gem- and Mineral-Collecting Sites," "The State Precious Stone—Emerald," and "The State Rock—Granite," and below that are acknowledgements of former Governor, James. B. Hunt, Jr.; former Secretary of the Department of Environment, Health and Natural Resources (now simplified to Department of Environment and Natural Resources), James B. Howes; and former Director of the Division of Land Resources and State Geologist, Charles H. Gardner. Finally, there is contact information for the North Carolina Geological Survey.

In between this and the leftmost text and map is a generalized "Geologic Time Scale for North Carolina" with the customary chronology of eons, eras, periods, and epochs, and brief remarks on "Geologic Events in North Carolina" for each epoch. Below that, in the extreme lower right, is a small and simple map of "Principal Mineral-Producing Areas," with fourteen entries presented in alphabetic code against the matrix of counties (unnamed) and postings of five of the state's major cities.

All of the above are part of this modest map's perimap, which is actually rather substantial and certainly relevant to interpreting the main cartographic text (just when was the late Proterozoic again?) and which purposefully points out the economic value of exploiting North Carolina's indigenous geologic resources. This perimap's dual purpose as both explanatory and promotional rhetoric is well achieved, but not the main focus of our discussion, so we flip back now to the crayon-colored "map side," the one we might display on our office walls and corkboards.

There, we now observe that there are thirty-two generalized geologic formations, with ten under the general heading of "Sedimentary and Metamorphic Rocks," six under the heading of "Intrusive Rocks," and another sixteen under the heading of "Sedimentary Rocks." The sedimentary and metamorphic rocks are presented in three age groupings from middle Proterozoic to early Paleozoic, the intrusive rocks in two age groupings of late Proterozoic and late Paleozoic age, and the sedimentary rocks in four age groupings from Triassic to Quaternary.

The first ten of these sport an array of bold colors, among them a vivid blue, orange, green, and yellow, along with a rich brown and deep ultramarine, and lie mostly in the western portion of the state, as well as the inner Piedmont and the Carolina Slate Belt. The intrusive rocks—in pink, fuscia, burgundy, and carmine—occupy much of the central Piedmont portion. The younger sedimentary deposits of the Sandhills and Coastal Plain in the east are presented in more pastel yellows, oranges, and yellow-oranges, along with greens and yellow-greens. The sedimentary rocks of this heading are, for the most part, lighter and paler than the others, and the igneous rocks of the second heading share an obvious kinship of hue, but a more rigorous logic of color is difficult to discern here. The map is a sort of circus of colors, with quite a few competing for attention simultaneously. And that's usually the case with geologic maps. To understand the logic of the geologic map, one has to *read the legend.*

First of all, there are the three major headings, differentiated primarily by their general geologic classes of origins: metamorphic, igneous, and sedimentary (although the mostly metamorphic class, which is labeled Sedimentary and Metamorphic Rocks, does include some relatively small occurrences of older sedimentary rocks and several metasedimentary formations). In the map, metamorphic and igneous rock appears mainly in broad linear belts (illustrated more simply in that black-and-white map on the reverse), which trend from southwest to northeast and reflect the alignment of ancient continental edges at the time of their collision about 500 million years ago, while the sedimentary formations and deposits hint of the positions of former but much more recent coastlines. The general arrangement is from west to east, with metamorphic rocks mostly in the western portion of the state, sedimentary rocks in the eastern, and igneous rocks more or less in the middle.

Under each of the three primary headings, formations are arranged by age in reverse chronological order. Sedimentary and metamorphic rocks are grouped into late Proterozoic to early Paleozoic, late Proterozoic, and middle Proterozoic. The upper end of that time frame is partly shared by the lower end of the time frame of the igneous rocks, which are divided into middle Paleozoic to late Paleozoic and late Proterozoic to middle Paleozoic (an overlap with the previous heading in this case of about 465 million years). The sedimentary rocks of the Coastal Plain are generally far younger and are grouped into Quaternary, Tertiary, Cretaceous, and Jurassic.

Finally, each heading, and each temporal division within each heading, is further subdivided into individual belts, groups, and formations based mainly on material composition, such as the metavolcanic rocks, metasedimentary rocks, and felsic gneiss of the Blue Ridge and inner Piedmont; or the meta-ultramafic rocks, metamorphosed gabbro and diorite, and syenite of the central Piedmont; or the limestones, sandstones, mudstones, marls, and conglomerates of the Coastal Plain.

These are the distinct colors that comprise the map, which can now be understood as not just a quasi-arbitrary painting of polygons but as an expression of a complex hierarchical taxonomy. Beneath the sign plane of the map lies a three-layered content plane, with a primary layer of content distinctions based mainly on process of origin (and reflected in the broad west-to-east sweep of the state's geology), a second layer differentiated by geological eras, and a third ultimately distinguishing individual semiotic elements according to composition and material. Expressed in a diagram, that might look something like figure 9.4.

This is and always has been the hallmark of the natural sciences, which define their most fundamental understandings of the world and what's in it through procedures of hierarchical multivariate deconstruction and classification. Maps of this kind are essentially spatializations of such taxonomies, with the geographic instantiations of their elements forming the maps' pertinent postings. The discovery of a new geological occurrence, like the discovery of a new species, emplaces a new content element at some level among the hierarchical layers of content planes, usually but not always the lowermost and most specific. In the case of this map, a well-established three-level taxonomy is compressed and spread across the cartographic sign plane of the state, a scientifically predigested presentation of evidential fact offered up for our own deconstruction and interpretation. And that's what all geologic maps are about, as are those of other scientific disciplines.

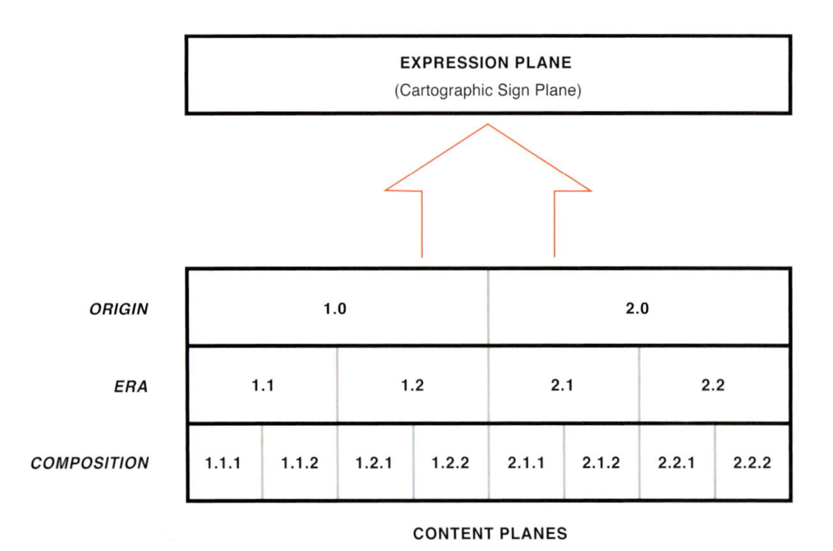

Figure 9.4
Diagrammatic representation of a multilevel, hierarchically organized content plane. Such a conceptual structure underlies not just this map, but nearly all maps of natural systems.

AND IN BETWEEN

Finally in this discussion, we have the *General Soil Map of North Carolina* produced by the Soil Conservation Service of the U.S. Department of Agriculture (figure 9.5). This is significantly larger than our previous map, although half the size of our first example, and rather more detailed. Its general appearance is quite different from the geologic map as well, with deeper more muted tones (mostly) organized by their memberships among the three major physiographic provinces of North Carolina, called in this map Blue Ridge Mountains, Southern Piedmont, and Southern Coastal Plain, from which there is also differentiated here a region of Atlantic Coast Flatwoods. The soils of the Blue Ridge Mountains appear in dark green and olive, those of the Southern Piedmont in pinks and salmon, those of the Southern Coastal Plain in warm dark brown and burgundy, and those of the Atlantic Coast Flatwoods in blues and greens and purple. Underlying the richly colored polygons is much of the usual set of topographic features: cities and towns, county boundaries, railways (but no roads), rivers and reservoirs, and all in black except for hydrography. There are no contours or any other expression of the land's relief.

The soil polygons themselves appear in the legend as sixty-six entries, organized by province and accordingly grouped into families of like colors. Each entry represents a "soil association" and we are informed in a small paragraph below the map's title that these have ". . . been adapted from sheets 10 and 11 of a Soil Map of the United States as associations of phases of subgroups" and that "Soil series names are tentative in portions of the state where detailed mapping is lacking." (To the uninitiated, that won't make much sense at all until we turn the sheet over and read that, "The soil associations have been named for two or three major soil series . . .")

Within each provincial grouping of soil associations, these are further grouped based on general soil characteristics and topographic setting, with anywhere from one to eighteen in each of these subgroupings. In the Blue Ridge, we have moderately shallow soils on steep ridges and mountains and moderately deep and deep soils of broad valleys and basins. In the Piedmont, we have slightly acid and neutral, very plastic soils, strongly acid and extremely acid plastic soils, and firm clayey soils on felsic rocks (granite or Carolina slates.) In the Coastal Plain, we have sandy and loamy soils of dissected uplands, loamy and clayey soils of smooth uplands, and loamy and clayey soils of lowlands and terraces of streams. And, in the Atlantic Coast Flatwoods, we find organic soils of shallow lake beds and river swamps, wet sandy soil of broad ridges, wet lowlands along stream valleys, sandy and loamy soils of uplands, and loamy and clayey soils of wet lowlands. Every soil association within each of these subgroupings shares one common and distinct color (fourteen in all) and, since individual associations do not for the most part have uniquely assigned colors (which would require sixty-six different colors and/or patterns), each carries, in the legend and within their postings on the map, a unique numeric identifier. (For example, number 63 is the Dunbar-Lynchburg-Goldsboro association, one of the sandy and loamy soils of uplands in the Atlantic Coast Flatwoods region.)

We can penetrate the map's content a bit more deeply if we flip the sheet over and examine the perimap material on the other side. In the upper left is a verbal legend of Soil Associations, Descriptions of Topography, and Selected Interpretations. This begins with the caveat that, "A general soil map does not show individual areas occupied by named soil series since it is impractical on a map of this size," and goes on to explicate the general terms given to surface slope classes ("level" means 0–3 percent, "sloping" means 3–15 percent, etc.) and to make it clear that,

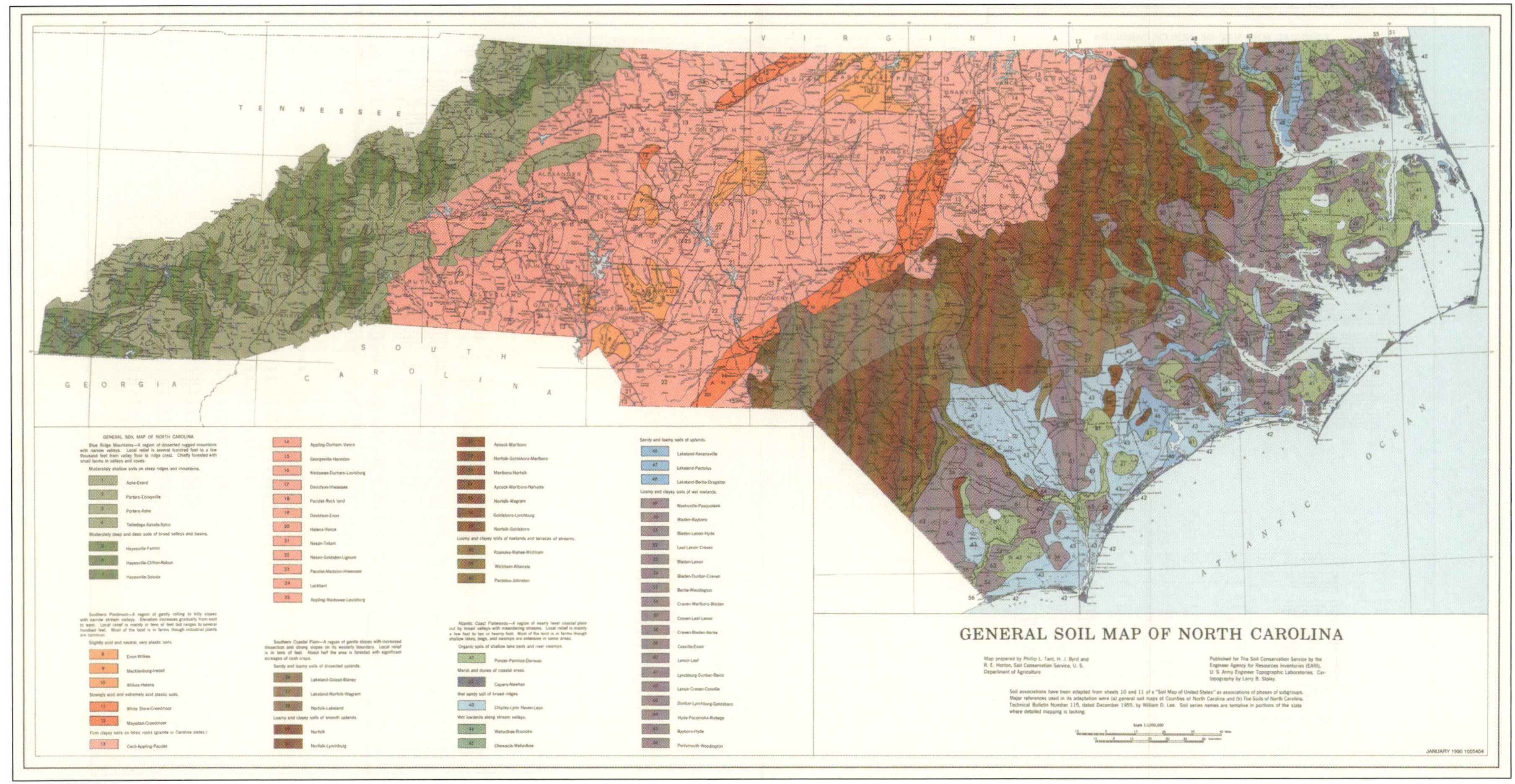

Figure 9.5
General Soil Map of North Carolina. This map, like the previous, presents only the uppermost levels of a complex hierarchical taxonomy. Map dimensions (WxH): 34in. 17in.; scale 1:1,000,000.

Courtesy of the USDA Soil Conservation Service, 1990.

> Interpretations of soil series named as major soils in associations are rated for depth to bedrock, shrink-swell potential, flooding; soil suitability for general agriculture as well as timber and pulp wood. Soil limitations for dwellings, shallow excavations and septic tank filter fields are also included.

The next two paragraphs appear under a heading simply called "Description" and clarify the soil units' ratings for shrink-swell potential and flood hazard.

The remainder of this verbal legend, which is about two-thirds of it, falls under the heading of Selected Interpretations, which includes soil suitability (good, fair, or poor suitability for tilled crops, pasture, tree fruits of the region, and timber and pulpwood) and soil limitations, which describes just what is meant by designations like "slight," "moderate," or "severe."

The entire remainder of this side is occupied by a very large table of soil associations corresponding with those appearing on the map side. As in its counterpart graphical legend appearing below the map (and of which this is clearly an extension), the table's entries carry the same numerical identifiers as the legend and are again organized by physiographic province.

The table for each province is in two parts. On the left, there are four columns: for map symbol (numerical identifier), soil association name, a capsule description of topographic setting and geologic substrate, and a list of minor soils (i.e., soil series) included in each soil association. In the Blue Ridge Mountains table, for example, the Hayesville-Fannin association (number 5) is found in settings characterized as "sloping to moderately steep mountain uplands; gneiss, schist and acid igneous rocks," and includes the Watauga, Ashe, and Chester soil series (curiously not the Hayesville series, which constitutes the Talladega-Saluda-Sylco association immediately above). The right-hand portion of the table enumerates, not soil associations, but their constituent soil series, with columnar entries for depth to bedrock, shrink-swell potential, flooding, suitability for general agriculture and for timber and pulpwood, and limitations for dwelling, excavations, and septic tank filter fields. And this goes on for a total of eighty-five soil series, with a few remaining unmentioned (presumably owing to insufficient information to fill the table's columns).

The lowermost layer of this map's content plane turns out to be the soil series, which are unexpressed in the map itself but are elaborated in this extended "legend" on the reverse. And if we hadn't flipped the thing over, we certainly would have missed this essential point. The soils series, which are explicitly endowed with pragmatic meaning, are assembled into soil associations in the map, and those are then grouped into topographic and material relationships, and finally into geographic and geologic relationships. The content plane of this map is structured very similarly to that of the *Generalized Geologic Map*. While the lowermost—and genuinely meaning bearing—layer of the content plane, the individual soil series, remains without direct cartographic expression, the higher layers of soil associations, topographic and material groupings, and provincial groupings are expressed quite clearly in the sign plane of this map. It's hard to imagine that any map of this type could function otherwise. A natural system of interest—like soils, geology, vegetation, whatever—is hierarchically deconstructed into its constituent components, insofar as we recognize them (which is a matter of no small debate among natural scientists) and then reassembled into familial or geographic content clusters in the map. That's why it's essential that content relationships at all levels be expressed somehow, through the paramap if necessary; otherwise the map would just be a chaotic catalog of individual and unrelated phenomena. Soils and geologic maps are visual localizations and structuralizations of complex systemic deconstructions of the natural systems within their prescribed domains.

OK . . . *now* we get it. The soil associations are the map's practical import; they are really what this map is about. If we want to start up a soybean farm in the Piedmont to furnish a burgeoning biodiesel industry in the state, then we know that suitability for agriculture on the Appling series (Wedowee-Durham-Louisburg association) is good; and if we want to excavate an extensive basement or parking garage for a large building in this same region, limitations for that purpose are severe on the Iredell series (Davidson-Enon association) but only moderate on the Mayodan Series (White Store-Creedmore association).

Of course, if we were really serious about such enterprises, then we'd consult the more detailed, large-scale county soils maps. So why such a broad and deep paramap for this map? Well, because maps of this type are not made just for the pleasure of their making, or viewing—although we could certainly appreciate that—but, since they're civic endeavors and ultimately taxpayer sponsored, there has to be some pragmatic rationale for their mandates in the first place. In the case of a state highways map, for example, that's explicitly about getting motorists surely and safely from one place to another or to a state's key attractions and, of course, to promote tourism. But for the vast majority of our culture, soil science and geology are far more obscure and opaque subjects than the all-American pastime of driving.[2] So the soils map must justify its existence clearly as an instrument of economic development, of the exploitation of the land essentially, just as the geologic map must justify its existence through the exploitation of mineral resources. Who but the genuinely initiate have any clue what the nomenclature and terminologies of such maps mean? If these maps are not about business at some level, then they would certainly remain well cloistered within the confines of their respective disciplines.

But those kinds of observations have been made many times, and are not really the point here. Our intention is simply to show how maps like the geologic map or soils map illustrate the systematic deconstruction of the natural world into recognizable and identifiable elements that can be spatialized as cartographic postings of relatively certain location and extent—or at least that's the presumption. As for our *State of North Carolina* topographic map, that clearly serves a similar purpose with respect to a more visible and more widely comprehensible world. But the topographic map does something rather different also. Having deconstructed a piece of the world into constituent systems, it goes on to reconstruct all that into an ostensibly whole and integral *synthesis of systems*,[3] a comprehensive physical landscape. (Yet, if that's the acknowledged intent of so-called topographic maps, it's hardly their exclusive domain.) Phenomenological deconstructions of the natural world would ultimately be of little use if they could not be subsequently reconstructed into larger understandings of that world.

And are there obvious kinships among these three maps? Of course there are. The general character of the state's three physiographic provinces, evident in the topographic map's shaded relief, is a reflection of the underlying geology and geomorphic history manifest in the *Generalized Geologic Map of North Carolina*. And the patterns of Coastal Plain swamplands there clearly echo those of the organic soils of shallow lake beds and river swamps and the wet lowlands along stream valleys presented in the *General Soil Map of North Carolina*. But, as one would expect, there are even more obvious similarities between the geologic and soils maps. The strongly acid and extremely acid soils of the White Store-Creedmoor and Mayodan-Creedmoor associations register perfectly with the Triassic bedrock of the Dan River Group and Chatham Group geologic units. Blue Ridge outliers of the Ashe-Evard moderately shallow soils of steep ridges and mountains overlie metamorphosed granitic outliers. The slightly acid and neutral, very plastic soils of the Enon-Wilkes, Mecklenburg-Iredell, and Wilkes-Helena associations correspond strongly with other granitic outliers, while the sandy and loamy soils of dissected uplands belonging to the Lakeland-Gilead-Blaney, Lakeland-Norfolk-Wagram, and Norfolk-Lakeland associations correspond with the Middendorf Formation of sand, sandstone, and clay. And other correspondences are apparent as well. After all, parent material—geologic substrate, generally speaking—is one of the primary determining factors in soil genesis. (The other four are relief, climate, organisms, and time.)[4]

Beneath the visible physiography of the topographic map lie distinctive patterns of soils and then geology, both systematically deconstructed here in the languages of their respective sciences. And that's not much different from what's happening on or above the surface either, like vegetation or even climate. This is the traditional basis of Western science—the deconstruction of some particular aspect of the world (the focus of a specific scientific discipline) into discreet recognizable terms, and its subsequent reconstruction into a systemic whole.

Systemic syntheses

We have previously pointed out how the signs operating within the map itself (the map's text as opposed to peritext or epitext) can usually be interpreted at three levels of integration.[5] At the *elemental* level, individual graphic marks within the map denote specific instances or occurrences of preformed conceptual types: a road or highway, river or stream, city or town, a political boundary of some

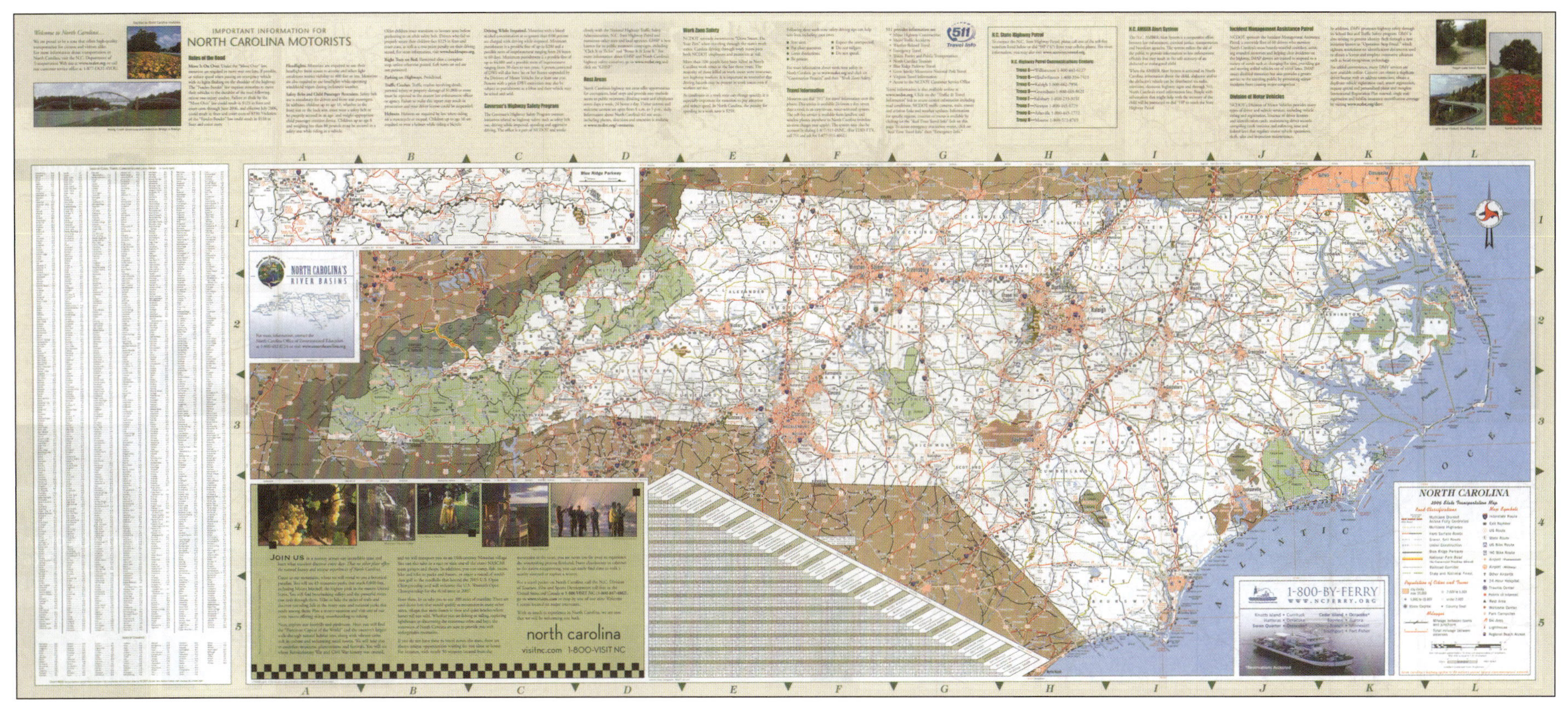

Figure 9.6
The *North Carolina 2006 State Transportation Map* synthesizes a number of systems: highways and roads, cities and towns, lakes and rivers, counties. Mostly, though, it's about pavement.

Source: North Carolina Department of Transportation, 2006.

kind, etc. These are the map's fundamental postings. At a somewhat higher level—the *systemic*—these are integrated into systems of like signs (supersigns), which is to say, like marks and like content elements: a network of roads and highways, a network of streams and rivers, a cluster or assemblage of municipalities, a matrix of counties or townships. At a still higher level—the *synthetic* (super-supersigns)—disparate sign systems, often embodying very different conceptual categories, exchange meaning among themselves in the potentially very complex spatial synthesis that we typically think of as a map. A road or railway lies where it is because it follows the course of a particular river; a county boundary is demarked along the course of another river or along the height-of-land that divides two river systems; a river and its tributaries flow where they do because of topographic constraints and opportunities; settlements and then towns and then cities rise along the rivers and then the railways and then the highways.

That's all readily evident in a map like the *State of North Carolina* topographic map. But that map, of course, is more than just a rendering of topography—it's a map of North Carolina, period, just like it says. No, there's no geology in that map (at least not explicitly shown), and there are no soils either (ditto), and vegetation is only partially indicated, and there's certainly nothing like, well, weather or climate, for example. The really tangible stuff, though—the stuff you know you can reach out and touch (except, perhaps, for the various territorial claims of the counties)—the things we take as more or less permanent within the time frames of our lives—*that's* there. And that's what the map is about: the state that we know and recognize faithfully reconstructed in cartographic form.

Sure, we can find there the town we live in and the county we live in and the way (if not the most current and detailed) to drive to the town where our friends live, or to Mount Mitchell or to the Outer Banks. It's all there, or about as all there as it needs to be or ever can be in such a map. As for the currency of the state's roads and highways on a map published in 1957 (and not revised since 1972), there's another map for that purpose. That map, of course, is the *North Carolina Transportation Map* (formerly *North Carolina Highway Map*), and every state has one (figure 9.6). North Carolina, like most states, publishes a new and up-to-date edition every year. The immediate function of those maps is up front and unambiguous: to assist motorists (and their families) in getting from their uncountable origins to their uncountable destinations efficiently and without incident. Since we have already addressed such maps' other functions elsewhere, there's certainly no point in doing that again here.[6] (The foremost of those other functions is, of course, the promotion of tourism to and within the sponsoring state. In fact, some states now offer their transportation maps only as detachable inserts in their voluminous state tourism guide books.) It's not our intention to address "highway maps" or "transportation maps" here at all, except to point out once again that the selection of the principal systems that participate in the ultimate synthetic presentation of almost any imaginable map is indeed a *selection of content in accordance with the map's polemical and social goals.*

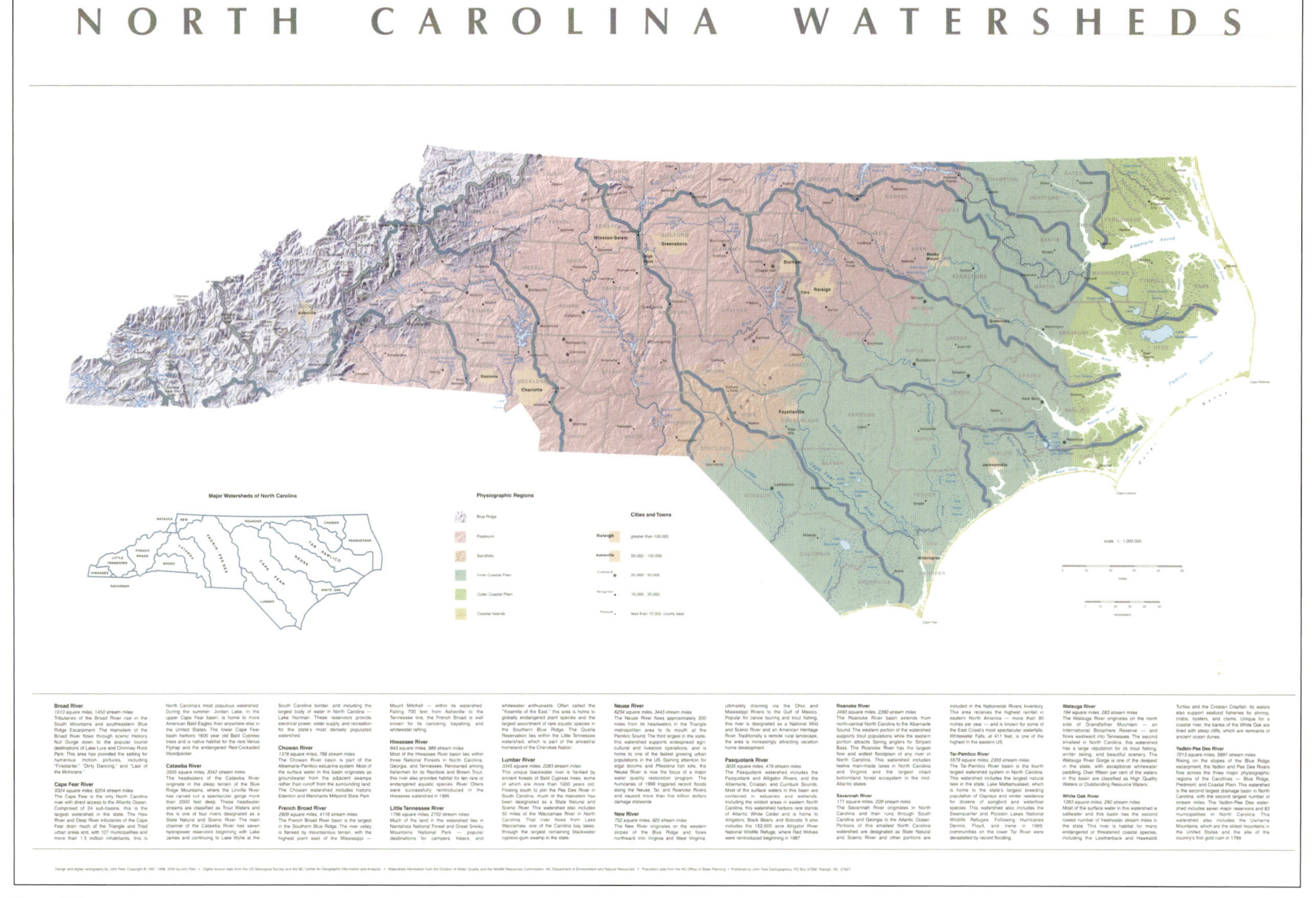

Figure 9.7
North Carolina Watersheds. There are no roads in this map. Instead, the map reader is encouraged to recognize an alternative geography of the state. Map dimension (WxH): 36in. x 24in.; scale 1:1,000,000.
Source: John Fels Cartographics, 2000. Data source: USGS.

That, of course, has been standard fare in cartographic textbooks for decades,[7] but always invoking something like "communication objectives" as the ultimate goals of an objective and appropriately balanced presentation of information instead of imbuing the map with something like a larger social agenda. But that's pretty old news now too, and not a point that needs to be belabored here.

What we're thinking is that the selection and content decisions intrinsic to any map are determined not just by the map's intended function or purpose, or even by some implicit or explicit social or political objective (the latter sometimes construed as a kind of cartographic deceit or deception), but more realistically that those decisions are fundamental and essential to the construction of alternative cartographic landscapes. Any highways map constructs a geography of its territory based on a clear and certain spatial/structural premise, an organizing principle—in that case, highways and roads, of course. That doesn't mean that highways and roads are more important than any other aspect of our state. Would anyone conceivably choose to visit the Grand Canyon or Yosemite, to travel to another state just because whatever other state had somehow a "better" highway system? We seriously doubt it.

So what is any content synthesizing map of any state really about anyway? Isn't it about offering up a particular spatial and conceptual framework for the social construction of that comprehensive space? Are there alternative or even competing views of just how that space could (should) be presented, understood, and socially constructed? How could there not be? And that rather long-winded preamble leads us to this case in point.

On the face of it, John Fels' *North Carolina Watersheds* map (figure 9.7) looks more than a little like a topographic map, and in many respects it is. There are the usual county boundaries, cities and towns, rivers and lakes and reservoirs. In lieu of contours, there is a vivid shaded relief (based on both elevation and surface illumination and extruded with a vertical exaggeration of five times to give the map a slightly oblique character), but that's not the most salient aspect of this map's first impression.

More apparent are the broad swaths of color painting the state's six physiographic regions, elaborated from the basic three physiographic provinces: the Blue Ridge (in a rich blue of course), the Piedmont and Sandhills (in subdued soils-like red and red-orange, respectively), and three divisions of eastern North Carolina (in green and yellow-green) into Inner Coastal Plain, Outer Coastal Plain, and Coastal Islands (i.e., the Outer Banks and more immediate barrier islands). Equally conspicuous are the broad blue-gray lines that delineate North Carolina's seventeen major watersheds, aligned roughly from northwest to southeast and following the general seaward slope of the state, perpendicular to prevailing geologic, pedologic, and physiographic trends. In the background are the light-gray lines demarking the county boundaries and an array of cities and towns, from "greater than 100,000" to "less than 10,000, county seat," with the larger incorporated municipalities appearing as pale tan-colored polygons.

To the lower left of the watersheds map itself, there is a smaller key map for the watersheds; and just right of that there is a legend for the six physiographic regions; and just right of that there is a legend for the cities and towns in five population classes. To the lower right of the watersheds map are scale bars in miles and kilometers. The map's title, in a widely spaced Optima type, stretches out above and across the lower part of the map sheet; there are eleven shallow columns of text giving the vital statistics (square miles and stream miles) for each watershed and pointing out their more noteworthy attributes.

All well and good, but there is something missing among all that —very obviously missing. *There are no roads.* There are no county roads, no state highways, no federal highways, not even interstates—*none.* This seems irreverent at least, maybe even sacrilegious, for a map of a state that takes such obvious pride in the extent and quality of its highway system. But that's exactly the point. The map's author puts it this way:

> *The North Carolina Watersheds* map was designed to promote a broader awareness of the environmental geography of North Carolina. The map portrays the three-dimensional topography of the state, its six physiographic regions, significant landforms, rivers and lakes, and major watershed boundaries. Cities, towns, and counties provide familiar references. Capsule descriptions furnish basic information and highlight the features of each watershed. Roads and highways are not shown since the essential purpose of the map is to encourage a different notion of geographic "connectedness"—one not based on automobiles and pavement but on the intrinsic connections of North Carolina's natural systems.

There isn't even a hint in this map that the major "Metrolina" cities of Raleigh, Durham, Greensboro, Winston-Salem, and Charlotte—and ultimately Asheville in the mountains and Wilmington on the coast—are connected by Interstates 40 and 85, even though that's probably the most widely comprehended aspect of the state's geography. What is evident, though, is that surface runoff (along with whatever else is in it) from Greensboro and Durham will ultimately find its way, via the Cape Fear River, to Wilmington or that wastewater discharges from Raleigh and Cary will find their way to New Bern via the Neuse River.

This is quite a different geography of the state than we've grown accustomed to and now take for granted as unquestionable fact. Indeed, this alludes in some ways to a much earlier state of North Carolina—a former time and place, one that preceded any integral road system or even railway system, when the rivers were the state's arteries of commerce and "geographic connectedness" was structured by water rather than pavement. If we could ignore the multiplicity and sprawl of all those present day cities and towns, this might be some kind of pre-Eisenhower North Carolina and, if we ignore the currency of the county boundaries, it's almost Colonial. But, of course, those are not to be ignored since, in the absence of all those roads and highways, these are the most familiar geographic touchstones remaining in this alternative and somewhat alien synthesis.

What a map—*any map*—has to say about the world, what particular "fact" it proposes to advance, is largely a matter of just which aspects of the world are allowed to participate in the map. A state of North Carolina structured by rivers and watersheds is not the same state as a North Carolina structured by roads and highways. They are different, very different, and they invoke very different concepts of place and very different social and political perspectives.

This one really blows us away

The next map we want to show you is, well . . . *awesome.* And that's just one of many superlatives we're inclined to heap upon it. But before we look at this, it's worthwhile looking at its predecessor.

That is the USGS's *Landforms of the Conterminous United States: A Digital Shaded-Relief Portrayal* by Gail P. Thelin and Richard J. Pike (figure 9.8). It's a large map and easily has enough visual presence to elicit quite a few "Wows" mounted on an office wall. The map is printed entirely in grayscale, with a solid black background that makes its computer-generated shaded relief look especially vivid. There are no states here, no cities, no highways, no rivers—nothing but the land surface itself, rarified as if in an X-ray image. This really is like an X-ray in some respects: anything accumulated on the land surface is just not there—just *seen through,* just *gone.*

There's very little perimap here as well. Below the map there are projection and scale statements (Albers equal-area conic projection, scale 1:3,500,000, one centimeter equals thirty-five kilometers) and there are the customary scale bars in miles and kilometers. Date of publication and USGS's modest credit are in the lower right, and in the lower left there is a paragraph of map notes that summarize some of the technical aspects of the map's production. In the extreme upper right is a small statement that reads, "Miscellaneous Investigations Series. Map I-2206. Explanatory pamphlet accompanies map."

That pamphlet is an 8.5 by 11 inch, sixteen-page booklet which, being a document apart, is, strictly speaking, considered epimap material but which does carry much of the perimap content that probably would have appeared surrounding the map if this were in *National Geographic* style.

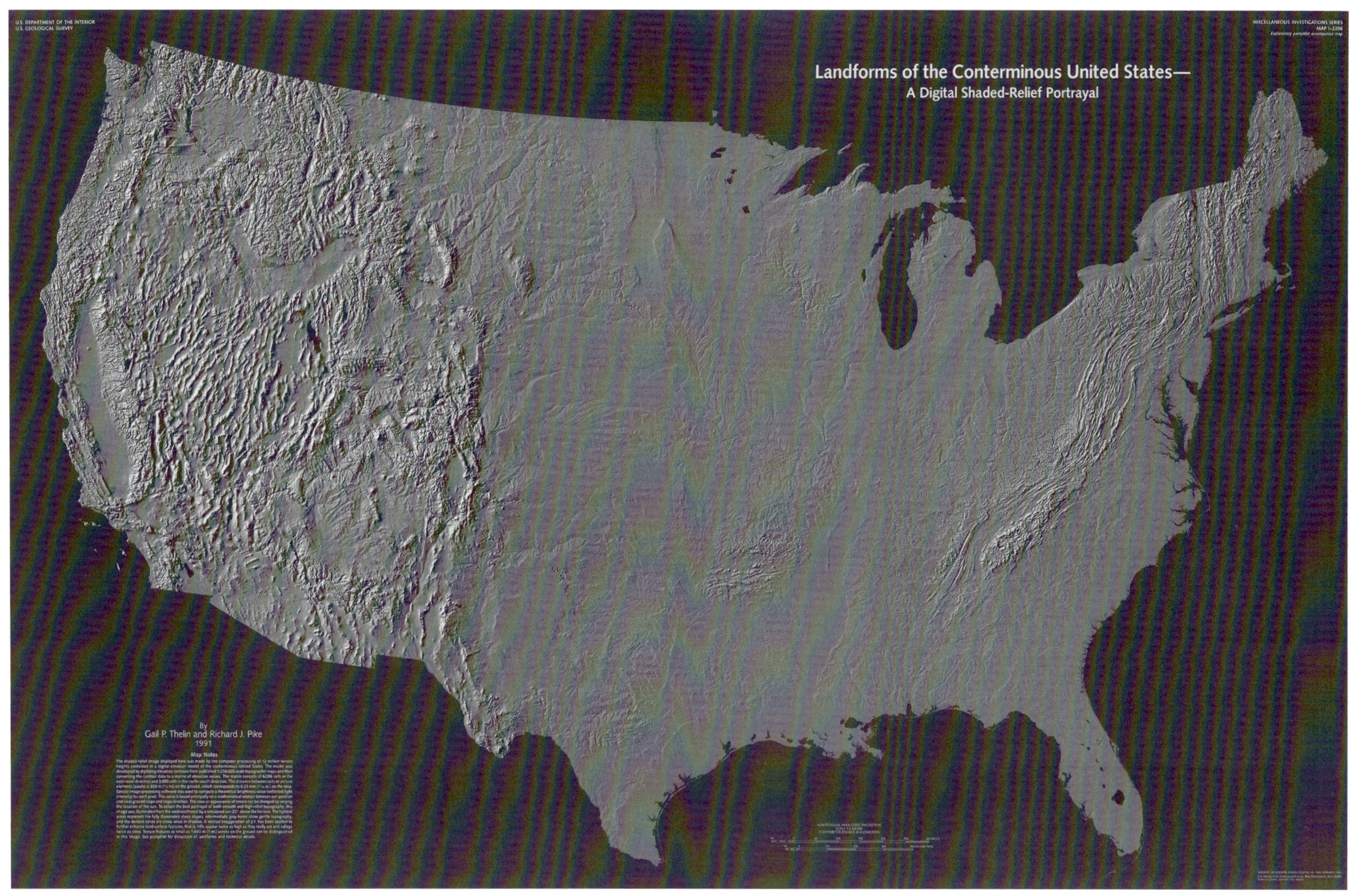

Figure 9.8
Landforms of the Conterminous United States, a striking high-tech rendering of the country's physiography. Map dimensions (WxH): 55in. x 35in.; scale 1:3,500,000.
Source: Courtesy of USGS, 1991.

The booklet begins with concise reviews of the history of landscape visualization in maps and the history (up to the map's 1991 date of publication) of automated relief shading. Those are followed by a brief discussion of the applications of digital landform maps and then by an overview of this map. What follows that is an enumeration of some landforms of interest that can be seen in the map, and then a brief discussion of future prospects. The final section of the booklet is simply titled "Technical Details" and includes discussions devoted to image processing and hillshading, source data, and computation and production. From this we learn, among other things, that this "reflectance map" consists of about 12 million data points, with each representing about 0.8 by 0.8 kilometers or about one-half mile square on the ground and measuring about 0.25 millimeters or 0.01 inches in the map (that previous map of *North Carolina Watersheds* consisted of about 3 million data points with each representing 300 feet by 300 feet on the ground); that these were reprojected in Albers equal-area conic; that the sun angle and elevation employed for relief shading were 300 degrees and 25 degrees respectively; and that five lithographic plates were required to print the map—one each for three different tonal ranges of the shaded relief and another two from a nondigital overlay "to ensure a deep black for the background."

Perhaps most importantly for the nontechnical, there are also included in the booklet three double-page reproductions of the map sheet (sans type, scale bar, or anything else), each with an overlay of linear symbols. The first of these delineates physiographic units at three hierarchical levels: divisions (with bolder outlines), provinces (designated by numeric codes), and sections (with finer outlines and designated by alphabetic codes). There are eight major divisions: Laurentian Upland, Atlantic Plain, Appalachian Highlands, Interior Plains, Interior Highlands, Rocky Mountain System, Intermontane Plateaus, and Pacific Mountain System. The physiographic divisions, in turn, consist of twenty-five provinces, which ultimately consist of eighty-five sections[8] (two of which, Continental Shelf and the Northern Section of the St. Lawrence Valley, do not actually appear in the map). The second of these maps gives us state boundaries, and the third a pointing out of twenty-seven significant geologic and geomorphic features that

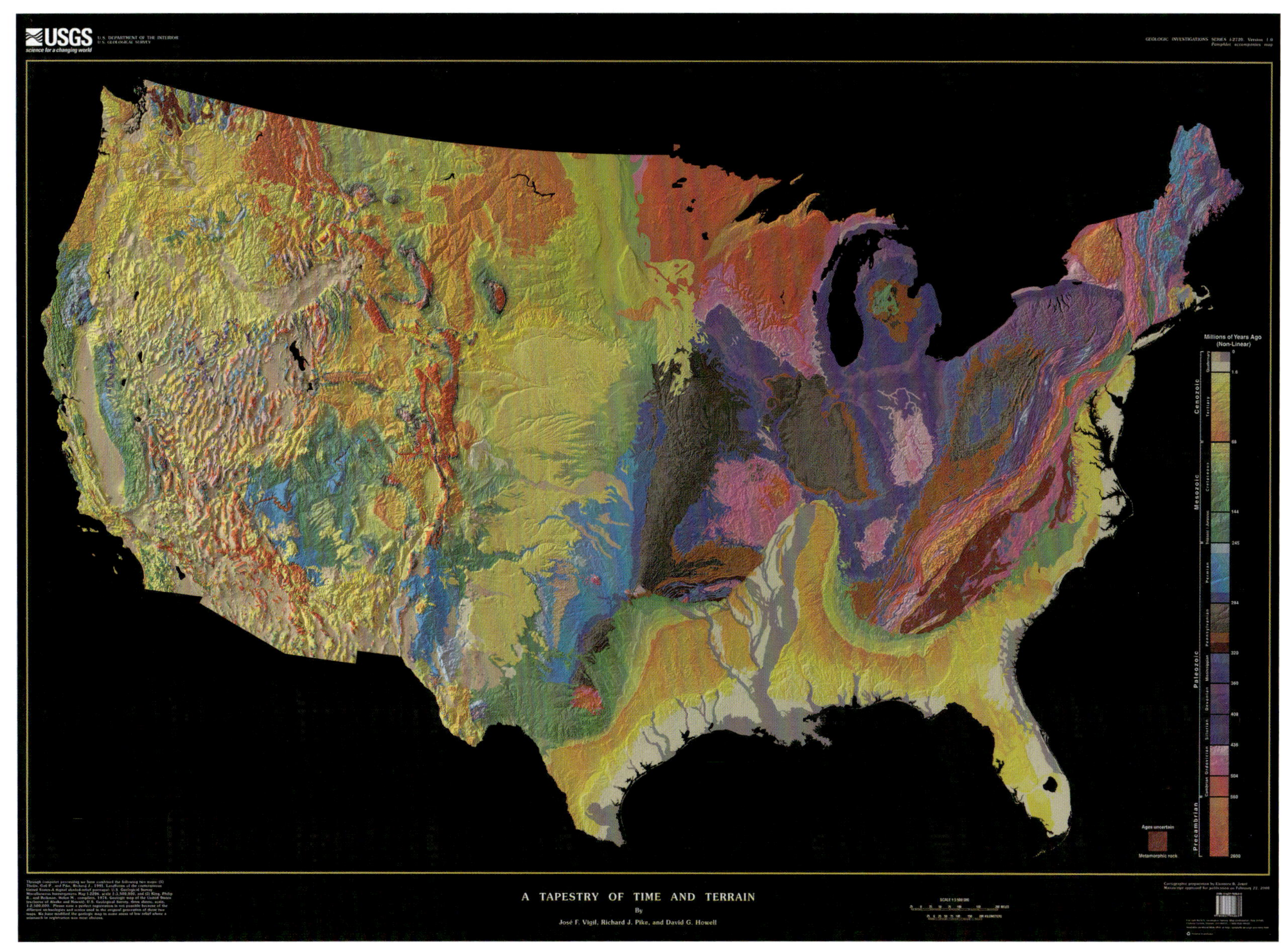

Figure 9.9
A Tapestry of Time and Terrain, the previous map combined with a map of geologic units. The information expressed here is more than the sum total of the information expressed in its two constituent maps. The whole really is greater than the sum of its parts. Map dimensions (WxH): 58in. x 42in.; scale 1,350,000.

Figures 9.9 and 9.10 courtesy of USGA, 2000.

are mentioned and listed in the booklet's text, as well as a delineation of the extent of Pleistocene glaciation.

Without these three maps, their larger counterpart would certainly be more difficult to interpret. Sure, some of the features here are almost impossible to miss, like the Appalachian Mountains, Mississippi Floodplain, Black Hills, Sierra Nevadas, or Sacramento Valley. Others, like Crowleys Ridge, Trail Ridge, the Clinton-Newbury and Garlock fault zones, or Olympic-Wallowa lineament, will probably escape or at least puzzle all but the most knowledgeable map readers.

That said, this is an extraordinary map. Even for those who don't bother to read the accompanying booklet and penetrate the map's content and process a bit, it still has to be quite a thrill. This map is so striking and so engaging that it's difficult not to feel that one is really seeing this country for the first time. It hardly even seems to be a map at all but an extraordinarily detailed super-photograph of some kind; there are no discrete cartographic postings, no labels, but it is a map nonetheless and most definitely not of photographic origin. (We'll deal with continuous-tone image maps in the next chapter.)

But as remarkable as this map is, it is not the ultimate object of this discussion. That would be USGS's *A Tapestry of Time and Terrain* by José F. Vigil, Richard J. Pike, and David G. Howell (figure 9.9). This map was published nine years after *Landforms of the Conterminous United States*, which appears here visually as an underlying base layer but which is actually an equal ingredient in the intriguingly titled *Tapestry*. This map is the same scale as its predecessor and the map sheet is just slightly larger. The shaded relief of *Landforms*—the "terrain" component of Tapestry—is overlain by a host of geologic units—the "time" component—simplified from King and Beikman's *Geologic Map of the United States*.[9] The geology is intensely colored in a "prismatic" series of fifty-two colors that indicate "just the geologic ages of rocks and surficial deposits." The layout is nearly identical to that of the previous map, except that the title is now below the map and there is a long (vertically) legend of color blocks along the right edge.

Precambrian rocks appear in several reddish oranges, and Paleozoic rocks in a series progressing from red through pink, violet, and blue violet to blue (Pennsylvanian rocks are brown and gray). Mesozoic rocks are dressed in greens and yellow greens. Cenozoic rocks are shown in a series from yellow orange to yellow (Tertiary) and in beige, tan, and a lighter gray (Holocene deposits). All this, and legend and scale bar, are again presented on an obsidian background that makes the map appear almost luminous. The "wow" factor here is even greater than for the previous map, as the colors practically leap off the page (and one feels they might well do so if not firmly anchored by the rock-solid shaded relief). This is one of the most striking and attention-grabbing maps we have ever seen. (We were both nearly speechless upon first seeing it . . . imagine that!) It is so stunning in appearance and complexity that at first glance it seems almost incomprehensible.

That could be rather frightening since, as with its predecessor, there is very little peritext present within the map sheet. Fortunately, like its predecessor, this map is also accompanied by an explanatory booklet that holds a wealth of paratextual information. This booklet is twenty-four pages and in a slimmer four-by-nine-inch format. Inside, there is a brief introduction and then a section titled "The Terrain," which reiterates, in more condensed form, much of the historical and technical discussion provided in the booklet accompanying *Landforms of the Conterminous United States*. The next section is titled "Geologic Time" and addresses "The second component of this tapestry . . ." This section discusses the four essential properties of geologic maps in general: classification by type of rock or surficial deposit, the ascription of earth's crustal materials to a specific environment or mode of origin, the identification of rock formations of distinctive materials and ages, and the arrangement of rock formations of different ages into a time sequence. Only the fourth and part of the third of those properties are of immediate relevance to this map, so the next paragraph goes on to clarify how King and Beikman's original 161 colors and patterns have been reduced to the fifty-two colors employed here based on geologic age alone and points out that the map "emphasizes time intervals defined by bedrock rather than surficial materials. In areas affected by continental glaciation, for example, ages shown are of underlying rocks, not the veneer of glacial (surficial) deposits." The third paragraph explains the subdivision of geologic time into eras, periods, and epochs, and the fourth briefly discusses the compilation of King and Beikman's map and cites a related 1976 report by King and another three by King and Beikman in 1974, 1976, and 1978.

The following section comprises the lion's share of the booklet and is devoted to "what the tapestry shows." The first paragraph offers a very general discussion of geologic concepts and principles, including *tectonic* and *magmatic* orogeny and subsequent erosion in a cycle of *dynamic equilibrium*. The second paragraph introduces concepts of *plate tectonics*, including *oceanic* (*basaltic*) and *continental* (*granitic*) *tectonic plates*, seafloor spreading and *subduction*, and the ancient supercontinents of *Rodinia* and *Pangaea*. (The words italicized above and below are those italicized in the booklet—"terms that may be unfamiliar to a general readership," as we had been told on the first page.)

The next paragraph prefaces the upcoming discussion of the map's features, and the one after that gives a very brief introduction to Fenneman's 1928 classification of *physiographic* regions of the U.S. and its three-tiered hierarchy of *division*, *province*, and *section*.

What follows is a short section on "General Features," such as the *active plate-margin* of the West, the *static midplate* of the remainder of the North American tectonic plate, and also *barrier islands*.

Next comes a section on "Specific Features"—forty-eight in all, including most of those mentioned in the *Landforms* booklet—presented in geologic chronology. For each of those features, there is a paragraph of intentionally nontechnical discussion. These are quite interesting and informative in themselves, but also found here is the real highlight of this publication.

That is an eight-panel double-page centerfold consisting of two full-color reproductions of the *Tapestry* with matching fold-over legends, scale bars, and, this time, graticule ticks in ten-degree increments. The map on the left is overlain with the same physiographic divisions shown in the previous booklet and the same alphanumeric key (shown in list form there on page 22). The map on the right is overlain with state boundaries and peppered with numbers referring to the textual descriptions of specific features. On the reverse of both maps are pages 11 and 12 of the text, the two legend panels, and another four panels of full-color enlargements of nine selected features or groups of features, which allow the reader to zoom in to those of most significant interest.

The booklet again closes with a short section on "Technical Details" explaining how the polygons of the differently projected and nonuniformly scaled geology component were scanned to convert them from vector to raster form and reprojected and rescaled to best fit the terrain component, how color and transparency choices were refined through trial-and-error, and what software and hardware were employed. The hierarchical listing of the physiographic regions, as well as acknowledgments and references, constitute the final four pages.

As cartographic syntheses go, this one is hardly complex in principle. Cartographic presentations often synthesize half a dozen to a dozen or more systems (thematic layers), but here we have only two: physiography and geologic time. And it's easy to think of the geology as overlain (or "draped") on the shaded relief—in fact the text puts it just that way—but in reality, of course, that's the other way around. Physiography is underlain by and, in large part, determined by the underlying geology. Perhaps it's best just to think of time and terrain as two equal ingredients in the fabric of the *Tapestry*.

But there's more to it than that. The shaded relief physiographic map, as we've seen, and the elaborately organized and colored geologic map both hold considerable interest in their own right but, synthesized together in one presentation as they are here, their collaboration gives rise to additional new meanings. The continuously

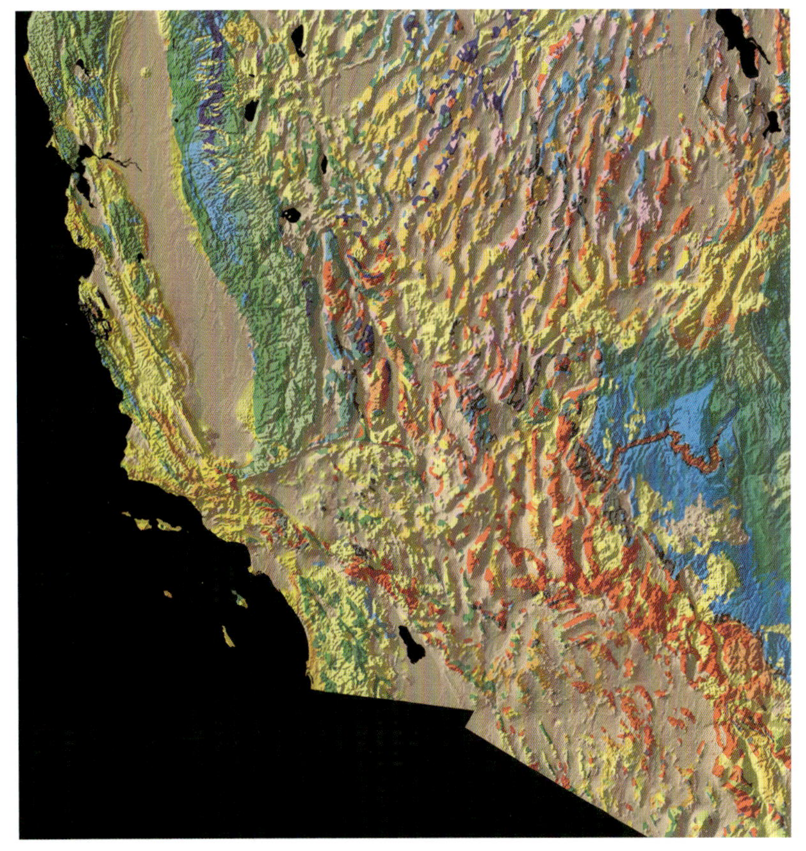

Figure 9.10
A detail of *Tapestry of Time and Terrain.* This says much, much more about California (or Nevada) than could be said by a physiographic or geologic map alone.

varying but undifferentiated physiography acquires new meaning when differentiated by geologic substrate, as geologic classifications acquire new meaning when their surficial expressions are made clear. (Just look at California in this map. The spatial synthesis of terrain and geology is nothing short of enlightening.) In such a spatial synthesis, all (here both) participating systems exchange meaning with one another, and this cannot help but generate new larger meanings. *The whole is more than the sum of its parts.* But that's essentially a definition of synthesis, and spatial synthesis is not only what maps do but something they do far better than any other conceivable medium.

A hierarchical deconstruction

Our next map is *Ecoregions of North America* (1997) by Robert G. Bailey of the USDA Forest Service (figure 9.11). It's a fairly large map, and also quite attractive. The main map image is of Ecoregion Provinces, and to the left are two smaller maps, of Ecoregion Domains of North America and Ecoregion Divisions of North America. All three maps are in Lambert azimuthal equal-area projection. A paragraph titled "Basis of Map Units" explains the following:

> This map shows ecosystems of regional extent. A hierarchical order is obtained by defining successively smaller ecosystems within larger ecosystems. At each successive level a different ecosystem component is assigned prime importance in the placing of map boundaries. *Domains* and *divisions* are based largely on the large ecological climate zones. Each division is further subdivided into *provinces* on the basis of macro features of the vegetation. Mountains exhibiting altitudinal zonation and having the climatic regime of the adjacent lowlands are distinguished according to the character of the zonation.

We are also told that "Three maps were used as a basis for compiling this map: *Ecoregions of North America*, 1981, by Robert G. Bailey and Charles T. Cushwa, *World Map of Present-day Landscapes*, published by Moscow State University, and *Ecoregions of the United States*, 1994, by Robert G. Bailey."[10] There is also, below the map's title, acknowledgment that the creation of this map was jointly supported by the U.S. Forest Service and The Nature Conservancy, Conservation and Science Division, and an indication that the map was revised in 1997.

The largest regional units, shown in a small map to the left, are the ecoregion domains, of which there are four: Polar (shown in blue), Humid Temperate (in green), Dry (in a salmon pink), and Humid Tropical (in violet). The Polar Domain occupies most of the continent north of southern Canada; the Humid Temperate Domain occupies the central and eastern United States as well as the west coast of the U.S. and Canada; the Dry Domain occupies the Rocky Mountains and the arid U.S. Southwest and northern Mexico; the Humid Tropical Domain occupies the rest of Mexico along with Central America, the Caribbean islands, and southern Florida.

In the small map below that, domains are subdivided into divisions, fifteen in all ranging from Icecap to Rainforest. A sixteenth legend block shows that Mountains with Altitudinal Zonation are overlain by a diagonal line pattern.

The main map further subdivides domains and divisions into ecoregion provinces. In a rather elaborate legend to the right, there are sixty-two of those in total, with thirteen under five subheadings in the Polar Domain (from Glacial ice to Tayga—tundra, high), twenty-two under eleven subheadings in the Humid Temperate Domain (Mixed deciduous—coniferous forest to Shrub or woodland—steppe—meadow), nineteen under eight subheadings in the Dry Domain (Coniferous open woodland and semi deserts to semi desert—open woodland—coniferous forest—alpine meadow), and another eight under four subheadings in the Humid Tropical Domain (Open woodlands, shrubs, and savannas to Evergreen forest—meadow or Paramus). Two more legend blocks show Riverine Forest in a dark green and, again, Mountains with Altitudinal Zonation (same as in the Ecoregion Divisions map). In this map, domain boundaries are demarked by broader black lines, division boundaries by somewhat thinner black lines, and the boundaries of the provinces by thinner lines still. Small black dots denote "representative climate stations." There is nothing else shown in the map except major lakes, national, state, and provincial political boundaries, and a ten-degree graticule.

Bailey's hierarchical classification of ecoregion domains, divisions, and provinces is not to be confused with Fenneman's physiographic classifications, Merriam's classification of life zones, Küchler's classification of potential natural vegetation, or Köppen's climatic classifications (later modified by Trewartha),[11] although there is a strong

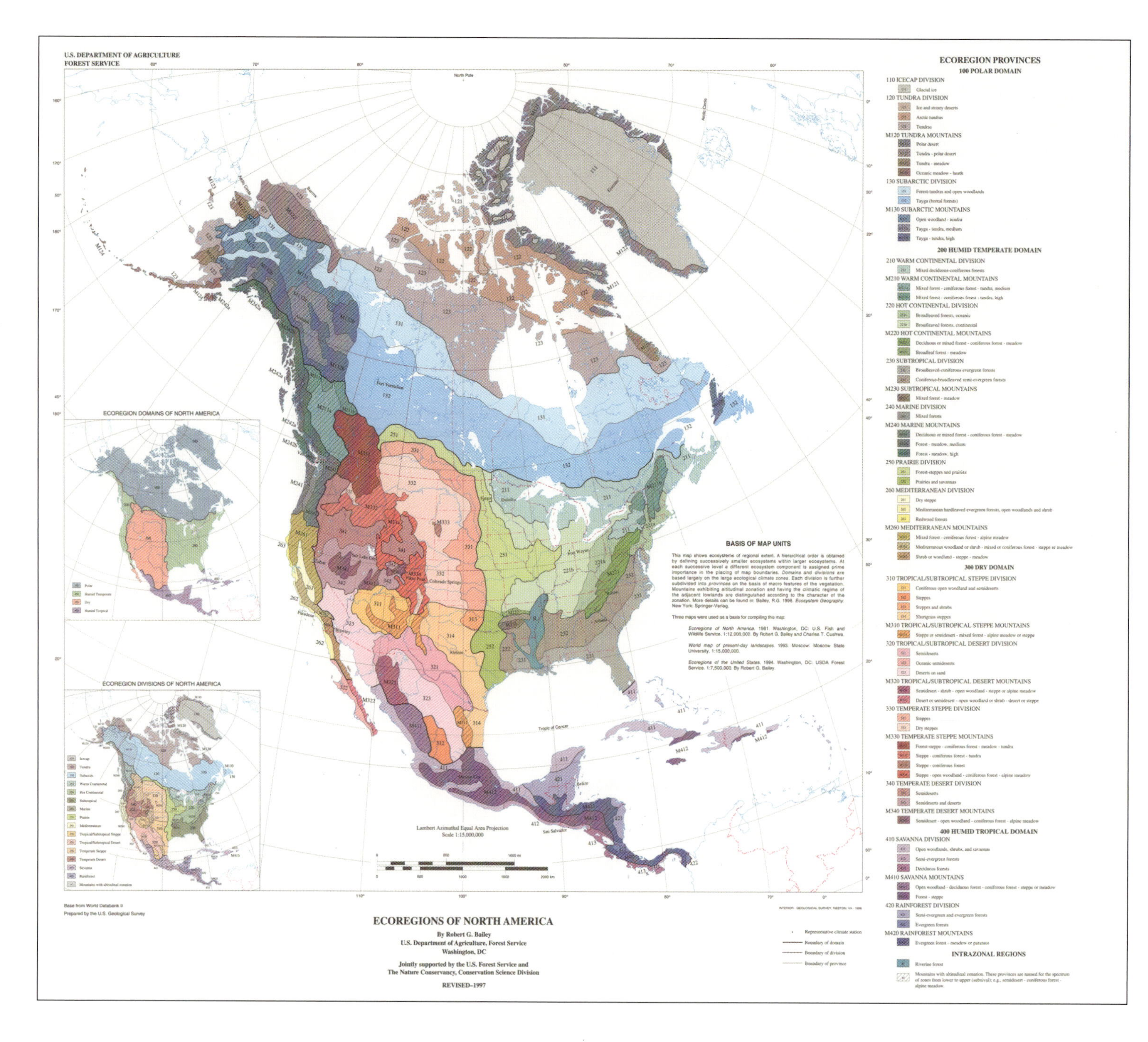

Figure 9.11
Ecoregions of North America. Underlying this very attractive map is, again, a complex, hierarchically organized content system, only the uppermost levels of which are expressed here. Map dimensions (WxH): 34in. x 31in.; scale 1:15,000,000.

Source: USDA National Forest Service, 1997.

admitted kinship with this last one. In his book *Ecosystem Geography*,[12] Bailey explains how domains and divisions "are based mostly on the large ecological climatic zones" and that divisions "are further subdivided into provinces on the basis of macrofeatures of the vegetation." In Bailey's system, the hierarchically scaled domains, divisions, and provinces constitute the macroscale level of the classification based mostly on macroclimatic differentiation. The subdivision of provinces into *landscape mosaics* is the mesoscale level of the classification and is based on landform differentiation. The further subdivision of landscape mosaics into *sites*, the microscale level of the classification, is based on edaphic-topoclimatic differentiation. The landscape mosaics (which Bailey also calls *mesoecosystems* or simply landscapes) are composed of linked sites and are "made up of spatially contiguous sites distinguished by material and energy exchange between them."[13] Landscape mosaics range in size from ten to several thousand square kilometers. He gives as an example Death Valley in California, where the mesoscale components of the mosaic are ranges and basins and the microscale sites are individual slopes within the mountain ranges.

This doesn't mean that Bailey's classification system can only be comprehended as a top-down structure—in contrast to the bottom-up geology and soils classifications—but he is very clear in stating that it *must* be created that way. Once created, however, this might also be regarded from the bottom up. After all, landscape mosaics consist of sites, and provinces consist of landscape mosaics, and divisions consist of provinces, and domains consist of divisions. This must be created from the top down because that is "the only way to ensure perfect spatial registration from any level to the one above." Still, any well-conceived and integral hierarchical system can be "read" from the bottom up as well as from the top down, and that is a fundamental test of its integrity and viability—hence the need for perfect spatial registration.

A well-conceived systemic hierarchy will likely appear simple in concept—and it should—but in principle and practice these can be far from simple. This one is no exception. In a presentation to a U.S. Army workshop in 2006,[14] Bailey enumerates no fewer than twenty essential principles to be observed in delineating ecosystem boundaries at various scales. These are as follows:

1. The series of ecoregions should express the changing nature of the climate over large areas.
2. Boundaries of ecoregions should coincide with certain climatic parameters.
3. Fine-scale climatic variations can be used to delineate smaller ecological regions.
4. Boundaries should capture the effect of mountains on climate.
5. A uniform pattern of mountain zonation is repeated over a climatic zone, which is the basic element in regionalizing mountain territories.
6. Ecoregional boundaries should delineate groups of upland sites with similar characteristics.
7. The mosaic of ecosystems found in major transitional zones (ecotones) should be delineated as separate ecoregions.
8. Context is often as important as content in mapping ecological regions.
9. Because subsystems can be understood only within the context of the whole, a classification of ecoregions begins with the largest units and successively subdivides them. *This is why higher level systems cannot be mapped (or refined) by aggregating lower level systems.*
10. The factors used to recognize ecoregions should be relatively stable.
11. Boundaries should circumscribe large, contiguous areas.
12. Potential vegetation, in contrast to actual, or real, vegetation, is useful in capturing ecological regions.
13. An understanding of the relationships between succession on identical landform positions in different climates is useful for establishing meaningful ecological regions.
14. Geologic factors may modify zonal boundaries.
15. Establishing a specific hierarchy of ecological boundaries should be based on understanding the formative processes that operate to differentiate ecoregions at various scales.
16. Criteria for setting ecoregion boundaries should be explicit in how ecoregions are identified on the basis of comparable likenesses and differences.
17. The limits of geographic ranges of species and races of plants and animals are not fully satisfactory criteria for determining the boundaries of ecoregions.
18. Ecoregions should have greater ecological relevance than large physiographic land units, or landform regions.
19. Ecoregion boundaries should have greater ecological relevance than watersheds.
20. The boundaries of ecoregions emerge from the study of spatial coincidences, patterning, and relationships of climate, vegetation, soil, and landform.

And at the end of the presentation there is a caveat that reads, "Finally, there is a limitation to all this. Understanding continental systems requires a grasp of the enormous influence that ocean systems exert on terrestrial climatic patterns, and thus the character and distribution of continental ecoregions."

One can imagine military eyes having long since glazed over at this point. Clearly, this is not so simple, even if it may have seemed so at first glance. On the face of it, the workings of *Ecoregions of North America* are not very different from the workings of our *General Soil Map of North Carolina*; it is an expression of a three-level, hierarchically organized content plane, with its uppermost level the ecoregion domains, the next down the ecoregion divisions, and the lowermost the ecoregion provinces. And again, there's more lurking below that unexpressed owing, inevitably, to the scale of the presentation. Here, those unexpressed still lower levels of latent content are the landscape mosaics and the individual sites, as was the case for soil series in our soils map, or individual geologic occurrences in our geology map. Without a doubt, this is a map that can only be really understood by penetrating and comprehending the vertical dimension of its multitiered content and not just the horizontal dimension of its spatial expression in the map.

But just what is this exactly? On the face of it, this appears to be another classification of some sort. Bailey seems to prefer the term "boundary delineation," but those boundaries delineate preformed content elements in some classification of the natural world, like Tundra or Tayga or Broadleaved Forest. Still, there has to be a natural system, or systems, behind all this, and Bailey is explicit regarding what he believes those are. At the uppermost level there is a system of climate; at a somewhat lower level there is a system of vegetation; and below that there are systems of topography and then soils and microclimate. Why are these different criteria the most meaningful at their respective scales, and why place climate at the top and topography near the bottom? After all, climate is hardly an independent entry point to all this, but the outcome of other geographic factors such as latitude, continental position, and altitude (this last falling under the rubric of topography) and all of these factors are interdependent and interactive anyway.

Interdependent and interactive. That's the essence of natural systems, which we can now imagine roiling and toiling and conjoining beneath the face of this map. Can all this really be so neatly teased apart? Of course not, but it's not the natural world that's being deconstructed here but our *concepts* of the natural world—climate, vegetation, topography, and so on. Is Bailey's approach the only possible approach? We seriously doubt that, but it does appear to make some sense of something that we're compelled to try to make sense of, even if that's a rather audacious and maybe capricious notion. And it does make for a great looking map.

Are these maps just guesses on paper?

Finally here we have a map of . . . well, of what exactly? The title says *Plant Associations of the Chattooga River Basin* (figure 9.12). The map is quite simple in layout. In the upper left are the map's title, scale bars

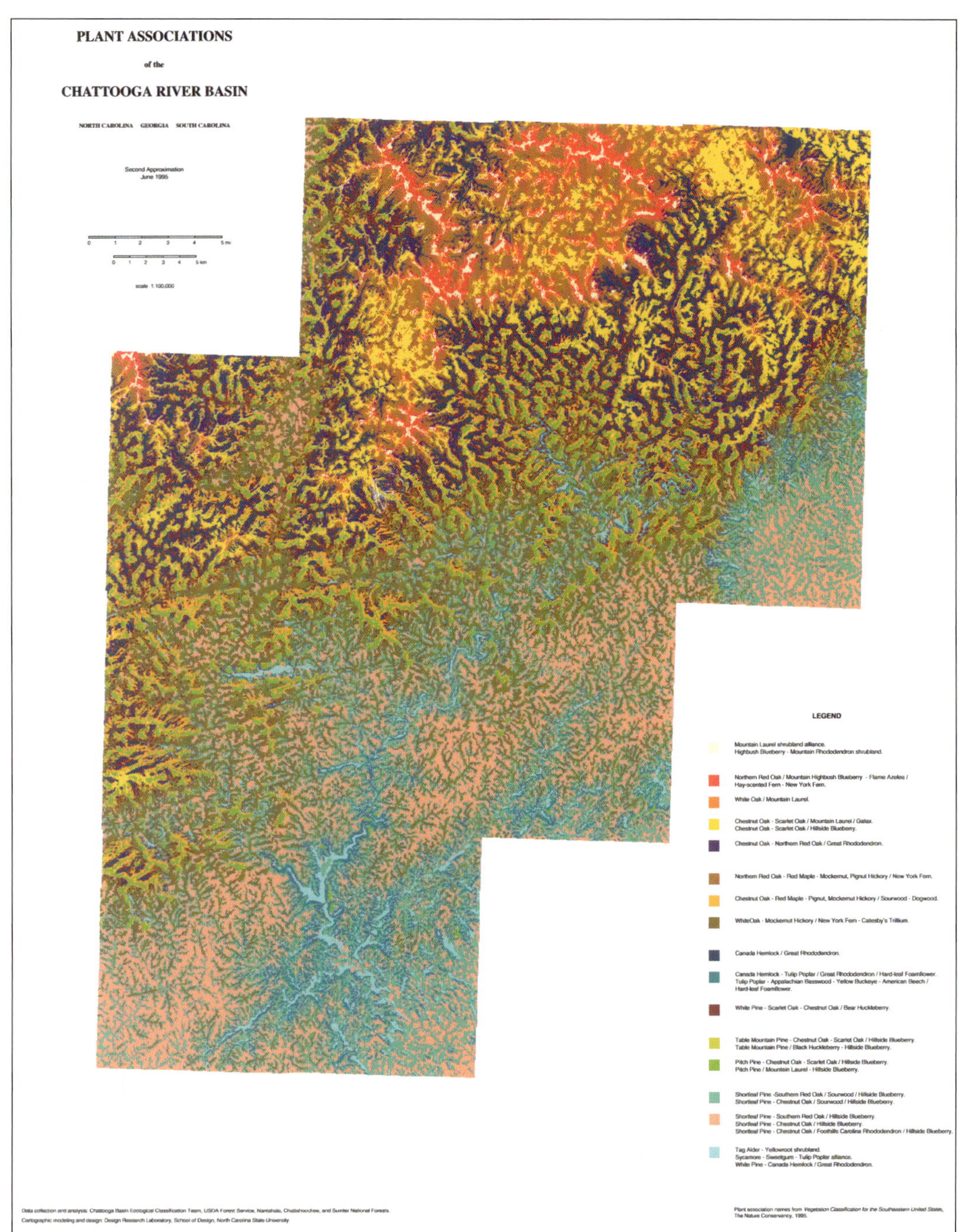

Figure 9.12
Plant Associations of the Chattooga River Basin. This is not a map of what is there, but of what is potentially there given the absence of natural or human disturbance. It is, in essence, a map of predictions and, as such, is inherently and inevitably speculative. Map dimensions (WxH): 24in. x 30in.; scale 1:100,000.

Source: John Fels and Steve Simon, 1995. Courtesy of the USDA National Forest Service.

and numerical scale, and a statement indicating that this version is the second approximation published in June 1995. In the lower right is a legend consisting of sixteen colored blocks, and a statement indicating that "Plant association names [are] from Vegetation Classification for the Southeastern United States, The Nature Conservancy, 1995." In the lower left is another brief statement telling us that data collection and analysis were done by the Chattooga Basin Ecological Classification Team, USDA Forest Service, and that cartographic modeling and design were done at North Carolina State University.

The legend is organized in six groupings with one to five members each. The first grouping consists of five color blocks for plant associations found mainly on upper elevation ridges, slopes, and drainages, from Mountain Laurel shrubland alliance to Chestnut Oak—Northern Red Oak/Great Rhododendron. The second grouping of three includes generally mid-elevation associations, from Northern Red Oak—Red Maple—Mockernut, Pignut Hickory/New York Fern to White Oak—Mockernut Hickory/New York Fern—Catesby's Trillium. The third grouping includes three Needleleaf-dominant associations, from Canada Hemlock/Great Rhododendron to White Pine—Scarlet Oak—Chestnut Oak/Bear Huckleberry. The next grouping of two is composed of Table Mountain Pine– and Pitch Pine–dominated associations typically found in the Chattooga Basin on lower elevation ridges and slopes, and the next grouping (also of two) includes Shortleaf Pine-dominated associations of the relatively dry, lower elevation uplands. The last legend block incorporates three associations common to riparian environments, from

Figure 9.13
A shaded relief image of the Chattooga Basin. The patterns of landforms visible here are strongly reflected in the map of potential vegetation.
Digital rendering by John Fels, 1995, from USGS DEM data.

Tag Alder—Yellowroot shrubland to White Pine—Canada Hemlock/ Great Rhododendron.[15]

Immediately apparent in the map, manifest in the patterns of vegetation, are the complex patterns of terrain typical in the region, from the red of the Northern Red Oak–dominated association on the uppermost slopes to the medium blue of the Canada Hemlock—Tulip Poplar–dominated association of the lower riverine and gorge slopes to the light blue of the Tag Alder, Sycamore, and White Pine—Canada hemlock–dominated associations of riparian settings. It is as if the terrain speaks of itself here *through* the various expressions of vegetation. And, in fact, it does, and for a very good reason. This is not a map of existing vegetation—that could have been much more easily obtained using remotely sensed data and image analysis and classification techniques. This is a map of *potential natural* vegetation. "Natural" means here, among other things, that the map does not include a category for agriculture, which is prevalent in the lower Chattooga Valley. "Potential vegetation" means that the map shows not just what is presently there but what in greatest probability could be expected to be there if the land were left undisturbed for a sufficient length of time (in the same sense as Bailey's point number 12 above).

This is predictive vegetation mapping (or PVM), a relatively new enterprise which, in the few decades of its existence, has seen its few practitioners employ a variety of analysis and modeling techniques, and computer simulations, to produce cartographic displays of just what vegetation (from plant species to plant communities—plant associations are somewhere in between) would most likely occur, or the probability or probable abundance of their occurrence, at various locations in the landscape.[16]

The technique employed in creating this vegetation map was *discriminant analysis*, which uses statistical methods to discriminate among a set of dependent variables (here vegetation types) on the basis of a set of independent variables (here a variety of terrain characterizations). One outcome of this is a *discriminant function* for each dependent variable (the plant associations). The discriminant function is a linear polynomial equation, which yields a value akin to probability for the dependent variable using values input for the independent variables. The resulting values for the dependent variables at any given location can then be used to determine just which of them is most likely to occur there.

In this effort, values for the independent, or predictor, variables were derived for more than two million data points in a raster digital elevation model (DEM) with a resolution of thirty meters. These variables included the original elevation values and derived values for slope steepness, slope aspect, concave, convex, and net slope curvature, terrain shape index and landform index (after McNab 1989 and 1993),[17] and landscape (slope) position, among others. In parallel with this phase of *digital terrain modeling*, there were two summers of field surveys in which existing vegetation was recorded at hundreds of randomly assigned locations. The observed vegetation types, together with the derived terrain variables for each of those locations, were employed in developing the discriminant functions, which were then applied to all data points to simulate probabilities for potential vegetation and ultimately produce a map of potential vegetation. Accuracy assessment was done by comparing model predictions with vegetation observed in the field at another set of randomly assigned locations.

Alright, that may be more than most readers really wanted to know, but it's essential to understanding the logic of this map. In terrain of significant relief, terrain characteristics such as elevation, slope, aspect, form, and position can have profound influences on vegetation composition at any and all locations. This is an indirect causality. Terrain characteristics affect factors such as microclimate, solar insolation, moisture regime, soil development, and nutrient availability, and those factors in turn affect potential vegetation. This notion (which we now pretty much take for granted) was first put forward by R. H. Whittaker in his seminal 1956 monograph, "Vegetation of the Great Smoky Mountains." Whittaker proposed that observed differences in vegetation at different locations in Great Smoky Mountains National Park were manifestations of species' responses to variations along *environmental gradients*. Whittaker identified two fundamental environmental gradients, temperature (based on elevation) and moisture, but we can now demonstrate a third influential nutrient gradient and possibly a fourth gradient of solar insolation. Some species prefer (which is to say, compete better on) warmer sites and some prefer cooler sites; some prefer drier sites and some prefer more moist sites; some require deep, nutrient-rich soils, while others can thrive on the poorest soils; some require a great deal of sunlight, while others are much more shade-tolerant.

And not only are these environmental gradients spatially complex, but they also interact with one another and are not independent of one another. The process of *gradient analysis*, as Whittaker called it, is predicated on two fundamental assumptions: that observed vegetation (barring natural or human disturbance, of course) is a manifestation of species' responses to underlying environmental gradients, and that basic environmental gradients (which cannot be directly measured over large areas) can be approximated by surrogate terrain variables that can be measured using digital terrain models.[18]

That might sound like a pretty big leap of faith and, whatever else it is, it certainly is that. Is this a map of what *could be* at any given location, or what *should be* there, or . . . just what? To say that this is a map of what could be there implies that it *could not be* elsewhere unless the map indicates that it is (we can't have more than one association at the same location), and that doesn't make any more sense than saying that a pin oak cannot be found outside its pink blob in the Golden Guide. And to say that this is a map of what should be there makes even less sense and unmasks the hubris inherent in any such conjectural mapping. (It should be there because we say so!) We know that this is not a map of what is there any more than a range map is but, like the range map, it doesn't explicitly admit that.

If we were to visit a site that this map identifies as, say, a "Northern Red Oak–dominated association," there's absolutely no guarantee that that's what we would find there. Of course, we wouldn't find palm trees or saguaro cacti, but we might find something other than what this map says we'll find. Maybe the site was logged or burned and is now in a successional phase. Maybe there is some anomaly of soils or geology not amenable to northern red oak. Maybe northern red oak was out-competed there by another similarly niched species or maybe its potentials are just unexpressed for reasons we can't possibly comprehend. A map like this can't account for chance.

And we're not really talking about northern red oak after all but about the Northern Red Oak/Mountain Highbush Blueberry—Flame Azalea/Hay-Scented Fern—New York Fern Plant association. Still, as with species, the vaporous notions of range and habitat are lurking here. The map in our Golden Guide says that the range of northern red oak is roughly from Maine to northern Mississippi (sans Coastal Plain) and westward to the Great Plains. The Chattooga River basin is clearly within that range, and we couldn't have a Northern Red Oak–dominated association if we didn't have any northern red oak in the first place. But the species is hardly ubiquitous in the Chattooga basin. On the contrary, this map confines that association to the steepest high-elevation slopes—in effect, the "habitat" of northern red oak within this part of its range. Does that mean we won't find northern red oak anywhere else? No. (Fels has a few on his property in Raleigh.) We just won't find the Northern Red Oak/Mountain Highbush Blueberry—Flame Azalea/Hay-Scented Fern—New York Fern plant association.

Or maybe we can't say that either. *Won't find* sounds too deterministic. Maybe we should say that we *shouldn't expect to find or are unlikely to find*. This is all beginning to make our heads swim, so let's think about just why that Northern Red Oak association got to be where it is in this map. (There was a reason for all that technical discussion earlier.) It got to be where it is because its discriminant function produced a higher score at those locations than the discriminant functions of the other plant associations—maybe not by much, but *this map has to choose.*

So what is this a map of *really*? Is it just a topographic map masquerading as a vegetation map? That's probably too simple an answer, but we find here a world deconstructed in terms of topography (figure 9.13) and reconstructed in terms of vegetation. When it comes down to it, this is really a map of—which is to say, spatialization of—a statistical function or, more precisely, twenty-eight statistical functions, the workings of each being visible only when it "out-competes" all the others (just as with the plant associations themselves). This is a map of predictions and those predictions are, in essence, best guesses, and some of those best guesses are better than others. We could illustrate that (and Fels has done this elsewhere) by making another map of the certainty of the predictions—something like the ratio of the highest to second-highest discriminant scores at each location. A relatively high number will indicate a relatively high degree of certainty. A number approaching one will suggest that we may as well flip a coin or, in more scientific parlance, that an ecotone has probably been revealed. Probably. This is all probabilistic, and there is no small debate among natural scientists as to whether a simulation approach such as this has any merit at all compared to traditional empirical observation. But how certain are those map postings that presumably are based on empirical observation? How certain are the ranges of those tree species or avian species? How certain are the boundaries of those geologic units or soil associations or ecoregions? No approach, GIS-based or field-based, can guarantee certainty in mapping large areas. There is always a leap of faith required.

In the end, we really *can't* map natural systems but only our conceptualizations of them, and that requires a lot of assumptions and a lot of hubris and, yes, a lot of guesswork to satisfy the cartographer's compulsion to draw hard-and-fast lines. The natural world is a very fuzzy concept in the first place, but no one wants a fuzzy map; such a map would be self effacing, even apologetic, and lacking the self-assumed credibility and authority we expect from maps. It's all probabilistic; it's all about making a best guess. Is this just a sophisticated way of groping in the dark? Maybe it is. Does this map have any right to present what it does as seemingly firm and certain fact? Maybe it doesn't . . . but that's what maps do.

Notes

1. Bull and Farrand, *The Audubon Society Field Guide to North American Birds, Eastern Region* (New York: Alfred A. Knopf, 1977), 558.

2. For some idea of just how complex the sciences of geology and soils are, see any good introductory textbook, like Strahler's *Physical Geology* (New York: Harper and Row, 1981) or Buol et al., *Soil Genesis and Classification* (Ames, Iowa: Iowa State University Press, 1989).

3. We discussed this idea in Wood and Fels, "Designs on Signs" and Wood, *Power of Maps.*

4. For an illuminating discussion of these soil forming factors, see Buol et al., *Soil Genesis and Classification*, 135–88.

5. Wood and Fels, "Designs on Signs" and Wood, *Power of Maps.*

6. Ibid.

7. See, for example, Robinson, et al., *Elements of Cartography.*

8. N. M. Fenneman and D. W. Johnson, *Physical Divisions of the United States* (Reston: USGS, 1946).

9. P. B. King and H. M. Beikman, *Geologic Map of the United State*s (Reston: USGS, 1974).

10. Bailey and Cushwa, *Ecoregions of North America* (Washington: USDA Forest Service, 1981); Moscow State University, *World map of present-day landscapes* (Moscow: Moscow State University, 1993); Bailey, *Ecoregions of the United States* (Washington: USDA Forest Service, 1994).

11. Fenneman, *Physiographic Divisions*; Merriam, *Life zones*; Küchler, *Potential natural vegetation*; Köppen, Grundriss der Klimakunde; and Trewartha, *Introduction to Climate.*

12. This book includes, in a pocket on the inside rear cover, two map plates: "Ecoregions of the Continents" and "Ecoregions of the Oceans." There is also a separately issued poster, *Ecosystem Geography*, which summarizes some of the main points of the book.

13. Ibid., 78, 80, 25.

14. Bailey, "Ecoregion Mapping and Boundaries."

15. With this protocol, replicated here exactly as in the map, the slash separates dominant overstory species (left) from dominant understory species (right) for each plant association. If more than one species is codominant at either level, then the main ones are listed, separated by dashes. The order of dominance or abundance at either level is from left to right. If there is a second slash and third species entry, then that entry identifies prevalent species in the lowermost (herb) layer.

16. For a comprehensive survey of the topic, see Franklin, "Predictive vegetation mapping."

17. McNab, "Terrain shape index" and "A topographic index."

18 This topic is pursued exhaustively in Fels's PhD dissertation, *Modeling and Mapping Potential Vegetation.*

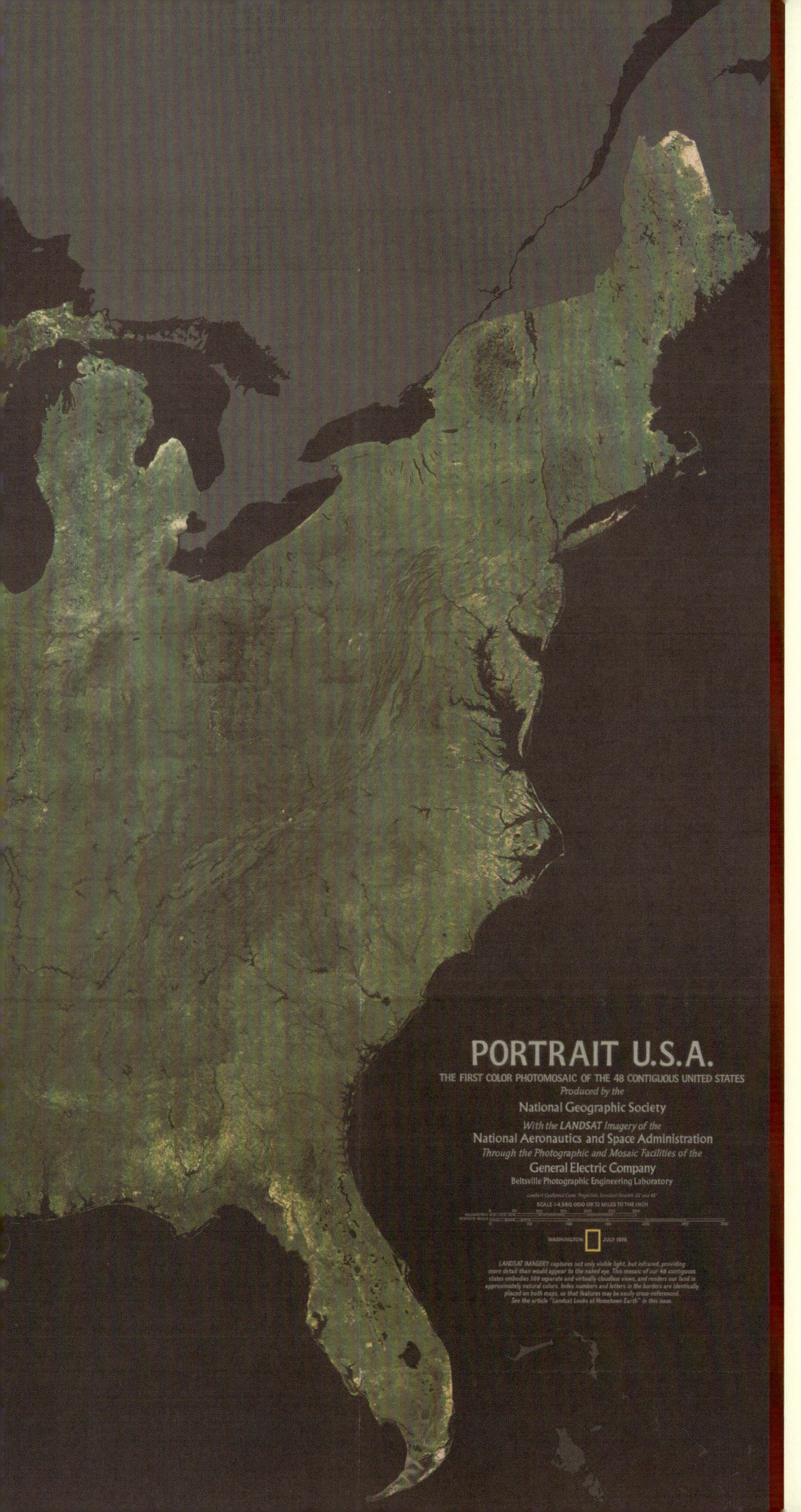

Ten

Nature as mystery

Map courtesy of Cartographic/National Geographic Image Collection.

Nature as mystery

The paradigmatic nature, the nature of science, is a nature known from within. It is a nature of relationships exposed, of methods uncovered, of secrets revealed. Or so it is claimed.[1] But there is another nature of science, one known from without. Or so it is claimed. As we shall see, this is less a nature of science than a nature of technology, and that by its nature it is unknown. It is a nature of unfathomable relationships, of methods merely guessed at, of secrets kept.

It is the nature of the earth seen from space.

Portrait U.S.A.

Here, for example, is "Portrait U.S.A" (figures 10.1 and 10.2).

Or, to give it its full title, "Portrait U.S.A./The First Color Photomosaic of the 48 Contiguous United States/Produced by the National Geographic Society/With the LANDSAT Imagery of the/National Aeronautics and Space Administration/Through the Photographic and Mosaic Facilities of the/General Electric Company/Beltsville Photographic Engineering Laboratory."

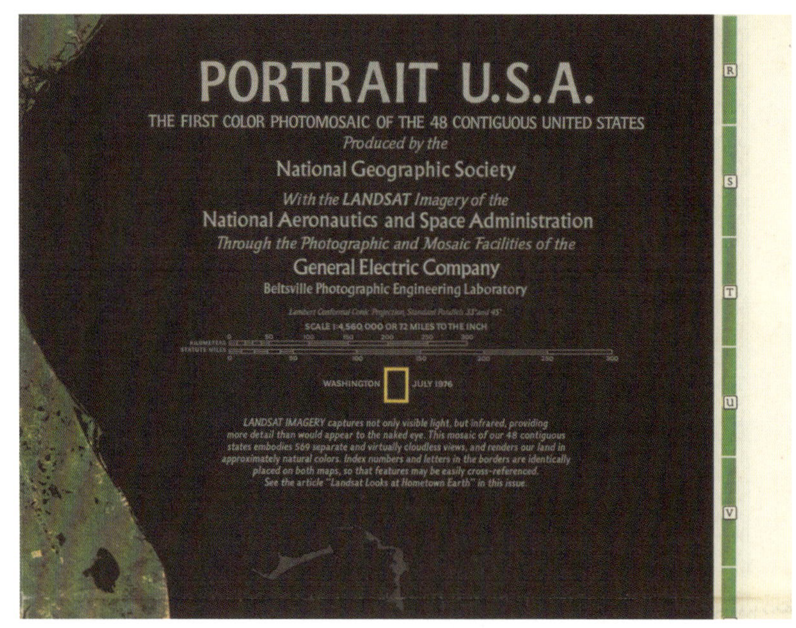

Figure 10.1
Cover of "Portrait U.S.A.," supplement to *National Geographic*, July 1976.
Figures 10.1–10.3 courtesy of Cartographic/National Geographic Image Collection.

The weight of objectifying resources is palpable. Beyond that of the National Geographic Society, NASA, and General Electric—holy concordance of educational nonprofit, government agency, and blue-chip corporation—we have that implied by *imagery, facilities, engineering, laboratory,* all this, together with the usual armamentarium of the projection's name (shored up by an indication of the standard parallels) and the scale (in four forms), on the cover fold in a silvery gray against a matte black background. The *National Geographic* doesn't get any more serious.

All this, on the one hand. On the other, *portrait.*

What is a portrait that it needs to be buffered by this massive show of authority? It is a drawing, a photograph, or a painting, of somebody, or of somebody's face. By extension, it's any close likeness of one thing to another; but the Latin roots are all about *exposing,* about *drawing forth,* about *revealing.* A portrait is not a snapshot. It is not a candid. It is not a mug shot. It is not a study, a draft, or a sketch. Even Ingres's portrait drawings, famously done at high speed, took a day to make.[2] Many photographic portrait sittings take this long. Most portraits take much longer. What is achieved with all this effort?

The living image, that's what's achieved.

A portrait is the very picture. It is a dead ringer, the spitting image. What does a portrait reveal? It reveals an essence. It unveils the real.

"Portrait U.S.A." *begs* to be taken as a portrait. We know it calls itself a portrait, but it observes other conventions of portraiture as well. Except for the (comparatively) discreet title block there is, in contradistinction to every other National Geographic map we've looked at, no other type.[3] Then, as though the U.S. were the sitter's face, it is subtly and richly rendered while, as though it were the sitter's clothing, the rest of the continent is given in a smooth, simple gray with the oceans—background—in black. The conventional Geographic-style political map is on the flip side (figure 10.3), with its pastel state borders, its pale blue oceans, inset maps, and gray scrim of type. Because their scales are identical, comparison is unavoidable and the differences between the maps reinforce the portrait-like qualities of the Portrait.

"Portrait U.S.A." dates from July 1976. Bannered in red across the top of the issue of the *Geographic* in which the map was enclosed is a piece of the epimap. This reads, "Supplement: The United States, plus a Bicentennial Portrait of the 'Lower 48'." "Portrait U.S.A." was a bicentennial gift to the nation. Red, white, and blue bunting waving across a photograph of a bald eagle proclaims, "This Land of Ours." The whole issue was devoted to the United States. An entire article, "A Satellite Makes a Coast-to-Coast Picture . . . ," was devoted to the making of the Portrait. A following article, ". . . and Becomes a Spectacular New Tool for Wildlife Research," was about an extension of the Portrait's technology into wildlife conservation (the technology made portraits but it was also useful!). The entire issue should probably be seen as the Portrait's epimap (figure 10.4).

Today, mosaics of the United States, even of the entire world, are commonplace, but thirty years ago they were sufficiently novel that this "First Color Photomosaic" made a splendid bicentennial salute to the nation. Computers may have been involved in the initial processing of the satellite signal, but in general the salute was a painstaking electro-optical collage of 569 "separate and virtually cloudless views" selected from 30,000 images taken during 3,500 orbits in the late spring, summer, and early fall between 1972 and 1975. These

Figure 10.2
"Portrait U.S.A" unfolded in all its full glory. Map dimensions (WxH): 42in. x 29in.

were printed through filters onto color film in the so-called false-color hues common to infrared photography. It took technicians four months to piece together portions of images into the sixteen regional mosaics—"almost 8,000 color prints contributed to the best possible color match and almost invisible join lines"—which were then rephotographed with a huge custom-made camera before being assembled into the final ten-by-sixteen foot photomosaic. Barry Bishop's "authoritative" article on the Portrait opened with a photograph of a General Electric scientist in front of this "billboard-size mural" holding what looks like a jar of paint in one hand and with the other, "blend[ing] join lines of two of the color prints."[4] In the process of converting the photomosaic into the Portrait, the infrared coloration was converted into "colors closer to those the eye would see," thus rendering "our land in approximately natural colors."

Ambiguity pervades this presentation. We have computers, but also collage. We have 30,000 images, but have selected just 569. We have 8,000 color prints, but technicians piecing them together. We have almost invisible join lines, but blend them together by hand. We have satellite measurements of radiation, but convert them into colors only approximately natural. On the one hand, there is science: satellites, scanning mirrors, banks of detectors, dish antennas, computer processing, magnetic tape, 569 images, 3,500 orbits, 8,000 color prints, 30,000 images, billions of pixels. On the other hand, there is art: it sorts, selects, pieces, colors, blends. On the one hand, we have the National Geographic, NASA, General Electric. On the other . . . what?

Is the Portrait a drawing? A photograph? A painting? It is not easy to say. It has elements of drawing. David Hockney has spoken about drawing with a camera,[5] and elements of drawing are also present in the projection. It may not be a photograph, but a photomontage is a "composite picture made from several photographs"[6]—it certainly is that—and David Hockney, again, has been making photo-collage portraits (figure 10.9). Is it a painting? Probably not, but there's that GE scientist painting/blending.[7] Is it a colorized photomontage/collage map?

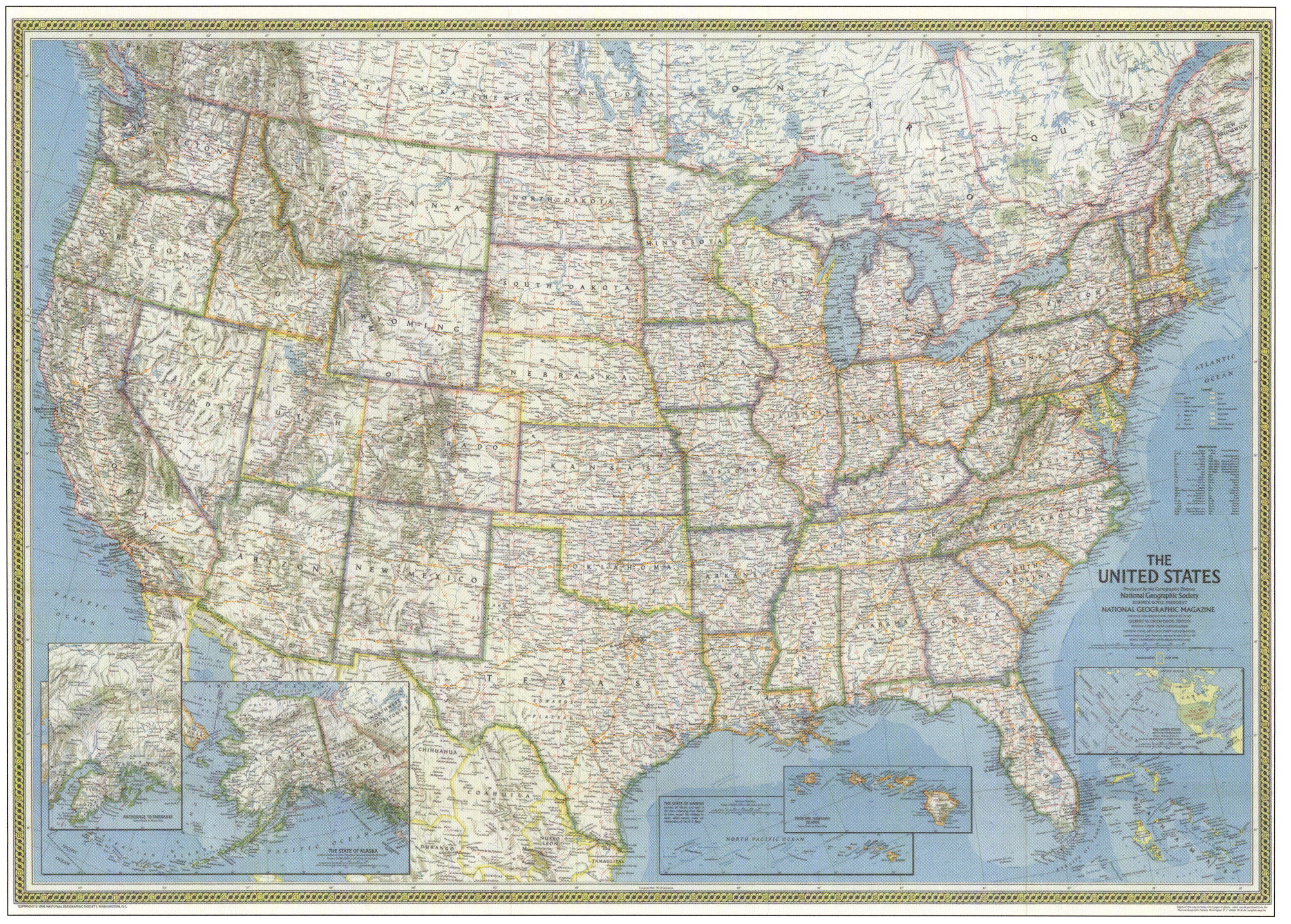

Figure 10.3
"The United States," found on the reverse side of "Portrait U.S.A." This political map of the United States in the National Geographic's house style is not only a map of the United States, but a key to the portrait on its other side. You may ask yourself, "What is that?" of a feature on the portrait. "The United States" can answer your question.

Evading this issue may perhaps be one reason the map was named "Portrait U.S.A.," but others can be imagined. There is a sense, for one thing, in which the country may be said to have sat for its portrait, for three long years in fact, with the "artist" eyeballing it 30,000 times, patiently trying to catch the country in its "unclouded" moments. The "artist" also chose the colors in which the country appears and worked hard to give the "face" its smooth and seamless "skin." The efforts point up the Portrait's interest in appearance, that is, in the way things look to the human eye, and this too in contradistinction to the interest of nearly every other map we've looked at. Most of the maps we've looked at—*most maps*—are about things that can't be "seen," often enough about things that don't even exist. Most maps advance arguments about property lines, next year's school districts, political boundaries, paths of storms and the flight paths of birds, the ranges of trees, geologic strata, ecological regions. Some of these things can be experienced but as conceptual, not perceptual constructs. Their appearance is rarely an issue.

It's true that this is not the first map we've referred to as a portrait. That was "Land of Living Fossils" about which we wrote that it was "as if the continent had sat for, perhaps even commissioned, this lush, gorgeous, almost tactile rendering." Yet that map did have type, and named cities, physical features, and political units; and so it provided, as Eduard Imhof insisted maps must, "the topographic, conceptual, and metric information which one expects from a map."[8] So did the inset physical map on "The Americas" (verso of "Bird Migrations in the Americas"). While we called this "one of those vegetationy-landformy maps," we concluded that it was sufficiently encumbered with map marks to be "less a portrait than as a paramap's caveat to the main map." One of the things, then, that marks "Portrait U.S.A." as a portrait is the limiting of the linguistic code

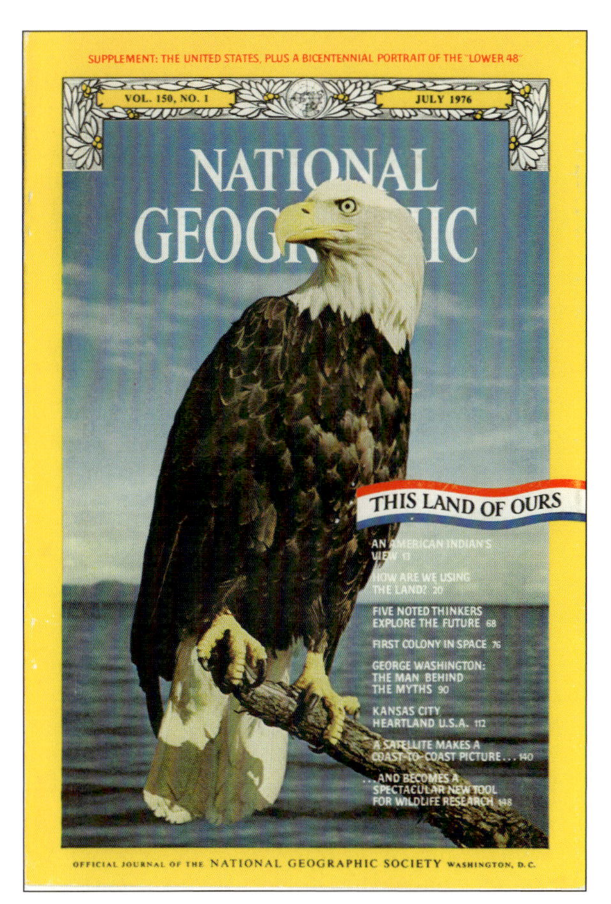

Figure 10.4
Cover of the July 1976 issue of *National Geographic*, for which "Portrait U.S.A." was a supplement. The entire bicentennial issue is devoted to the U.S. and thus should be seen, in its entirety, as the epimap of the Portrait.

Courtesy Norbert Rosing/National Geographic Image Collection.

Figure 10.5
A quintessential portrait: Jan van Eyck's *Cardinal Niccolò Albergati*, 1432, oil on wood. The cardinal's face is subtly and richly rendered, his robe less so, and the setting not at all, being replaced by a contrasting color. The use of optical aids to achieve paintings like this may explain these characteristic differences, but as conventions these differences have endured and characterize all the portrait maps we look at in this chapter.

Jan van Eyck, Cardinal Niccolò Albergati, 1432. Painting, oil on wood.

to the title block, because this shackles the map's ability to raise existence claims on the behalf of rivers and mountains, deserts and plains, cities and states, that is, to raise the conceptual world that maps help bring into being.

The metaphysical truth in appearance

Why make such a map? In a way this is tantamount to asking why make a portrait, a picture, any image? Images function in many ways. They play powerful roles in religious and secular ritual. Images preserve, they commemorate, they disseminate. They arouse. They evoke responses.[9] Fundamental to many of these functions is a belief in the truth of appearance. This amounts to a belief that appearance reveals deep truths, for instance, in the finest portraits, those fired with psychological insight and emotional empathy, truths perhaps inaccessible to other forms of analysis. This is the point we wanted to make when we said that portraits revealed an essence, that they unveiled the real (figure 10.5). For instance, the art historian E. H. Gombrich has written of the portrait heads of ancient Egypt that, "One sees that the sculptor was not trying to flatter his sitter, or to preserve a fleeting impression. He was concerned only with essentials, [and] the observation of nature and the regularity of the whole are so evenly balanced that [the heads] impress us as being lifelike and yet remote and enduring." Speaking of the *Mona Lisa*, Gombrich says that what strikes us is "the amazing degree to which Lisa looks alive. She really seems to look at us and have a mind of her own. Like a living being she seems to change before our eyes and looks a little different every time we come back to her." In Rembrandt's great portraits Gombrich says that, "we feel face to face with real people, we sense their warmth, their need for sympathy and also their loneliness and suffering. Those keen and steady eyes that we know so well from Rembrandt's self-portraits must have been able to look straight into the human heart."[10] It is these kinds of truths that portraits aspire to and that "Portrait U.S.A.," in its own way, promises. The reason for making such a map, then, is because it offers truths about the United States that the political map on its flip side can't, offers truths that can't be reached by the usual methods of science and cartography.[11]

What are these truths? One way to approach this is to ask, if it is a portrait, what kind of portrait it is? Does it flatter the country, or show it warts and all? Is it fanciful or sober and plain? Is it impressionistic or illusionistic? Is it an idealization or is it realistic? The Portrait looks realistic, but a typical definition of realism, "giving a truthful impression of actuality as it appears to the normal human consciousness,"[12] raises doubts whether it could be. One of these doubts would be whether the United States as the Portrait presents it is the sort of thing a normal human consciousness could ever actually experience. The epimap—the July 1976 issue of the *Geographic*—would suggest the United States may not be such a thing. Gilbert Grosvenor's editor's column begins with the following:

> When I first saw the unique and splendid satellite portrait of the 48 states that accompanies this Bicentennial issue, I found myself absorbed in its detail—the clean-cut filigree of Lake Powell and the Grand Canyon; the undulating veins of the Mississippi; the gentle folds of the Appalachians.
>
> Yet there is an even larger picture, an even more compelling one. For this land of endless diversity, of city, forest, field, plain, river, and mountains, of complex ethnic differences, of goals and purposes as various as our people, is as the Founding Fathers had hoped: *E pluribus unum*—out of many, one.

Perhaps, though it's the *pluribus* that preoccupies the *magazine*. The issue opens with a photographic portfolio. Across successive double-page spreads, the Pacific sweeps onto a rocky coast, deer browse in the dappled peace of a Virginia wood, storm clouds ride the Grand Canyon's mottled breadth, bison huddle in a Wyoming blizzard, and autumn leaves brighten the Colorado Rockies. This is not a picture of one, but pictures of many. Six stories follow—we get an American Indian's view, there's a story on "Kansas City: Heartland U.S.A.," one on how we're using the land—before we get to the two stories about the Portrait and wildlife conservation. None of these pages begins to suggest how a normal human consciousness would take in the United States as a whole. But isn't it in fact the function of mapping to bring into discourse what *can't* otherwise appear to the normal human consciousness? What in this case has been annealed from 569 separate satellite views, and smoothed, and blended, and colorized? The mere idea of mapping is unrealistic.[13]

This aside, is there any other sense in which the Portrait gives a "truthful impression of actuality"? Flaunting the apparent richness of its detail, the clean-cut filigree, the undulating veins, the gentle folds, the map strains to elicit a positive response, but *in actuality* . . . there would be clouds. Contrast "Portrait U.S.A." with this image from the National Geographic Society's 1985 *Atlas of North America* (figure 10.6). Though it's captioned "Face of the Continent" , this is not a portrait. It is not a montage, not a collage, not an assemblage of images. It is a snapshot. Though color has been added, the United States here looks nothing like the glowing emerald that it appears to be in "Portrait U.S.A." But then here it's fall, October 9, 1980, in fact, an actual day and not a collage of summer days culled from the summers of three years; and though October 9, 1980, was a remarkably clear day, fog nonetheless mantles the California coast and clouds drift over Texas, east-central Florida, and the Appalachians. Over northern New England and the northern Great Plains the skies are gray. Outside the United States the clouds are dramatically more prominent, as they usually are over the United States,[14] and the clouds invest the image with a swirling dynamism that animates every photograph ever taken of the planet from this altitude. The clouds acknowledge the actuality that the earth has an atmosphere, that this atmosphere is vigorous, frequently violent, and in fact the sole reason the Portrait can be as green as it is.

In actuality there are clouds, half the time it is night, oftentimes there is snow, the green sweeps north in the spring, the brown slips south in the fall, and everywhere are signs of human occupance.[15] There is in fact *no* sense in which "Portrait U.S.A." may be said to give a truthful impression of actuality as it appears to the normal human consciousness. On the one hand, this no more than acknowledges that the Portrait is a map which, as Tom Patterson has recently had occasion to recall, is inescapably an idealization and inherently abstract.[16] But idealization can take many forms, and in selecting to show terrain and vegetation on an imaginary, cloudless, summer day, the Portrait declares its commitment to a particular idealism. Despite the detail (which could suggest an almost illusionistic naturalism), the Portrait's idealism is deeply classical, rooted in both Aristotelian and Platonist practices. In an Aristotelian framework, idealism refers to the practice of "reproducing the best of nature but improving it and perfecting it, by eliminating the inevitable imperfections of particular examples."[17] This is just what the Portrait does, reproducing the best of nature—the U.S.A.—but improving it by painting it exclusively in the hues of summer, and perfecting it by eliminating 29,431 images caught on imperfect, cloudy days. In Platonist, or at least Neo-Platonist thought, idealism refers to works of art that "directly mirror forth the essence of the *Idea*," that unchanging reality supposed to stand behind the imperfect copies accessible to perception.[18] It is this unchanging reality that the Portrait has rendered, the *essential* U.S.A. that endures through the ceaseless permutations of night and day, the seasons, and history.

These are of course the very traits that for Gombrich characterized the Egyptian portrait heads of the Old Kingdom, neither flattery nor the preservation of a fleeting expression, but the solemnity and simplicity of the essentials, and an observation of nature and a regulation of the whole so evenly balanced that the heads seemed lifelike yet remote and enduring. These are also traits Gombrich associated with the Greek sculptors of the Early Classical, when "the artist was not out to imitate a real face with all its imperfections but . . . shaped it out of his knowledge of the human form."[19] Such touchstones of classical idealism—knowledge of form, solemnity, simplicity, lifelike yet remote and enduring—Gombrich finds in the work of later artists too, in Poussin, for example, and Cézanne and, we would like to imagine, in the Portrait as well.[20] *Isn't* this the sense we get from the Portrait, not the momentary turmoil of "The Face of the Continent," but the solemnity and simplicity of the land and its vegetation? Not the ceaseless turning of the year, but an observation of nature regulated (smoothed and colored) to reveal a knowledge of its physical form; lifelike, but caught in an eternal summer; real, but unchanging?

The truth the map gives us, then, is a classical one, an integrated unity, an organic whole, Grosvenor's *unum* in fact, untroubled by, rising above, the *pluribus* of physiography, ethnicity, and goals and purposes "as various as our people." This enduring singleness absorbs them all. This steadfast land, this emerald reality, is as firm in its solid solemnity, is as dignified, is as remote as the bald eagle on the maga-

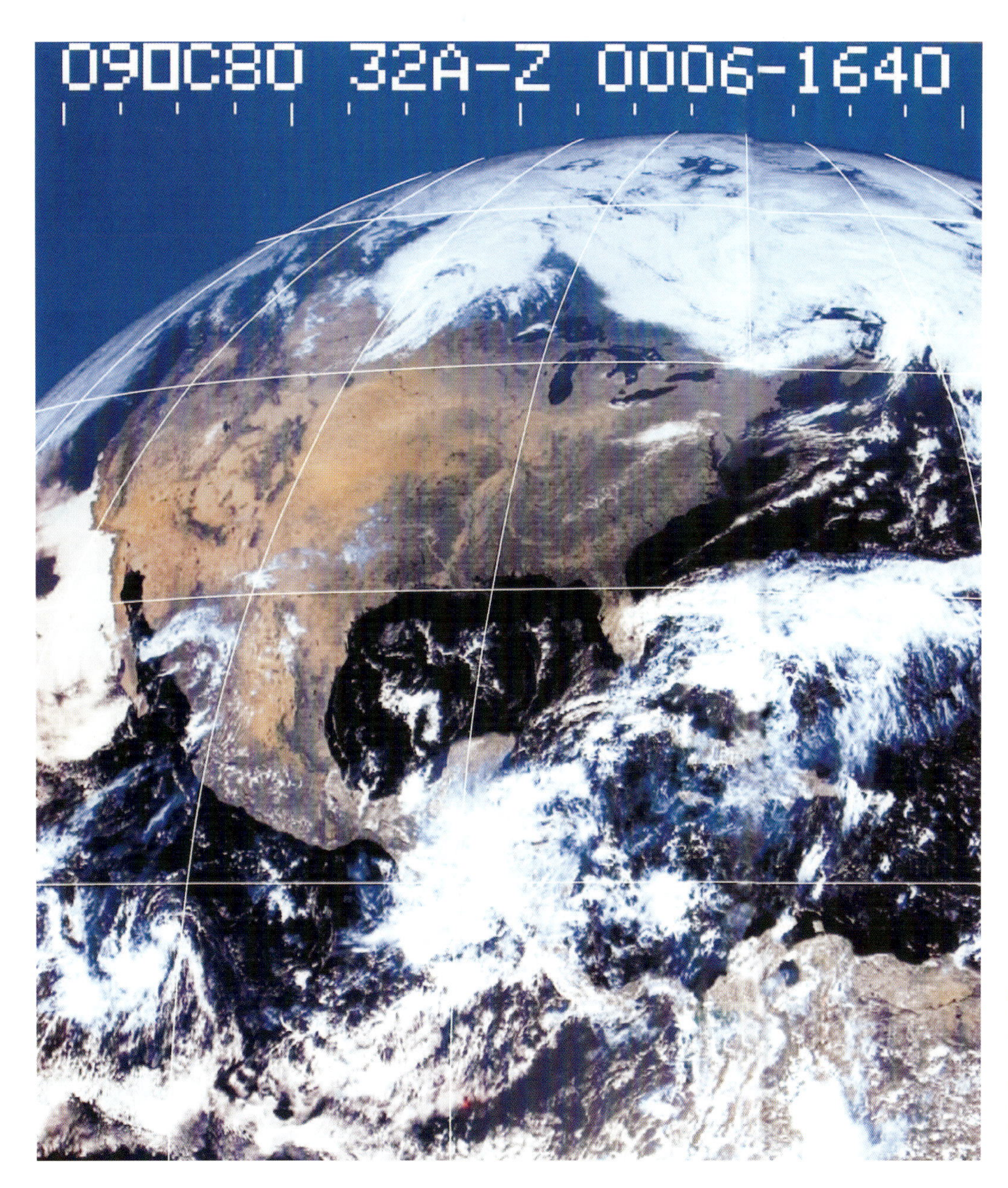

Figure 10.6
"Face of the Continent," a satellite photo of North America, from National Geographic Society's *Atlas of North America: Space Age Portrait of a Continent.* Everything about this image contributes to its sense of urgent dynamism, from the clouds, to the lines of the graticule rushing toward the pole, to the alphanumeric code at the top of the frame.
Courtesy of Ng Maps/National geographic Image Collection.

zine's cover, "living symbol of our land."[21] *The map is the eagle.* Isn't this why "Portrait U.S.A." gives us *not* the U.S.A. but the "Lower 48"; gives us the compact mass of its contiguous parts, not the nation in its fragmentary divisions, its largest state separated from the rest by a foreign country, its island state, fragmentary in itself, by half an ocean? These are included, as a matter of course, on the map's flipside, called simply The United States, where all *fifty* states are displayed in the full absurdity of their historical shapes and sizes, tiny Rhode Island, gargantuan Texas, the idiosyncrasies of the states' shapes in the east, the comparative regularities of their shapes in the west.[22]

The disappearance of history . . .

What is inscribed in these fragmentations, in these idiosyncrasies? History is inscribed in them, the reality of the way we have written ourselves across the land in time, which in the name of an imaginary unity has been all but erased from the Portrait—Alaska gone, Hawaii not there, none of the states demarcated. But then . . . why not? *Nature itself has been edited,* and in precisely the same way, the clouds blown away, winter denied, and fall, and spring, the sweet night obliterated, every trace of time extracted, as if nature—in reality—existed outside time, outside history. With its wealth of detail, its photographic verisimilitude, its escutcheon plated with the names of an educational nonprofit organization, a government agency, and a blue-chip corporation, the Portrait insists on this truth.

David Hockney feels that the advantage of photography is that people believe the photograph: "They believe," Hockney says, "it was made in a certain way, and they believe the photographer was there." He goes on:

> Last Christmas I was in London, and Mark Hatworth-Booth had come round to get me to talk on the tape recorder about Bill Brandt's photographs, specifically those in the

Figure 10.7
"Lenin addresses the troops," Moscow, May 5, 1920. Original photograph, including Trotsky and Kemenev.
Figures 10.7 and 10.8 courtesy of the David King Collection, London.

Figure 10.8
"Lenin addresses the troops," Moscow, May 5, 1920. Collaged and retouched photograph in which Trotsky and Kemenev have been replaced by stairs, an example of what David Hockney refers to as Stalinist collage. (Courtesy of the David King collection, London.)

book *Literary London*. Many of the photographs are very striking—there's one of Top Withens in Yorkshire, which is a very dramatic photograph. I was looking at this for quite a while, and I started asking him some questions about it. I said, there's one thing I've noticed that's strange—there's a very bright light in the sky back there, but the grass in the foreground has also got a very bright light coming towards it here. He must have used a flash of some sort, otherwise where was the light coming from? And he said: ah, the sky is from another negative.

Well, this horrified me, and I suggested this was Stalinist photography. It was a collage, really, but there was no evidence of it being a collage. There's nothing wrong with collage at all, but it should be quite clear that one thing is stuck on top of another. This photograph was not like that, and so people would assume it had been made from a single image. When you can tell that the sky is from another day and yet you pretend that it's not, then I think you can talk about Stalinist photography.[23]

What Hockney is referring to was the habit under Stalin of retouching and restructuring photographs with scalpel and gluepot, to make once important and often famous politicians and other figures disappear.[24] The first and probably most famous example involved the excision of Trotsky and Kamenev from a 1920 photograph of Lenin addressing the troops in Moscow,[25] but this was but one of thousands of similar falsifications (figures 10.7 and 10.8). "With a sharp scalpel an incision could be made along the leading edge of the image of the person or object adjacent to the one who had to be removed. With the help of some glue, the first could simply be stuck down on top of the second. A little paint or ink was then carefully brushed around the edges and background of the picture to hide the joins."[26]

Because the Portrait's reality as a mosaic is spelled out right there under the title—even if it is in minuscule type—it would be going too far to characterize "Portrait U.S.A." as a Stalinist image. It's worth noting, however, that titles are hardly what Hockney had in mind. Hockney was talking about clear evidence *within* the image that one part of it was stuck down on another:

The reason that Stalinism works in photography is that we do believe what is there in front of us. When Trotsky is next to Lenin, and then he's taken out, the picture suggests that Trotsky was not there at all. Painting is not the same. You can paint a picture of Lenin making a speech and never put Trotsky there, as though you never noticed him. But the camera is not like this, and so you're back to the point that what you depict should be in front of you. So these techniques seem deceitful to me.[27]

The difference between photography and painting to which Hockney alludes is founded in the assumption that what appears in a photograph was *present* to the eye of the photographer, or in the case

Figure 10.9
"Place Furstenburg, Paris, August 7, 8, 9, 1985," a David Hockney photo collage. Notice, in contradistinction to the retouched "Lenin Addresses the Troops" of figure 10.8, how obvious it is here that different photographs have been glued down on top of each other. Notice too the way this builds time into the image, not only through its self-evident construction, but in the movements Hockney had to have made in order to take the photographs from which the image is composed (for example, the graffiti that from a fixed perspective would have been *behind* the tree). Note as well that Hockney acknowledges in his very title the three days it took to take these pictures, unlike "Portrait U.S.A." or the other portrait maps we'll examine in this chapter.
Courtesy of David Hockney Number 1 U.S. Trust.

of the Portrait the eye in the sky. It is this presence alone that gives a photograph its evidential quality, and that justifies the Portrait's claim to its name.[28] Signs within the Portrait that acknowledged the collaging would, it must have been imagined, undo this pretense, and thus reduce the image to the status of a map or a painting, in any event, to the realm of the imaginary.

No, for it to really count, the evidence of the collage has to be *in the structure of the image*. It has to be quite clear that one thing is stuck down on top of another, as here in this photo collage of the Place Furstenburg in Paris that Hockney made in 1985.[29] "It's our movement that tells us we're alive," Hockney says, as here it is his evidently shifting gaze, his panning from side to side, and the tilting of his head that animates what in other hands would be a snapshot.[30] In other works, it has been his subjects that have moved, as in his well-known portrait of Billy Wilder where we can follow Wilder strike a match and light a cigar, in what must have been a series of characteristic gestures. In "Fredda Bringing Ann and Me a Cup of Tea Los Angeles April 1985" both Hockney and his subject are in motion.[31] Anne Hoy adds that the movement that interests Hockney

> . . . reveals character or challenges commonplaces of pictorial representation. Both traits of movement are addressed in his photocollage portraiture. As in traditional portraits Hockney's subjects generally sit, in their homes or the artist's studio. But collage obviates the traditional need to distill a characteristic expression and pose into a single image. Collage documents individual fleeting movements of face, hands, and body, then permits a synthesis through choice and juxtaposition of individual prints.[32]

Unstated, though perhaps implied, is the fact that the collage builds time back into the image, as in "Place Furstenburg" where we can practically follow the movement of Hockney's eyes. Piecing the collage together, picture by picture, Hockney felt that he was drawing, that linking the photographs together was about his hand and what his hand was doing, and that the choices he was making were essentially like drawing.[33] With the time built back into the portrait, the static character vanishes—even from an image of an urban square—that "remote and enduring" quality that Gombrich so appreciated in those Egyptian portrait heads.[34] This quality is exchanged for a vibrant immediacy that recalls, in its very different way, the dynamism that animates the "Face of the Continent."

There is no reason a collage of satellite images can't look like a collage. Here, for example, is "Images from Space," a collage the National Geographic published in 1990 in the sixth edition of its *Atlas of the World* (figure 10.10). "A continental mosaic," the caption reads, "begins with a review of hundreds of images to find 20 to 50 suitable scenes, usually from the same season and time of day. These are registered to a base map and overlap at first."[35] The registration keeps "Images from Space" from possessing the sense of drawing that animates Hockney's collages, but then, unlike Hockney, the National Geographic didn't consider this a final product, just a step in the process leading to the usual inert, synoptic view. In the end "computer technicians adjust the data, reducing color variations for a

Figure 10.10
"Images from Space," an unadjusted photomosaic of North America from the *National Geographic Atlas of the World, Sixth Edition.* Its caption notes that, "A continental mosaic begins with a review of hundreds of images to find 20 to 50 suitable scenes . . . These are registered to a base map and overlap at first," fulfilling completely Hockney's requirement for a self-evident collage. Notice how dynamic this image feels, with the photos "pasted down" and the clouds within the photos.
Courtesy of Ng Maps/National Geographic Image Collection.

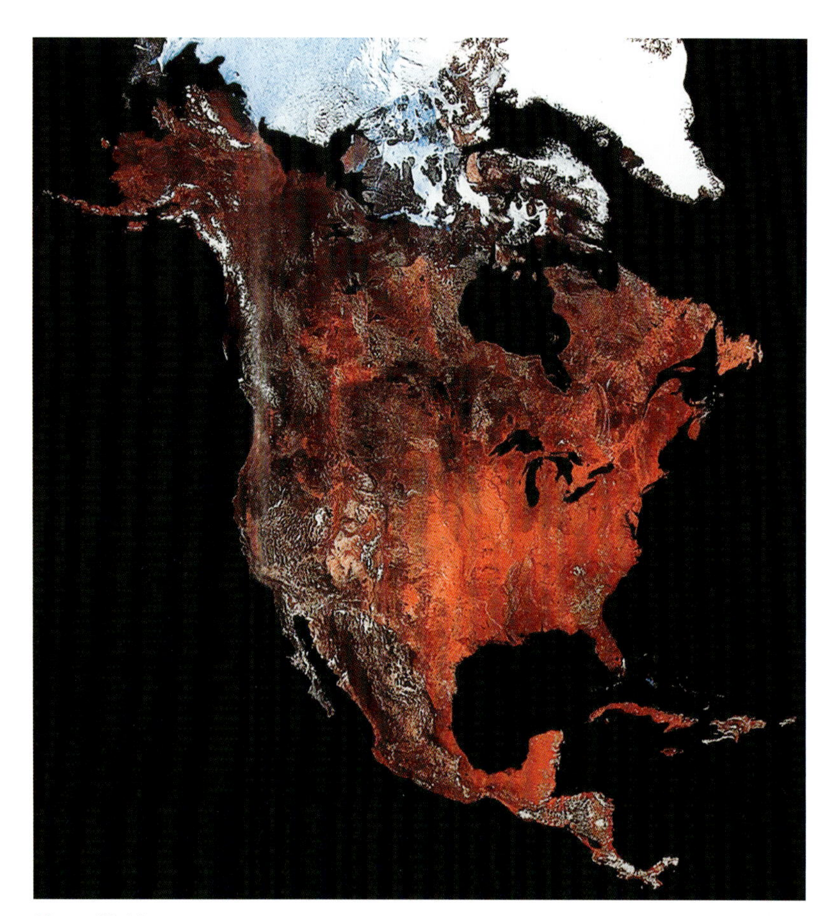

Figure 10.11
"North America," a false-color photomosaic also from the *National Geographic Atlas of the World, Sixth Edition.* This is what resulted from the unadjusted "Images from Space" collage of figure 10.10. Notice, in comparison, how lifeless this appears once computer technicians have adjusted the data and reduced color variation to produce "a coherent image."
Courtesy of General Dynamics.

coherent image," which amounts to the distillation of "a characteristic expression and pose into a single image." Any evidence of collage has been adjusted away.

"North America" (figure 10.11), the result of this process, was printed false-color on black, to establish itself, unarguably, as another portrait.[36] Spread across a full page and a half, it introduces the *Atlas*' section on North America. What exactly was gained by all this trimming and adjusting? Compare "North America" with "Images from Space" and "Face of the Continent." A snapshot, "Face of the Continent" captures a moment of the earth's history in all its excitement and particularity. It may or may not be a map, but the superimposed graticule suggests that it may be, and the adjacent thumbnail, covering the same territory in an identical projection, certainly is. A collage, "Images from Space" captures a bunch of moments, yet thanks to the registration of its images it remains a map despite the unruliness. Because the images have yet to be cleaned up and trimmed, they exude a lot of the same kind of energy the snapshot does (there are lots of clouds) *plus* all of the energy involved in registering and piecing them together (which can be read straight off the collage).

None of this energy survives in "North America," which floats against its deep black background as remote and untouchable as an exquisite jewel on a velvet cloth beneath the glass of a jeweler's counter. With every sign of motion trimmed, smoothed, and blended away,[37] North America comes across as inert as a piece of petrified wood which, with the ferric-oxide tonality of its apparent agatization, it could easily pass for. Something about it seems so precious you're almost afraid to touch it, as you'd be afraid of dropping it if the jeweler let you handle it. As solid as it seems, there's something brittle about it too, something fragile. Some of this has to do with the fineness of the detailing—which shrieks delicacy—but some of it comes from the inky void the continent's suspended in, the way it's just *snapped off* at Colombia, its detachedness, its defenseless isolation. Zoom in to the edges of the continent: there's nothing beyond it. It's all alone.

. . . LEADS TO A NATURE OUTSIDE OF TIME

But North America isn't alone. It doesn't end at the beach. In fact further on in the atlas are maps of the oceans that show a wide Continental Shelf—actually a more or less inundated reach of the Coastal Plain—stretching off into the Atlantic. Those oil platforms in the Gulf? They're standing on this shelf; and the fishing fleets of the Grand Banks are fishing it. There aren't any edges in nature because in nature things are never alone. Nothing is isolated like this. Everything is connected.[38]

Then what's the point of this severing? Of this hard edge, this isolation, this inky blackness that whispers only of outer space? It is, of course, some kind of proposition. We've seen that "Portrait U.S.A.," by wrenching the United States from the grip of history, proposed the existence of an essential U.S.A., one capable of playing the role of the Founding Fathers' *unum* in the *Geographic's* cartographic construction of a coherent U.S. history. Surrounding the states with gray and black helped condense the map's illusionistic detail into a solid, sempiternal mass. The cost was that of wrenching nature from the flow of time. "Portrait U.S.A." gave us a United States sans clouds, sans night, sans seasons, sans almost everything.

Just as sans clouds, sans night, sans seasons, sans almost everything as the Portrait, "North America" can only give us a *continent* outside history. But with no national myth to support, to what end? Only, as "Portrait U.S.A." itself implied, that nature itself exists outside history. In fact, "North America" proposes an idealized nature, one as remote and enduring as Gombrich's portrait heads, nature as the ultimate touchstone of the true, the foundation, the rock upon which everything else *must* be erected. This justifies its location in the *Atlas'* section on North America: a*t the beginning*. Correspondingly, the text opens: "The land is ancient."[39]

The land is ancient. It is ancient because it has endured. It has endured because it *is* the foundation. Pictured here—mapped, by satellite—is a nature that all but directly mirrors forth the essence of the *idea* of nature. Outside time, this is as close to the Platonistic ideal of nature as one could hope to get. This nature could never be threatened by humankind, any more than it could stoop to threaten humankind. This nature is as far beyond spectacle as it is above being cuddled. It is not a nature that can be collected, any more than it can be taken apart to learn how it works. It is a nature than can barely be mapped (indeed, it barely *is* mapped).[40]

Describe, label, name this nature? *Only in your dreams.*

It is precisely the transcendental quality of this—it *has* to be capitalized—Nature that explains the contradiction at the heart of these portrait maps. Maps are loquacious. It is their nature to babble. Maps are collocations of propositions. "*This is there,*" maps say, again and again. On the flipside of "Portrait U.S.A." is a map, "The United States," that is stuffed with *thises*: *this* is a state, *this* is an island, *this* is a river it says (again and again), and furthermore *this* is called Mississippi, and *this* is called Oahu, and *this* is called St. Johns. So dense is it with this linguistic apparatus it barely has ink for the nature that consumes the Portrait: of landform, only a dusting barely visible beneath the type; of vegetation, only swamps. On the pages of the *Atlas* that succeed "North America," map after map is just as garrulous as "The United States" is: *these* are mountains, the maps say, and *this* point we call the North Pole, and *here* is an area affected by acid rain, and *this* is an ocean.

It is the nature of maps to label and name. It is the most potent source of their power to socialize the world. It is the nature of maps, but not of portrait maps. Portrait maps label nothing, they name nothing. Indeed, to what could labels or names attach themselves, to what could they refer? For as they do without type, the portrait maps also dispense with signs. Or at least they dispense with more than one sign, the signified of which is a nature whose signifier is a subtly modulated, endlessly variegated surface. It is a surface that promises an infinitude of signs but that finally offers only itself, whole, intact. If you thought you saw something on it—the Grand Canyon, say, as Grosvenor did on "Portrait U.S.A."—then this is only because *you* provided the content and you carved the signifier from its surface (delimited it, decided where the feature began and ended, extracted an icon: *clean-cut filigree*; and labeled it, gave it a name: *Grand Canyon*). Those intimate with the United States and with North America may be able to do this repeatedly (*and that undulating vein is the Mississippi and those gentle folds are the Appalachians*), but the map itself is dumb. Indeed, only *because* the map is dumb can it speak of nature. Otherwise it would be babbling, like most maps, about no more than this river or that, this or that plain. But nature is a *this* that subsumes *all* these other *thises*.

In effect, portrait maps are wholly constituted of single super-supersigns, super-supersigns that ferociously resist decomposition into supersigns or their constituent elemental signs. As you'll recall from the last chapter, an elemental sign (such as a contour line) is formed whenever an element from the plane of signification (say a brown line) is linked to the content plane (say altitude) by a code. As you'll also recall, supersigns are aggregations of elemental signs that demand to be read as a single syntactic entity, as contour lines demand to be read as a topographic surface. At the synthetic level, again, supersigns enter into alliances in which they offer meaning to each another in the genesis of an embracing geographic icon. With this super-supersign we have a map image which, when integrated with perimap elements at the presentational level (title, legend, scale), congeals into discourse.[41] But in portrait maps it's as though the content plane hovers above the plane of signification like a ghost. It promises but never quite makes contact, and so never manifests elemental signs or, of course, supersigns; nevertheless, it bleeds out at the synthetic level as nature, an indecomposable super-supersign. In the case of "North America" we have red, brown, blue-white, and gray coded for vegetation, desert, ice, and urban.[42] Though forming a bizarre class and surpassingly general, these could work as elemental signs (a mark, red, linked to a concept, vegetation), but on "North America" these signifiers so fade, mix, merge, blend, and grade into each other (for instance, the "brownish red" representing boreal forests in Canada) that they can't be disentangled to sediment elemental signs. Without elemental signs for composing syntactic entities, there can be no supersigns. Yet at the synthetic level the variations in color come together as a super-supersign for nature.[43] For "North America" this is pretty much it, for in common with other portrait maps it lacks much by way of a perimap. ("Portrait U.S.A." is exceptional in adorning itself even with a scale and the name of its projection).

Among other things most portrait maps do without are graticules, grids, indexes, indeed any of the locative apparatus usually exploited to distinguish the sign plane of the map from other sign planes. (With its alphanumeric index, "Portrait U.S.A." is unusual in this regard as well.) But absent this, what kind of posting can our single super-supersign participate in? *The same as any other sign.* When a content element is expressed in the sign plane of the map, it is simultaneously given a location in that sign plane. At that moment, the *this* is no longer *a this* but *that this*. So by being expressed on the sign plane of the map, the portrait maps' contents cease being nature and become American Nature and North American Nature.[44] *This* (nature), "Portrait U.S.A." says, is *there* (the U.S.A.); *this* (nature), "North America" says, is *there* (North America). Since every such proposition is an equivalence, this is simultaneously to say that the U.S.A. is nature and that North America is nature, which amounts to saying (doesn't it?), that the U.S.A. is natural, that North America is natural.

A more straightforward example of Barthes's "naturalization of the cultural" would be hard to imagine.[45]

An unnamed portrait in natural-looking colors

The title page of the *National Geographic Atlas of the World, Sixth Edition* contains—in addition to the title—a "global portrait" printed in "natural-looking colors" on the same black background that "North America" was printed on. It doesn't say it's a "global portrait" on either of the two pages across which it's spread, or on the preceding half-title page where it appears in reduced size printed on white.[46] As with other portrait maps—as with other portraits of any kind—everything that doesn't emanate from the simple appearance of the subject is kept to a minimum. Here this means no title, legend, scale, projection name, north arrow, graticule, grid ticks, index, neat line, or type, in a word, none of the elements usually taken to indicate a map. No wonder they don't call it a map.

They call it a portrait two pages later, in a small block of type that dangles beneath Gilbert Grosvenor's name, the name appended to a column headed "Preface." Here's what it says in that dangling block of type:

> The global portrait on the preceding pages, produced in natural-looking colors, is the first virtually cloud-free satellite image of the planet.
>
> This portrait was created by artist Tom Van Sant and scientist Lloyd Van Warren of NASA's Jet Propulsion Laboratory from visible and infrared data recorded between 1986 and 1989 by National Oceanic and Atmospheric Administration satellites. Orbiting at an altitude of 850 kilometers, the satellites scanned the surface in four-square-kilometer sections, or pixels.
>
> Data from different times of the year were acquired to ensure the best lighting and maximum vegetation. A computer then converted the data into images. Van Sant reviewed the entire world and selected the best data for the final composite image, which comprises 35 million pixels. Geographic and elevation data bases were used to verify and enhance drainage and relief.
>
> Van Sant chose colors that would give a realistic view of the physical world. Gray-brown areas along coasts represent silt discharges of great rivers, algae blooms, or upwellings of cold, deep water.[47]

Little of this need detain us here. As with "Portrait U.S.A." there's the tension between art and science, but here it's explicitly balanced between an artist and a scientist. On the part of science we have the national agencies, the large numbers, the satellite scanners, the computer conversion; on the part of art, the reviewing, the selecting, the assigning of colors. That this was old hat by the time the *Atlas* was published may be gauged by the matter-of-fact description of the process of creation. In 1976, this took an entire eight-page article. Here in 1990, it's reduced to the four paragraphs we've just quoted, and two others that open a *Geographic* blurb for the new atlas. Attached to these is a three-page foldout of the global portrait. There it's headlined, "First-of-a-kind portrait from space."[48]

Indubitably this is a portrait.

And because it is, it gets away with *so* much. Most astonishingly, no title. That is, no name, no descriptive heading, no nothing to say what it is. Two pages later, with no heading of their own, four small, italicized paragraphs. Every other map in this atlas of hundreds of maps has at least a title, even if it is no more than a name, "Pitcairn Island," for example, or "Southern Africa," or "Bucharest, Romania." Some have descriptive material built into them, as "Physical Map of Europe" does, to distinguish it from "Europe," which has more political information on it, or "The Geography of Europe: Environmental Stress," which has a complicated legend and five paragraphs of descriptive text as well. Even "The World" has a title, as does the "Physical Map of the World," which sort of looks like the global portrait.

Though the "Physical Map of the World" has type on it which, of course, the global portrait doesn't.

No type. No title. Nothing, really, to say what it is. And yet, floating beneath it, a part of the title of the *Atlas* the portrait adorns, the words—in enormous type, letters nearly an inch and a half high—THE WORLD.

Oops, our bad: of course it's the world. *Why would it need a title?* And that's the idea here, that this *is* the world. But the world as it *really* is, not the world as humans want to imagine it with their possessiveness and squabbling, their political boundaries and presumptuous names (as though the Pacific needed a name). No, that world is displayed on a three-page foldout further on in the atlas, and it's got a title: "The World." Flip "The World" over and you've got the "Physical Map of the World," same size, same projection, but without the human culture, that is, without the political boundaries, the cities. Wait, there's another absence: there's no bathymetry on the "Physical Map of the World" either (whereas there is on "The World") and the continents have been vignetted into the oceans.

But if this is the physical map, what's the unnamed portrait? Not the cultural world, not the physical world, can it be *the natural world*? And if it is, what's the difference between the natural world and the physical world?

Actually, that's easy. The physical world is decomposable into elemental signs. Here the content plane *makes contact* with the plane of signification, and the contact sediments elemental signs: here are oceans, continents, islands, mountains, plains, deserts, basins, rivers, and other physical features (falls, for example, capes). Many are even named: Guiana Highlands, Qatara Depression, Deccan Plateau. The rest of the map apparatus is also present. The projection is named. The scale is given in three forms. There's a graticule. There's a border.

Because there's no pretense here that this is anything but a map, the Arctic Ocean appears as an ocean. That is, it's the same blue as the rest of the oceans. This is not the case on the unnamed portrait where "Van Sant chose colors"—as the explanatory text pointed out—"that would give a realistic view of the physical world." In this case this means that the Arctic Ocean is an icy white. So what? Well, the physical map categorizes and names the Arctic Ocean as an ocean. The unnamed portrait does neither. Instead it offers us what "Portrait U.S.A." offered us, a truth that the physical map can't, a truth beyond the usual methods of science and cartography. Again, it offers us nature, whole, but in contrast to the robust green nature of "Portrait U.S.A.," or the perdurable ferric-oxide nature of "North America," what we're given here is a pale washed-out nature, a fragile nature, a nature practically on life-support.

Since in the indecomposable super-supersigns that comprise these portrait maps it is variations in color that signify, the differences among the maps largely reduce to differences in coloration. (Shape also signifies but essentially to distinguish subjects: the United States, North America, the world.) In "Portrait U.S.A." nature came dressed in "approximately natural colors." In "North America" nature appeared in false color. The unnamed portrait offers us nature in "natural looking colors" or, in the words of a piece of the epimap, "colors so true to life and details so crisp that one can imagine zooming in to find a neighbor busy in his garden."[49] Coming from a journal with pretensions of scientificity, this is a remarkable claim. It implies the fulfillment of Jean Baudrillard's nightmare prophecy of the representational imaginary finally culminating in and being "engulfed by the cartographer's mad project of an ideal coextensivity between the map and the territory."[50] The pretense exalted by the *Geographic* is that the portrait is a photograph, that nothing stands between you and the world it records in perfect detail except the size of the page it's reproduced on, that you could fall into it, that it would get larger and larger, that there would be more and more detail. It's a vision inaugurated by the iconic sequence in Michelangelo Antonioni's film *Blow-Up* in which the photographer, Thomas, repeatedly enlarges a photograph in search of a corpse he believes must be concealed in the background. It's become a mandatory trope in any spy film, the protagonist leaning over the computer terminal where the able technician in a kind of *resolution orgy* continually zooms in until we can see the face of the villain.[51]

Of course, there's no zooming in to the unnamed portrait map—each pixel is four square kilometers—and if you could zoom in, rather than finding your neighbor busy in his garden, you might find that his house had yet to be built since, given the image-surfing required to keep this map cloud-free, adjacent pixels can be separated by as many as four years. The unnamed portrait is a collage, not a photograph; its promise of unending resolution is illusory; and the true-to-life colors have been picked by an artist to paint not reality on the ground—again, half the time it's night, there are seasons, and so on—but an idealized vision of the world that is uniquely the artist's. In fact, thinking about the unnamed portrait as a collage may miss the point: when it comes to color, which is most of it in a portrait map, it's nothing but a painting.[52]

Painting nature

This is not a condemnation but a characterization. It's one we make to undermine the hyperbole that has stuck to every one of these portraits. From the beginning each has been praised for its extraordinary realism, a sequence of panegyrics rendered in "colors so true to life and details so crisp that one can imagine zooming in to find a neighbor busy in his garden." To think about these portraits as paintings is to directly query their claims to objectivity; and so to complicate with assertions of their subjectiveness their claims to realism. But it is also to open the door to a discussion, not of the billions of pixels, the satellites, the computers, but of that most painterly of questions, the choice of color.

Where Van Sant, the artist who colorized the unnamed portrait, has had little to say about the grounds for his color choices, Tom Patterson has written about the grounds for his color choices at length. A practicing mapmaker whose *Land Cover Portrait/Coterminous United States* we turn to now (figure 10.12), Patterson has also surveyed the color choices made by such acknowledged masters as Eduard Imhof, Heinrich Berann, Richard Edes Harrison, Hal Shelton, and Tibor Toth, each celebrated for the realism and beauty of his maps.[53] None of these, according to Patterson, followed a common set of precepts. Imhof and Harrison leaned toward what Patterson characterizes as "conventional colors," while Berann preferred "the uninhibited end of the color spectrum." Shelton fell somewhere in between, as does the still-practicing Toth. Patterson's traversal renders it undeniable that what we're talking about is painting. He foregrounds the brushes, the mediums, the supports. He gives Toth's color formulas. (Deciduous forest is mixed from 83 drops of Liquitex cadmium red medium, 56 drops of Liquitex pthalocyanine blue, and 751 drops of Liquitex cadmium yellow medium.) Even the computer, Patterson's tool of choice, is treated in painterly terms (color palettes, custom brushes). At the same time Patterson's traversal renders incontrovertible the degree to which the choice of colors is a matter of the mapmaker's individual taste, notwithstanding ceaseless chatter to the contrary about naturalism, intuitiveness, and memorability.

Patterson's *Land Cover Portrait/Coterminous United States* was produced by the U.S. National Park Service as part of an ongoing effort to make its maps "more inviting and understandable."[54] The agency makes maps for nearly four hundred parks which range from the Caribbean to Alaska and the South Pacific, and which are visited by some 300 million people a year. Many of these visitors are inexperienced map readers and non-English speakers, facts that the Park Service has interpreted as requiring less abstract and more realistic maps, especially of mountainous terrain and natural landscapes; and which has led the agency to experiment with cartographically realistic map design.[55] This work has led the agency away from both hypsometric layer tinting and satellite imagery to land-cover data and shaded relief, and *Land Cover Portrait* is an essay in the application of these techniques.

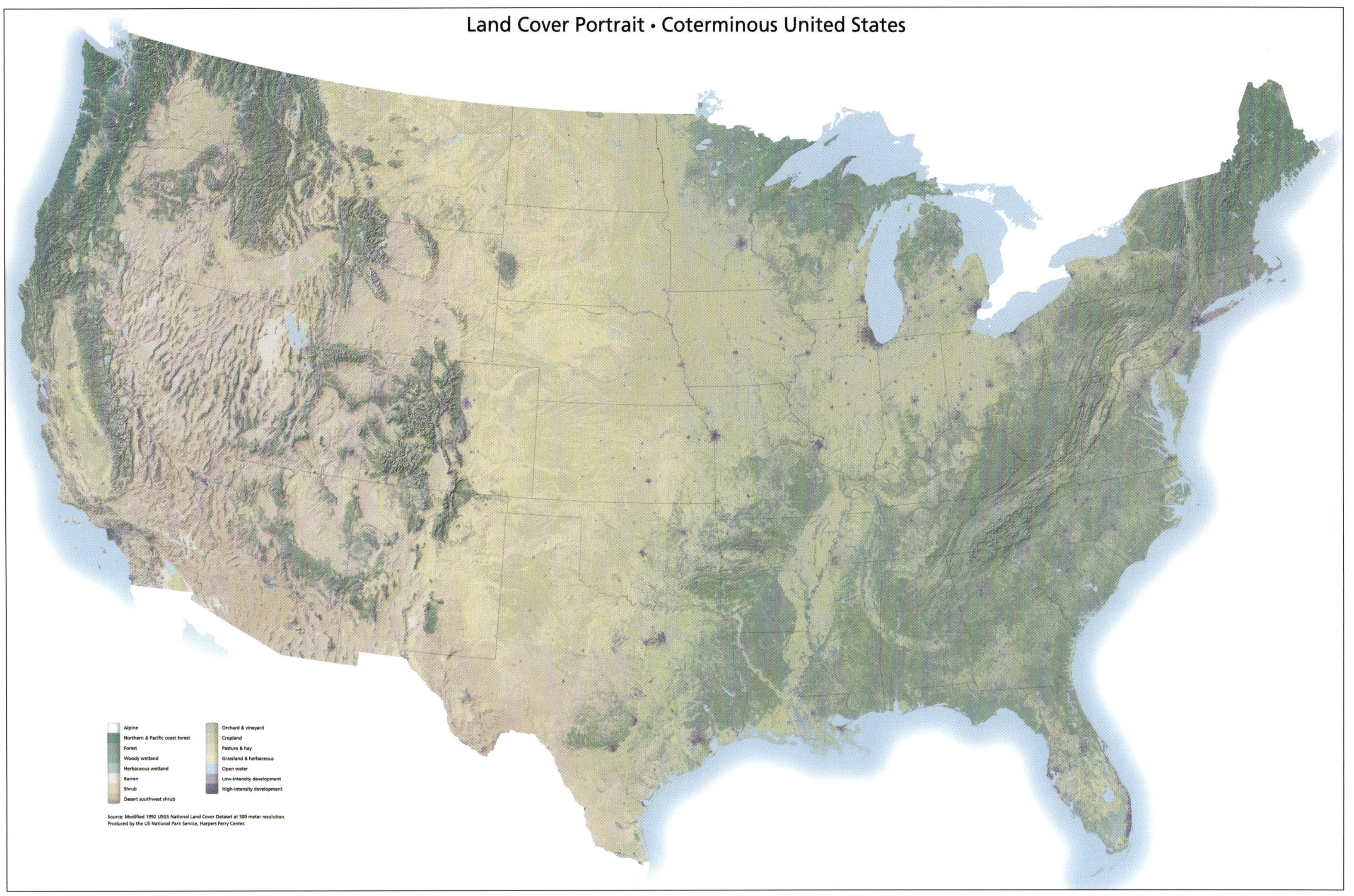

Figure 10.12
Land Cover Portrait/Coterminous United States. This exquisite map was produced by the U.S. National Park Service, Harpers Ferry Center, by a team led by Tom Patterson, as part of an effort to create more realistic maps. Rather than using satellite photo data like other portrait maps in this chapter, it relies on classified land-cover data produced by the USGS. The map is, however, classically idealistic. (1992)
Courtesy of the U.S. National Park Service; data from USGS.

In contrast to the unnamed portrait—and to a greater or lesser degree the rest of the portrait maps—*Land Cover Portrait* is completely upfront about being a map: it has a title, a fifteen-item legend, a note about sources, a production credit, and an extensive technical epimap. It even displays (unobtrusively) state borders. Because it makes no bones about being a map, *Land Cover Portrait* raises the problem posed by its ascription as a portrait in an unusually straightforward fashion. What is it that is being claimed for this map?

Well, again, that the map is a portrait, that is, an image of reality, an image, as we know, supposed to be real in ways other images aren't. Among other things, it's supposed to be immediately accessible, which as we know is an important Park Service goal.[56] What Patterson means by accessible is simple: the image is supposed to look like, is supposed to appear to be the thing itself. This is not the case with something like a USGS topo quad where a mountain, for example, appears as a set of nested contour lines. When you look up at the mountain on the land you don't see these lines, and so the mountain on the quad doesn't look like, doesn't appear to be the mountain on the land. Another example of the same sort of thing shows up with hypsometric layer tinting. On the maps of California in many atlases, Death Valley is green. This is because green is the color most hypsometric tinting schemes assign to elevations lower than a thousand feet, but green isn't a color you see much of when you look out over Death Valley. In a wholly different register, take a close look at a satellite image, one printed in natural, not false colors. In the photo, there's a lot of color variation you don't see when you look at the land. This points to variations in things like soil moisture or mineral composition, things you can't see that were picked up by the satellite in the nonvisible ranges of the electromagnetic spectrum.[57]

By replacing contours and hypsometric tinting with shaded relief, and satellite photo coverage with land-cover data, which has already been both classified and generalized, *Land Cover Portrait* resolves these conflicts between the map and the world so that, Patterson insists, what you see is what you see. While he's acknowledged that maps "are idealized representations of the Earth and are inherently abstract," Patterson has also insisted that, "On the other hand, most of us would agree that some maps appear more realistic and are more intuitively comprehensible than others." He continues:

> For example, a shaded relief map with terrain represented by softly modulated light and shadows appears more realistic than a contour map with a multitude of isolines connecting points of equal elevation value. The ersatz realism one sees on a map—which I will call "cartographic realism"—is not reality, but is instead a graphical representation. The snow on the ground outside my window as I write this is real. The white coloration representing snowy mountain peaks on the topographic map of the Alps on my wall is cartographic realism.[58]

This seems comparatively straightforward—ideologically prejudiced toward visual appearance, perhaps, but straightforward—but as usual the devil's in the details.

That snow, for example, so real outside Patterson's window, and so realistic on his map of the Alps, turns into an instance of "meteorological interference" on a natural-color Landsat image Patterson and Kelso critique for containing "graphical elements inconsistent with cartographic design goals."[59] Meteorological interference is, for these realists, a major problem with satellite imagery:

> Because clouds on average cover 64 percent of Earth's surface (54 percent of land areas) at any given time, the odds of finding satellite images completely free of clouds and their shadows are slim at best. Even one small cloud on an image requires a cartographer to make a difficult choice—an exercise in cartographic situational ethics. Is it best just to leave the unsightly blemish on the image, or is it proper to quickly remove it with the Clone Stamp (Rubber Stamp) tool in Photoshop? After all, who would notice or object? Other meteorological interferences encountered on satellite images include snow-covered ground, frozen water bodies, smog, smoke plumes from wildfires, and lowland flooding. Such undesirable traits plague many of the satellite images available online for free, which are provided by organizations that monitor the environment and natural disasters. When using satellite images as backdrops on maps, boring is better.[60]

One has to imagine that Stalin's censors had similar thoughts about the unsightly blemishes (discredited politicians) and undesirable traits (inadequately large crowds) that appeared in the images they had to work with, blemishes and traits that lacking Photoshop's Clone Stamp they had to use a sharp knife and a bit of glue to deal with. Who, they too doubtless wondered, would notice or object? Object or not, what we have to observe is that there is nothing the slightest bit realist about these procedures, which amount to as pure an exemplification of Aristotelian idealism as one could hope to find, the inevitable imperfections of particular examples—and presumably smoke plumes from pulp plants would be Rubber Stamped out as well—eliminated in the name of "boring is better," a thoroughly classical predisposition.

This is not, however, the idealism dominant in *Land Cover Portrait* for which Patterson eschewed satellite imagery for derivative land-cover data, specifically the USGS's National Land Cover Dataset (NLCD). Patterson and Kelso describe this as a type of categorical land-cover data:

> With categorical land cover data, each pixel represents a sampled area on the ground and receives a classification as one type of land cover or another. For example, if the contents of a 30 x 30-meter sample of NLCD were 51 percent shrub and 49 percent evergreen forest, then the sample receives the shrub assignation entirely—winner takes all. What categorical land cover lacks in subtlety it makes up for in quantity. The millions of pixels that comprise these data when reduced in scale blend land cover colors together smoothly, a desirable trait on natural-color maps.[61]

This is to say that the map's smooth color-blends derive from a kind of majoritarianism which, while out of sync politically with Platonist doctrine, nevertheless results in thoroughly ideal Neo-Platonist land-cover types.

Land Cover Portrait displays fifteen of these types (or colors—since each type is assigned a map color, Patterson thinks about the system as a palette): alpine, northern and Pacific Coast forest, shrub, open water, high intensity development, and so on. Patterson derived his types from the twenty-one level-two categories of the NLCD. These had in turn been derived from the land-cover and land-use categories—over a hundred of them—of the USGS's Anderson Land Use and Land Cover Classification System. (If the Anderson categories had been single malt scotches, the categories of *Land Cover Portrait* would be blends of blends.)

The transformation of the NLCD categories into the Portrait palette involved both aggregation and reclassification of categories *and* the creation of new ones (for instance, alpine), for each of which Patterson chose a color. In contrast to the bright and distinct colors of the NLCD, "the colors chosen for the natural-color palette were complementary and representative of natural environments;" except, that is, for the blue chosen for open water (dictated by graphical pragmatism), the muted purples assigned to low and high intensity development ("unnatural colors for unnatural information"), and the white assigned to . . .

Yes, the white assigned to *what* exactly? We've seen that Patterson thinks that white representing snowy mountain peaks on topographic maps of the Alps is an example of cartographic realism, and that white representing snow-covered ground on satellite images is an example of meteorological interference. This might have led us to suppose that on the cartographically realist *Land Cover Portrait* white would have been assigned to snowy mountain peaks, but,

> On natural-color maps the appearance of white (snow) in lofty mountain areas tells readers that these areas are higher and colder than adjacent lowlands. In the continental U.S., however, the NLCD category *perennial ice/snow* occupies only scattered tiny areas in the Cascades and northern Rockies. To give high western mountains the emphasis they deserve, the palette contains a new category called *alpine*. It encompasses all areas above timberline and slightly lower in select places, such as the snowy and rugged Wasatch range of Utah that barely reaches timberline. Because the elevation of timberline varies depending on latitude, continentality, and other factors, a DEM and biogeography references proved essential for delineating alpine areas. The procedure involved reclassifying all *perennial ice/snow, barren, shrubland, and herbaceous/grassland* as *alpine* for areas above the documented timberline elevation of each mountain range.[62]

What? Can this possibly mean that everything above timberline—and in selected places *below it*—is white?

Really, is there anything even remotely realist about this?

No, as with every other portrait we find ourselves here in the grip of an absolute idealism, one committed to painting a map of an idealized nature (at the very least a nature in which high western mountains get the emphasis they deserve). Despite the desire to bring "to the printed map a colorized portrait of the landscape that closely approximates what people see in the natural world around them;" despite the psychological insight that people living in mid and high latitudes "tend to associate green with the color of vegetation, brown with aridity, and white as the color of snow;"[63] despite the folksy comparison of the snow outside his window with the white on a map of the Alps; in the end Patterson assigned white to barrens, to shrubland, and to herbaceous/grasslands (as well as to perennial ice and snow), just as long as they grow above the tree line (except here and there below it too). What exactly does the tree line have to do with "the colors humans observe everyday in nature"? Nothing, really. It's just that in Patterson's Nature altitude trumps color, even on natural-color maps.

Perhaps it would be as well to give up the idea of natural color. The "overarching goal" of *Land Cover Portrait* "was to achieve a soft impressionistic portrayal,"[64] and certainly this is how many people want to imagine nature, soft and impressionistic. It's the blockbuster nature you find on the posters in art museum gift shops with their carefully cropped Monets and Pissarros and Renoirs, with their meadows and their blossoming trees and their river walks. Probably impressionism is a far better term than realism for the cartography Patterson practices, for Patterson seems to be fighting on the terrain of the map the very fight the Impressionists fought on the terrain of painting, a fight not over how nature is to be portrayed but how, through its

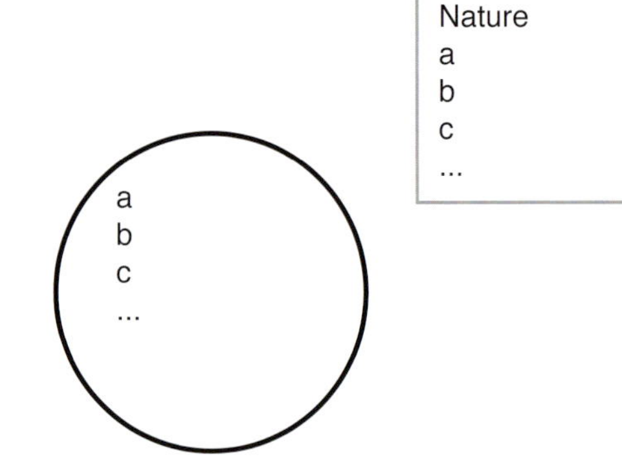

Figure 10.13
The cognitive structure of *Land Cover Portrait/ Coterminous United States.* The simplicity of this diagram reflects how little is going on here as the portrait map retreats from propositional discourse toward the purely pictorial. The elements of the image, its shape, orientation, and characteristic variations in tone, are organized by two frames (or idealized cognitive models), the lower 48 and nature. With its indecomposable super-supersign, its refusal of the linguistic code, and its limited perimap, this is all there is to it. Contrast this structure with that of "Australia Under Siege," "Natural Hazards," or even "Portrait U.S.A.," which with its political map of all fifty states on the flip side, would at least have had that to engage. You look at a portrait map and it looks back at you, returning nothing you haven't brought with you in the form of frames or idealized cognitive models. The portrait map is . . . just barely a map.

construction of nature, the bourgeoisie is to understand itself.[65] Patterson is not alone here. No matter how caught up they are in the technology they exploit, all the mapmakers we've looked at in this chapter want to declare the nature of their humanity by taking a stand on the side of nature, on the side of the clean-cut filigree of the Grand Canyon, on the side of the undulating veins of the Mississippi, and the gentle folds of the Appalachians; on the side of the cloudless summer day, perfect, unchanging through the unceasing permutations; on the side of the solemnity and simplicity of the enduring land; on the side of a nature whole, intact, barely tainted (if at all) by the unnatural colors assigned to the low and high intensity development that is the only *hint*—beyond the maps themselves—of the culture, with its rockets, satellites, cameras, dish antennas, transmission lines, computers, pine plantations, railroads, pulp plants, paper mills, chemical works, and so forth and so on, that permits these maps of nature to be made.

The maps are the mark of a certain schizophrenia with its characteristic *withdrawing* from reality. It's unlikely that the nature the maps propose ever existed. Certainly it does not today. But then maybe that's why these mapmakers are so desperate to insist on its continuing existence.

Notes

1. It's an almost reflexive formulation. Here's Andrew Pickering summarizing a chapter of his on the invention of the bubble chamber: "It was knowledge of the inner workings of the bubble chamber . . . rather than of the inner workings of nature. The latter, however, is what we are accustomed to regard as exemplary of scientific knowledge," in *Mangle of Practice*, 68. Pickering is not approving of this usage, which can arise only in a representational idiom.

2. The speed with which Ingres drew his portraits is an essential consideration in David Hockney's thesis about how they were made. See Hockney's *Secret Knowledge: Rediscovering the Lost Techniques of the Old Masters* (New York: Viking Studio, 2001), 20–33 and passim. We'll be returning to Hockney soon.

3. At least within the simple green and white border. In the margins are the usual copyright and ordering information.

4. Bishop, "Landsat Looks at Hometown Earth," *National Geographic* (July, 1976), 140–7. It may be doubted that General Electric's Tony Rossi is actually retouching the mural here. Surely he is merely posing for photographer Joe Bailey. The ascription "authoritative" used to describe Bishop's article comes from a General Electric advertisement for its photographic and mosaic facilities in the magazine's unpaginated front matter. General Electric would go on to form EOSAT, a joint venture with Hughes Aircraft, to commercialize Landsat.

5. Hockney claims that the assembly of photos into collages involves precisely those skills and habits characteristic of drawing. A 1981 show of his photocollages was called "Drawing with a Camera." See Hockney's *On Photography* (New York: André Emmerich Gallery, n.d.). The publication is not dated, but it's the transcription of a lecture Hockney gave at the Victoria & Albert Museum, London, in November 1983. Since then Hockney has "written" about photography a lot, usually in the form of taped conversations. See, for example, Hockney *On Photography: Conversations with Paul Joyce* (New York: Harmony Books, 1988) (not to be confused with *On Photography*), which is an extended meditation on time and space in painting and photography. Hockney's *That's Way I See It* (San Francisco: Chronicle Books, 1993) pursues these themes into the realm of perception generally.

6. The Portrait is both a photomontage, generally understood as involving manipulation during processing, and a photocollage, generally understood as involving the cutting and reassembling of photographs, though the distinction is not universally accepted. See Ades, *Photomontage* (London: Thames and Hudson, 1986), 15–17 and passim for an extended discussion.

7. Again, we hope he's posing, but this photograph does open the essential epimap. The photo is directly adjacent to a textual subhead that reads "Billions of 'Pixels' Beamed to Earth," and the juxtaposition encapsulates the ambiguity infesting the map.

8. Imhof, *Cartographic Relief Presentation*, 337. Imhof makes this remark apropos of his well-known "map as a landscape painting" of the region around the Walensee in Eastern Switzerland. In a recent traversal of this remark of Imhof's, Tom Patterson tries to contextualize it as an admonition merely intended to keep impressionable students from wasting their time trying to achieve such effects ("Getting Real," see especially pp. 45–46), but Imhof is actually quite doctrinaire: "The end result [of such naturalism] is not a map," he writes, because it doesn't provide what a map calls for: topographic, conceptual, and metric information. Though Imhof did title his painting "A Map of the Area around the Walensee," in his discussion Imhof points out that he "deliberately dispensed with any cartographical conventions," and refers to his effort as "landscape painting," "plan view painting," a "painting experiment," "plan-view landscape pictures," and so on. He says, "It is a painting," and he adds, "The end result is not a map." (It's worth comparing Imhof's painting with the 1:25,000 Walensee sheet of the *Landeskarte der Schweiz* to appreciate Imhof's point of view.)

9. For one profound reading of the functions of images, see David Freedberg's *The Power of Images: Studies in the History and Theory of Response* (Chicago: University of Chicago Press, 1989).

10. Gombrich, *Story of Art*, 12th ed. (London: Phaidon, 1972) 33, 227–8, 332. These sentiments of Gombrich's are staples in surveys of art history and may be replicated ad infinitum. Compare, for example, Janson and Janson, *History of Art*, 5th ed. (New York: Abrams, 1997): an Egyptian sculptural portrait is "almost overpowering in its three-dimensional firmness and immobility" (p. 68); the *Mona Lisa* "seemingly embodies a quality of maternal tenderness that was to Leonardo the essence of womanhood. Even the landscape in the background, composed mainly of rocks and water, suggests elemental generative forces" (p. 456); and in Rembrandt's self-portraits "the bold pose and penetrating look bespeak a resigned but firm resolve" (p. 584). Gombrich fully appreciated the traditionalism of *The Story of Art*. See his characterization in the posthumous *Preference for the Primitive: Episodes in the History of Western Taste and Art* (London: Phaidon, 2002), 8–9.

11. For a magazine aspiring to the mantle of science, this is sufficiently problematic to swath the Portrait with as many attestations of science as possible. The desire in this bicentennial issue to do more than science can, but without slipping into pseudoscience, is responsible for the slippery, ambiguous presentation.

12. Pulled from the lead sentence of the entry "realism" in Preminger, ed., *Princeton Encyclopedia of Poetry and Poetics* (Princeton: Princeton University Press, 1974), 685. On page 43 of his "Getting Real: Reflecting on the New Look of National Park Service Maps," (*Cartographic Perspectives* 43 [Fall 2002]: 43–56),Tom Patterson works with a definition from *Webster's New World Dictionary*: "The picturing in art and literature of people and things as it is thought they really are, without idealizing." We could multiply these. The topic is enormous, the literature immense. Suffice it to say we are not concerned here with the subtleties of philosophical realism, nor with issues of *scientific realism* (see Leplin, ed., *Scientific Realism* [Berkeley, Calif.: University of California Press, 1984], for an introduction).

13. This is what complicates what Patterson, "Getting Real," 44, wants to call "ersatz" or "cartographic realism."

14. During his first orbital space flight, John Glenn observed that, "I was surprised at what a large percentage of the track was covered by clouds." He found most of the United States under a blanket of cloud. See Voas, "John Glenn's Three Orbits in *Friendship 7*: A Minute-by-Minute Account of America's First Orbital Space Flight," *National Geographic* (June, 1962): 792–827. Quotation on p. 806.

15. See our treatment of a similar image in the third chapter of *Power of Maps*, 48–69 ("Every Map Shows This . . . But Not That").

16. Patterson, "Getting Real," 43. What he said is, "all maps (and even many remotely sensed images) are idealized representations of the Earth and are inherently abstract."

17. From the "ideal" entry in Osborne, ed., *Oxford Companion to Art* (Oxford: Oxford University Press, 1970), 555. The reference is to the *Poetics*, where Aristotle speaks of poetry quite broadly, encompassing the epic, tragedy, comedy, dithyrambic poetry, and most flute and lyre playing, all as imitative arts, and into which he occasionally drags painting as well. See section 1448a in any standard edition.

18. Osborne, *Oxford Companion to Art*, 555–6.

19. Gombrich, *Story of Art*, 57.

20. Kenyon Cox, who wrote at the turn of the last century, was the great apostle of classicism in American art. Introducing a reprint of Cox's most important texts (*What Is Painting?: "Winslow Homer" and Other Essays* [New York: Norton, 1988]), Gene Thornton writes, "To Cox, classicism meant not so much depending on Greek or Renaissance models as achieving the close union of representation and abstraction that the ancient Greeks were the first to achieve and that was achieved again by the great masters of the Italian Renaissance. To him, a classical American art would be an art as clear, orderly and accessible as the art of the Greeks, even when it dealt realistically with the everyday life of a world that the ancient Greeks never knew or imagined" (pp. xvi–xvii) or, again, "Usually he means that the painter, starting from nature, has eliminated from his subject everything that is not characteristic and essential, and has recreated it on canvas free from all irrelevant detail" (p. xvi), both of which could serve to describe the ambitions of the GE, NASA, and National Geographic mapmakers.

21. So described in the caption, *National Geographic*, July 1976, p. 1.

22. It certainly wasn't because photomosaics weren't available. At the same time the Geographic published its Portrait, NASA published *Mission to Earth: Landsat Views the World*. Spread across the front endpapers is Space Portrait

U.S.A./The First Color Photomosaic of the Contiguous UNITED STATES/ Produced by GENERAL ELECTRIC COMPANY/Beltsville Photographic Engineering Laboratory/From 569 Landsat Satellite Images/In Cooperation with/The National Geographic Society and/National Aeronautics and Space Administration. Its caption describes it as, "The first false color mosaic of the conterminous United States," and indeed here it is red and blue and black and white. Across the book's endpapers, however, is "the first precision black and white photomosaic of the coterminous United States," this prepared by the Soil Conservation Service for NASA's Goddard Space Flight Center with mosaics of Alaska and Hawaii on the page just prior to the endpapers. Mosaics of the whole United States had been made, they just looked clunky and awkward, as if a leg and a hand had been painted separately from the rest of the body, and the portrait consisted of three framed fragments. This is not the way to present an image of the *unum*. The problem with "Portrait U.S.A." is not that it's ideological—that's unavoidable—but that it's exclusively ideological.

23. Hockney, *Hockney On Photography*, 44–45.

24. King, *The Commissar Vanishes: The Falsification of Photographs and Art in Stalin's Russia* (New York: Henry Holt, 1999), 9. The falsification of photographs was part of a broader falsification of history generally.

25. King reproduces nine variants of this image (ibid., 66–71).

26. Ibid., 13.

27. Hockney, *Hockney On Photography*, 45. The degree of this difference between painting and photography may be measured by Goethe's remarks to Johann Eckermann in response to Eckermann's objection that the light in a Rubens painting came from two opposing directions, "which is quite contrary to nature." "If it is contrary to nature," Goethe answered, "I also say that it is *higher* than nature. I say that it is the bold hand of the master, whereby he demonstrates in a brilliant way that art is not entirely subject to natural necessity, but rather has laws of its own." Glossing this passage, Siegfried Kracauer adds, "A *portrait painter* who submitted entirely to 'natural necessity' would at best create photographs" (all in Kracauer's 1927 essay "Photography" in *The Mass Ornament: Weimar Essays* [Cambridge, Mass.: Harvard University Press, 1995]).

28. That this is the case is due not merely to the photochemical "fixing" of whatever stands before the photographer's lens either, but to what Alan Trachtenberg has termed "this uniquely modern conflation of vision and knowledge": "The emulsified image made an extraordinary difference in how reality was understood; the emulsion was capable of rematerializing as an image, almost instantaneously, in defiance of time, space, and mass. On the threshold of cinematography, then, the realm of photography was already established as the realm of free-floating, emulsified images. Not 'pictures,' properly speaking, photographs performed their work in society in similar ways, *as depictions of the real*" (our emphasis, from Trachtenberg's "Photography/Cinematography" in *Before Hollywood: Turn-of-the-Century Films from American Archives* [New York: American Federation of the Arts, 1986] the quoted text from p. 75.) In his *Techniques of the Observer: On Vision and Modernity in the Nineteenth Century* (Cambridge, Mass.: MIT, 1990), Jonathan Crary sites this valorization of the photographic aura in the nineteenth century's much earlier formation of the observer.

29. "Place Furstenburg, Paris, August 7, 8, 9, 1985," appears in Los Angeles County Museum of Art, *David Hockney: A Retrospective* (New York: Abrams, 1988), 247.

30. Ibid., 58.

31. "Billy Wilder Lighting His Cigar Dec. 1982," ibid., 221. "Fredda," ibid., 64

32. "Billy Wilder Lighting His Cigar Dec. 1982" appears in Los Angeles County Museum of Art's *David Hockney*, 221, where, incidentally, Hockney's portrait collage of Bill Brandt and his wife, Noya, can be seen on p. 59. Quotes from pp. 58-59. Anne Hoy contributed an essay to the volume called "Hockney's Photocollages" (55–66).

33. This sentence strings together phrases from several pages of Hockney's *On Photography* (not to be confused with *Hockney on Photography*).

34. Hockney attributes this static quality to the fact that "the best portrait photographs are those that capture in a fraction of a second a period of time that looked as though it had been longer. Yet this also results in a certain static aspect to the face." Ibid., 27.

35. *Atlas of the World*, vi. Note that with improved technology, the number of images sorted through has shrunk from thousands to hundreds.

36. Ibid., plate 14 (these numbers refer to left–right openings). Actually the *Atlas*, and the text referred to in the next sentence, call "North America" an "image" (but then, see below, it appears in an atlas that opens with a Van Sant-Warren natural-color "portrait"). There are a couple of paragraphs about the construction of "North America" in an "On Assignment" column on the last page of the November 1990 issue of *National Geographic* (unpaginated back matter). Chris Chiesa and his colleagues at the Environmental Research Institute of Michigan "spent months writing software and reviewing hundreds of images recorded by NOAA weather satellites," to pick the fifty "clear-weather images" used "to create the clearest ever mosaic" of North America. Piecing the images together "was a lot like putting together a jigsaw puzzle" and "color variations were blended, leaving a seamless image." A lot has changed since the construction of "Portrait U.S.A.," but the language used to describe the maps and their construction is not one of them. The accompanying photo of Chiesa sitting at his terminal in front of a mural-size version of the map similarly recalls that of GE's Tony Rossi blending join lines on the mural-size mosaic of the U.S.A. We note, though, that in 1990 Chiesa was not doing this with a jar of paint in his hand, but at the keyboard of his computer.

37. Actually the image contains clouds—for instance, over Central America (it's only the "clear*est* ever mosaic")—but since the obvious referent for white is Arctic and alpine snow and ice, the image's few clouds fail to read as clouds, especially with their siblings over the oceans blacked out. The interesting question is, why include the Arctic ice, to say nothing of the Chukchi Peninsula, especially when the continent's so brutally truncated at Colombia?

38. For a book-length treatment of this theme, see Wood's *Five Billion Years*.

39. Ibid., 14 (left). Most of this text is devoted to a breathless recitation of the continent's human history, though it does conclude with an acknowledgment that "Meanwhile nature changes the land, as volcanoes erupt and earthquakes shudder. The California coast grinds northward at three meters a century—a powerful reminder that this vast continent, although ancient, is still very much alive." A necessary reminder too, since the image makes it look so dead.

40. Indeed Tom Patterson, an advocate, claims that, "Compared to conventional maps, realistic maps are, undeniably, dumbed down—users have to grapple with fewer abstractions, and intelligence is commonly defined as the ability to think abstractly. However, by avoiding the use of abstract symbolization, realistic maps have the potential to communicate more efficiently to a greater number of users," "Getting Real," 44.

41. We first argued this case in "Designs on Signs," 88–90, but also see *Power of Maps*, especially 132–3.

42. This code is spelled out in a small block of italicized type below the text block to the left of the image. It also refers to a blend: "Boreal forests in Canada are brownish red." One of our points would be that you can find brownish red all over the map. The code fails to assign content to white (usu-

ally clouds?), blue (which, in the Bahamas and San Francisco Bay is unlikely to be ice), or black (which would be an unusual choice for water). But then, the code is purely indicative and not meant to be taken too seriously.

43. Another way of saying this would be to note that the map image of North America consists of but a single layer.

44. Note that we have understood as *there* what, on another map, might as easily be taken for a *this*. In "North America," which is without other locative indication, this course seems almost unavoidable. In "Portrait U.S.A.," as just noted, an index does run around the border, and Canada and Mexico are also indicated in dark gray. In that case, we could describe the *there* without referring to the United States, and the map would consist of two postings with coincident *theres* (Nature is *there* and North America is [also] *there*).

45. See Barthes, *Mythologies*, 121–31 (especially, 127–31), but also elsewhere in Barthes' writings, indeed, almost everywhere.

46. *Atlas of the World*, i and ii–iii.

47. Ibid., v.

48. Grosvenor, "New Atlas Explores a Changing World," *National Geographic* (November, 1990): 126–9.

49. From the same "On Assignment" column on the last page of the November 1990 issue of *National Geographic* (unpaginated back matter) that we cited earlier apropos Chris Chiesa and his work on "North America."

50. Baudrillard, "Simulacra and Simulations: Disneyland," in *Social Theory: The Multicultural and Classic Readings*, ed. Charles Lemert (Boulder, Colo.: Westview Press, 1993), 524–29, the quotation on p. 525.

51. During the 1960s, Antonioni was the subject of a minor publishing industry. For a critical perspective, see Ian Cameron and Robin Wood's *Antonioni*. Rev. ed (New York: Praeger, 1971); for one more balanced, see Tudor "Antonioni, Michaelangelo" in *World Film Directors: Volume II 1945–1985*, ed. John Wakeman (New York: H. W. Wilson, 1988). In 1981, Brian de Palma made *Blow Out*, an homage to *Blow-Up*, in which he replaced the photograph with a sound recording. In 1982, Ridley Scott made *Blade Runner*, in which he replicated Antonioni's image enlargement but on a computer. Since then we've been inundated with films in which critical evidence is secured by using computers, typically accompanied by Tinkerbelle-like magic-wand music, to "blow up" sights or sounds which have been recorded onto grain-free media. The acme, or nadir, appeared in 1992 in Philip Noyce's *Patriot Games* where we were treated to real-time, infrared, reconnaissance satellite video of special forces killing people in a night raid on a terrorist training camp. It's not just about spying: it's a kind of *resolution pornography*.

52. "Every portrait that is painted with feeling is a portrait of the artist, not of the sitter," Oscar Wilde had his character, Basil Hallward, say in *The Picture of Dorian Gray*. This is certainly the case here. For more on Van Sant's maps, see Wood's *Power of Maps*, where the entire third chapter is turned over to their deconstruction (pp. 48–69); as well as Wood, Kaiser, and Abramms's *Seeing Through Maps*, 2nd ed. (Amherst: ODT, 2006) where Van Sant's maps are discussed on pp. 64–73. There we conclude that a Van Sant map is a painting: "The planet sits for its portrait, but not before a camera. Or rather the painter does use a camera but he uses the camera the way a painter uses a brush. The painter sets the pose (without clouds), arranges the light (no nighttime shadows), selects the season (summer), and then at the console of his computer, picks realistic colors" (p. 72). Recent versions of Van Sant's map contain not 35 but 90 million pixels, each one kilometer square, suggesting that they too may be examples of resolution porn.

53. Patterson does this in a valuable trilogy of papers that begs to be turned into a book. His tribute to Berann is in "View From On High: Heinrich Berann's Panoramas and Landscape Visualization Techniques for the U.S. National Park Service," (*Cartographic Perspectives* [Spring 2000]: 38–65); his tribute to Hal Shelton is in "Getting Real;" and his tribute to Imhof, Shelton, Berann, Harrison, and Toth in his and Kelso's "Hall Shelton Revisited: Designing and Producing Natural-Color Maps with Satellite Land Cover Data" (*Cartographic Perspectives* [Winter 2004]: 28–55 and 69–80). The papers also constitute a ringing declaration of cartographic realist principles, and while we dissent from even the idea, we commend Patterson's willingness to lay them out in such detail. They're essential reading.

54. Patterson and Kelso, "Hall Shelton Revisited," discusses the creation of this map. An acknowledgement (p. 52) underscores the extent to which the map was a team effort. We refer to it as Patterson's as a matter of convenience.

55. Patterson, "Getting Real," 43.

56. Recall that for Kenyon Cox (*What Is Painting?*), "a classical American art would be an art as clear, orderly and accessible as the art of the Greeks, even when it dealt realistically with the everyday life of a world that the ancient Greeks never knew" (our emphasis). Patterson is thoroughly in tune with these idealist precepts.

57. All these examples—there are many others—have been advanced by adherents of natural-color mapping. These can all be found in Patterson and Kelso, "Hall Shelton Revisited."

58. Patterson, "Getting Real," 43–44.

59. Patterson and Kelso, "Hall Shelton Revisited," 38. The natural-color Landsat image Patterson and Kelso critique appears in black and white on p. 38 and in color on p. 72.

60. Ibid., 37–38.

61. Ibid., 39.

62. Ibid., 40.

63. Ibid., 27, 34.

64. Ibid., 40.

65. The critical text here is T. J. Clark's *The Painting of Modern Life: Paris in the Art of Manet and His Followers* (Princeton: Princeton University Press, 1984), the whole book, but especially for us his arguments about the role of nature in the identity of the bourgeois and the role of landscape painting in the negotiations surrounding that identity: "There was a struggle being waged in these decades for the right to bourgeois identity. It was fought out quite largely in the forms the new city had brought to perfection: the squares, the streets, and the spectacles. The crowds on the riverbank on Sunday afternoon, all moving about in identical dress, all eager to be seen, were engaged in a grand redefinition of what counted as middle-class . . . To have access to nature be the test of class is to shift the argument to usefully irrefutable ground: the bourgeoisie's Nature is not unlike the aristocracy's Blood: what the false bourgeois has is false nature, *nature en toc, la nature des environs de Paris*; and beyond or behind it there must be a real one, which remains in the hands of the real bourgeoisie," and so on (pp. 155–6). The problem for painting was to define this nature which, in fact, was being obliterated by French industrialization. While Manet confronts the resulting untidiness, Monet turns away from it, "preferring to focus on what the scene still offers of pleasure or nature in undiluted form" (p. 180), which is precisely what the creators of the maps in this chapter have done. See the related arguments in Raymond Williams' *The Country and the City* (New York: Oxford University Press, 1973).

PROPOSED GREAT SMOKY MOUNTAINS NAT
GREAT SMOKY MOUNTAINS
PISGAH
NATIONAL FOREST
NANTAHALA NATIONAL
YELLOW CREEK MOUNTAIN
SEVIERVILLE
MARYVILLE
Rockford
Gatlinburg
Cades Cove
Thunder Head
Spencer Cabin
Gregory Bald
Parson Bald
Shuck Stack
Welch Bald
Mt Mingus
Indian Gap
Chimney Tops
Le Conte
Bull Head
Roundtop
Blanket Mt.
Clingmans Dome
Andrews Bald
Bryson
Henderson Springs
Tuckaleechee Cove
Sugarland Mt.
Proctor
Siler's Bald
Cold Spring Knob
High Rocks
Bee Knob
Little Bald
Mt Collins
Sharptop Mtn
Rattlesnake Cliff
Millstone Gap
Chilhowee

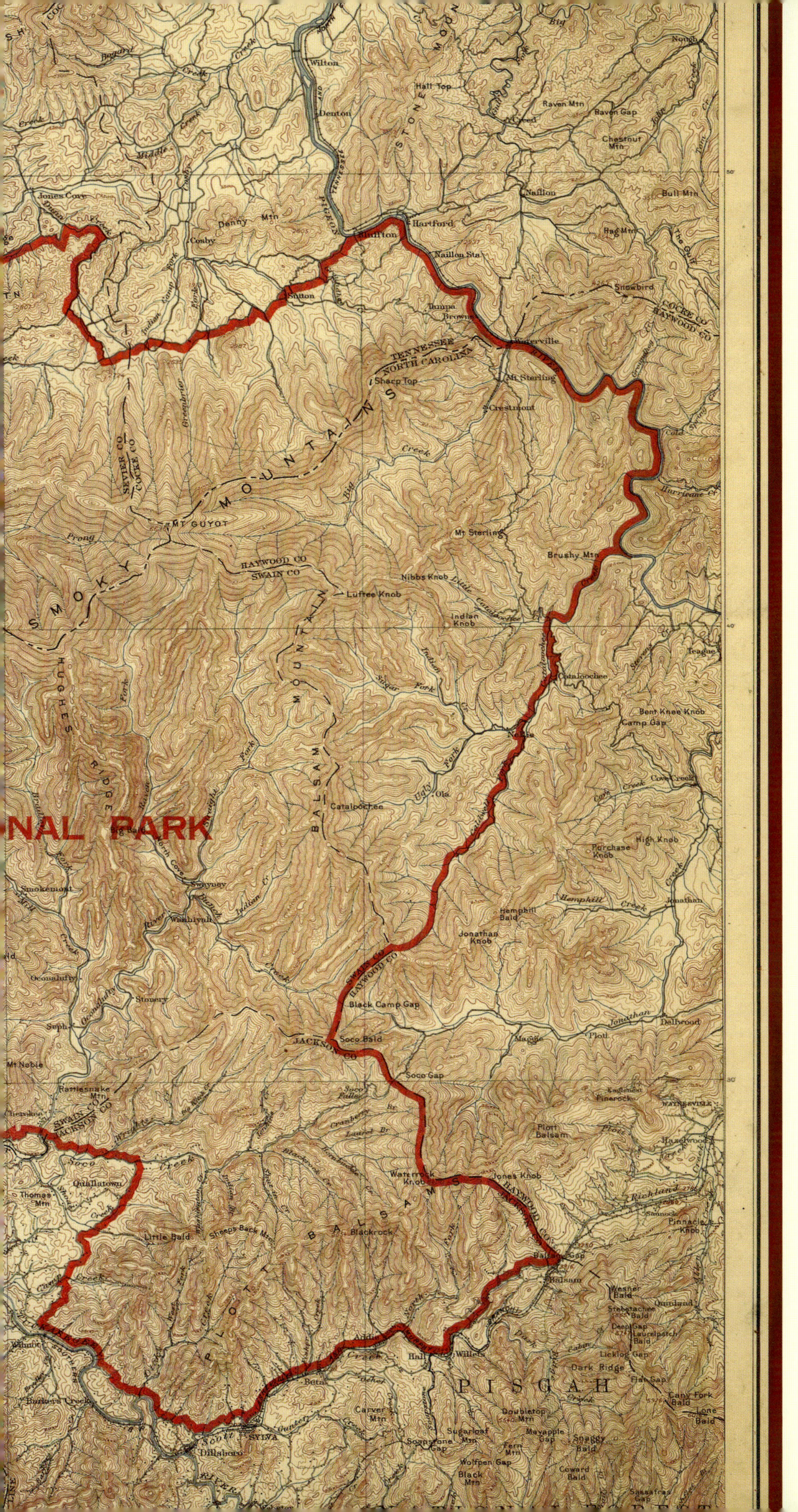

Eleven

Nature as park

Map courtesy of USGS, 1926; and the Library of Congress, Geography and Map Division.

Nature as park

Here's an alternative vision—another that claims to show the world as it is. This is a topographic survey sheet. It's the one we opened the book with, the United States Geological Survey's *Cove Creek Gap* map (figures 11.1 and 11.2). Unlike the portrait maps we've been looking at, where nothing is named, where little is even classified, here nothing exists that has not been classified, and little enough that has not been named.

Figure 11.1
Here is a detail from the USGS *Cove Creek Gap* topographic map, a different detail from the one with which we opened our book. Compared to portrait maps, which name nothing, topographic maps can't stop naming things: hills, mountains, churches, rivers, park borders.

Figures 11.1 and 11.2 courtesy of USGS, 1997.

The garrulousness of the topographic survey

The names lie heavy on the land: Mossy Branch, Matthews Branch, Upper Double Branch, Lower Double Branch, Palmer Creek, Cataloochee Creek. Between the branches and creeks run ridges, Middle Ridge to Canadian Top, Jesse Ridge to Bald. Beyond these rise Cooks Knob, Noland Mountain. There is Bratcher Gap and Davidson Gap. There is Short Bunk. There is Little Cataloochee Church and Palmer Chapel. There is Palmer Creek Trail, Caldwell Cemetery, and Cataloochee Ranger Station.

After the reticence of a portrait map the garrulousness of a topographic map can stupefy. Beyond the creeks, branches, ridges, tops, knobs, and mountains, beyond the gaps, churches, chapels, trails, cemeteries, and ranger stations, there are ruins. There are houses. There is a light duty road and a secondary road. The elevation is scored in forty-foot intervals. It is spelled out every two hundred feet. Bald Top stands 3,960 feet above the National Geodetic Vertical Datum of 1929.

If the hand of man is light, even invisible in the portrait maps, in this corner of what is in fact Great Smoky Mountains National Park that hand is omnipresent. If in the portrait maps the content plane hovered above the plane of signification like a ghost, here the two are inseparably fused. If the portrait maps at the level of synthesis were constituted of a single, indecomposable super-supersign, nature, here Great Smoky Mountains National Park, Pisgah National Forest, and Pisgah Game Land are readily decomposed into systems of rivers, landforms, transportation, each of which is no less readily decomposed into its elemental signs, creek, contour, road. Below this lie only characteristic and connotation, for neither of which was there room in the overstuffed super-supersigns of the portrait maps.

What is stuffed here is the content plane. If you ask, "What is it?" the map returns an answer. It's a gauging station, it's a river, it's a dam. It's an abandoned mine, it's a tunnel, it's a transmission line. The iconic code of the Geological Survey is capacious. It gathers to itself well over two hundred categories of things, organized, on the Survey's "Topographic Map Symbols" Web site (figure 11.3), alphabetically: bathymetric features, boundaries, buildings and related features, coastal features, contours, control data and monuments, glaciers and permanent snowfields, land surveys, marine shorelines, mines and caves, and so and so on. Each of these can be unpacked. Out of coastal features tumbles foreshore flat; coral or rock reef; rock, bare or awash, dangerous to navigation; group of rocks, bare or awash; exposed wrecks; depth curve, sounding; breakwater, pier, jetty, or wharf; seawall; and oil or gas well, platform.[1]

Categories proliferate in the linguistic code where content distinctions overlooked in the iconic code are made. None of those creeks, branches, ridges, tops, knobs, and mountains, for instance, appears in the iconic code which of landforms knows only contours, cuts, fills, and continental divides, and of waterways only washes, streams, and rivers. Geographic nomenclature would be impoverished without its ability to designate. Without its creek, trail, or chapel, what would we make of the Palmers crowding the lower left-hand corner of our map? How would we distinguish Bald Top from Bald Gap? For that matter, what would we make of Short, Middle, and Lower Double? At the same time the names nail the types into specificity. It's not any gap but Davidson Gap. Out of it flows Davidson Branch, which flows into Palmer Creek. Palmer Creek flows into Cataloochee Creek, which flows into Walters Lake pooled behind Walters Dam on the Pigeon River. Under the pressure of the names—Davidson, Palmer, Cataloochee, Walters, Pigeon—the landscape spalls into uncounted places.

The map can't stop talking. But despite babbling of Upper Double Branch and the Pigeon River, of Caldwell Cemetery and

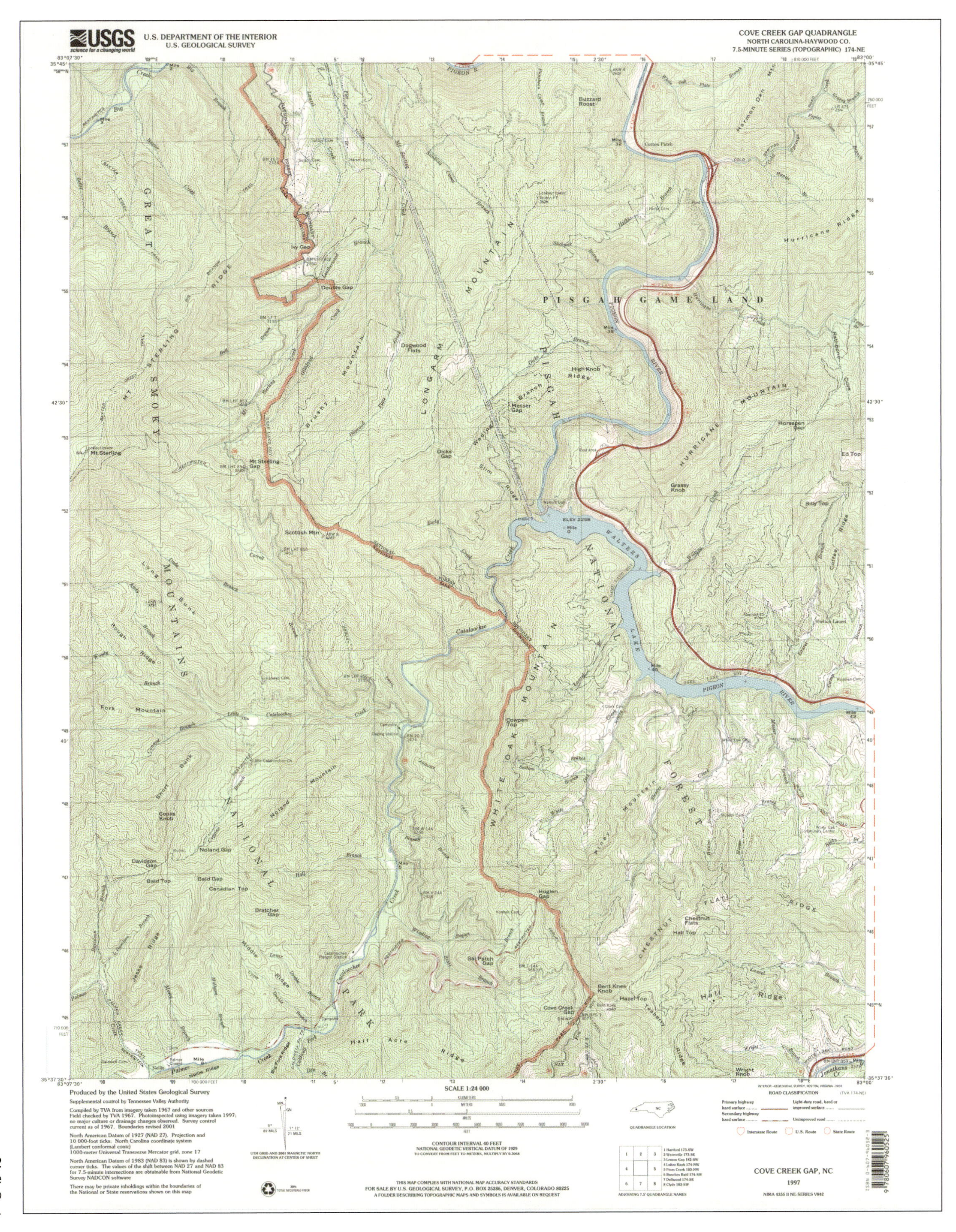

Figure 11.2
The *Cove Creek Gap* quadrangle map in its entirety. Map dimensions (WxH): 22in. x 27in.

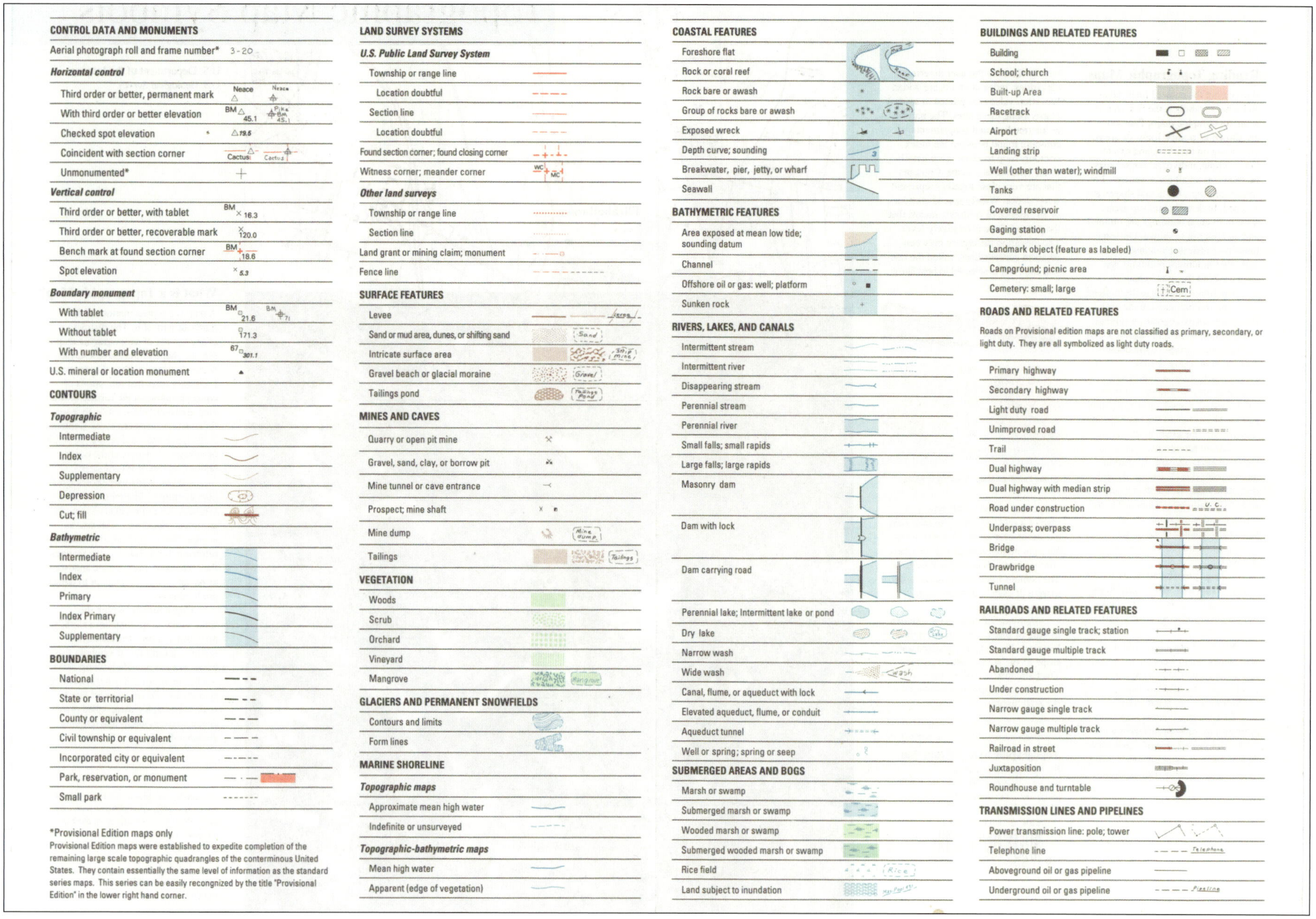

CONTROL DATA AND MONUMENTS

Aerial photograph roll and frame number* 3-20

Horizontal control

Third order or better, permanent mark

With third order or better elevation

Checked spot elevation

Coincident with section corner

Unmonumented*

Vertical control

Third order or better, with tablet

Third order or better, recoverable mark

Bench mark at found section corner

Spot elevation

Boundary monument

With tablet

Without tablet

With number and elevation

U.S. mineral or location monument

CONTOURS

Topographic

Intermediate

Index

Supplementary

Depression

Cut; fill

Bathymetric

Intermediate

Index

Primary

Index Primary

Supplementary

BOUNDARIES

National

State or territorial

County or equivalent

Civil township or equivalent

Incorporated city or equivalent

Park, reservation, or monument

Small park

*Provisional Edition maps only

Provisional Edition maps were established to expedite completion of the remaining large scale topographic quadrangles of the conterminous United States. They contain essentially the same level of information as the standard series maps. This series can be easily recognized by the title "Provisional Edition" in the lower right hand corner.

LAND SURVEY SYSTEMS

U.S. Public Land Survey System

Township or range line

Location doubtful

Section line

Location doubtful

Found section corner; found closing corner

Witness corner; meander corner

Other land surveys

Township or range line

Section line

Land grant or mining claim; monument

Fence line

SURFACE FEATURES

Levee

Sand or mud area, dunes, or shifting sand

Intricate surface area

Gravel beach or glacial moraine

Tailings pond

MINES AND CAVES

Quarry or open pit mine

Gravel, sand, clay, or borrow pit

Mine tunnel or cave entrance

Prospect; mine shaft

Mine dump

Tailings

VEGETATION

Woods

Scrub

Orchard

Vineyard

Mangrove

GLACIERS AND PERMANENT SNOWFIELDS

Contours and limits

Form lines

MARINE SHORELINE

Topographic maps

Approximate mean high water

Indefinite or unsurveyed

Topographic-bathymetric maps

Mean high water

Apparent (edge of vegetation)

COASTAL FEATURES

Foreshore flat

Rock or coral reef

Rock bare or awash

Group of rocks bare or awash

Exposed wreck

Depth curve; sounding

Breakwater, pier, jetty, or wharf

Seawall

BATHYMETRIC FEATURES

Area exposed at mean low tide; sounding datum

Channel

Offshore oil or gas: well; platform

Sunken rock

RIVERS, LAKES, AND CANALS

Intermittent stream

Intermittent river

Disappearing stream

Perennial stream

Perennial river

Small falls; small rapids

Large falls; large rapids

Masonry dam

Dam with lock

Dam carrying road

Perennial lake; Intermittent lake or pond

Dry lake

Narrow wash

Wide wash

Canal, flume, or aqueduct with lock

Elevated aqueduct, flume, or conduit

Aqueduct tunnel

Well or spring; spring or seep

SUBMERGED AREAS AND BOGS

Marsh or swamp

Submerged marsh or swamp

Wooded marsh or swamp

Submerged wooded marsh or swamp

Rice field

Land subject to inundation

BUILDINGS AND RELATED FEATURES

Building

School; church

Built-up Area

Racetrack

Airport

Landing strip

Well (other than water); windmill

Tanks

Covered reservoir

Gaging station

Landmark object (feature as labeled)

Campground; picnic area

Cemetery: small; large

ROADS AND RELATED FEATURES

Roads on Provisional edition maps are not classified as primary, secondary, or light duty. They are all symbolized as light duty roads.

Primary highway

Secondary highway

Light duty road

Unimproved road

Trail

Dual highway

Dual highway with median strip

Road under construction

Underpass; overpass

Bridge

Drawbridge

Tunnel

RAILROADS AND RELATED FEATURES

Standard gauge single track; station

Standard gauge multiple track

Abandoned

Under construction

Narrow gauge single track

Narrow gauge multiple track

Railroad in street

Juxtaposition

Roundhouse and turntable

TRANSMISSION LINES AND PIPELINES

Power transmission line: pole; tower

Telephone line

Aboveground oil or gas pipeline

Underground oil or gas pipeline

Figure 11.3
"Topographic Map Symbols," an extensive legend that serves as a reference for all USGS topographic maps.

Courtesy of USGS.

Little Cataloochee Church, of *this* branch and *that* river, of *this* cemetery and *that* church, the map says nothing of nature, little enough of the natural. This is a characteristic of the topographic quad that emerges from the Survey's mission which, in short, is to identify "What there is in this richly endowed land of ours which may be dug, or gathered, or harvested, and made part of the wealth of America and of the world, and how and where it lies."[2] This produces a phenomenological inventory in the form of a map, the heterogeneity of which reflects our opportunistic, flexible attention. It is this that drives the iconic code to jumble together not only mines and caves, control data and permanent snowfields, but railroads and related features, rivers, lakes, and canals, submerged areas and bogs, transmission lines, and vegetation. Whatever our attention has seized on has been tossed onto the map so that under vegetation we find woods, scrub, and mangrove, but also orchard and vineyard. Yet we don't find rice fields. Along with marshes, swamps, and land subject to inundation, they're included as submerged areas and bogs. Under the pressure of the shifting interests of industry, client agencies of government, and the general public, the code has evolved as the USGS has struggled to augment it consistent with its own commitments to permanence, cost control, and legibility.[3] The code's heterogeneity makes sense only as a historical assemblage.

Yet only rarely are topo quads consulted as palimpsests for the economic (or other) history that may be found there. Instead they're turned to for . . . well, here's the description of our quad that you'll find at the Great Smoky Mountains Association Online Bookstore and

Figure 11.4

The cover of *Great Smoky Mountains* is composed of a simple title and photograph. With the first and second folds, credit citations and introductory text are added.

Figures 11.4–11.6 courtesy of National Park Service and U.S. Department of the Interior, 2001.

Great Smoky Mountains
National Park
Tennessee and North Carolina
National Park Service
U.S. Department of the Interior
No place this size in a temperate climate can match Great Smoky Mountains National Park's variety of plant and animal species. Here are more tree species than in northern Europe, 1,500 flowering plants, dozens of native fish, and more than 200 species of birds and 60 of mammals. International Biosphere Reserve and World Heritage Site designations have recognized this remarkable biological diversity and the cultures humans wrested from its abundance. The National Park Service mission is to preserve this natural and cultural heritage unimpaired for this and future generations. Most of the park is now managed as wilderness.
The Cherokee described these mountains as *shaconage*, meaning "blue, like smoke." They farmed the land and built log homes. The Cherokee tried to adapt to the Europeans, but the newcomers took their land. During the 1790s white settlement began in the lowlands and climbed the hills as eastern farmland became scarce and commercial agriculture migrated to the Midwest. The Eastern Band of Cherokees now lives on its reservation next to the national park. Most tribe members are descendants of those not forcibly removed in the 1830s.
Alarmed at commercial logging threats to the forests, Congress authorized the park in 1926. Established in 1934, this was among the first national parks assembled from private lands. The states of North Carolina and Tennessee, private citizens and groups, and schools contributed money to purchase these lands for donation to the Federal Government.
Great Smoky Mountains

Gift Shop: "This is the official U.S. Geological Service 7.5 minute topographic map, the most accurate map available for hikers and horseback riders who really want to know where they are and what they're seeing."[4] Not, note, "want the map to look like what they're looking at"—which would be a test for Patterson's cartographic realism—but want the map to help them "know where they are and what they're seeing," that is, the name and the nature of the thing or things they're looking at. "What's that?" the hiker asks, and the map answers: "It's Noland Mountain. It's 3,920 feet high. You're at Mile 5 on Cataloochee Creek almost 1,400 feet below it."

By assigning white to its alpine class, this is more or less what Patterson's map did too: "Look," it said, "these are high western mountains, okay?" In the end it's what all maps do, say what's there. It's the source of their power to socialize the world, even portrait maps. It's just that portrait maps have one big thing to say ("this is nature"), while topographic survey sheets have a million little ones ("this is Big Fork Ridge, this is Caldwell Fork Trail, this is . . ."). We're not exaggerating when we say a million either. Edward Tufte estimates that large scale topographic maps contain 150 million bits of data,[5] and if it doesn't make sense to equate all of these with individual instances of saying what's there—with individual postings—the number of things such maps say is still very large.

The Park Service sees things differently

To appreciate how large, it may help to compare the Cove Creek Gap quadrangle with the smaller-scale map the National Park Service provides visitors to Great Smoky Mountains National Park. Its structure should be familiar. On the front there's a poster. The map's on the back. The words "Great Smoky Mountains" run up the cover-fold white on black along a strip of trees—in their fall colors—that climb into the eponymous "smoke." The first fold doubles this strip in height (these are "great" mountains indeed) and gives us, discreetly, the full name of the park, the states it's in, and the names of the institutions responsible for it—to whose oversight we owe any pleasures the park

Figure 11.5
Great Smoky Mountains unfolded to its full extent reveals a poster featuring five vignettes and an abundance of animals.

might give us—the National Park Service and the U.S. Department of the Interior (figure 11.4). The sheet then accordions out to reveal three paragraphs of introductory text and, taking up most of the page, vignettes of the Great Smoky Mountains forest types (spruce-fir, northern hardwood, cove hardwood, hemlock, and pine-and-oak). Each is crammed with plants and animals dominated by fur and feathers: a chipmunk, a bobcat, a white-tailed deer, a red fox, a red squirrel, a wild hog, a ruffed grouse, an owl, a turkey, a hooded warbler, a scarlet tanager, and so on; with a box turtle here and a brook trout there and here a katydid almost invisible against the foliage and there a moss spider little more than a smudge on the paper (figure 11.5). The equation is straightforward: Great Smoky Mountains National Park = rhododendron and magnolia, black bears and ravens = nature.

The other side is headlined “Discovering Diversity in the Smokies” (figure 11.6). This features photos of forest, waterfalls, a log cabin, Cherokees, “mountain people,” a basket, some flowers, and a couple of maps. The smaller of these locates the park in the interstate, federal, and state highway systems. The other, a map of the park, is the largest thing on the page. Cognitively the map is base. Its “park” spreads to the vignettes on the other side via optimization, whose “nature” then floats up to suffuse the park; the map in this way lending its authority to, spatializing the name Great Smoky Mountains National Park, even as it spatializes the “nature” illustrated in the vignettes (figure 11.7).

Note that there are only nine items on the map’s legend compared to the couple of hundred on the legend of the topographic map.

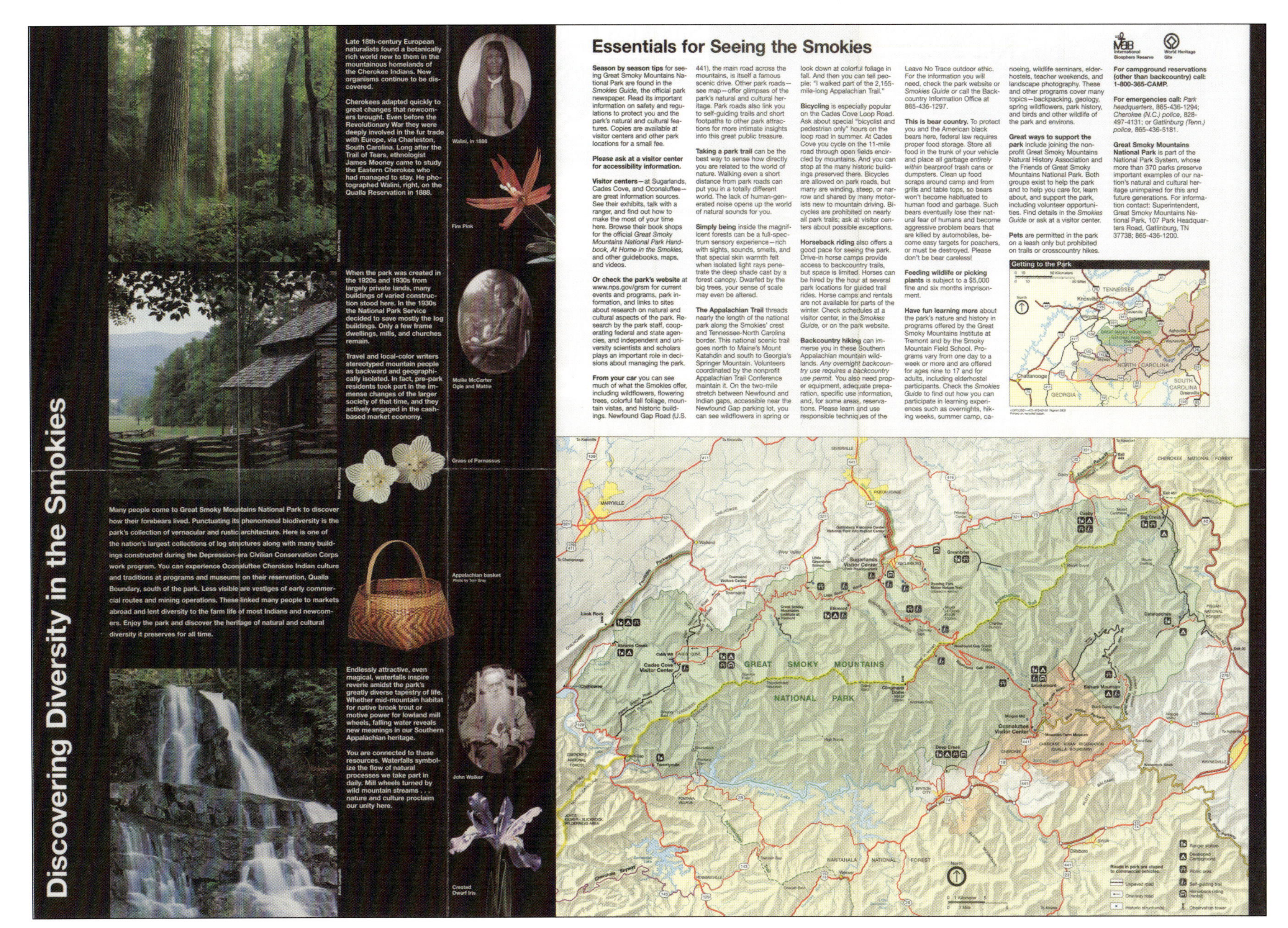

Figure 11.6
The reverse side of the Great Smoky Mountains pamphlet (titled "Discovering Diversity in the Smokies") includes a map of the park, a small overview highway map (called "Getting to the Park"), and more tourist/marketing information presented with photos and copy.

While there are symbols deployed on the map that are not included in the legend, the number is not large even taking these into account. This is a park map and its concern is with park things: ranger stations, campgrounds, picnic areas, self-guiding trails, historic structures. Here the topography, substantially generalized, is rendered as shaded relief, with a concomitant loss of information. Together with the contours, most of the place names have disappeared too.[6] It's not that there's no room for the names—there's plenty—but that the map is torn between the need to provide a roadmap and the desire to paint a portrait, and as we know there's no room on portraits for names. Hence the clash between the straightforward abstractions of the signs for roads, ranger stations, and campgrounds, and the ersatz realism of the relief.

Captured in this clash is the nature of the park itself, a place both for people to visit and for nonhuman things to thrive undisturbed. This clash is particularly well-focused in Great Smoky Mountains National Park. As an International Biosphere Reserve and home to more tree species than exist in all of northern Europe, to 1,500 kinds of flowering plants, to more than two hundred species of birds, to sixty of mammals, and to dozens of native fish, much of the park is managed as wilderness. But as a World Heritage Site containing the nation's largest collection of log buildings (among other things), the park is also managed as

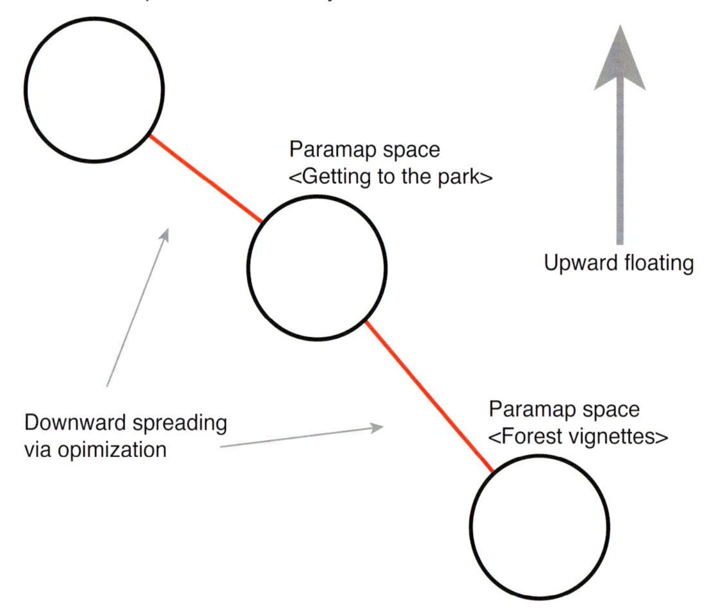

Figure 11.7
The cognitive structure of *Great Smoky Mountains* / "Discovering Diversity in the Smokies" assumes the main map on the poster side is base and spreads, via optimization, to the smaller map, "Getting to the Park," and to the vignettes of forest types on the other side, which are thereby spatialized. The vignettes naturalize everything through floating.

a museum that attracts more than twice as many visitors as any other national park in the country, in excess of ten million last year alone.

The park resolves this conflict by keeping most of its visitors on or near the roads. As text just above the main map points out, "From your car you can see much of what the Smokies offer, including wildflowers, flowering trees, colorful fall foliage, mountain vistas, and historic buildings." The text points out that U.S. 441 is itself a famous scenic drive, and adds that, "Park roads also link you to self-guiding trails and short footpaths to other park attractions," which obviate the need to venture into what on the map is green shaded relief. Since everyone using the map is in a car, the map has no need to name Short Bunk or Little Cataloochee Creek since neither is visible from the road, and indeed what the map posits is that beyond the road the park is one vast expanse of nature, the very nature sketched in the vignettes on the poster side.[7] This is a nature, it says in the first paragraph of introductory text, that it is the National Park Service's mission to preserve unimpaired for this and future generations.[8]

Another way the Park Service fulfills this mission is by penalizing poaching. In a little block of text next to the Cove Hardwood Forest vignette it says,

> Poachers with guns, dogs, and traps illegally kill bears and deer inside the park. More park bears are shot if they leave the park's protection to get food. Many are killed only for the gall bladder, considered an aphrodisiac in Asia. Plants are poached too. Ginseng roots are dug and sold illegally for their supposed "vital" nature.

Next to this text is a photograph of a ginseng root. The example, of course, is the one we opened our book with. When we did so, however, it was only to illustrate our contention that what makes a map a map is not the posting per se (*ginseng* grows *here*) but the fact that the map links postings together (ginseng grows here + Great Smoky Mountains National Park is here) to create and convey authority over territory (so that to pick ginseng in the park is to risk a year in prison[9]).

The maps behind the park

At this point, though, we can see that the map is much more deeply involved in the story than this, can see that it has taken map after map after map—each of them creating and conveying authority over territory—to bring this state of affairs into being. As we know now, maps are implicated in the construction of the very idea of nature that the park was created to preserve. When we speak of preserving nature today, especially "unimpaired for future generations," we're not talking about sticking chunks of it—a rock, a piece of coral, a stuffed bird, a dried flower—into a cabinet of curiosities. We're not talking about systematically storing samples of it in a natural history museum either. We're not even talking about nurturing examples in botanical gardens, arboretums, and zoos. What we're talking about is fencing off pieces of the earth's surface—some of them quite sizeable—and keeping at bay the forces that would change them, developers, for instance, and lumber and mining companies, ranchers, poachers.[10]

We've seen that the spatialization of nature this implies has taken centuries to bring about, and has been achieved by natural scientists interested in plants, in animals, in what would become geology, in the tides and the winds, in weather and its forecasting, in the night sky, indeed in every facet of what today we think about as the natural world. Over the years these scientists found such increasing utility in casting their thoughts—their data, their hypotheses, their conclusions—in the form of maps that it's now hard for us to think about nature without them.[11] From range maps in birders' books, through the maps that tracked Katrina, to maps of the new hiking trails through your favorite park, our lives are pervaded by them. It's this whole culture of a spatialized nature realized in maps that stands behind our park's interest in ginseng and its ability to penalize poachers.

The first nature we looked at was the nature threatened by man. It was this very nature that Great Smoky Mountains National Park was created to preserve. The self-serving language of the park map's introductory text says, "Alarmed at commercial logging threats to the forests, Congress authorized the park in 1926," though in fact by the time Congress acted the logging had been going on a long time and even the effort to create the park was forty years old. The problem for Congress was that the land proposed for Great Smoky Mountains was owned by thousands of people, all of whom would have to be bought out, unlike Yellowstone, for instance, or Yosemite, which had been carved out of public lands. Congress had a longstanding policy of not appropriating funds for park land, and its 1926 authorization of the park was entirely conditional on its advocates' ability to buy the land. In the end, John D. Rockefeller, Jr., contributed $5,000,000, the legislatures of North Carolina and Tennessee each $2,000,000, and a million was raised from smaller donors or, as our map puts it, "The states of North Carolina and Tennessee, private citizens and groups, and schools contributed money to purchase these lands for donation to the Federal Government." In 1933, Congress kicked in a final $2,000,000, and in 1934 it formally established the park.[12]

Behind the creation of the park stretched a trail of maps that successively realized the great mountain forests as a resource to be possessed, as a threatened resource, and as a park. Drawing on precedent efforts in the 1850s and 1860s by Arnold Guyot and others, the U.S. Geological Survey began publishing maps of the area in 1885 (figure 11.8). As you may recall from our chapter on possessable nature, in 1884 Charles Sargent published the first maps of U.S. tree distributions. In their 1897 report, *Timber Trees and Forests of North Carolina,* Gifford Pinchot and William Ashe published the first maps of state tree distributions.[13] In 1909—which turned out to be the peak year for lumbering in the Smokies—the U.S. Forest Service published *Forest Conditions/Western North Carolina.* This map laid out not only where lumbering had already taken place but, by implication, where it was going to. During the long fight for the park many other maps were produced in the debates over its location. The last of these was the Geological Survey's 1926 *Proposed Great Smoky Mountains National Park, North Carolina-Tennessee.* Over the next five years, the USGS produced a set of maps at the then unusually large scale of 1:24,000 for land acquisition and park planning. Later the Survey published still other maps, including a shaded relief map of the park in 1950, and the 1997 sheet we opened this chapter with. The Park Service began its mapping in 1931, publishing its first map of the park in 1940. Since then it's produced many maps, including the 2001 sheet we've illustrated.

Unavoidably, the successive maps tracked changes in the region, but because maps are realized as systems of signs, and signs are formed when an element from the plane of signification is linked to an element in the content plane, what the maps mostly track are changes in the content plane, that is, changes in the way we think about trees. The earlier maps, those of Guyot and the Geological Survey, don't think about trees at all. The trees simply aren't there for them. They're not on the content plane at all.

With the maps of Sargent, and Pinchot and Ashe, the opposite is true: there is little on the content plane except for trees, or not trees precisely but species—*nature,* if you recall our discussion of the range maps—of which the range is its spatial analogue. You may also recall our conclusion that range maps were tokens of a possession rarely entered into except when used in the control of nature; and, while the maps in the report of Pinchot and Ashe were not produced by the timber industry itself, they were very much produced in its spirit.[14] Here is their "Map showing economic distribution of yellow poplar" (figure 11.9), where the distinction between "few merchantable trees" and "merchantable timber" is the distinction between the eastern part of the state where most of the poplar had already been sawed (for crates, trucking boxes, and the like) and the mountains where, despite the 18 million board feet of poplar sawed west of the Blue Ridge in 1892 alone, Pinchot estimated that another 500 million board feet of poplar still stood.[15]

Yet it is also true that of the report's thirty-eight maps (posting the distribution of sixty-two species), only five were of these "economic distributions," and in fact Pinchot felt it necessary to *excuse* the failure to

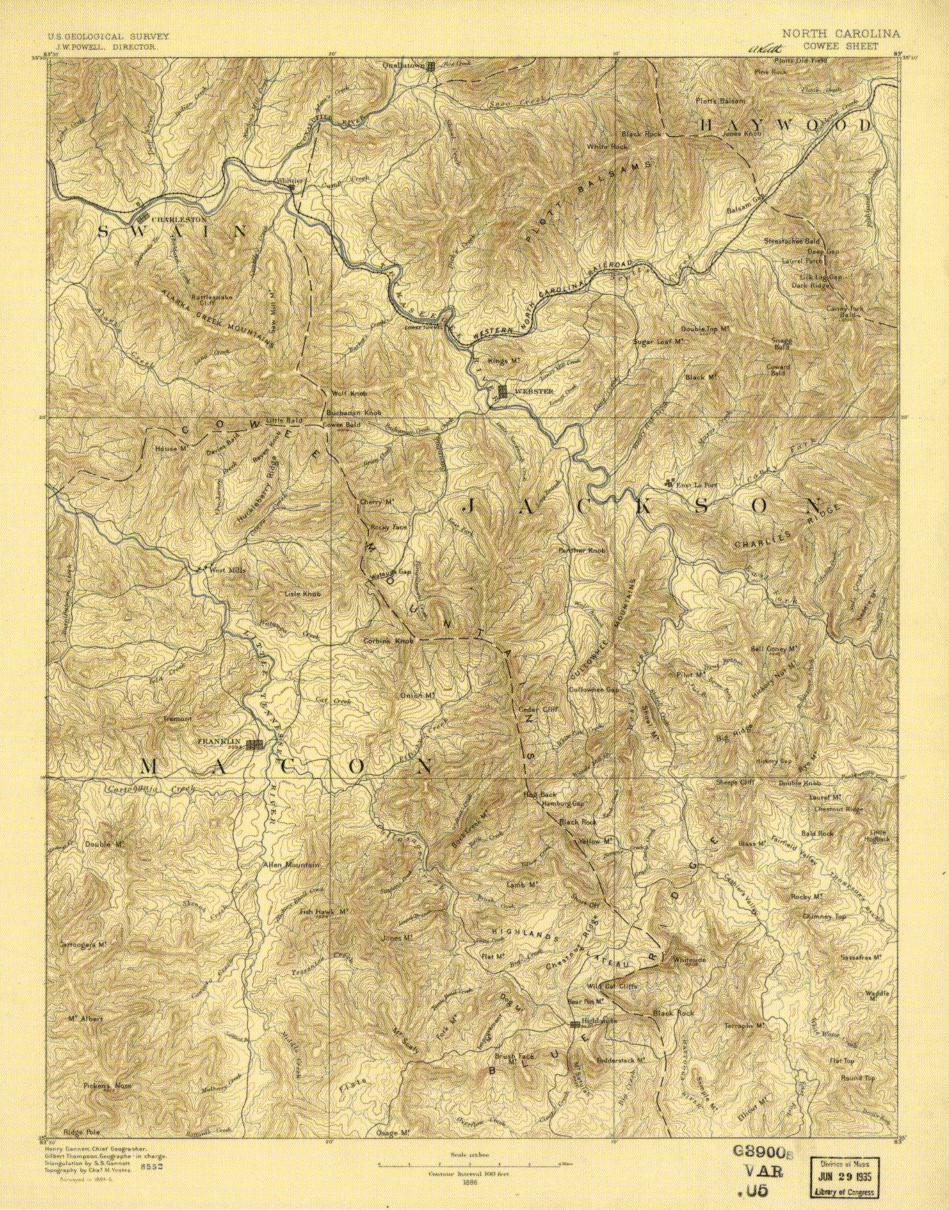

Figure 11.8
The oldest USGS maps of the areas that were eventually incorporated into Great Smoky Mountains National Park were of Cowee in 1885 (pictured here), Knoxville and Nantahalah in 1895, and Mount Guyot in 1893, all at a scale of 1:250,000. None of them, remarkably from our current perspective, paid the slightest attention to the tree cover, though all mapped a possessable nature.

Courtesy of USGS, 1885, and the Library of Congress, Geography and Map Division.

confine his and Ashe's attention to trees of economic interest: "The forest flora of no other State is more varied," he wrote, "nor in many ways so interesting as this. Partly for this reason a close restriction to those species which are of present value for timber has not been maintained."[16] In his descriptions of the high mountain forests Ashe waxed almost lyrical—"The cover of these forests has scarcely been broken, the tops of the trees presenting a nearly uniform surface throughout, the crowns closely interlocking and forming a dense shade"—and he was explicit about the rarity and the importance of the large stands of virgin forest in the Smokies.[17] Both men would go on to work for the Forest Service—Pinchot as its first chief forester—and both made

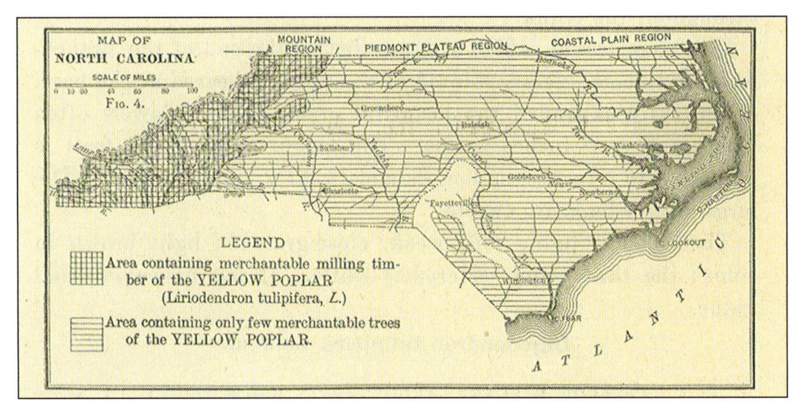

Figure 11.9
Map of North Carolina showing the economic distribution of yellow poplar. In contradistinction to the USGS maps of the area, which paid no attention to the tree cover at all, this map by Pinchot paid attention to little but the trees, though solely from an economic perspective: the amount of merchantable timber left to harvest. That is, this too was a map of a possessable nature.

Source: "Timber Trees and Forest of North Carolina" by Gifford Pinchot and William Ashe, 1897.

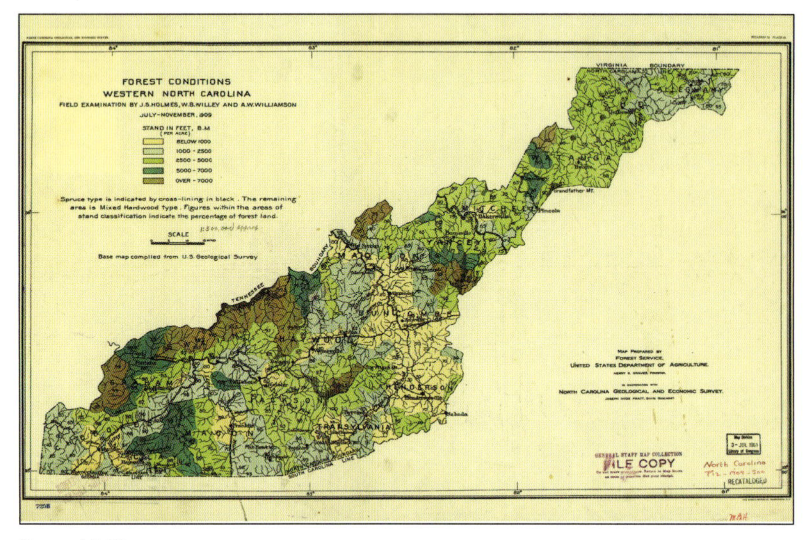

Figure 11.10
Forest Conditions of Western North Carolina. This map derives from the possessable nature tradition responsible for Pinchot's "Map showing economic distribution of yellow poplar" of figure 11.9; but with Ashe's likely involvement, it very probably also pointed to the threat facing the virgin stands of the high mountains, precisely where efforts to establish a park were already twenty years old, that is, toward a threatened nature. Map dimensions (WxH): 26in. x 15in.

Source: U.S. Forest Service and the North Carolina Geological and Economic Survey, 1909. Courtesy of the Library of Congress, Geography and Map Division.

significant contributions to the development of the National Forest system.[18]

Though there can be little question that the nature caught in their maps is a possessable one, explicitly so with the merchantable timber, the attitude behind the maps is complicated, and Pinchot's and Ashe's work presses toward both a knowable, scientific nature and one verging on awe. It was this latter nature, at the time increasingly brought into public consciousness by the writings of Emerson, Thoreau, Marsh, and Muir, and the paintings of Thomas Cole, Frederic Church, and Albert Bierstadt, that made it possible to begin thinking about timber as a threatened virginity rather than as an inexhaustible fecundity.[19] The U.S. Forest Service's *Forest Conditions of Western North Carolina* (figure 11.10) is every bit as ambiguous—as complicated in its motivations—as the range maps of Pinchot and Ashe. At one level, it's a map of the amount of timber per acre, in board measure, present in the state's western counties; but this forces the conclusion that timbering will have to move into the high mountains, into the very regions, that is, where Ashe had underlined the rarity and importance of the virgin forest and where, for the previous twenty years, effort had been increasingly sharply focused on establishing a park.[20] Although Ashe's name is not on the map, it is hard to imagine he had nothing to do with its production, especially since it was issued the year Ashe left his job as forester with North Carolina's Geological and Economic Survey and began working fulltime for the U.S. Forest Service. Given that the map was produced by the Forest Service in cooperation with the North Carolina Geological and Economic Survey this is even more likely.

If ultimately the range maps posted a possessable nature and *Forest Conditions* a threatened one—the very term "Conditions" implies the threat—then the Geological Survey's *Proposed Great Smoky Mountains National Park* (figure 11.11) posted a saved one, although the saving had to be postponed for another eight years, and even then was hardly inviolable.[21] This is an interesting map. It's based, as it says in the lower left-hand corner, on "U.S. Geological Survey topographic maps surveyed in 1885–1905"; so on the one hand, it's consumed the history of topographic mapping that began with Guyot. Yet in that it's a map of the proposed park, it has also consumed the history of tree mapping that began with Sargent and proceeded through Pinchot and Ashe and *Forest Conditions* to morph into the maps proposing the park, including this one. What makes it so interesting is the complete absence of trees from the map.

Everything at a certain level of generality

Think about it. It was the great forests everyone wanted to save, yet the map proposing the park that would save them doesn't even bother to post them. But then, how could it have? To begin with, the USGS didn't begin posting vegetation until well after 1905, so there wasn't any on the maps on which *Proposed Great Smoky Mountains National Park* was based.[22] But had vegetation been posted, what exactly would it have been? Recall the content categories that tumbled out of vegetation: woods, scrub, mangrove, orchard, vineyard. Clearly the only applicable category is woods, but what are woods? Woods, or woodland as the category is sometimes called, is "an area of normally dry land containing tree cover or brush that is potential tree cover. The growth must be at least 6 feet tall and dense enough to afford cover for troops."[23] This description can be further unpacked with respect to density, areas, clearings, and woodland boundary accuracy. For instance, woods (and scrub) are mapped (it says with respect to density), "if the growth is thick enough to provide cover for troops or to impede foot travel." The Survey takes this to be the case "if the average open-space distance between the crowns is equal to the average crown diameter,"[24] and as we know from Ashe, the crowns in the virgin stands of the Great Smokies were actually interlocking.

That is, except for a cove or two and the areas of proposed parkland being ferociously clear-cut to beat the establishment of the park, the whole thing should have been a smooth, uniform green, all of the whole proposed park, the national forests around it, and most

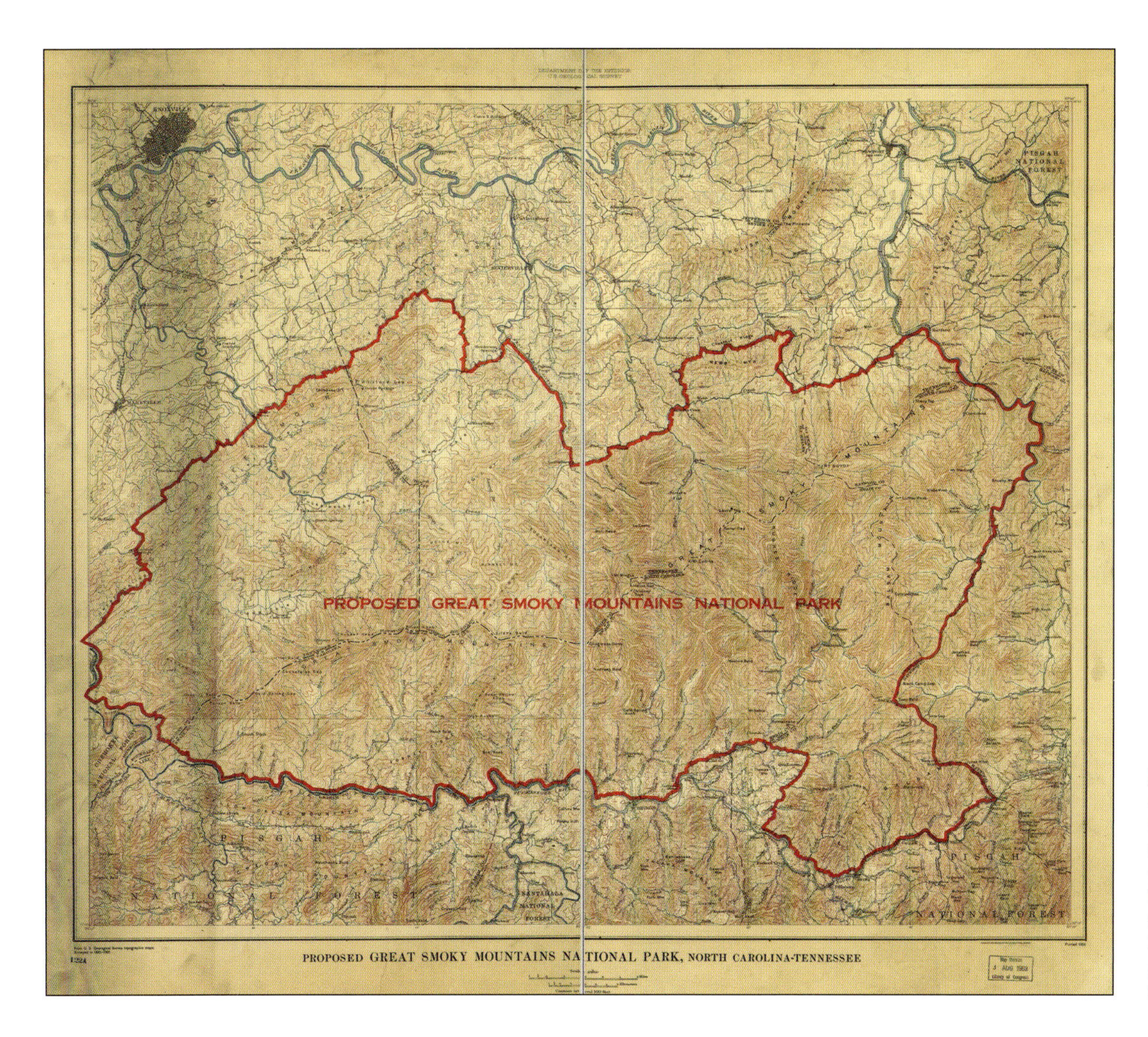

Figure 11.11
Proposed Great Smoky Mountain National Park, North Carolina-Tennessee. This is a map of nature saved, or about to be saved, from the threat mapped in *Forest Conditions of Western North Carolina*, and implicit, from today's perspective, in the earlier economic maps of Pinchot and Ashe. Map dimensions (WxH): 29in. x 23in.

Courtesy of USGS, 1926, and the Library of Congress, Geography and Map Division.

other adjacent lands, but without distinguishing dense, old-growth, virgin forest from younger second growth, or even more recently reforested clear cuts. The map would have been green, like *Cove Creek Gap* is green. Were it not for its fat red border, the proposed park would have been as hard to tell from its surroundings *with* the vegetation overlay as without it, and this fact points to what parks actually are and to what Nature—or nature—has become. After all, Great Smoky Mountains National Park is not a *natural* entity, but a legal, a social one. That is, the park is not a collection of trees, shrubs, and other wildlife, which would be no more than a forest, virgin or not. No, the park is a system of expectations, rules, regulations, and laws. The park is a way of *relating* to trees, shrubs, and other wildlife; and its expectations, rules, regulations, and laws are organized first and foremost to preserve the *natural* against encroachments of and destruction by the *cultural*.[25]

What does this mean? There are photographs that can make what it means clear to the dimmest eyesight. At the moment we're looking at a photograph of the western boundary of Yellowstone National Park where it abuts the Targhee National Forest. It's like two different worlds: one, Yellowstone, is forested; the other, the national forest, is not. As the caption says, "Twenty-five years of heavy logging have devastated the Targhee which is now dominated by large clear cuts ringed by strips of standing trees" (and which inappropriately enough the Survey's topographic quadrangles continue to show in woodland green). Here's another photo caption:

> A fence marks the boundary of Hakalau Forest National Wildlife Reserve, established high on the windward slope of Mauna Kea to preserve the area's 36 species of bird, including 7 endangered native species. The fence keeps out feral cattle and pigs, both descendants of animals introduced by Europeans, and alien plants. To the left of the fence is land that has been overgrazed by cattle; to the right, in the reserve, cattle have not grazed in five years, and the grasses and ferns are growing more luxuriously. Eventually the park managers hope to reestablish the slope's original forest."[26]

This is what is implied by the red line on *Proposed Great Smoky Mountains National Park*: the end of harvesting, mining, grazing; the beginning of a fight against nonnative species of all kinds; and, where necessary, management to maintain and reestablish the original forest.

These days such intentions are spelled out in elaborate decades-long strategic campaigns broken up into long-term goals that are implemented as five-year plans. There are any number of these for Great Smoky Mountains National Park and they cover everything from public education and outreach to the control of invasive animals.[27] In the latter case, the plan is to get one invasive species under effective control by 2008:

> Although there are several non-native animal and insect populations in [the park], we have identified 11 which are

"invasive"; these include European wild hogs, feral cats, balsam woolly adelgid, hemlock woolly adelgid, beech bark scale, fire ants, gypsy moth, Chinese jumping worm, Asian tiger mosquito, rainbow trout, and European starling. From 2004 through 2008, we plan to remove 1,050 wild hogs, treat 3,000 acres for hemlock woolly adelgid infestation, repeatedly treat 5 acres for balsam woolly adelgid infestation, and remove rainbow trout from 1.5 miles of stream. Fire ants will be treated as nests are found. Although we are working on several invasive populations, we plan to "effectively control" one.[28]

There is a similar plan to control fifty species of invasive exotic plants, another to monitor key vital signs of natural resources, and so on. The human effort committed to keeping nature natural in Great Smoky Mountains National Park is enormous.

What this amounts to is the fact that nature today must be managed if it is to stay natural, managed by humans within prescribed boundaries that show up as lines on maps (albeit maps the woolly adelgid can't read). In fact, this is the *only* way nature shows up on the U.S. Geological Survey's topographic quadrangles, as otherwise undifferentiated areas within boundaries drawn with a dash and a dot and a dash, usually fringed on the park side in red. Promiscuously sweeping together rivers and canals, mines and caves, marsh and rice field, bare rock and exposed wreck, the Survey's code otherwise calls out nothing at all as natural. In the inventory maintained by the Survey "of what there is that may be dug or gathered or harvested, and how and where it lies," nature manifests itself solely as that which may not be dug or gathered or harvested, which may at most be visited, preferably from the confines of a car: take only memories, or soon we'll not even have those to pluck.

We have come full circle, back to the threatened nature where we started, only to find it saved, preserved against the dangers that confronted it. It's as though having spatialized nature, having constructed it as this extended thing, we realized the only way we could keep it was to give it up, and that the only way we could give it up was to be kept from it, was to lock it away behind a wall of permits and prohibitions, which allow it to be maintained in a condition that conforms to the idea of nature we've constructed.[29] Here the map works like a photograph after all, isolating the natural, preserving it against the ravages of history, but only by enabling a system of rules that permits no more communion than that vouchsafed by the eyes and then only for a glimpse through the windshield of a car.

Better yet, the screen of your TV.

Perhaps best of all on the flip side of a map. And perhaps this *is* where nature is most accessible, at once preserved and displayed, in both its particularities and its spatialities. Certainly this is true of the park map. From your car your chance of seeing a bobcat or a brook trout, a black bear or a box turtle is remote. It's not even very good from one of the park trails, no mater how completely you may be immersed there, in the words of the map, "in a full-spectrum sensory experience—rich with sights, sounds, smells, and that special skin warmth felt when isolated light rays penetrate the deep shade cast by a forest canopy." But there they are on the flip side of the map—the map which spreads these creatures out across the eight hundred square miles of the park—the box turtle and the black bear, the brook trout and the bobcat, a little unnaturally crowded together perhaps, but at least naturally posed. One of you can even read about them to the other as you speed back to Gatlinburg or on to Asheville and the Econo Lodge or the Days Inn.

This is also true of the first map we looked at in any detail, "Australia: Land of Living Fossils." There they were—all those animals! all that fur and feathers!—for whom "Australians have set aside 3 percent of their nation," preserved "for the rare creatures of a land that time forgot." This 3 percent shows up on the map proper—how else?—as a little park here (Cape Arid National Park) and a little park there (Little Desert National Park), and here a bigger one (Great Victoria Desert Nature Reserve) and there a bigger one still (Tanami Desert Wildlife Sanctuary), within dashed lines—most of them sort of square—and green, even the desert parks. The wildlife pictured on the poster side seeps through the page to make the green = nature, as the map seeping back forces each of the animals to signify Australia: koala = Australia, kangaroo = Australia, platypus = Australia.

It's the final nature, isn't it, the nature of parks? Where the nature that maps helped bring into being—of neotropical fauna and the world of flowers, of Mount Everest and the Grand Canyon—finally retreats, to dwindle into gated preserves outside whose (carefully mapped) borders the bears looking for food are slaughtered, the forests are felled, cattle graze, and roads reign. It's said that maps are important tools in the fight to preserve nature, but maybe all that's true is that we've mapped nature to death.

Notes

1. While legends are traditionally a piece of a map's perimap, the U.S. Geological Survey's "Topographic Map Symbols" is so elaborate it's a piece of every quadrangle's epimap.

2. In the words of A. S. Hewitt, congressman of New York, who wrote the legislation authorizing the Geological Survey in 1879. We have taken minor liberties with the phrasing. See Thompson, *Maps for America: Cartographic Products of the U.S. Geological Survey and Others* (Washington, D.C.: U.S. Geological Survey, 1979), v.

3. Ibid., 27.

4. We found this at www.smokiesstore.org.

5. Tufte, *Envisioning Information* (Cheshire, Conn.: Graphics Press, 1990), 49.

6. The conflict between contours and shaded relief is not inherent. In fact in 1950, the USGS published a map of the park at a scale of 1:125,000 that displayed both and was also thick with place names. As we stressed in the last chapter, the decision to delete information is made in the name of a certain idealism. The first Park Service map we know of that displays the park in shaded relief dates to 1981 and is at a scale of about 1:168,000. Though this displays plenty of names, what it's really dense with are trails, each segment of every one of which has its length indicated. Names and other data really start disappearing from the Park Service maps with the 1:300,000 1991 version.

7. Or as one guide book puts it, "If you stay in your car, you'll miss most of the park," which is exactly what the park wants (Hargan, *The Blue Ridge and Smoky Mountains* [Woodstock, Vt.: Countryman Press, 2002], 268).

8. This language comes from the 1926 Act of Congress that authorized the park. "The Park's purpose," the act says, is "to preserve exceptionally diverse resources and to provide for public benefit from and enjoyment of those resources in ways which will leave them basically unaltered by human influences." The relationship of this language to the park's mission statement is discussed in the *2005 Strategic Plan: Great Smoky Mountains National Park, October, 2005–September 30, 2008.*

9. As we pointed out in a footnote in our first chapter, poaching in the national park carries a maximum sentence of a year; though, as Burkhard Bilger reports in "Wild Sang," no one has ever received this sentence.

10. Poaching is not a trivial concern. Between 1991 and 2001, 11,654 pounds of illegally harvested ginseng roots (an estimated 3,496,200 plants) were seized in Great Smoky Mountains National Park alone, which is soon to publish the conclusions of an exhaustive study of this problem. Since April 1, 2003, North Carolina has required that even wild ginseng plants collected outside the park be at least five years old and have three prongs (leaves) or, in the absence of leaves, at least four discernible bud scars plus a bud on the rhizome. The new regulation also requires harvesters to plant the seeds of any harvested plants within one hundred feet of where ginseng is located in the wild (see the Rules on Ginseng Collection and Trade in North Carolina, North Carolina Administration Code, Title 2, 48F).

11. In a related domain, Tom Koch has recently traversed the parallel spatialization of disease in his compendious and lavishly illustrated *Cartographies of Disease: Maps, Mapping, and Medicine* (Redlands, Calif.: ESRI Press, 2005). Koch begins his survey in the late seventeenth century, but as we know the groundwork was laid earlier. See Findlen's *Possessing Nature,* as well as David Freedberg's more recent *The Eye of the Lynx: Galileo, His Friends, and the Beginnings of Modern Natural History* (Chicago: University of Chicago Press, 2002), with its wealth of illustrations, many in color.

12. Dr. Chase P. Ambler, of Asheville, made the first suggestion for a national park in the southern Appalachians. From 1889 to 1905 he directed a strongly supported campaign that nevertheless failed to win congressional approval. With the failure of this effort, interest turned to the creation of national forests, where the lumbering at least could be controlled, and these achieved some success. The successful effort to establish the park was initiated in 1923 by Willis Davis, of Knoxville, working with Stephen Mather of the then only recently formed National Park Service. The story is easily turned into a hero saga, whether about Ambler, Davis, Mather, or John D. Rockefeller, Jr. See, among many others, Arthur Stupka's *Great Smoky Mountains National Park: National History Handbook Series No. 5* (Washinton, D.C., Government Printing Office, 1960), 67–68; Norman T. Newton's *Design on the Land: The Development of Landscape Architecture* (Cambridge, Mass.: Harvard University Press, 1971), 536; and the Library of Congress's "Mapping the National Parks: Maps of Great Smoky Mountains National Park" Web site compiled by Patricia Molen van Ee. The strength of Newton's synoptic (if canonic) version is that it's embedded in the larger history of the national parks movement, and this within the still larger history of the parks movement generally. The strength of van Ee's Web site is its cartographic focus and wealth of illustrations.

13. In chapter 8, we alluded to the sixteen colored range maps Sargent attached as a special portfolio to his *Report on the Forests of North America* (Washington, D.C.: Government Printing Office, 1884). Sargent's report contained an additional thirty-nine black-and-white maps of forests, including a "Map of North Carolina Showing the Distribution of the Pine Forests with Special Reference to the Lumber Industry" (which was pretty much confined to the Coastal Plain). Acknowledging and building on Sargent's contributions, in 1897 Gifford Pinchot and William W. Ashe published "Timber Trees and Forests of North Carolina" *North Carolina Geological Bulletin,* no. 6 (1897). The maps, amounting to the first state atlas of species ranges ever published, all fall in Pinchot's essay on timber trees, which he considered species by species. Ashe wrote about the state's forests, which he considered region by region. The combination of the systematic and regional perspectives made for a compelling portrait not only of North Carolina's trees but of the profession of forestry.

14. The bulletin had been planned as part of the North Carolina forest exhibition at the Chicago World's Fair. Pinchot tells us that work on the bulletin had begun before the Fair's opening in 1893, but he chooses not to explain the delay in publication.

15. Sargent had previously drawn the same distinction, noting both that, "A larger proportion of the pine forests of the coast has been destroyed in North Carolina than in any of the other southern states," and that, "The inaccessibility of this mountain region has protected these valuable forests up to the present time" (Report on the Forests, 515).

16. Pinchot and Ashe, "Timber Trees and Forests," 13. This too had been observed by Sargent who wrote, "The forests of North Carolina were once hardly surpassed in variety and importance by those of any other part of the United States" (Report on the Forests, 515).

17. Pinchot and Ashe, "Timber Trees and Forests," 219–22.

18. In a very interesting paper, "Visual Harvest: Ambiguity in the U.S. Forest Science at Home and in the Tropics, 1890–1925" (Annual Meeting, History of Science Society, Milwaukee, 2002), Hannes Toivanen argues that early leaders of American forestry such as Bernhard Fernow and Gifford Pinchot believed in industrial capitalism, including lumbering, "but sought to control the damage to nature that shocked the voting public." Fernow and Pinchot positioned forestry to "mediate between private and public interests in a way that promised a win-win situation [and] perceived forestry as a special variant of scientific management that required . . . particular technologies of representation—statistics and maps—to demonstrate their claims about American nature" (all p. 2). Toivanen stresses the distinction between this approach and the "romanticizing preservationism of John Muir," especially Fernow's belief that because "preservationists were helpless in the face of industrial capitalism" only forestry could control the lumber industry (p. 4). It is precisely this tension we feel in Pinchot and Ashe's work and in their maps which, Toivanen insists, "were the material technology of the systems analysis they advanced."

19. As early as 1873, Franklin Hough had read a paper to the American Association for the Advancement of Science called "On the Duty of Governments in the Preservation of Forests." Classic discussions of the growing idea of nature as an essential national interest during this period include Henry Nash Smith's *Virgin Land* (New York: Knopf, 1950); Leo Marx's *Machine in the Garden* (New York: Oxford, 1964); and Barbara Novak's *Nature and Culture: American Landscape and Painting, 1825–1875* (New York: Oxford, 1980).

20. Kirk Fuller points out that reports made by Ashe in conjunction with the U.S. Bureau of Forestry in 1903–1904, "aroused public attention to the rapid rate at which the southern mountain slopes were being denuded and the resultant permanent injury" (in his "History of North Carolina State Parks" in Histories of Southeastern State Park Systems [Associations of Southeastern State Park Directors, 1977]). In his important history, *Americans and Their Forests: A Historical Geography* (Cambridge: Cambridge University Press, 1989), Michael Williams concludes that the debate over forest depletion slid easily and naturally into a debate about forest preservation and conservation (p. 394). This map embodies that slide.

21. The story of the Little River Lumber Company is illustrative. In 1925, the company agreed to sell its 77,000 acres if it could retain logging rights for fifteen years. The Tennessee National Park Commission agreed, and the company denuded the virgin forests of the Little River drainage by 1939, the last five years within the borders of the established park. In the words of Jim

Hargan, *Blue Ridge*, "The logging company had purchased a huge tract of old-growth forest, removed every tree, left a destroyed land bereft of economic value, and sold it to the government for a 50 percent profit" (301). For a very different perspective, in which the company heroically advocates for the park, see Robin Bible's "Stringtowns: Early Logging Communities in the Great Smoky Mountains" *Forest History Today* (Spring 2002): 29–32.

22. According to Arthur Krim—curator of the 2005 Harvard Map Library exhibition *Early USGS Topographic Maps: Cartographic Portraits of the American Frontier, 1878–1893*, the Survey didn't begin routinely posting vegetation until 1917–1918. Matthew Edney, director of The History of Cartography Project at the University of Wisconsin, believes he's seen the vegetation overlay as early as 1910 on selected Maine quads. Even this, though, would postdate by five years the mapping of the Great Smokies on which Proposed Great Smoky Mountains National Park was based.

23. Thompson, *Maps for America*, 70. In *New Nature of Maps*, J. B. Harley cites this description of Thompson's as "a trace of the military mind" (pp. 39–40); while in *How to Lie with Maps* (Chicago: University of Chicago Press, 1996), Mark Monmonier refers to it as "perhaps the most subtle example of military influence on civilian topographic maps," following a historical sketch of military influence on the Geologic Survey (p. 127). We are less convinced that the relationship is this straightforward, especially given the late date for the application of the vegetation overlay, though it's conceivable the specification had been written in the 1880s and not applied until technological developments in printing made it feasible. In any case, as stated in Thompson, the description is neither a "trace" nor particularly "subtle." According to the Army's own *Map Reading and Land Navigation*, green "identifies vegetation with military significance, such as woods, orchards, and vineyards" (Department of the Army, FM 21–26, Washington, D.C., 1987, p. 3–8).

24. All Thompson, *Maps for America*, 70–71.

25. Leila Harris and Helen Hazen give us a nuanced, or perhaps merely ambivalent, review of this issue in their recent "Power of Maps: (Counter) Mapping for Conservation" *ACME: An International E-Journal for Critical Geographers* 4, no. 1 (2006): 99–130.

26. Dobson, *Conservation and Biodiversity* (New York: Scientific American Library, 1996), 197, 175.

27. There are thirty-eight Long-term Goal Performance Targets included in the 2005 Strategic Plan (see note 8 above), not all of which have plans to achieve them.

28. Ibid., where this is Park Goal Number Iao10

29. In *Walden*, Thoreau says, "Not till we are lost, in other words, not till we have lost the world, do we begin to find ourselves, and realize where we are and the infinite extent of our relations."

Bibliography

Abramson, Howard. *National Geographic: Behind America's Lensn the World.* New York: Crown Publishing, 1987.

Ackerman, Jennifer. "Cranes: The Long Road Home." *National Geographic,* April 2004: 38–57.

Ades, Dawn. *Photomontage.* London: Thames and Hudson, 1986 [1976].

Allen, Richard Hinckley. *Star Names: Their Lore and Meaning.* New York: Dover Books, 1963 [1899].

Alpha, Tau Rho, Janis S. Detterman, and James M. Morley. *Atlas of Oblique Maps: A Collection of Landform Portrayals of Selected Areas of the World.* USGS Miscellaneous Investigations Series I-1799. Reston, Va.: U.S. Geological Survey, 1988.

Amarel, Saul. "On the Mechanization of Creative Processes," *IEEE Spectrum* 3 (April 1966): 112–14.

Anderson, Edgar. *Plants, Man and Life.* Berkeley: University of California Press, 1952.

Annerino, John. *Canyons of the Southwest: A Tour of the Great Canyon Country from Colorado to Northern Mexico.* Tucson, Ariz.: University of Arizona Press, 1993.

Baigell, Matthew. *Albert Bierstadt.* New York: Watson-Guptill, 1981.

Bailey, Robert G. *Ecosystem Geography.* New York: Springer-Verlag, 1996.

Bakhtin, M. M. *Speech Genres and Other Late Essays.* Austin, Tex.: University of Texas Press, 1986.

Barthes, Roland. *The Fashion System.* New York: Hill and Wang, 1983 [1967].

———. *Image – Music – Text.* New York: Hill and Wang, 1977.

———. *Mythologies.* New York: Hill and Wang, 1972 [1957].

———. "The Plates of the Encyclopedia." in *New Critical Essays.* New York: Hill and Wang, 1980 [1972].

———. *S/Z.* New York: Hill and Wang, 1974 [1970].

Bataille, Georges. "The Language of Flowers." In *Visions of Excess: Selected Writings, 1927-1939.* Minneapolis, Minn.: University of Minnesota Press, 1985.

Baudrillard, Jean. "Simulacra and Simulations: Disneyland." In *Social Theory: The Multicultural and Classic Readings,* edited by Charles Lemert, 524-529. Boulder: Westview Press, 1993.

Baughman, Mel, ed. *Reference Atlas to the Birds of North America.* Washington, D.C.: National Geographic Society, 2003.

Bennett, Donald P., and David A. Humphries. *Introduction to Field Biology.* 2nd ed. London: Edward Arnold, 1974.

Bertin, Jacques. *Semiology of Graphics: Diagrams, Networks, Maps.* Madison, Wis.: University of Wisconsin Press, 1983 [1967].

Bible, Robin. "Stringtowns: Early Logging Communities in the Great Smoky Mountains." *Forest History Today,* Spring 2002, 29–32.

Bilger, Burkhard. "Wild Sang: Rangers, Poachers, and Roots that Cost a Thousand Dollars a Pound." *The New Yorker,* July 15, 2002, 38–47.

Bishop, Barry C. "Landsat Looks at Hometown Earth." *National Geographic,* July 1976, 140–47.

Bixel, Patricia, and Elizabeth Turner. *Galveston and the 1900 Storm.* Austin, Tex.: University of Texas Press, 2000.

Black, Jeremy. *Maps and History: Constructing Images of the Past.* New Haven, Conn.: Yale University Press, 1997.

———. *Maps and Politics.* Chicago: University of Chicago Press, 1997.

Blaut, J. M. *The Colonizer's Model of the World.* New York: Guilford, 1993.

Boime, Albert. *The Magisterial Gaze: Manifest Destiny and American Landscape Painting, c. 1830-1865.* Washington, D.C.: Smithsonian, 1991.

Bonner, John Tyler. *The Scales of Nature.* New York: Harper and Row, 1969.

Boomgaard, Peter. *Frontiers of Fear: Tigers and People in the Malay World, 1600-1950,* New Haven, Conn.: Yale, 2001.

Boucher, Norman, ed. *Bird Lover's Life List and Journal.* Boston: Museum of Fine Arts, 1992.

Boyle, Bill. *My First Atlas.* London: Dorling Kindersley, 1994.

Bramwell, Anna. *Ecology in the 20th Century: A History.* New Haven, Conn.: Yale, 1989.

Breeden, Stanley, and Kay Breeden. *Australia's South East: A Natural History of Australia.* Vol. 2. Sydney: Collins, 1972.

Brockman, C. Frank. *Trees of North America: A Field Guide to the Major Native and Introduced Species North of Mexico.* New York: Golden Books, 1968.

Bryan, C. D. B. *The National Geographic Society: 100 Years of Adventure and Discovery.* Rev. ed. New York: Abrams, 1997.

Bull, John, and John Farrand, Jr. *The Audubon Society Field Guide to North American Birds, Eastern Region.* New York: Alfred A. Knopf, 1977.

Bullard, E. C., J. E. Everett, and A. G. Smith. "The Fit of the Continents around the Atlantic." In *A Symposium on Continental Drift,* P. M. S. Blackett, E. C. Bullard, and S. K. Runcorn, eds. *Philosophical Transactions of the Royal Society,* A258 (1965): 41–51.

Buol, S. W., F. D. Hole, and R. J. McCracken. *Soil Genesis and Classification.* Ames, Iowa: Iowa State University Press, 1989.

Burnham, Jr., Robert. *Burnham's Celestial Handbook: An Observer's Guide to the Universe Beyond the Solar System.* Rev. ed., 3 vols. New York: Dover, 1978 [1966].

Burnie, David. *Eyewitness Books: Birds.* New York: Knopf, 1988.

Camerini, Jane. *Darwin, Wallace and Maps.* PhD diss., University of Wisconsin, Madison, 1987.

———. "Evolution, Biogeography, and Maps: An Early History of Wallace's Line," *Isis* 84 (1993): 700–27.

———. "*The Physical Atlas* of Heinrich Berghaus: Distribution Maps as Scientific Knowledge." In *Non-Verbal Communication in Science Prior to 1900*, edited by R. G. Mazzolini, 479-512. Florence: Olschki, 1993.

Cameron, Ian, and Robin Wood. *Antonioni.* Rev. ed. New York: Praeger, 1971.

Campbell, Neil A. *Biology.* 4th ed. Menlo Park, Calif.: Benjamin/Cummings, 1996.

Carr, Gerald L. *Bierstadt's West.* New York: Gerald Peters Gallery, 1997.

Charrand, Mark R. *The National Audubon Society Field Guide to the Night Sky.* New York: Knopf, 1991.

Clark, T. J. *The Painting of Modern Life: Paris in the Art of Manet and His Followers.* Princeton: Princeton University Press, 1984.

Cohen, Michael P. *The History of the Sierra Club, 1892-1970.* San Francisco: Sierra Club, 1988.

———. *The Pathless Way: John Muir and American Wilderness.* Madison, Wis.: University of Wisconsin Press, 1981.

Collins, Henry Hill. *Complete Field Guide to American Wildlife.* New York: Harper and Row, 1959.

Cooper, James Fenimore. *The Pathfinder.* New York: Signet, 1961.

Cosgrove, Denis. *Apollo's Eye: A Cartographic Genealogy of the Earth in the Western Imagination.* Baltimore: Johns Hopkins University Press, 2001.

———, ed. *Mappings.* London: Reaktion Books, 1999.

Cosgrove, Denis, and Stephen Daniels, eds. *The Iconography of Landscape.* Cambridge: Cambridge University Press, 1988.

Cox, Kenyon. *What Is Painting?: "Winslow Homer" and Other Essays.* New York: Norton, 1988.

Crary, Jonathan. *Techniques of the Observer: On Vision and Modernity in the Nineteenth Century.* Cambridge, Mass.: MIT Press, 1990.

Critchfield, Willis, and Elbert L. Little, Jr. *Geographic Distribution of Pines of the World, Miscellaneous Publication 991.* Washington, D.C.: United States Department of Agriculture, 1966.

Cronon, William: *Changes in the Land: Indians, Colonists, and the Ecology of New England.* New York: Hill and Wang, 1983.

———, ed. *Uncommon Ground: Rethinking the Human Place in Nature.* New York: W. W. Norton & Co., 1996.

Dancygier, Barbara, and Eve Sweetser. *Mental Spaces in Grammar: Conditional Constructions.* Cambridge: Cambridge University Press, 2005.

Davis, Derek S., and Sue Brown, eds. *The Natural History of Nova Scotia.* 2 vols. Nova Scotia: Nova Scotia Museum/Department of Education and Culture, 1996.

Del Casino, Vincent, and Stephen Hanna. "Mapping Identities, Reading Maps: The Politics of Representation in Bangkok's Sex Tourism Industry." In *Mapping Tourism*, edited by Hanna and Del Casino, 161–85. Minneapolis, Minn.: University of Minnesota Press, 2005.

Dellenbaugh, Frederick S. *A Canyon Voyage: The Narrative of the Second Powell Expedition down the Green-Colorado River from Wyoming, and the Explorations on Lands, in the Years 1871 and 1872.* New Haven, Conn.: Yale University Press, 1926.

Dent, Borden. *Cartography: Thematic Map Design.* 3rd ed. Dubuque: William Brown Publishers, 1993.

Derrida, Jacques. *Of Grammatology.* Baltimore, Md.: Johns Hopkins University Press, 1976 [1967].

———. *The Truth in Painting.* Chicago: University of Chicago Press, 1987 [1978].

De Roos, Robert. "The Flower Seed Growers: Gardening's Color Merchants." In *National Geographic*, May 1968, 720–38.

De Santillana, Giorgio, and Hertha von Dechend. *Hamlet's Mill: As Essay on Myth and the Frame of Time.* Boston: Gambit, 1969.

De Saussure, Ferdinand. *Course in General Linguistics.* New York: McGraw-Hill, 1966 [1916].

Devlin, John, and Grace Naismith. *The World of Roger Tory Peterson.* New York: Times Books, 1977.

Dobson, Andrew P. *Conservation and Biodiversity.* New York: Scientific American Library, 1996.

Duncan, Wilbur H., and Marion B. Duncan. *Trees of the Southeastern United States.* Athens, Ga.: University of Georgia Press, 1988.

Du Toit, Alexander L. *A Geological Comparison of South America with South Africa.* Washington, D.C.: Carnegie Institution of Washington, 1927.

———. *Our Wandering Continents: An Hypothesis of Continental Drifting.* Edinburgh: Oliverand Boyd, 1933.

Dutton, Clarence E. *Tertiary History of the Grand Canyon District, With Atlas.* Washington, D.C.: U.S. Government Printing Office, 1882.

Eco, Umberto. *A Theory of Semiotics.* Bloomington, Ind.: Indiana University Press, 1976.

Edney, Matthew. *Mapping an Empire: The Geographical Construction of British India, 1765-1843.* Chicago: University of Chicago Press, 1997.

Elias, Thomas S. *The Complete Trees of North America: Field Guide and Natural History.* New York: Grolier, 1980.

Encyclopedia Britannica. 11th ed., vol. XXI. Physiologus, 1910–1911.

Erickson, R. O. "The *Clematis fremontii* var. *riehlii* population in the Ozarks." *Annals of the Missouri Botanical Garden* 32 (1945): 413–60.

Espinosa, Alvaro F., Wilbur Rinehart, and Marie Tharp. *Seismicity of the Earth, 1960-1980.* Washington, D.C.: U.S. Navy Office of Naval Research, 1982.

ESRI. *ESRI Map Book.* Vol. 19. Redlands: ESRI Press, 2004.

Euler, Robert C., and A. Trinkle Jones. *A Sketch of Grand Canyon Prehistory.* Grand Canyon, Ariz.: Grand Canyon Natural History Association, 1979.

Farber, Paul Lawrence. *Discovering Birds: The Emergence of Ornithology as a Scientific Discipline, 1760-1850.* Baltimore: Johns Hopkins University Press, 1997 [1982].

———. *Finding Order in Nature: The Naturalist Tradition from Linnaeus to E. O. Wilson.* Baltimore, Md.: Johns Hopkins University Press, 2000.

Fauconnier, Gilles. *Mappings in Thought and Language.* Cambridge: Cambridge University Press, 1997.

Fauconnier, Gilles, and Eve Sweetser, eds. *Spaces, Worlds and Grammar.* Chicago: University of Chicago Press, 1996.

Fauconnier, Gilles, and Mark Turner. *The Way We Think: Conceptual Blending and the Mind's Hidden Complexities.* New York: Basic Books, 2002.

Foucault, Michel. *Discipline and Punish: The Birth of the Prison.* New York: Vintage, 1979 [1975].

———. "The Eye of Power," in *Power/Knowledge*, edited by C. Gordon, 146–65. New York: Pantheon, 1980.

Findlen, Paula. "The Formation of a Scientific Community: Natural History in Sixteenth Century Italy." In *Natural Particulars: Nature and the Disciplines in Renaissance Europe*, edited by Anthony Grafton and Nancy Siraisi, 369–400. Cambridge, Mass.: MIT Press, 1999.

———. *Possessing Nature: Museums, Collecting, and Scientific Culture in Early Modern Italy.* Berkeley: University of California Press, 1994.

Fowells, H. A., compiler. *Silvics of Forest Trees of the United States, Agriculture Handbook 271.* Washington, D.C.: United States Department of Agriculture, 1965.

Frank, Andre Gunder. *ReOrient: Global Economy in the Asian Age.* Berkeley, Calif.: University of California Press, 1998.

Freedberg, David. *The Eye of the Lynx: Galileo, His Friends, and the Beginnings of Modern Natural History.* Chicago: University of Chicago Press, 2002.

———. *The Power of Images: Studies in the History and Theory of Response.* Chicago: University of Chicago Press, 1989.

Fuller, Kirk. "History of North Carolina State Parks." In *Histories of Southeastern State Park Systems,* 126. Association of Southeastern State Park Directors, 1977.

Fussell, Paul. *Class: A Guide Through the American Status System.* New York: Simon & Schuster, 1983.

Gaffney, B. M. "Roots of the Niche Concept." *American Naturalist* 109 (1973): 490.

Gall, James. "Use of Cylindrical Projections for Geographical, Astronomical, and Scientific Purposes." *Scottish Geographical Magazine* 1 (1885): 119–23.

Galilei, Galileo. *Sidereus Nuncius or The Sidereal Messenger,* trans. by Albert Van Helden. Chicago: University of Chicago Press, 1989 [1610].

Gallup Organization. *Geography: An International Gallup Survey 1988.* Princeton: Gallup, 1988.

Genette, Gerard. *Paratexts: Thresholds of Interpretation.* Cambridge: Cambridge University Press, 1997 [1987].

———. *Palimpsests: La littérature au second degré.* Paris: Seuil, 1982.

George, Wilma. *Animals and Maps.* Berkeley, Calif.: University of California Press, 1969.

Glacken, Clarence. *Traces on the Rhodian Shore: Nature and Culture in Western Thought from Ancient Times to the End of the Eighteenth Century.* Berkeley, Calif.: University of California Press, 1967.

Glen, William. *The Road to Jaramillo: Critical Years of the Revolution in Earth Science.* Stanford, Calif.: Stanford University Press, 1982.

Godlewska, Anne. "The Napoleonic Survey of Egypt: A Masterpiece of Cartographic Compilation and Early Nineteenth Century Fieldwork." *Cartographica Monograph 25* (1988).

Godlewska, Anne, and Neil Smith, eds. *Geography and Empire.* Oxford: Blackwell, 1994.

Goffman, Erving. *Frame Analysis: An Essay on the Organization of Experience.* Cambridge, Mass.: Harvard University Press, 1974.

Gombrich, E. H. *The Preference for the Primitive: Episodes in the History of Western Taste and Art.* London: Phaidon, 2002.

———. *The Story of Art.* 12th ed. London: Phaidon, 1972.

Gore, Rick. "A Bad Time to be a Crocodile." *National Geographic,* January 1978: 90–115.

Gorman, James. "So Many Field Guides You Need a Guide." *New York Times,* July 20, 2004: D3.

Gould, James L., and Carol Grant Gould, eds. *Life at the Edge: Readings From Scientific American Magazine.* New York: Freeman, 1989.

Gould, Stephen Jay. *I Have Landed: The End of a Beginning in Natural History.* New York: Harmony Books, 2002.

Grosvenor, Gilbert. "New Atlas Explores a Changing World," *National Geographic,* November 1990, 126–29.

Hacking, Ian. *The Emergence of Probability.* Cambridge: Cambridge University Press, 1975.

———. *The Social Construction of What?* Cambridge, Mass.: Harvard University Press, 1999.

Hansard, Peter, and Burton Silver. *What Bird Did That? A Driver's Guide to Some Common Birds of North America.* Berkeley, Calif.: Ten Speed Press, 1991.

Hargan, Jim. *The Blue Ridge and Smoky Mountains.* Woodstock, Vt.: Countryman Press, 2002.

Harley, J. B. *The New Nature of Maps: Essays in the History of Cartography.* Baltimore, Md.: Johns Hopkins University Press, 2001.

Harper Collins. *The Harper Atlas of World History.* New York: Harper Collins, 1992.

Harris, Leila, and Helen Hazen. "Power of Maps: (Counter) Mapping for Conservation." *ACME: An International E-Journal for Critical Geographers* 4, no. 1 (2006): 99–130.

Harrison, Colin, and Alan Greensmith. *Birds of the World, Eyewitness Handbooks.* New York: Dorling Kindersley Publishing, 1993.

Harvey, P. D. A. *The History of Topographical Maps: Symbols, Pictures and Surveys.* London: Thames and Hudson, 1980.

———. *Medieval Maps.* Toronto: University of Toronto Press, 1991.

Hays, Samuel B. *Conservation and the Gospel of Efficiency: The Progressive Conservation Movement, 1890-1920.* Cambridge, Mass.: Harvard University Press, 1959.

Hess, Harry H. "History of ocean basins" in *Petrological Studies,* edited by A. E. J. Engle, 599–620. New York: Geological Society of America, 1962.

Hockney, David. *Hockney on Photography: Conversations with Paul Joyce.* New York: Harmony Books, 1988.

———. *On Photography.* New York: André Emmerich Gallery, n.d.

———. *Secret Knowledge: Rediscovering the Lost Techniques of the Old Masters.* New York: Viking Studio, 2001.

———. *That's the Way I See It.* San Francisco: Chronicle Books, 1993.

Holmes, Arthur. *Principles of Physical Geology.* Sunbury-on-Thames: Thomas Nelson and Sons, 1944.

Hui, T. Keung. "Social Studies Squeeze." [Raleigh] *News and Observer,* August 13, 2003: 1B, 9B.

Hutchins, Edwin. *Cognition in the Wild.* Cambridge, Mass.: MIT Press, 1995.

Hutchinson, G. Evelyn. *An Introduction to Population Ecology.* New Haven, Conn.: Yale, 1978.

Imhof, Eduard. *Cartographic Relief Presentation.* Berlin and New York: Walter de Gruyter, 1982 [1965].

Janson, H. W., and Anthony F. Janson. *History of Art.* 5th ed. New York: Abrams, 1997.

Jay, Martin. *Downcast Eyes: The Denigration of Vision in Twentieth-Century Thought.* Berkeley, Calif.: University of California Press, 1993.

Jeunesse, Gallimard, Claude D. Delafosse, Sylvaine Perols (illustrator). *Atlas of Plants. A First Discovery Book.* New York: Scholastic, 1996.

Johnson, Douglas H., and Alan B. Sargeant. "Impact of Red Fox Predation on the Sex Ratio of Prairie Mallards," *Wildlife Research Report* 6. Washington, D.C.: Fish and Wildlife Service, United States Department of the Interior, 1977.

Johnson, Hugh. *The Principles of Gardening.* New York: Simon & Schuster, 1979.

Kant, Immanuel. *Observations of the Feeling of the Beautiful and Sublime.* Berkeley: University of California Press, 1960 [1763].

Kay, Paul. "The Inheritance of Presupposition." *Linguistics and Philosophy* 15 (1992): 333–81.

Kessel, B. "Distribution and Migration of the European Starling in North America." *Condor* 55 (1953): 49–67.

King, David. *The Commissar Vanishes: The Falsification of Photographs and Art in Stalin's Russia.* New York: Henry Holt, 1999 [1997].

Klonsky, Milton, ed. *Speaking Pictures: A Gallery of Pictorial Poetry from the Sixteen Century to the Present.* New York: Harmony, 1975.

Koch, Tom. *Cartographies of Disease: Maps, Mapping, and Medicine.* Redlands, Calif.: ESRI Press, 2005.

Kracauer, Siegfried. "Photography." In *The Mass Ornament: Weimar Essays,* 43–63. Cambridge, Mass.: Harvard University Press, 1995 [1927].

Lakoff, George. *Women, Fire, and Dangerous Things.* Chicago: University of Chicago Press, 1987.

Lakoff, George, and Mark Johnson. *Philosophy in the Flesh: The Embodied Mind and Its Challenges to Western Thought.* New York: Basic Books, 1999.

Larsen, Eric. *Isaac's Storm.* New York: Crown Publishing, 1999.

Latour, Bruno. *Pandora's Hope: Essays on the Reality of Science Studies.* Cambridge, Mass.: Harvard University Press, 1999.

———. *The Pasteurization of France.* Cambridge, Mass.: Harvard University Press, 1988.

———. *Science in Action.* Cambridge, Mass.: Harvard University Press, 1987.

———. *We Have Never Been Modern.* Cambridge, Mass.: Harvard University Press, 1993.

Latour, Bruno, and Woolgar, Steve. *Laboratory Life: The Social Construction of Scientific Facts.* Los Angeles: Sage, 1979.

Law, John, and Michael Lynch. "Lists, Field Guides, and the Descriptive Organization of Seeing: Birdwatching as an Exemplary Observational Activity." In *Representation in Scientific Practice,* edited by Michael Lynch and Steve Woolgar, 267–99. Cambridge, Mass.: MIT Press, 1990.

Le Pinchon, Xavier, Jean Francheteau, and Jean Bonnin. *Plate Tectonics: Developments in Geotectonics, No. 6.* New York: Elsevier, 1973.

Leplin, Jarrett, ed. *Scientific Realism.* Berkeley, Calif.: University of California Press, 1984.

Levy, David. *Skywatching* (The Nature Company Guides). New York: Time-Life Books, 1994.

Little, Jr., Elbert L., et al. *Atlas of United States Trees, Volumes 1-6, Miscellaneous Publications Nos. 1146, 1293, 1334, 1342, 1361,* and *1410.* Washington, D.C.: United States Department of Agriculture, 1971–1981.

———. "Important Forest Trees of the United States." In *Trees: The Yearbook of Agriculture, 1949,* edited by Alfred Stefferud, 763–814. Washington, D.C.: United States Department of Agriculture, 1949.

———. "Mapping Ranges of the Trees of the United States." *Rhodora* 53 (1951): 195–203.

Long, Kim. *Squirrels: A Wildlife Handbook, Johnson Nature Series.* Boulder, Colorado: Johnson Books, 1995.

Longwell, Chester R., and Richard F. Flint. *Introduction to Physical Geology.* New York: John Wiley & Sons, Inc., 1955.

Longyear, William. *Advertising Layout.* New York: Ronald Press, 1946.

Los Angeles County Museum of Art. *David Hockney: A Retrospective.* New York: Abrams, 1988.

Loveland, Thomas, et al. "Map Supplement: Seasonal Land-Cover Regions of the United States," *Annals of the Association of American Geographers* 85, no. 2 (1995): 339–55.

Lutz, Catherine A., and Jane L. Collins. *Reading National Geographic.* Chicago: University of Chicago Press, 1993.

Lyell, Charles. *The Principles of Geology.* Vol. 2. London: John Murray, 1832.

MacEachren, Alan. *How Maps Work: Representation, Visualization, and Design.* New York: Guilford Press, 1996.

March, James A., and Richard A. Hunt. "Mallard Population and Harvest Dynamics in Wisconsin." *Technical Bulletin No. 106.* Madison: Department of Natural Resources, 1978.

Margulis, Lynn. *Symbiotic Planet.* New York: Basic Books, 1998.

Margulis, Lynn, and Dorian Sagan. *Microcosmos: Four Billion Years of Microbial Evolution.* New York: Summit, 1986.

Margulis, Lynn, and Karlene V. Schwartz. *Five Kingdoms: An Illustrated Guide to the Phyla of Life on Earth.* 2nd ed. New York: Freeman, 1988.

Martin, Martha Evans, and Donald Howard Menzel. *The Friendly Stars: How to Locate and Identify Them.* New York: Dover, 1964 [1907, Martin].

Marx, Leo. *The Machine in the Garden.* New York: Oxford, 1964.

McHarg, Ian L. *Design With Nature.* Garden City, N.Y.: Doubleday, 1969.

McPhee, John. *Annals of the Former World.* New York: Farrar, Straus, and Giroux, 1998.

———. *In Suspect Terrain.* New York: Farrar, Straus, and Giroux, 1982.

Menard, H. W. *The Ocean of Truth.* Princeton: Princeton University Press, 1986.

Menzel, Donald. *Field Guide to the Stars and Planets.* Houghton Mifflin, 1963.

Mohlenbrock, Robert, and Hohn Thieret. *Trees: A Quick Guide Reference to Trees of North America.* (Macmillan Field Guides.) New York: Macmillan, 1987.

Monastersky, Richard. "The Rise of Life on Earth." *National Geographic,* March 1998, 54–81.

Monmonier, Mark. *Air Apparent: How Meteorologists Learned to Map, Predict, and Dramatize Weather.* Chicago: University of Chicago Press, 1999.

———. *Drawing the Line.* New York: Henry Holt, 1995.

———. *How to Lie with Maps.* Chicago: University of Chicago Press. 1996.

Morell, Virginia. "The Variety of Life." *National Geographic,* February 1999, 6–87.

Munns, E. N. *The Distribution of Important Forest Tress of the United States, Miscellaneous Publication No. 287.* Washington, D.C.: United States Department of Agriculture, 1938.

National Geographic Society. *Atlas of North America: Space Age Portrait of a Planet.* Washington, D.C.: National Geographic Society, 1985.

———. *Millennium in Maps, Physical Earth.* Washington, D.C.: National Geographic Society, 1998.

———. *National Geographic Atlas of the World.* 6th ed. Washington, D.C.: National Geographic Society, 1990.

———. *National Geographic: Index 1888-1988.* Washington, D.C.: National Geographic Society, 1989.

Newton, Norman T. *Design on the Land: The Development of Landscape Architecture* Cambridge, Mass.: Harvard University Press, 1971.

Novak, Barbara. *Nature and Culture: American Landscape and Painting, 1825-1875.* New York: Oxford, 1980.

Nyerges, Alexander Lee. *In Praise of Nature: Ansel Adams and Photographers of the American West.* Dayton, Ohio: Dayton Art Institute, 1999.

Obata, Chiura, Janice Tolhurst Driesbach, and Susan Landauer. *Obata's Yosemite: The Art and Letters of Chiura Obata from His Trip to the High Sierra in 1927.* El Portal, Calif.: Yosemite Association, 1993.

Oreskes, Naomi, ed. *Plate Tectonics: An Insider's History of the Modern Theory of the Earth.* Boulder, Colo.: Westview Press, 2001.

Osborne, Harold, ed. *The Oxford Companion to Art.* Oxford: Oxford University Press, 1970.

Patterson, Tom. "Getting Real: Reflecting on the New Look of National Park Service Maps." *Cartographic Perspectives* 43 (Fall 2002): 43–56.

———. "A View From On High: Heinrich Berann's Panoramas and Landscape Visualization Techniques for the U.S. National Park Service," *Cartographic Perspectives,* Spring, 2000, 38–65.

Patterson, Tom, and Nathaniel Vaughn Kelso. "Hall Shelton Revisited: Designing and Producing Natural-Color Maps with Satellite Land Cover Data," *Cartographic Perspectives,* Winter 2004, 28–55 and 69–80.

Peirce, Charles S. "Logic as Semiotic: The Theory of Signs." In *Philosophical Writings of Peirce,* edited by Justus Buchler, 98–119. New York: Dover, 1955.

Peters, Arno. *Die Neue Kartographie/The New Cartography.* Klagenfurt and New York: Universitätsverlag and Friendship Press, 1983.

Peterson, Roger Tory. *Peterson First Guide to Birds of North America.* Boston: Houghton Mifflin, 1986.

Peterson, Roger Tory, Guy Mountfort, and P. A. D. Hollom. *A Field Guide to the Birds of Britain and Europe.* 4th ed. Boston: Houghton Mifflin, 1983.

Petrides, George A. *A Field Guide to Eastern Trees, Eastern United States and Canada, Peterson Field Guides.* Boston: Houghton Mifflin, 1988.

Pickering, Andrew. *The Mangle of Practice: Time, Agency, and Science.* Chicago: University of Chicago Press, 1995.

———. "Objectivity and the Mangle of Practice." *Annals of Scholarship* 8 (1991): 409–25. Reprinted in Allan Megill, ed., *Rethinking Objectivity.* Durham: Duke, 1994, 109–25. Also incorporated into chapter six of Pickering's *The Mangle of Practice: Time, Agency, and Science.*

Pickstone, John. *Ways of Knowing: A New History of Science, Technology and Medicine.* Chicago: University of Chicago Press, 2001.

Pinchot, Gifford, and William W. Ashe. "Timber Trees and Forests of North Carolina," *North Carolina Geological Bulletin No.* 6., 1897.

Pomian, Krzysztof. *Collectors and Curiosities: Paris and Venice, 1500-1800.* London: Polity, 1990.

Portinaro, Pierluigi, and Franco Knirsch. *The Cartography of North America, 1500-1800.* New York: Crescent Books, 1987.

Powell, James Lawrence. *Mysteries of Terra Firma: The Age and Evolution of the Earth.* New York: The Free Press, 2001.

Powell, Major John Wesley. *Exploration of the Colorado River of the West, and Its Tributaries, Explored in 1869, 1870, 1871, and 1872.* Washington, D.C.: U.S. Government printing Office, 1875.

———. *The Exploration of the Colorado River and its Canyons.* New York: Dover, 1961.

Preminger, Alex, ed. *Princeton Encyclopedia of Poetry and Poetics.* Princeton: Princeton University Press, 1974.

Radford, Albert E. et al. *Manual of the Vascular Flora of the Carolinas.* Chapel Hill, N.C.: University of North Carolina Press, 1968.

Raff, A. D., and R. G. Mason. "Magnetic Survey off the West Coast of North America, 40° N Latitude to 52° N Latitude." *Bulletin of the Geological Society of America* 72(1961): 1267–270.

Rapoport, Eduardo H. *Areography: Geographical Strategies of Species.* Oxford: Pergamon, 1982.

Rey, H. A. *The Stars.* Boston: Houghton Mifflin, 1952.

Ricklefs, Robert E. *Ecology.* 3rd ed. New York: Freeman, 1990.

Robbins, Chandler S., Bertel Bruun, and Herbert Zim. *A Guide to Field Identification of Birds of North America.* New York: Golden Press, 1966.

Robinson, Arthur. "Arno Peters and His New Cartography." *American Cartographer* 12, no. 2 (1985): 103–11.

———. *Early Thematic Mapping in the History of Cartography.* Chicago: University of Chicago Press, 1982.

———. "Rectangular World Maps – No!" *The Professional Geographer* 42, no.1 (1990): 101-104.

Robinson, Arthur H., and Barbara Bartz Petchenik. *The Nature of Maps: Essays Toward Understanding Maps and Mapping.* Chicago: University of Chicago Press, 1976.

Robinson Arthur H., Joel L. Morrison, Phillip C. Muehrcke, A. Jon Kimerling, and Stephen C. Guptill. *Elements of Cartography.* 6th ed. New York: Wiley, 1995.

Rogoff, Irit. *Terra Infirma: Geography's Visual Culture.* London: Routledge, 2000.

Romm, James. "A new forerunner for continental drift." *Nature* 367 (1994): 407–08.

Rothenberg, Tamar Y. "Voyeurs of Imperialism: *The National Geographic Magazine* before World War II." In *Geography and Empire,* edited by Anne Godlewska and Neil Smith, 155–72. Oxford: Blackwell, 1994.

Sargeant, Charles. *Report on the Forests of North America.* Vol. 9 of the *Tenth Census of the United States.* Washington, D.C.: Government Printing Office, 1884.

Schank, Roger. *The Connoisseur's Guide to the Mind.* New York: Summit Books, 1991.

Schank, R. C., and R. P. Abelson. "Scripts, Plans and Knowledge." In *Thinking: Readings in Cognitive Science,* eds., P. N. Johnson-Laird and P. C. Wason. Cambridge: Cambridge University Press, 1977.

Schopf, J., William. "The Evolution of the Earliest Cells." *Scientific American,* September 1978: 110–38.

Schulten, Susan. *The Geographical Imagination in America, 1880-1950.* Chicago: University of Chicago Press, 2001.

Seuss, Eduard. *The Face of the Earth.* Translated by Hertha Sollas. Oxford: Oxford University Press, 1904–1924.

Shapin, Steven, and Simon Schaffer. *Leviathan and the Air-Pump: Hobbes, Boyle, and the Experimental Life.* Princeton: Princeton University Press, 1985.

Simon, Herbert. *Sciences of the Artificial.* 2nd ed. Cambridge, Mass.: MIT Press, 1981.

Sloss, Peter W. *Surface of the Earth.* Boulder, Colo.: NOAA/National Geophysical Data Center, 1996.

Smith, Henry Nash. *Virgin Land.* New York: Knopf, 1950.

Smith, Melvin T. *The Colorado River: Its History in the Lower Canyons Area.* Provo, Utah: Brigham Young University, 1972.

Snyder, George Sergeant. *Maps of the Heavens.* New York: Abbeville, 1984.

Stanley, Steven M. *Life and Earth Through Time.* 2nd ed. New York: Freeman, 1989.

———. *Exploring Earth and Life Through Time.* New York: Freeman, 1993.

Strahler, Arthur N. *Plate Tectonics.* Cambridge: Geo Books, 1998.

Studer, Jacob. *Studer's Popular Ornithology: The Birds of North America.* New York: Harrison House, 1977 [1881].

Stupka, Arthur. *Great Smoky Mountains National Park: Natural History Handbook Series No. 5.* Washington, D.C.: Government Printing Office, 1960.

Sturken, Marita. "Desiring the Weather: El Niño, the Media, and California Identity," *Public Culture* 13, no. 2 (2001): 161–89.

Sunquist, Mel, and Fiona Sunquist. *Wild Cats of the World.* Chicago: University of Chicago Press, 2002.

Takenichi, H., S. Uyeda, and H. Kanamori. *Debate About the Earth: Approach to Geophysics through Analysis of Continental Drift.* San Francisco: Freeman, Cooper and Co., 1967.

Taylor, F. B. "Bearing of the Tertiary Mountain Belt on the Origin of the Earth's Plan." *Geological Society of America Bulletin* 21 (1910): 179–226.

Terborgh, John. *Diversity and the Tropical Rainforest.* New York: Scientific American, 1992.

———. *Where Have All the Birds Gone?* Princeton: Princeton University Press, 1989.

Thompson, Morris M. *Maps for America: Cartographic products of the U.S. Geological Survey and others.* Washington, D.C.: U.S. Geological Survey, 1979.

Thurman, Judith. "String Theory: Ready-to-Wear in Paris, a Versace Retrospective in London." *The New Yorker* (November 4, 2002): 107.

Toivanen, Hannes. "Visual Harvest: Ambiguity in the U.S. Forest Science at Home and in the Tropics, 1890-1925." Annual Meeting, History of Science Society, Milwaukee, 2002.

Toulmin, S. E. *The Uses of Argument.* Cambridge: Cambridge University Press, 1958.

Trachtenberg, Alan. "Photography/Cinematography." In *Before Hollywood: Turn-of-the-Century Films from American Archives,* 73–79. New York: American Federation of the Arts, 1986.

Tudor, Andrew. "Antonioni, Michelangelo." In *World Film Directors: Volume II, 1945-1985,* edited by John Wakeman, 59–69. New York: H. W. Wilson, 1988.

Tufte, Edward. *Envisioning Information.* Cheshire, Conn.: Graphics Press, 1990.

———. *Visual Explanations: Images and Quantities, Evidence and Narrative,* Cheshire, Connecticut: Graphics Press, 1997.

Turnbull, David. *Maps Are Territories: Science Is an Atlas.* Geelong, Victoria: Deakin University Press, 1989.

———. *Masons, Tricksters and Cartographers: Comparative Studies in the Sociology of Scientific and Indigenous Knowledge.* London: Routledge, 2000.

Tuttle, Merlin D. *America's Neighborhood Bats.* Austin, Tex.: University of Texas Press, 1988.

Tyning, Thomas F. *Guide to Amphibians and Reptiles, Stokes Nature Guides.* Boston: Little, Brown, 1990.

Udvardy, Miklos D. F. *The Audubon Society Field Guide to North American Birds (Western Region).* New York: Knopf, 1977.

U.S. Geological Survey. *This Dynamic Planet: World Map of Volcanoes, Earthquakes, Impact Craters, and Plate Tectonics.* Denver, Colo.: U.S. Geological Survey, 1994.

Voas, Robert. "John Glenn's Three Orbits in *Friendship 7*: A Minute-by-Minute Account of America's First Orbital Space Flight," *National Geographic,* June 1962, 792–827.

Warner, Deborah J. *The Sky Explored: Celestial Cartography 1500-1800.* New York: Alan R. Liss, 1979.

Webb, Thomas W. *Celestial Objects for Common Telescopes.* London: Longmans, Green, 1917 [1859].

Webster's Third New International Dictionary of the English Language, Unabridged. Springfield, Mass.: G & C Merriam Company, 1961.

Wegener, Alfred. *The Origin of Continents and Oceans.* Translated by John Biram. New York: Dover, 1966.

Wetzler, Brad. "Base Camp Confidential: An Oral History of Everest's Endearingly Dysfunctional Village." *Outside,* April 2001.

White, T. H. *The Bestiary, A Book of Beasts.* New York: Putnam, 1954.

Whittaker, R. H., and S. A. Levin, eds. *Niche: Theory and Application, Benchmark Papers in Ecology.* Vol. 3. Stroudsburg, Pa.: Dowden, Hutchinson and Ross, 1975.

Whittaker, R. H., S. A. Levin, and R. B. Root. "Niche, habitat and ecotope." *American Naturalist* 107 (1973): 321–38.

Whitfield, Peter. *The Image of the World: 20 Centuries of World Maps.* San Francisco: Pomegranate Artbooks, 1994.

Whitney, Stephen. *The Audubon Society Nature Guide to Western Forests.* New York: Knopf, 1985.

Williams, Michael. *Americans and Their Forests: A Historical Geography.* Cambridge: Cambridge University Press, 1989.

Williams, Raymond. *The Country and the City.* New York: Oxford University Press, 1973.

———. *Keywords: A Vocabulary of Culture and Society.* New York: Oxford University Press, USA, 1976.

Willinsky, John. *Empire of Words: The Reign of the OED.* Princeton: Princeton University Press, 1994.

Wilson, Edward O. *The Diversity of Life.* New York: Norton, 1992.

Wilson, J. Tuzo, ed. *Continents Adrift: Readings from* Scientific American. San Francisco: W.H. Freeman and Company, 1972.

———. *Continents Adrift and Continents Aground: Readings from* Scientific American. San Francisco: W.H. Freeman and Company, 1976.

Wolfe, David. *Tales From the Underground: A Natural History of Subterranean Life.* Cambridge, Mass.: Perseus, 2001.

Wollen, Peter. "Mappings: Situationists and/or Conceptualists." In *Rewriting Conceptual Art,* edited by Michael Newman and Jon Bird. London: Reaktion, 1999.

Wood, Denis. *Five Billion Years of Global Change: A History of the Land.* New York: Guilford, 2004.

———. "*The Mapmakers* by John Noble Wilford," *Cartographica* 19 (Autumn & Winter 1982): 127–31.

———. "Maps and Mapmaking." In *Encyclopedia of the History of Science, Technology and Medicine in Non-Western Cultures,* edited by Helaine Selin. Dordrecht, Boston: Kluwer Academic, 1997.

———. "Now and Then: Comparisons of Ordinary Americans' Symbol Conventions with Those of Past Cartographers." *Prologue: The Journal of the National Archives* 9 (Fall 1977): 151–61.

———. "P. D. A. Harvey and Medieval Mapping: An Essay Review," *Cartographica* 31 (Autumn 1994): 52–59

———. *The Power of Maps.* New York: Guilford Press, 1992.

Wood, Denis, and John Fels. "Designs on Signs: Myth and Meaning in Maps," *Cartographica* 23 (Autumn 1986): 54–103.

Wood, Denis, Ward Kaiser, and Bob Abramms. *Seeing Through Maps.* 2nd ed. Amherst: ODT, 2006.

Woodward, David. "Medieval *Mappaemundi.*" In *The History of Cartography: Volume One: Cartography in Prehistoric, Ancient, and Medieval Europe and the Mediterranean,* edited by J. B. Harley and David Woodward. Chicago: University of Chicago Press, 1987.

Wright, J. K. *Human Nature in Geography.* Cambridge, Mass.: Harvard University Press, 1956.

Zavarzin, George. "Cyanobacteria mats in general biology." In *Microbial Mats: Structure, Development and Environmental Significance.* Vol.35 of *NATO ASI Series G: Ecological Sciences,* edited by Lucas Stal and Pierre Caumette, 443–52. Berlin: Springer-Verlag, 1994.

Zim, Herbert S., and Robert H. Baker. *Stars: A Guide to the Constellations, Sun, Moon, Planets, and Other Features of the Heavens.* New York: Golden Press, 1951.

Zim, Herbert S., and Alexander C. Martin. *Trees: A Guide to Familiar American Trees.* Rev. ed. New York: Golden Press, 1956 [Simon & Schuster, 1952].

List of Key Maps

In order of first appearance

Cove Creek Gap, NC. U.S. Geological Survey, Reston, Va., 1997. (Figures 1.1, 1.18, 11.1, 11.2,)

"Australia Under Siege/Australia." Supplement to *National Geographic,* July 2000. National Geographic Society, Washington D.C. (Figures 1.4, 4.1–4.10)

The Peters World Map. Arno Peters. Version shown here published by ODT, 2005. (Figure 1.5)

Seasonal Land-Cover Ecoregions. U.S. Geological Survey, Reston, Va., 1993. (Figure 1.8)

"The Relentless Season." 2004 storm tracks map from WRAL-TV's *2005 Stormtracker* publication, 2005. (Figures 1.12, 5.3)

"Mt Everest." Supplement to *National Geographic,* November 1988. National Geographic Society, Washington D.C. (Figures 1.13, 6.16, 6.17)

"The World of Flowers." Ned Seidler. In "The Flower Seed Growers: Gardening's Color Merchants" by Robert de Roos, *National Geographic,* May 1968: 720–38. (Figures 1.14, 7.1)

A Tapestry of Time and Terrain. Jose F. Vigil, Richard J. Pike, and David G. Howell. Geologic Investigations Series 1-2720, U.S. Geological Survey, Reston, Va., 2000. (Figures 1.16, 9.9, 9.10)

"Portrait U.S.A./The United States." Supplement to *National Geographic,* July 1976. (Figures 1.17, 10.1, 10.2, 10.4)

"Australia: Land of Living Fossils/Australia." Supplement to *National Geographic,* February 1979. National Geographic Society, Washington D.C. (Figures 3.1–3.7, 3.11–3.13)

"Hawaii." Part of the "Making of America" series. Supplement to *National Geographic,* November 1983. National Geographic Society, Washington D.C. (Figure 4.15)

"If you Evacuate." Hurricane evacuation map from WRAL-TV's *2005 Stormtracker.* (Figure 5.2)

Hurricane Tracking Map from WRAL-TV's *2005 Stormtracker.* (Figure 5.4)

"Natural Hazards of North America/Great Disasters: Nature in Full Force." Supplement to *National Geographic,* July 1998. Photographs by Tad Nichols (courtesy of Northern Arizona University), Jim Wells, Alan J. Oxley, Frank Oberle, John Lopinot, Howard Bluestein, J. Boissinot (courtesy of CP Photo Assignment Services), Robert Eplett, Warren Faidley; artwork by Rob Wood. National Geographic Society, Washington D.C. (Figures 5.9–5.15)

"The Earth's Fractured Surface." Supplement to *National Geographic,* April 1995. National Geographic Society, Washington D.C. (Figures 5.18, 5.19)

Cantino Planisphere. Anon., 1502. Courtesy of Biblioteca Estense, Modena IT. (Figure 5.20)

"Agnese World Map." In *Portolan Atlas* by Battista Agnese, 1544. Courtesy of The Huntington Library, San Marino Calif. (Figure 5.21)

"Ortelius World Map." In *Theatrum Orbis Terrarum* by Abraham Ortelius, 1595. Courtesy of The Huntington Library, San Marino Calif. (Figure 5.22)

World Ocean Floor. John Fels Cartographics, 2006. Image from ETOPO2 data, NOAA/NGDC. (Figure 5.27)

Three-Dimensional Anaglyph of the Earth. Brian Davis, Paul Morin, and Monte Ramstad, U.S. Geological Survey, Reston, Va., 2003. Also reproduced in *ESRI Map Book,* vol. 19, p. 16. (Figure 5.30)

Qblique Map of Yosemite Valley. Tau Rho Alpha, U.S. Geological Survey, Reston, Va., 1991. (Figures 6.4, 6.5)

Map of Yosemite Valley, Shaded Relief Edition. U.S. Geological Survey, Reston, Va., 1958, revised 1970. (Figure 6.6)

Satellite Orthoimage of Yosemite Valley. U.S. Geological Survey, Reston, Va., 1986. (Figures 6.7, 6.8)

Grand Canyon National Park and Vicinity. U.S. Geological Survey, Reston, Va., 1962. (Figures 6.9, 6.10)

Grand Canyon National Park and Vicinity – Shaded Relief Edition. U.S. Geological Survey, Reston, Va., 1972. (Figure 6.11)

Trails Illustrated Map of Grand Canyon National Park, Arizona, USA. National Geographic Maps, Evergreen, Colo., 1987, revised 2003. (Figures 6.12–6.15)

"Everest 50/Everest." Supplement to *National Geographic,* May 2003. (Figure 6.18, only "Everest" side shown)

Satellite Image Map of Antarctica. Miscellaneous Investigations Series Map I-2560, U.S. Geological Survey, Reston, Va., 1996, revised 2000. (Figure 6.19)

"Africa." In *My First Atlas* by Bill Boyle. London: Dorling Kindersley Books, 1994. (Figure 7.3)

Esbtorf World Map. Created at the Kloster Ebstorf monastery, 1241–1245. Image from Martin Warnke. (Figures 7.4, 7.5)

"Schegel's Versuch Elner Schlangen-Karte (The Distribution of Snakes)." In *Heinrich Berghaus' World Atlas,* 1845. David Rumsey Collection. (Figure 7.6)

"Geographical Divisions and Distribution of Carnivora." In *A. K. Johnston's*

Physical Atlas, 1856. David Rumsey Collection. (Figure 7.7)

"Distribution of Animals in Zones." In *Frye's Grammar School of Geography,* 1895. (Figure 7.8)

"Plants and Animals." In *Maury's Manual of Geography,* 1898. (Figure 7.9)

Map of the Arrowhead Country, Minnesota, Quetico Park, New Ontario and Nipigon. Railroad tourist map, Canadian National Railway, date unknown. (Figure 7.12)

"Bird Migration in the Americas/The Americas." Supplement to the *National Geographic,* August 1979. (Figures 7.13-7.16)

"Bird Migration: Eastern Hemisphere." Supplement to the *National Geographic,* April 2004. National Geographic Society, Washington D.C. (Figure 7.18)

State of North Carolina. U.S. Geological Survey, Reston, Va., 1957, rev. 1972. (Figures 9.1, 9.2)

Generalized Geologic Map of North Carolina. North Carolina Geologic Survey, Raleigh N.C., 1991, reprinted 1996. (Figure 9.3)

General Soil Map of North Carolina. USDA Soil Conservation Service, Washington D.C., 1990. (Figure 9.5)

North Carolina Transportation Map. North Carolina Department of Transportation, 2006. (Figure 9.6)

North Carolina Watersheds. 3rd ed. John Fels Cartographics, 2000. (Figure 9.7)

Landforms of the Conterminous United States. Gail P. Thelin and Richard J. Pike, U.S. Geological Survey, Reston, Va., 1991. (Figure 9.8)

Ecoregions of North America. R. G. Bailey, USDA Forest Service. Rocky Mountain Research Station, 1997. (Figure 9.11)

Plant Associations of the Chattooga River Basin. John Fels and Steven Simon, USDA Forest Service, Southeastern Region, Asheville, 1995. (Figure 9.12)

Shaded relief image of Chatooga Basin. John Fels Cartographics, 1995. (Figure 9.13)

"Face of the Continent." Satellite photo of North America from *Atlas of North America: Space Age Portrait of a Continent,* 10–11, National Geographic Society, Washington D.C., 1985. (Figure 10.6)

"Images from Space." Unadjusted photomosaic of North America from *National Geographic Atlas of the World,* 6th ed., vi, National Geographic Society, Washington D.C., 1990. (Figure 10.10)

"North America." False color photomosaic of North America from *National Geographic Atlas of the World,* 6th ed., 14–15, National Geographic Society, Washington D.C., 1990. (Figure 10.11)

Land Cover Portrait/Coterminous United States. Tom Patterson et al., U.S. National Park Service, Harpers Ferry Center, 1992. (Figure 10.12)

Great Smoky Mountains. National Park Service and U.S. Department of the Interior, 2001. (Figures 11.4–11.6)

Cowee Quadrangle. U.S. Geographic Survey, Reston, Va., 1886. Courtesy of Library of Congress, Geography and Map Division. (Figure 11.8)

Forest Conditions of Western North Carolina. The U.S. Forest Service and the North Carolina Geological and Economic Survey, 1909. Courtesy of Library of Congress, Geography and Map Division. (Figure 11.10)

Proposed Great Smoky Mountains National Park, North Carolina–Tennessee. U.S. Geographic Survey, Reston, Va., 1926. Courtesy of Library of Congress, Geography and Map Division. (Figure 11.11)

Denis Wood holds a master's degree and PhD in geography from Clark University. He taught at the College of Design at North Carolina State University for a quarter of a century and curated the Smithsonian's "Power of Maps" exhibition. He's best known for his path-breaking book *The Power of Maps* (1992, with John Fels). Other books include *Home Rules* (1994, with Robert Beck), *Seeing Through Maps* (2001, with Ward Kaiser), *Five Billion Years of Global Change* (2004), and *Making Maps* (2005, with John Krygier). He has also published more than a hundred articles in a wide variety of journals.

Dr. John Fels holds a master's degree in graphic design and PhD in forestry from North Carolina State University. He has worked as a professional cartographer with the Ministry of Natural Resources, Province of Ontario, and as a freelance cartographic designer and consultant. For twelve years he served on the faculty of the School of Natural Resources at Sir Sandford Fleming College, Ontario, where he developed the core curriculum in design for the cartography and GIS programs. He is currently a consultant in geospatial modeling and analysis and an adjunct associate professor in the College of Natural Resources at North Carolina State University. He is the coauthor of *The Power of Maps*.